Jean-Dominique Deuschel
Andreas Greven (Eds.)

Interacting Stochastic Systems

With 17 Figures

Jean-Dominique Deuschel
Technische Universität Berlin, Fakultät II
Institut für Mathematik
Straße des 17. Juni 136
10623 Berlin, Germany
e-mail: deuschel@math.tu-berlin.de

Andreas Greven
Universität Erlangen
Fachbereich Mathematik und Physik
Mathematisches Institut
Bismarckstr. 1 1/2
91054 Erlangen, Germany
e-mail: greven@mi.uni-erlangen.de

Mathematics Subject Classification (2000): 60xx, 92Bxx, 62B35, M27004, P19013, S1400
Physics and Astronomy Classification Scheme (2003): 02

ISBN 978-3-642-06196-7   e-ISBN 978-3-540-27110-9

Springer is a part of Springer Science+Business Media
springeronline.com

© Springer-Verlag Berlin Heidelberg 2010
Printed in Germany

Cover design: Erich Kirchner, Heidelberg

Printed on acid-free paper 41/3142YL – 5 4 3 2 1 0

Table of Contents

Interacting Stochastic Systems

The Parabolic Anderson Model
Jürgen Gärtner, Wolfgang König

Random Spectral Distributions
Friedrich Götze, Franz Merkl

Branching Processes in Random Environment –
A View on Critical and Subcritical Cases
Matthias Birkner, Jochen Geiger, Götz Kersting

Part III Stochastic Analysis

Thin Points of Brownian Motion Intersection Local Times
Achim Klenke

Coupling, Regularity and Curvature
Karl-Theodor Sturm

Part IV Applications of Stochastic Analysis in Finance, Engineering and Algorithms

---

Random Dynamical Systems Methods in Ship Stability: A Case Study
Ludwig Arnold, Igor Chueshov, Gunter Ochs

Analysis of Algorithms by the Contraction Method: Additive and Max-recursive Sequences
Ralph Neininger, Ludger Rüschendorf

Introduction

Jean-Dominique Deuschel[1] and Andreas Greven[2]

[1] Fakultät II - Mathematik und Naturwissenschaften, Institut für Mathematik,
Technische Universität Berlin, Straße des 17. Juni 136, D-10623 Berlin
Tel. (0049 30) 314-25193, Fax: (0049 30) 314-21695
deuschel@math.tu-berlin.de

[2] Mathematisches Institut, Bismarckstr. 1 ½, D-91054 Erlangen,
Tel. (0049 9131) 85-22454, Fax (0049 9131) 85-26214

This volume collects original work and reviews on work in the field of probability which took place within the framework of the DFG-Schwerpunkt: Interacting stochastic systems of high complexity. This research network started in May 1997 and was funded till May 2003. In this network between 20–30 (depending on the 2-year periods of grant renewal) groups from probability, statistical physics and mathematical statistics within Germany were active. An extensive international collaboration was an essential part of this network and here particularly intense contacts were built up between the DFG-Schwerpunkt and EURANDOM, a European institute for research in stochastics, which was founded in 1997. This partnership reached from joint workshops and colloquia to collaborations on specific projects.

The key scientific idea which was behind the research network was to explore and develop the connections between research in infinite dimensional stochastic analysis, statistical physics (Gibbs measures, random media), spatial population models from mathematical biology, complex models of financial markets or of stochastic models with interacting components in other sciences as for example in climatology. At the time the proposal was written such connections were about to emerge and become part of the proposals, other connections developed as the work in the network proceeded.

An example of these newly emerged connections is the appearance of Neveu's branching process in the theory of spin glasses as key model in the area of random media. In this particular area the connections planned in the proposal were more in the line of the application of renormalization methods and ideas common in the statistical physics community in the field of stochastic population models, a topic which also flourished during the time of the network. Other examples are the applications of the theory of random dynamical systems, part of the original proposal, which shifted from the development of the theoretical framework and questions of evolutionary biology to treating questions of the stability of ships.

The topics covered by the research in the "Schwerpunkt" were grouped in the following general themes:

- Interacting particle systems in statistical physics,
- random media and homogenization,
- branching models in population genetics,
- stochastic methods for the analysis of financial derivatives,
- consistency and efficiency of Monte Carlo algorithm,
- stochastic analysis: Dirichlet forms, partial stochastic differential equations and theory of large deviations.

This original plan for and the original structure of the network underwent rapidly changes as the scientific work proceeded. The theory of random Schrödinger operators became part of the random media section of the network, the theory of spin glasses led to a closer connection between random media and interacting particle systems in statistical physics, the theory of Dirichlet forms and large deviation merged with both these fields, besides financial markets and Monte Carlo algorithms new applications came into sight like sequence alignment or modelling of the climate using stochastic analysis and random dynamical systems. All these developments benefitted from the regular yearly meetings of the network, the workshops and summer schools.

In line with this development the "Schwerpunkt" participated in various joint efforts with other mathematical areas and with other disciplines in the sciences. Highlights were a symposium on entropy jointly organized by three mathematical "Schwerpunkte" together with the Max-Planck-Institute for the Physics of Complex Systems (Dresden) and workshops on mathematical biology jointly with the Max-Planck-Institute for Evolutionary Anthropology and the Max-Planck Institute for Mathematics in the Sciences, both in Leipzig. Over the whole period efforts were made to co-sponsor a series of workshops on climatology. In addition to that a quite successful series of workshops on statistical physics and on stochastic analysis (both jointly with EURANDOM) were organized supporting the collaboration between these institutions, within the network and with the international community of probabilists.

The contribution in the volume will clearly show the extensive international collaboration taking place in order to achieve the scientific progress made as the project was running. Through these activities it was possible to provide for a substantial number of young scientists the possibility to participate in workshops and joint projects with internationally leading scientists. This way the goal was reached to create in the field of stochastics in Germany a group of young researchers which are internationally competitive.

We describe next in more detail what happens in the different chapters of this volume and also give some references to work in the network which is not described in this volume.

The first chapter treats stochastic models from statistical physics and the closely related topic of processes in random media. The contributions to this field concentrate on two main topics. The first topic evolves around the Gibbs measure while the second had its focus on random Schroedinger operators, the parabolic Anderson model and spectra of random matrices. This chapter illustrates how modern stochastic methods can be very successfully applied in a very broad range of models of statistical physics and random media. The applications range from quantum electrodynamics, to quantum field theory and from magneto-hydrodynamics, to quantum anharmonic crystals and random Schrödinger operators.

One of the central themes is *Gibbs distributions* and their properties. In the formalism of thermodynamics, Gibbs distributions describe the equilibrium states with respect to some given interaction. Some of the challenging questions deal with existence and uniqueness of the corresponding states. In the case of non uniqueness we have a phase transition, this corresponds to a strong interaction or the low temperature regime. In this regime the analysis is very delicate and various technique have been developed to study the properties in this regime and to establish the phase transition.

One technique relies on long range interactions of the mean field type, so called Kac models characterized by a parameter describing the range of interaction, see the paper of Bovier and Külske. For lattice dimensions greater than three, one can show the existence of ferromagnetic order Gibbs states for a range of temperatures that is uniform in the Kac parameter. The proof is based on a a coarse-graining procedure or renormalization argument, once the long-range random field model has been reduced to an effective short range contour model.This analysis sheds light into more realistic models with local interactions.

Another interesting application of Gibbs measures deals with quantum anharmonic crystals. The idea is to represent the crystal with the help of Euclidean Gibbs measures on loop lattice, and is developed in the work of Albeverio and Roeckner. The main result here is the existence and uniqueness of the Gibbs measure for sufficiently small values of the mass. The technique is based on an integration by parts formula, which gives an alternative description of Gibbs measures. In particular the Dobrushin uniqueness criterion, exponential decay of correlations are derived.

Gibbs measures on the space of continuous trajectories relative to Brownian motions as a model of quantum electrodynamic is the objective of Betz, Lörinczi and Spohn. Here the existence and path properties of such measures are investigated. In particular the existence of phase transition and a functional central limit theorem are proved using cluster expansion techniques.

Another model of quantum field is modelled by a Markov jump process with minimal jump rates, the so-called Bell process in the paper of Georgii and Tumulka. Here global existence is obtained for manifolds with boundaries.

The second main theme of the first section deals with *random media* which appears in the context of spatially inhomogeneous systems. One of the standard method used in studying spatially disordered system is what is called homogenisation. This means that under rescaling the system with random coefficients converges to a PDE with deterministic coefficients obtained by "properly averaging" random coefficients. However in many important situations, random systems cannot be adequately described by a deterministic approach. This is for instance the case for localization effects for the electron transport in disordered media given by the Schrödinger operators with random potentials. They are used to model quantum aspects of disorder electronic systems like unordered alloys, amorphous solids or liquids. In this range of questions belongs also the analysis of spectral properties of random matrices, a field in rapid development.

The main question in this setting concerns the phenomena of *localization* and *delocalization*. The localized phase corresponds to a pure point spectrum with only bounds states prohibiting transport of current, whereas the delocalized phase with absolutely continuous spectrum exhibits scattering states with electron transport. While periodic potentials have absolutely continuous spectrum, disordered systems have a tendency to localize states, the so-called Anderson localization.

Although situations with localized states are well known, so far delocalization has not been settled for genuine, i.e. translation invariant, random models. The object the work of Böcker, Kirsch and Stollmann is to investigate the spectrum and the density of states in the context of nonstationary random potentials with a delocalized phase. In particular different models with sparse random potentials and random surface models are presented.

While the previous paper deals with a discrete model, the paper of Leschke, Müller and Warzel is concerned with the spectrum of continuous models for amorphous solids. In this setting, translation invariance is assumed, and the paper analyzes the density of states and corresponding Lifshitz tail in localized regime for both generic Gaussian and Poissonian random potentials.

The paper of Gärtner and König studies the parabolic Anderson model, that is with the Cauchy problem for the heat equation with random potentials on the lattice. Unlike the two previous papers which dealt only with the spectrum, this paper focuses on the intermittency of the solutions. Intermittency takes place in the localized regime, when one expects that the solution of the system develops pronounced spatial structures on islands located far from each other. The aim of the paper is to give a mathematical description of this phenomena, which differs depending on the quenched setting with given "frozen" random media or the annealed setting, averaged over the media. In particular the paper illustrates the four different universal classes of possible asymptotics. Both time independent and time dependent random potentials such as the one generated by a Poisson field are considered.

The last paper of this section deals with the topic of random matrices. In this area quite some development took place while the research network was active and the group around F. Götze started working in this field. In this contribution Götze and Merkl review results on the limit distributions and the rate of convergence for the spectra of certain ensembles of random matrices. The ensembles include the Wigner anf general unitary ensemble (GUE), the Laguerre ensembles and the circular unitary ensemble (CUE). Connections to quantuum field theory and distribution of the zeros of the zeta function is also illustrated.

A central aim of the research in the network was besides the application of stochastic analysis to further develop the foundations of stochastic analysis. Some of this work developed in close connection with certain applications as for example the work of Alberverio and Röckner found in the chapter on statistical physics models some of the work however focussed on structural aspects, which is collected in section 3, which we discuss next, except the contribution concerned with the concept of random dynamical systems, this will be discussed in context of the last chapter.

The paper of Sturm investigates the space of probability measures over a Euclidean or Riemannian space. One of the objective introduce nonlinear diffusion equations describing the flow of the distribution of interacting stochastic particles. These diffusions can be characterized as the gradient flow of an appropriate free energy functional. Moreover the barycenter map is shown to be contractive in metric spaces with nonpositive curvarture. This allows to generalize the classical law of large number numbers to this abstract setting. Also contraction properties for the heat semigroup on a Riemanian manifold with its relations to bounds on the Ricci curvature are discussed.

In the work of Deuschel, a wide class the qualitative properties of interactive diffusion processes in time and space are studied. The aim of the paper is to present a technique, the random walk representation which allow to express the space-time covariances in terms of the Green function of a random walk in random environment. In particular existence of invariant measures, convergence rate to equilibrium and various aging phenomena can be derived for various models which in the language of mathematical physics are described as massless, see also the references to effective interface models in this paper.

The work by Klenke and Mörters is concerned with the intersection local time of two independent Brownian paths in $\mathbb{R}^d$ for $d = 2$ and 3. This intersection set has been studied for typical points by its Hausdorf dimension and Hausdorf gauge function. The work here is concerned with the question whether there are thick or thin points in S with local dimension lower or higher than typically expected. The results disprove some assertions made by the multi-fractal formalism from statistical physics.

Besides the application of stochastic analysis in models from statistical physics a central goal of the network was to apply stochastic methods in models from mathematical biology. The main focus was on evolutionary questions in a broad sense. The chapter on population models exhibits very well the spectrum of tasks mathematicians face modelling and analysing the evolution of populations. This spectrum reaches from the effort to model a biological phenomenon appropriately in mathematical terms accessible to statistical methods, the development of the statistical theory up to the final task the construction of a transparent mathematical theory for the qualitative behavior of stochastic population models.

The building block of population models are stochastic systems in which the reproduction and the migration of individuals in geographic space can be described. A paradigmatic process is the so-called branching random walk and its relatives super random walk and super Brownian motion which arise by considering continuum limits in the mass, i.e. replace individuals by smaller and smaller masses with at the same time increasing number of individuals, and rapid reproduction, respectively continuum limits in space replacing for example the lattice Z^d as geographic space by $\mathbb{R}^d$.

The research on the qualitative properties of such models has been focussed on a number of directions. First try to understand not only the size and distribution of the population at given times, but try to understand the complete history of such individuals and their ancestors. This means besides location and type of individuals record their whole *genealogy* and describe qualitative properties of this object. The second direction has been to find the classes of models which lead when considered on large space and time scales to the same qualitative behavior. The key word here is *universality*. This is a theme which connects the theory of these biologically motivated systems with the corresponding attempts in statistical physics and in particular systems in inhomogeneous media. The third direction of great importance in applications is to incorporate in population models mechanisms representing mutation, recombination and selection which are responsible for many effects observed in nature. This leads to many new and difficult problems and in particular requires a close connection to the methods of infinite dimensional analyses in order to be able to construct the appropriate mathematical models and to develop methods to tackle these highly nonlinear systems.

The analysis of population models evolves now on three levels. On the mathematical level where properties of these models are derived on the basis of rigorous mathematical reasoning, on the level of a statistical theory which allows the choice of parameters in those models on the basis of data and finally the level on which the probabilist tries to adapt models he can handle both mathematically and computationally to the situations and the questions raised by the biologist. The latter process of balancing mathematical tractability with the search for realistic models is a very delicate task due to the complexity of the biological situation with often overwhelming amounts of data versus the mathematical restriction as to which nonlinear effects can still be rigorously analysed.

The level of mathematical theory is here represented by two articles, the contribution of Birkner, Geiger and Kersting and the contribution of Greven.

The first one shows how insight on branching populations can be gained by focussing on the whole genealogical tree and how new results can be obtained this way even on questions one can raise on the basis of population size alone. More examples of such results can be found in the references in this chapter. The contribution by Greven deals with a mathematically rigorous approach to the problem of understanding the universality classes of neutral population models and desribes the progress which has been made in passing from single type to multitype population models. On the level of modelling the contribution by Fleiss ner, Metzler, von Haeseler and Wakolbinger describes the progress which has been made in understanding the procedure of sequence alignment based on stochastic models.

Besides this work in population models documented here research on population models in random environment was pursued by Greven, Klenke and Wakolbinger in a series of papers ([GKW99, GKW01, GKW02]) and by Fleischmann and his international collaborators ([DFMPX]). The main focus of these papers is the rigorous construction of catalytic branching processes and the determination of its longtime behavior in its dependence on the dimension of the geographic space in which the population lives. This work gained substantially by ideas developing in the circle working on random media problems motivated by statistical physics. Here the fundaments have been laid to proceed to models where medium and process interact as for example for the process called mutually catalytic branching ([CDG]).

In addition to the themes mentioned the research network was active in a broad field of applications reaching from climatology, stability of ships, stochastics in financial markets to the study of algorithms. The last chapter of this book tries to give an impression of this work.

A concept with a broad range of applications is the concept of a random dynamical system. This concept has been extensively studied by the group around Arnold in Bremen and Imkeller and Scheutzow in Berlin. In this volume this work is represented on a more theoretical level by contribution of Chueshov, Scheutzow and Schmalfuß, by Hermann, Imkeller and Pavlyukevich and a contribution on the applied side by Arnold, Chueshov and Ochs. The first contribution focusses on continuity properties of the inertial manifold in stochastic retarded equations in the regime of small delay times and in the regime of small noise. Another topic relevant in the sciences and falling in the framework of random dynamical systems is what is called stochastic resonance of a random dynamical system. Here the article by Herrmann, Imkeller and Pavlyukevich contributes by analysing a one-dimensional diffusion driven by a Brownian motion with variance ε and a double well potential providing the drift. Here the theory of large deviation and couplings with Markov chains allowing for metastable states play a central role to study the qualitative behavior of this system.

The contribution of Arnold, Chueshov and Ochs uses the concept and the theory of random dynamical systems to analyse in a case study a nonlinear model under the point of view of stochastic stability. The concrete case chosen here is the motion of a ship. The model arises by describing the sea as a stationary random field and formulating the equations of motions from which then a simpler nonlinear random differential equation is extracted for the roll motion. This random dynamical system is then analysed.

A key question in financial markets is how to model and deal with the possibility of a crash. In the contribution of Korn and Menkens the concept of the worst-case portfolio optimization is described and analysed under the point of view of the expected utility. The analysis is focussed on model assumptions relevant for an insurer.

The contribution of Neininger and Rüschendorf on algorithms of divide and conquer type has a quite different flavour. In the center of this work is analysis of additive recursive sequences for which the contraction method is exploited, which is a useful tool for the analysis of stochastic algorithms. The technique allows to derive quite general limit theorems.

Further work not covered here concerns the application of stochastic analysis in climatology, we refer the reader to the proceedings book of Imkeller and von Storch ([IvS01]). The stochastics of financial markets was investigated by the group around Föllmer using stochastic analysis. For example interacting Markov chains, a theme, which is present in most of the other topics of the Schwerpunkt, are applied to the theory of the stochastics of financial markets by Föllmer and Horst in ([FH01]).

References

[CDG]   T. Cox, D.A. Dawson, A. Greven, *Mutually catalytic super branching random walks: large finite systems and renormalization analysis*, to appear in Memoirs of the AMS in 2004.

[DFMPX] D.A. Dawson, K. Fleischmann, L. Mytnik, E.A. Perkins and J. Xiong, *Mutually catalytic branching in the plane: Uniqueness*, Ann. l'Institut Henri Poincaré, Probab. Statist. Vol 39, 135-191 (2003).

[FH01]   H. Föllmer, U. Horst, *Macroscopic and microscopic convergence of Markov chains on infinite product spaces*, Stochastic Process. Appl., Vol. **96 (1)**, 99-121, (2001).

[GKW99] A. Greven, A. Klenke, A. Wakolbinger, *The longtime behavior of branching random walk in a catalytic medium*, EJP, Vol **4**, paper 12, 80, pages (electronic), (1999).

[GKW01] A. Greven, A. Klenke, A. Wakolbinger, *Interacting Fisher-Wright Diffusions in a Catalytic Medium*, Prob. Theory and Rel. Fields, Vol. **120**, No. 1, 85-117, (2001).

[GKW02] A. Greven, A. Klenke, A. Wakolbinger, *Interacting diffusions in a random medium: comparison and longtime behavior*, Stochastic Process. Appl., Vol. **98**, 23-41, (2002).

[IvS01] P. Imkeller, J.-S. von Storch, *Stochastic Climate Models*, Birkhäuser: Basel (2001).

Part I

Stochastic Methods in Statistical Physics

Coarse-Graining Techniques for (Random) Kac Models*

Anton Bovier[1] and Christof Külske[2]

[.] Weierstraß-Institut für Angewandte Analysis und Stochastik
Mohrenstrasse 39, 10117 Berlin, Germany
bovier@wias-berlin.de
[.] Mathematisches Institut, Technische Universität Berlin
Strasse des 17. Juni 136, 10623 Berlin, Germany
kuelske@math.tu-berlin.de

Summary. We review our recent results on the low temperature behavior of Kac models. We discuss translation-invariant models and the Kac version of the random field model. For the latter we outline, how various coarse-graining techniques can be used to prove ferromagnetic ordering in dimensions $d \geq 3$, small randomness, and low temperatures, uniformly in the range of the interaction.

MATHEMATICAL SUBJECT CLASSIFICATION (2000): 82B44, 82B20, 82B28

1 Introduction

Mean field theory is one of the standard tools of statistical mechanics to get a fast first insight into the behaviour of a complex interacting system. The most prominent example in this context is of course the famous van der Waals theory for the liquid vapour transition. However, mean field models have some undesirable "non-physical" properties, in particular they give rise to non-convex thermodynamic potentials, and as a consequence non-monotonous relations between intensive variables and their conjugate fields. These pathologies can be ad-hoc cured by the Maxwell construction, which simply consists in replacing the non-convex potentials by their convex hulls. To make sense of this procedure as an asymptotic theory for realistic physical models, Kac et al. [17] proposed a model with long, but finite, range interactions (of the form $J_\gamma(r) \equiv \gamma^d J(\gamma r)$, with J a function with bounded support or rapid decay), known as the Kac model. Taking the infinite volume limit for such a model first, and then considering the limit as the range of interactions tends to infinity while appropriately rescaling the interaction strength, one then recovers mean field theory. The most precise and complete form of this asymptotic relation was later proven by Lebowitz and Penrose [23]. They showed that the rate function for the total mean magnetization in the Kac model converges, in the limit of infinite interaction range, to the convex hull

---

* This work was supported on part by the DFG priority program 1033

of the corresponding rate function in the Curie-Weiss model. Such results were later recovered for more complicated mean field models, such as the Curie-Weiss-Potts model (see e.g. [16] for a survey).

While these results show that mean field theory can provide reasonable free energies, the issue remains whether finer result, and in particular the phase structure of the model on the level of the Gibbs measures is properly represented. Obviously this cannot in general be true, since mean field theory does not see the influence of dimensionality. How is this reflected in the properties of the Gibbs measures of the corresponding Kac models? If we take, for instance the Kac version of the ferromagnetic Ising model, it is clear that for any value of the parameter γ, if $d = 1$, there exists a unique Gibbs state, while the mean field model has two extremal Gibbs states if the inverse temperature, β, is smaller than the critical values 1. The question to what extend a refined analysis of the Gibbs measures of the Kac models allows one to see some trace of the mean field phase transition was addressed in a seminal paper by Cassandro, Orlandi, and Presutti [13]. They showed that on suitably chosen mesoscopic scales the local mean magnetization under the Gibbs measure of the Kac-Ising model is concentrated sharply near the values of the two mean field magnetisations $\pm m^*(\beta)$ (if γ is small). Moreover, typical magnetization profiles are constant near one of the two values over lengths of the order of $\exp(\gamma^{-1}\beta)$.

In higher dimension, one would expect that the convergence of the Gibbs measures to the mean-field limit is even more straightforward since now more than two extremal states can exist already for finite γ. A natural conjecture would be that the critical temperature $\beta_c(\gamma)$ in all such models should converge to that of the mean field model, as $\gamma \downarrow 0$. Such a result was proven in the case of the Ising model in $d \geq 2$ only rather recently in Cassandro, Marra, and Presutti [9] and Bovier and Zahradník [10]. Both proofs rely very heavily on the spin flip symmetry of the model, and therefore do not generalize easily to a wider class of systems.

Studying Kac versions of models with more complex interactions is actually very attractive since one can hope to take advantage of the fact that the model is "close" to a mean field model in some sense. This may allow, for instance, to estimate the local entropy of configurations close to candidate ground states. This has lead us to be interested in Kac versions of certain types of disordered spin systems whose mean field versions can be rigorously analysed. A particularly attractive model in this respect are the Hopfield models. On the level of large deviation results à la Lebowitz-Penrose, this program has been carried out successfully in [2]. Pushing the analysis towards the Gibbs measures, however, proved much harder. In $d = 1$ it was shown in [3] that local profiles again concentrate near the admissible mean field values, and lower bounds on the typical length over which these remain constant were given. Some finer results were later proven in the simpler case of a random field Kac-Ising model by Cassandro, Orlandi and Picco [12].

The analysis of higher dimensional disordered models has proven to be a major challenge due to the surprising lack of adequate techniques. The development of such tools has been a central theme of the research project that is being reported here. Let us mention that parallel to our efforts, the idea of taking advantage of Kac interactions has been very successfully implemented in a different context, namely that of the liquid vapour transition in a particle system in continuous space. Defining a model with a rather particular form of a four body Kac-interaction, Lebowitz, Maazel, and Presutti [22] have given the first proof of such a phase transition in a single type particle system in continuous space.

The basic difficulties encountered in the analysis of the Gibbs measures of Kac models at low temperatures comes from the fact that the basic methods of low temperature expansions are all devised for models with predominantly strong and short range interaction. In such a situation it is possible to devise perturbative expansions around a single "ground state configuration". In Kac models, where the interactions are very weak, but long range, such an expansion must invariably fail (unless $\beta = O(\gamma^{-d})$). As all methods to analyse the phase structure of lattice models rely on such expansion techniques, there one of the first tasks is the development of suitable methods of low temperature expansion in such a situation. This was done in [11] (and also in [22]).

In the present paper we review our results about the Gibbs measures of (mostly) long-range models. The most difficult part of the analysis concerns the proof of the existence of ferromagnetic order in the random field Kac model for a range of temperatures that is uniform in the Kac parameter γ, for dimensions $d \geq 3$. We cannot give full proofs here but we will explain the various coarse-graining procedures that are involved. This is done in a sequence of steps. Our main philosophy is the *reduction of the long-range random field model to an effective discrete short range contour model* on large enough scales. The latter can then be treated by the renormalization group techniques of Bricmont and Kupiainen. This strategy was already successfully applied to prove ferromagnetic ordering in the continuous spin random field model in [18].

To start, in Section 2 we describe how to define contours in long-range models. The contour definition we give is somewhat subtle [11],[15], but it is needed, even in translation-invariant models, for a satisfactory understanding of the low-temperature phases. These contours will be thick on the order of $1/\gamma$. In Section 3 we review the known results about short-range random field models. We first outline of the RG-treatment of the short range random field Ising model in [8], see also [6]. Then we sketch how the continuous spin random field model can be reduced to the discrete case, as a first successful application of the *reduction* strategy.

In Section 4 we give our main result about the random field Kac Ising model and provide some ideas about the key steps of the proof. The coarse-graining to the nearest neighbor model builds on the contour techniques of section 2, but there is some new difficulty that necessitates the introduction of an additional coarse-graining on an even larger scale than the range of the interaction.

2 Translation-Invariant Long-Range Models

An important technical tool for the analysis of the phase structure of spin systems are reformulations in terms of contour models and convergent cluster expansions. In the case of Kac models, there are some fundamentally new effects that required a substantial reformulation of the classical theory. In [11] this was developed for a rather broad class of predominantly "ferromagnetic" models with weak long range interactions. The challenge here is to develop convergent expansions for a range of temperatures that is *independent of the range of the interactions*, but depends only on the "total strength" of the dominant part of the attractive interaction.

Specifically, we consider Hamiltonians of the form

$$H(\sigma) = \sum_{\{i,j\}} \Phi^\gamma_{i-j}(\sigma_i, \sigma_j) + \sum_i U(\sigma_i) \tag{1}$$

with σ_i taking values in a finite set Q. The interaction kernel Φ^γ_{i-j} is assumed to have finite range $1/\gamma$, to be roughly constant on its range and to verify $\sum_k \|\Phi^\gamma_k\|_\infty = O(1)$. We don't assume any symmetry under permutation of the spin-values.

In this context [11] developed a contour representation for the partition function that allows to compute relevant physical quantities in terms of convergent expansions for a range of temperatures that is uniform in γ as γ tends to 0. As expanded on in [24], this allows to extend the full power of the Pirogov-Sinai theory to such model, and in particular to give a rigorous analysis of the phase diagram in this range of temperatures.

In the remainder of this section we will explain the main ideas in a simplified context that will be relevant for the application in the random field Kac model.

2.1 $1/\gamma$-Contour Model Representation

The main step of the proof is a reformulation of the long-range model in terms of a suitable contour model. We call it $1/\gamma$-contour model because the resulting contours will be thick on the scale of the range of the interaction. Remember that our main emphasis lies on the fact that we are able to treat models without permutation symmetry under the spins. However, for the sake of a transparent explanation of the basic features that are due to the long-range nature, let restrict ourselves just to the simplest non-trivial long-range model, i.e. let us specialize to the Ising model with spin-flip symmetric translation-invariant interaction. We remark that the formulae to be discussed here have immediate random generalizations to the random field Kac model, to be discussed in Section 4.

So, let us look at the Hamiltonian

$$H(\sigma) = -\sum_{\substack{\{i,j\} \\ i \neq j}} J_\gamma(i-j)\sigma_i\sigma_j \tag{2}$$

The spin variables $\sigma = (\sigma_i)_{i \in \mathbb{Z}^d}$ take values in $\{-1,1\}^{\mathbb{Z}^d}$. We take the simplest possible choice for the couplings being the indicator function of a cube of sidelength $R = \gamma^{-1} \in \mathbb{N}$ and the energy-difference due to the pair interaction of flipping one spin in a sea of plusses is exactly equal to one, that is

$$J_\gamma(i) = \frac{1_{U_R^\bullet}(i)}{|U_R^\bullet|} \qquad \text{where} \qquad U_R^\bullet = \{j \neq 0, |j| \leq R\} \tag{3}$$

The main observation is that, in contrast to a short range model, it is impossible to expand the model around the perfect plus- resp. minus-groundstates. Applying the standard PS-theory for short-range models would only yield ferromagnetic ordering for temperatures that decrease with the range of the interaction tending to infinity. This problem is cured (very roughly) in the following way: 1) Replace the perfect infinite volume plus- or minus-groundstate by sets of perturbed plus- or minus-like configurations ("ensembles") that are characterized by a low density of wrong signs. These ensembles can be treated by high-temperature expansions. 2) Define the contours as the complement of those regions, that is those regions of space where the spins are not plus- resp. minus-like. Show that these contours obey Peierls estimates and their ensembles can be treated by methods known from short-range Pirogov-Sinai theory. While this is the correct main idea, a contour definition in terms of just one density constraint following the above procedure literally is a little too simplistic, and needs some modification.

Contours of a Configuration and Peierls Bounds We would like to include a correct definition of a contours and low-density ensembles for the Ising model, although it seems technically slightly involved at first sight. The version of the contour definition coincides essentially with the one given in [11]; however we present the version given in [15] that is also used in [7].

Fix a density threshold $\delta > 0$ and a sign $s = \pm 1$. We call i a (δ, s)-*correct point* of a configuration σ if

$$\#\{j : j - i \in U_R^\bullet, \sigma_j \neq s\} \leq \frac{\delta}{2}|U_R^\bullet| \tag{4}$$

Note that this definition makes no reference of the sign at the site i itself; it is only a property of the configuration around it. So, let us define the *cleaned configuration* $\bar{\sigma} = \bar{\sigma}(\sigma)$ corresponding to σ by setting

$$\bar{\sigma}_i(\sigma) = \begin{cases} s & \text{if } i \text{ is } (\delta, s) - \text{correct for } \sigma \\ \sigma_i & \text{else}, \end{cases} \tag{5}$$

We denote by $I_\delta(\sigma)$ the set of δ-incorrect point of the configuration σ. This set would be a natural first guess for the support of the contours to be defined. This is clear because deviation from almost homogeneous spin-configurations should come at a high energetic price. However, in order to arrive at a successful decomposition of the partition function into low-density ensembles and contours a little more care is needed.

First of all, as usual in short-range PS-theory, a *contour* $\Gamma = (\underline{\Gamma}, \sigma_{\underline{\Gamma}})$ is by definition a pair given by a connected subset $\underline{\Gamma} \subset \mathbb{Z}^d$, called the *support*, and a spin configuration $\sigma_{\underline{\Gamma}}$ on this subset. Here we will define supports of contours that are thick of the order of the range of the interaction. To be explicit, partition $\mathbb{Z}^d$ into disjoint cubes of side length $l > 2R$ whose centers lie of the sites of a square sub-lattice of $\mathbb{Z}^d$, and assume that $l = \nu R$, $\nu > 2$. Denote the set of all subsets of $\mathbb{Z}^d$ that can be written as unions of such cubes (with fixed sublattice!) by $C^{(l)}$. Denote by $[A]_R = \{j, |j - i| \leq R, i \in A\}$ the R-neighborhood of a set A.

Now, take a second density threshold value $\tilde{\delta} < \delta$.

Definition 2.1 *Let σ be a spin-configuration. We call the connected components of*

$$I^*(\sigma) := \bigcap_{\substack{M \in C^{(l)}, \, M \supset [I_\delta(\sigma)]_R \\ M \supset [I_{\tilde{\delta}}(\sigma_M \bar{\sigma}_{M^c})]_R}} M \tag{6}$$

the support of the contours of the configuration σ. A system of contours $\overline{\Gamma} = \{\Gamma_1, \ldots, \Gamma_n\}$ is admissible if there is a spin-configuration σ such that $\overline{\Gamma}$ is the associated system of contours.

Denote by $\Lambda_s(\sigma)$ the set of (δ, s)-correct points. Then one has that $d(\Lambda_+(\sigma), \Lambda_-(\sigma)) > l$ and $\mathbb{Z}^d = I^*(\sigma) \cup \Lambda_+(\sigma) \cup \Lambda_-(\sigma)$. It is not too difficult to see that the intersection is over non-empty sets M and $I^*(\sigma)$ is in fact the smallest set M having the three properties appearing in the last intersection. This is done explicitly in [15] Lemma 2.2. ff.

If we want to control the finite volume Gibbs measures with plus boundary conditions we have to look at the partition function in finite volume Λ with plus boundary conditions. In the following the summation over all spin-configurations will be split in the following way. 1) Sum over all possible sets $I^*(\sigma)$, and the possible spin configurations on those sets. 2) Conditional on those sets and spin-configurations sum over all spin-configurations that are compatible with them.

This procedure gives the following decomposition into a sum over compatible contours and partition functions of *restricted low-density* ensembles.

$$\sum_{\sigma_\Lambda} e^{-\beta H_\Lambda^+(\sigma_\Lambda)} = \sum_{\overline{\Gamma}} \left(\prod_\Gamma \rho(\Gamma) \right) Z_{\Lambda_+}^{+,r}(+_{\Lambda^c}, \sigma_{\overline{\Gamma}}) \, Z_{\Lambda_-}^{-,r}(\sigma_{\overline{\Gamma}}) \tag{7}$$

For the partition functions of the restricted ensembles one has the explicit formula

$$Z_\Lambda^{s,r}(\sigma_{\Lambda^c}) = \sum_{\sigma_\Lambda} \left(\prod_{i \in [\Lambda]_R} 1_{i \text{ is } (\delta,s) \text{ correct for } \sigma}} \right)$$

$$\times \exp\left(2\beta \sum_{\substack{\{i,j\} \cap \Lambda \neq \emptyset \\ i \neq j}} J_\gamma(i-j) 1_{\sigma_i \neq s} 1_{\sigma_j \neq s} - \beta \sum_{i \in \Lambda} 1_{\sigma_i \neq s} \right) \tag{8}$$

They depend on boundary conditions imposed by the spin-configurations $\sigma_{\overline{\Gamma}}$ on the contours (and the plus boundary condition outside of Λ). The contour activities $\rho(\Gamma)$ simply collect the terms of the interactions that are only between sites located on the support of the contour. (Here the interaction between contour-sites and the cleaned configuration put on the sites outside of the contour is included, too.)

A main virtue of the contour-definition is that the following Peierls estimate holds.

Lemma 1. *There is a dimension-dependent constant $c = c(\tilde{\delta})$ such that*

$$\rho(\Gamma) \leq e^{-c(\tilde{\delta})\beta|\underline{\Gamma}|} \tag{9}$$

Note the important fact that the volume $|\underline{\Gamma}|$ is always a multiple of l^d and so we have a rather strong suppression. How can we understand that this volume-suppression is always on the scale of the range of the interaction? It is because once there are incorrect points for some density, there are even Const R^d of them, for a slightly lowered density, and each of these contributes an amount of energy that is bounded below by a positive constant. We don't give a full proof here.

Cluster Expansion of Restricted "Low Density" Ensembles The next important step of the preparation of the model is the following: The restricted partition function (8) can be written as a polymer partition function

$$Z_\Lambda^{s,r}(\sigma_{\Lambda^c}) = \sum_{(P_1,\ldots,P_n)_{cp}} \prod_{i=1}^n \tilde{w}_{P_i} \tag{10}$$

with polymers P_i that interact only via volume constraints. The logic to achieve this is as follows: Expand the interaction term between any pair of sites i, j in (8) first. We note that each site with the wrong sign $-s$ is energetically suppressed, due to the second term in the second exponential in (2.9). So, there is sufficient energetic suppression of a configuration as soon as there are not too many of such spins cooperating, so that the pair-inteaction term is not too big. This is shown using the indicator functions in the first brackets. Next, the indicator functions describing the density-constraint are expanded. This is needed in order to produce a gas of non-interacting polymers. Their activities are then shown to obey nice estimates and allow for a cluster expansion.

Because of their geometry these polymers were given the name *galaxies* in [11]. To be a little more precise, the polymers are of the form $P = (t_1, \ldots, t_l, N_1, \ldots, N_k)$ where t_i are trees connecting the sets N_i. The trees t_i arise from the high-temperature expansion in the restricted ensembles. The sets N_i are connected components of the form $\cup_j B_R(j)$. They arise from the expansion of the non-local density constraints. One has for the polymer activities $\tilde{w}_P$ of galaxies P the estimates

$$\sum_{P:V(P)\ni i} |\tilde{w}_P| e^{\mathrm{const}\,\beta|V(P)|} \le \varepsilon \tag{11}$$

where $V(P) = \bigcup_t V(t) \cup \bigcup_i N_i$.

From (11) follows that one may now perform a cluster expansion of (10) and exponentiate the restricted partition function. We stress again at this point that the same constructions work for non-symmetric models with finite state space.

The resulting representation is then the starting point for the further treatment of such model by short-range Pirogov-Sinai type techniques. This is explained in [24].

3 Random Short Range Models and Coarse-Grainings

To achieve our final goal and treat the Kac random field Ising model we will find an effective short-range model that describes the long range model on a sufficiently large scale. For this short-range model ferromagnetic order is well-established [8] and can be carried over to the original model. This philosophy of "reduction to an effective short-range Ising type model" was already devised in [18] to treat the nearest continuous spin model. This will be shortly discussed in Subsection 3.2. However, at first we must turn to a review of the short-range model itself.

3.1 The Random Field Ising Model

So, let us consider the random field Ising model (with symmetric non-degenerate distribution) and nearest neighbor interaction whose Hamiltonian is given by

$$H[\eta](\sigma) = -\sum_{\langle i,j \rangle} \sigma_i \sigma_j - \sum_i \eta_i \sigma_i \tag{12}$$

Here the $(\eta_i)_{i \in \mathbb{Z}^d}$, are i.i.d. symmetrically distributed random variables that satisfy the probabilistic bound

$$\mathbb{P}[\eta_i \geq t] \leq e^{-\frac{t^2}{2\sigma^2}} \tag{13}$$

where the $\sigma^2 \geq 0$ governing the smallness of the random variables has to be sufficiently small.

We are then interested in the random Gibbs measures

$$\mu[\eta](d\sigma) \propto e^{-\beta H[\eta](\sigma)} d\sigma \tag{14}$$

where $d\sigma$ is the symmetric product measure on the spin configurations σ. (Of course, this is to be understood as a solution to the DLR-equations written down for fixed random field configuration η.)

For this model it was proved in [1] that there is unicity of the Gibbs measure in 2-dimensions, at any fixed temperature, for $\mathbb{P}$-a.e. η. This is in contrast to the case of the model without disorder, which shows that the introduction of arbitrarily weak random perturbations can destroy a phase transition. It shows that randomness can potentially alter the behavior of the system in a fundamental way, and cannot always be treated as a small perturbation. distribution energies in finite martingale techniques We remark that their (martingale-)method was later applied by [5] to show the non-localization of interfaces in random environments in the framework of certain models for interfaces without overhangs in space dimensions less than $3 = 2 + 1$.

In the three or more dimensional random field Ising model, for small disorder, and small temperature, however, disorder does *not* destroy the ferromagnetic ordering. Here, Bricmont and Kupiainen showed in their famous paper [8] that there exist Gibbs measures $\mu^+[\eta]$ (and $\mu^-[\eta]$), which, for typical magnetic field configuration η, describe small perturbations around a plus-like (respectively a minus-like) infinite-volume ground state. A plus-like ground state looks like a sea of pluses with rare islands of minuses in those regions of space where the realizations of the magnetic fields happen to be mostly oriented to favor the minus spins. We remark that the result of [8] was a nice example where a question that was truly under debate among theoretical physicists could be settled by mathematicians.

Contour Model Representation of Short-Range Random Field Model – Renormalization Group The method they used to control the Gibbs measures, the so-called 'renormalization group', is a multiscale method that consists in a successive application of a certain coarse-graining/rescaling procedure. This is necessary because there is no simple Peierls-condition for this model (say around the all-plus state.) The individual steps are controlled by expansion methods and probabilistic estimates of the undesirable event that regions of exceptionally large magnetic fields occur. This has to be done for all hierarchies occurring. This method is conceptually beautiful but technically hard to implement. It was later also applied by [6] to show the stability of certain interface models in dimensions $d + 1 \geq 4$. behavior

What is the Bricmont-Kupiainen renormalization group in a little more detail? First of all, one has to find a representation of the model as a short-range contour model. Here a contour is again a pair $\Gamma = (\underline{\Gamma}, \sigma_{\underline{\Gamma}})$ of a support and a spin-configuration on the support. Then the RG-transformation of the model is carried out in the contour representation in finite but arbitrarily large volume, for every fixed realization of the disorder η. This RG-transformation roughly consists of two steps: Fix an integer L describing the blocklength. First step: Integration of small contours with diameter smaller than $L/4$. The resulting expression for the partition function contains only a sum over the remaining large contours with new interaction produced by the small-contour sum. Second step: Do a coarse-graining of the remaining large contours. This eliminates the "wiggles" of the remaining large contours so that the resulting model can then be rescaled. The result is a model that lives again on a lattice with the original lattice spacing.

It is the virtue of Bricmont and Kupiainen to have found a representation of the model that is stable under this transformation for all hierarchies, along with a suitable type of bounds. Running the RG an iterative proof shows that a) temperature effectively becomes smaller and b) that randomness becomes smaller and thus more irrelevant on large hierarchies. This translates into the fact that the formation of large contours is very unlikely. This proves ferromagnetic ordering. The same basic structure underlies the work on the stability of SOS-interfaces by the authors.

Now, let us provide some more details about the representation of the contour model that is "BK-renormalizable", that is invariant under the renormalization procedure sketched above. These details will be needed to understand some key features of the proof for the Kac random field model sketched in Section 4. For a spin configuration and a collection of contours $\overline{\Gamma}$ we denote by $\sigma_i(\overline{\Gamma})$ the spin configuration that equals σ_i on $\overline{\Gamma}$ and equals s for $i \in \Lambda_s$. Denote by $Z_\Lambda[\eta]$ the partition function of the model after an arbitrary number of steps of the renormalization group transformation. Then, for each realization of η, and after each finite number of steps one always has the representation

$$Z_\Lambda[\eta] = \sum_{\overline{\Gamma}} \left(\prod_\Gamma \rho[\eta](\Gamma) \right) \tag{15}$$

$$\times \exp\left(\sum_{i \in \Lambda} S_i(\eta) \sigma_i[\overline{\Gamma}] + \sum_{C \in \Lambda_+(\overline{\Gamma})} S_C^+(\eta) \right.$$

$$\left. + \sum_{C \in \Lambda_-(\overline{\Gamma})} S_C^-(\eta) + \sum_C W_C^{\overline{\Gamma}}(\eta) \right)$$

Here $S_i(\eta)$ is a local random field at the site i, which depends on the original random field η only in a stricly local way. It obeys a bound of the type (13) with a new "renormalized (upper bound on the) variance" σ^2. The quantity σ^2 depends on the hierarchy and decreases with application of the RG. $S_C^\pm(\eta)$ are nonlocal random fields, depending on *connected sets* C on the lattice of at least 2 points. They are exponentially suppressed in the volume,

$$|S_C^\pm(\eta)| \le e^{-\text{const } \tilde{\beta}|C|} \tag{16}$$

with some hierarchy-dependent constant $\tilde{\beta}$ and keep the symmetry of the model, $S_C^+(\eta) = -S_C^-(-\eta)$. When we will try to apply the renormalization group strategy to the long range model this geometric structure of the C 's being *connected sets* will become important and cause some additional complexities.

The $\rho[\eta](\Gamma)$ are contour activities, depending on the realization of the random fields. *Essentially* they obey deterministic Peierls bounds $0 \le \rho[\eta](\Gamma) \le e^{-\text{const } \beta|\underline{\Gamma}|}$, β being a "renormalized inverse temperature". However the Peierls bound just stated holds only for contours in regions of space where the underlying randomness was not "too big". A failure of the simple Peierls bound may happen in the so called large field region. Without being explicit about this let us mention that it is one important part of the proof to show that this large field region is very exceptional and becomes less and less important under RG. Finally, $W_C^{\overline{\Gamma}}[\eta]$ are nonlocal interactions between the contours; they are nonzero only for connected sets C that intersect the support of the contours and obey deterministic upper bounds of the same type as the non-local random fields. They are just correction terms whose creation under RG can't be avoided.

The hard technical work of the RG-analysis then consists in showing that this form is invariant, with $\beta \uparrow \infty$, $\tilde{\beta} \uparrow \infty$, $\sigma^2 \downarrow 0$ and the bad region dying out under successive application of the RG.

3.2 The Continuous Spin Random Field Model

In the context of disordered systems the continuous spin version corresponding to the random field model is an important model to study. In the physics literature is no less popular than the Ising model itself, so it is interesting to see whether ferromagnetic ordering can be proved for this model, too. Now the spin variables m_i take values in $\mathbb{R}$ and the formal Hamiltonian for a spin-configuration $m \in \mathbb{R}^{\mathbb{Z}^d}$ in the infinite volume is given by

$$H[\eta](m) = \frac{q}{2} \sum_{\langle i,j \rangle} (m_i - m_j)^2 + \sum_i V(m_i) - \sum_i \eta_i m_i \tag{17}$$

where the first summation extends over all pairs of nearest neighbors $\langle i,j \rangle$. The finite volume Gibbs measures are then obviously given by taking the exponential of the negative finite volume restriction of (17) as the non-normalized Lebesgue-density. The potential V has a symmetric double-well structure. We will stick to the most popular choice which is a polynomial of fourth order. Let us choose a scaling where the potential has unity curvature in the minima $\pm m^*$ that is

$$V(m_i) = \frac{\left(m_i^2 - (m^*)^2\right)^2}{8m^{*2}} \tag{18}$$

and investigate the Gibbs measures for $q \geq 0$ sufficiently small and $q\,(m^*)^2$ sufficiently large. The latter quantity gives the order of magnitude of the minimal energetic contribution to the Hamiltonian (17) caused by neighboring spins in different wells. Thus it corresponds to a Peierls constant. Moreover we impose a fixed uniform bound on $|\eta_i|$, independent of σ^2. This is for technical reasons.

Since we are dealing with continuous degrees of freedom a contour definition and a direct application of the renormalization group in the framework of Bricmont and Kupiainen is not immediate, and if possible would entail a huge amount of technical work. Moreover one might be afraid that the additional continuous degrees of freedom could lead to a possible loss of order.

In this context we show that there is in fact a 'ferromagnetic' phase transition, in dimensions $d \geq 3$, for sufficiently small σ^2 (meaning small disorder), sufficiently large $q(m^*)^2$, and not too big $q(m^*)^{\frac{2}{3}}$ (controlling the 'anharmonicity' of the minima, as it can be seen from the proof). We prove the following: The [random] Gibbs-probability (w.r.t. to the finite volume-measure with plus-boundary conditions) of finding the spin left to the positive potential well is very small, uniformly in the volume, on a set of realizations of η of a size [w.r.t $\mathbb{P}$] of at least $1 - e^{-\frac{\text{const}}{\sigma^2}}$. The precise statement is found in Theorem 1 p.1272 of [18]. For more information and explanation we refer to the introduction of [18]. Let us however mention the following: The particular form of the potential as a fourth order polynomial is of no importance, as well as the requirement of uniform boundedness on the random fields and the restriction to nearest neighbor couplings in the Hamiltonian (instead of finite-range interactions) could be given up.

Reduction to Short-Range Contour Model
– Coarse-Graining in Spin-Space The novelty of the proof is the use of a stochastic mapping of the continuous spins to their sign-field (independently over the sites). We choose it such that the probability that a continuous spin m_x is mapped to its sign is given by $\frac{1}{2}(1 + \tanh(am^*|m_i|))$. (Here a is a parameter close to one that needs to be tuned in a useful way.) The image measure of a particular sign-configuration then gives the approximate weights of finding continuous spins in the neighborhood of the potential wells indexed by these signs. Using a suitable combination between high temperature and low temperature expansions it is shown that the resulting model has the form

of an Ising model with exponentially decaying interactions. (These expansions are related to those used by [Za00] in the translation-invariant context where however, due to the lack of positivity, no probabilistic interpretation can be given.) This can be seen as a 'single-site-coarse-graining'-method. Next, having constructed the Ising-system, it can be cast into a contour model representation for which the renormalization group of [BrKu88] can be used. We remark that a lot of technical work is needed to implement this idea of the reduction to a discrete model. Still, while doing so, there is still a great deal of work saved which was already done on the level of the discrete RG.

For the readers interested in the theory of generalized Gibbs measures we also remark that this mapping is really compatible with the infinite volume limit in the sense that the infinite system under consideration is mapped to a proper infinite volume *Gibbs* measure of an Ising model (see Theorem 2 of [18], p.1273). So, this stochastic map also provides an interesting example of a 'coarse-graining without pathologies'. This means that the coarse-graining produces no 'artificial' non-local dependencies in the conditional expectations of the resulting measure.

Let us just mention at this place that, in contrast to that, disordered systems frequently provide a source of various non-Gibbsian measures when we look at them jointly on the product space of disorder variables and spin variables. This is known as the so-called Morita-approach in theoretical physics. For more rigorous research on this we refer to [19, 20, 21] and the references therein.

4 The Random Field Kac Ising Model

Let us finally turn to the discussion of our main result. Consider the model with Hamiltonian

$$H_\gamma[\eta](\sigma) = -\sum_{\substack{\{i,j\} \\ i \neq j}} J_\gamma(i-j)\sigma_i\sigma_j - \sum_i \eta_i\sigma_i$$

Then we prove the following.

Theorem 4.1 *Assume that $d \geq 3$. Assume that η_i are symmetrically distributed i.i.d. and $|\eta_i| \leq \delta_{RF}$. Then there is $\gamma_0 > 0$ and $\beta_0 < \infty$, such that for all $\gamma \leq \gamma_0$ and $\beta \geq \beta_0$ we have*

$$\mathbb{E}\left(\mu^+_{\beta,\gamma}[\eta](\sigma_i = 1)\right) > \frac{1}{2}.$$

We will sketch two steps of the proof. In the first step, described in 4.1 we will obtain a contour-representation that is quite analogous to the one described for the translation-invariant Kac-model in Subsection 2.1. Unfortunately the result is not yet good enough to yield a formulation of the model allows for the RG-treatment. We will then briefly indicated what is the problem and the cure in Subsection 4.1.

4.1 $1/\gamma$-Contour Model Representation

Let us do the decomposition into contours and low-density ensembles without any reference to the configuration of the the random field η.

The first main point is the following decomposition that generalizes (1) to the random situation

$$Z_{\gamma,\Lambda}^{\mathrm{Kac}}[\eta] \equiv \sum_{\sigma_\Lambda} e^{-\beta H_\Lambda^+[\eta](\sigma_\Lambda)} \tag{19}$$

$$= \sum_{\overline{\Gamma}} \left(\prod_\Gamma \rho(\Gamma) \right) \exp\left(\beta \sum_{i\in\Lambda} \eta_i \sigma_i[\overline{\Gamma}] \right) Z_{\Lambda_+}^{+,r}[\eta_{\Lambda_+}](+_{\Lambda^c}, \sigma_{\overline{\Gamma}})\, Z_{\Lambda_-}^{-,r}[\eta_{\Lambda_-}](\sigma_{\overline{\Gamma}})$$

Here the contour-activities are identical to those in the non-random case described in (8) but the restricted partition functions acquire the modified form

$$Z_\Lambda^{s,r}[\eta_\Lambda](\sigma_{\Lambda^c}) = \sum_{\sigma_\Lambda} \left(\prod_{i\in[\Lambda]_R} 1_{i \text{ is } (\delta,s) \text{ correct for } \sigma} \right) \tag{20}$$

$$\times \exp\left(2\beta \sum_{\substack{\{i,j\}\cap\Lambda\neq\emptyset \\ i\neq j}} J_\gamma(i-j) 1_{\sigma_i\neq s} 1_{\sigma_j\neq s} - \beta \sum_{i\in\Lambda}(1+2s\eta_i) 1_{\sigma_i\neq s} \right)$$

They can be written as partition functions for a polymer gas in the very same way as for the translation-invariant model. Of course, now the polymer activities will depend on the random fields. Under the condition of uniform boundedness of the random fields with δ_{RF} sufficiently small there is however still sufficient suppression of the polymer-activities. Performing the cluster expansion for these partition functions we obtain the representation

$$Z_{\gamma,\Lambda}^{\mathrm{Kac}}[\eta] = \sum_{\overline{\Gamma}} \left(\prod_\Gamma \rho(\Gamma) \right) \tag{21}$$

$$\times \exp\left(\beta \sum_{i\in\Lambda} \eta_i \sigma_i[\overline{\Gamma}] + \sum_{\substack{\mathcal{C} \\ V(\mathcal{C})\subset\Lambda_+(\overline{\Gamma})}} w_{\mathcal{C}}^+(\eta) + \sum_{\substack{\mathcal{C} \\ V(\mathcal{C})\subset\Lambda_-(\overline{\Gamma})}} w_{\mathcal{C}}^-(\eta) + \sum_{\mathcal{C}\mathrm{icp}\,\overline{\Gamma}} w_{\mathcal{C}}^{\overline{\Gamma}}(\eta) \right)$$

The symbols $\mathcal{C}$ denote clusters of the type of polymers defined below (10), and $V(\mathcal{C})$ is just the union of the volumes $V(P)$ over the polymers P the cluster is made of. From a uniform version of (11) than follows that the corresponding fields obey the uniform exponential estimate

$$\sum_{\mathcal{C}:V(\mathcal{C})\ni i} \sup_\eta |w_{\mathcal{C}}^{+,-,\overline{\Gamma}}(\eta)| e^{\mathrm{const}\,\beta V(\mathcal{C})} \leq \mathrm{Const} \tag{22}$$

At first glance this looks close to the form (15) that is invariant under the Bricmont-Kupiainen renormalization group, but it is not! This is due to the fact that the clusters $\mathcal{C}$ are made of polymers whose geometric structure is quite different from the connected sets occurring for the short-range random field model. We will see that problems are occurring if we perform one additional RG-step of the Bricmont-Kupiainen renormalization. The problems are due to the fact that in the RG-invariant formulation of the model *connected* sets appear as indices of the non-local contributions of the field while we are dealing here with possibly spread out geometric objects. Remember that the clusters are made of polymers which contain trees arising from the expansion of the long-range interactions. The simplest and "most spread-out" polymer occurring is just a string composed of bonds whose length are of the order of the range of the interaction R. The most straightforward idea to create corresponding connected sets would therefor be to perform one additional blocking-step with blocks of side-length of the order R. However, as we will see below this won't provide us with the desired bounds that allow for further application of the RG for a range of inverse temperatures β that is uniform in the range of interaction. To explain the problem and its cure a little more in detail let us be more general and see what happens under a blocking with block-length LR, L being a constant to be determined.

4.2 *LR*-Blocking

Let us denote by x a block on the RL-lattice and put for the *local part* of the effective random field associated to this block the trial definition

$$S_x(\eta) := \sum_{i \in x} \beta \eta_i + \sum_{\substack{\mathcal{C}:V(\mathcal{C}) \subset x \text{ or} \\ d(\mathcal{C}) < LR/4}} \frac{w_{\mathcal{C}}^+(\eta) - w_{\mathcal{C}}^-(\eta)}{2n(\mathcal{C})} \qquad (23)$$

where $n(\mathcal{C})$ is the number of blocks on the RL-lattice that are intersected by the volume of the cluster $V(\mathcal{C})$. This is a straightforward and natural construction in the RG-treatment.

The remaining sums over clusters of galaxies have to be blocked by infection. They give rise to a non-local small field term S_K that is indexed by connected subsets K of adjacent blocks on the RL-lattice.

$$S_K^\pm(\eta) := \sum_{\substack{\mathcal{C}:\mathcal{C} \mapsto K \text{ and} \\ d(\mathcal{C}) \geq (RL)/4}} w_{\mathcal{C}}^\pm(\eta) \qquad (24)$$

where we sum over those clusters $\mathcal{C}$ for which K is the minimal set of adjacent RL-blocks that contains their volume $V(\mathcal{C})$.

The local term poses no difficulties, it is the non-local problem that is difficult. Indeed, to treat the first, we note that the variation of the function $S_x(\eta)$ w.r.t. change of a single random field is at most $\beta\delta_{RF} + e^{-\text{const }\beta}$. Since $S_x(\eta)$ depends only on η_i in a set of sites that is of the order $(RL)^d$ one gets by a well-known martingale estimate the desired Gaussian upper bound of the form (13)

$$\mathbb{P}\left[S_x \geq t\right] \leq e^{-\text{const}\,\frac{t^2}{(\beta\delta_{RF}+e^{-\text{const}\,\beta})^2(RL)^d}} \tag{25}$$

for all $t \geq 0$.

Let us now try to get an exponential bound of the type (16) for the non-local part (24), where $|C|$ has to be replaced by the rescaled volume, which is just the number of blocks $n = |K|/(RL)^d$. The best we can do is to estimate

$$|S_K^\pm| \leq \sum_{i \in K} \sum_{\substack{C:V(C) \ni i\text{ and }C \mapsto K \\ \text{and }d(C)\geq (RL)/4}} |w_C^\pm(\eta)| \leq n(RL)^d e^{-\text{const}\,\beta Ln} \tag{26}$$

Here the factor L in the exponential stems from the fact that the "worst" clusters that are contributing to the sum have a minimal volume of the order L. This is already seen by looking at polymers made of trees with bonds with width of the order R. The prefactor $n(RL)^d$ counts the number of "anchoring points" i of such clusters. This bound is not unnatural because the cluster-sum corresponds to a sort of free energy which should be of the order of the volume, measured on the original scale. This shows that this bound can not easily be improved. Now, the problem is that the prefactor causes this bound to be non-uniform in R.

However, for any fixed L, the exponential dominates the prefactor as soon as we demand that $\beta \geq \text{Const}\log R$. In fact, in this range of temperatures the full preparation of the model to accommodate for the short-range RG is easy and we get ferromagnetic ordering without too much difficulty.

However, the reason why we made L explicit in the above formulas is that we want to make it R-dependent according to $L(R) = \text{Const}\log R$.

Then, for $\beta \geq \beta_0$ sufficiently large, where β_0 is *uniform in* R, we get the desired estimate

$$|S_K^\pm| \leq e^{-\text{const}\,\beta L(R)n} \tag{27}$$

The present argument only gives an outline why we are forced to deal with an additional length-scale. It remains to show that a model (15) with desired bounds can be rigorously derived, incorporating the influence of contours whose diameter is smaller than $RL/4$. There is a fair amount of technical work involved. The details and additional arguments can be found in [7].

References

1. M. Aizenman and J. Wehr, "Rounding effects on quenched randomness on first-order phase transitions", Commun. Math. Phys. **130**, 489 (1990).
2. A. Bovier, V. Gayrard, and P. Picco, "Large deviation principles for the Hopfield model and the Kac-Hopfield model", Prob. Theor. Rel. Fields **101**, 511-546 (1995).
3. A. Bovier, V. Gayrard, and P. Picco, "Distribution of overlap profiles in the one-dimensional Kac-Hopfield model", Comm. Math. Phys. **186**, 323–379 (1997).

4. A. Bovier and C. Külske, "A rigorous renormalization group method for interfaces in random media", Rev. Math. Phys. **6**, 413–496 (1994)
5. A. Bovier and C. Külske, "There are no nice interfaces in $(2 + 1)$-dimensional SOS models in random media", J. Stat. Phys. **83** (1996), 751-759 (1996).
6. A. Bovier and Ch. Külske, A rigorous renormalization group method for interfaces in random media, Rev. Math. Phys. **6**, 413-496 (1994).
7. A. Bovier, C. Külske, *Phase Transition in the three dimensional random field Kac model*, in preparation
8. J. Bricmont, and A. Kupiainen, "Phase transition in the 3d random field Ising model", Commun. Math. Phys. **116**, 539-572 (1988).
9. M. Cassandro, R. Marra, and E. Presutti, "Corrections to the critical temperature in 2d Ising systems with Kac potentials", J. Stat. Phys. **78**, 1131-1138 (1995).
10. A. Bovier and M. Zahradník, "The low-temperature phase of Kac-Ising models". J. Stat. Phys. **87**, 311–332 (1997).
11. A. Bovier and M. Zahradník, "Cluster expansions and Pirogov–Sinai theory for long range spin systems", Markov Proc. Rel. Fields **8**, 443-478 (2002).
12. M. Cassandro, E. Orlandi, P. Picco, "Typical configurations for one-dimensional random field Kac model", Ann. Probab. **27**, 1414–1467 (1999)
13. M. Cassandro, E. Orlandi, and E. Presutti, "Interfaces and typical Gibbs configurations for one-dimensional Kac potentials", Prob. Theor. Rel. Fields **96**, 57-96 (1993).
14. A. De Masi, E. Orlandi, E. Presutti, and L. Triolo, "Glauber evolution with Kac potentials, I. Mesoscopic and macroscopic limits, interface dynamics", Nonlinearity **7**, 633-696 (1994); Glauber evolution with Kac potentials. II. Fluctuations. Nonlinearity **9**, 27–51 (1996); Glauber evolution with Kac potentials. III. Spinodal decomposition. Nonlinearity **9**, 53–114 (1996).
15. S. Friedli S., C.-E. Pfister, "Non-Analyticity and the van der Waals limit", preprint (2003), to appear in J. Stat. Phys.
16. H. Kesten and R. Shonmann, "Behaviour in large dimensions of the Potts and Heisenberg model", Rev. Math. Phys. **1**: 147-182 (1990).
17. M. Kac, G. Uhlenbeck, and P.C. Hemmer, "On the van der Waals theory of vapour-liquid equilibrium. I. Discussion of a one-dimensional model" J. Math. Phys. **4**, 216-228 (1963); "II. Discussion of the distribution functions" J. Math. Phys. **4**, 229-247 (1963); "III. Discussion of the critical region", J. Math. Phys. **5**, 60-74 (1964).
18. C. Külske, "The continuous spin random field model: Ferromagnetic ordering in $d \geq 3$", Rev.Math.Phys., **11**, 1269-1314 (1999)
19. C. Külske, "Weakly Gibbsian representations for joint measures of quenched lattice spin models" Probab. Theory Related Fields 119 (2001)
20. C. Külske, "Analogues of non-Gibbsianness in joint measures of disordered mean field models" J. Stat. Phys. **112**, 1079–1108 (2003)
21. C. Külske, A. Le Ny, F. Redig, "Relative entropy and variational properties of generalized Gibbsian measures", Eurandom preprint 2002-035, accepted for publication in Annals of Probability
22. J. L. Lebowitz, E. Presutti, "Liquid-vapor phase transitions for systems with finite-range interactions", J. Stat. Phys. **94**, 955–1025 (1999)
23. J. Lebowitz and O. Penrose, "Rigorous treatment of the Van der Waals Maxwell theory of the liquid-vapour transition", J. Math. Phys. **7**, 98-113 (1966)

24. M. Zahradník, "Cluster expansions of small contours in abstract Pirogov-Sinai models", Markov Process. Related Fields **8**, 383–441 (2002)

Euclidean Gibbs Measures of Quantum Crystals: Existence, Uniqueness and a Priori Estimates

Sergio Albeverio[1,3,4], Yuri Kondratiev[2,5],
Tatiana Pasurek[2,3], and Michael Röckner[2,3]

- Institut für Angewandte Mathematik, Universität Bonn,
  D-53155 Bonn, Germany
  albeverio@uni-bonn.de
- Fakultät für Mathematik, Universität Bielefeld, D-33615 Bielefeld, Germany
  {kondrat,roeckner}@mathematik.uni-bielefeld.de
  pasurek@physik.uni-bielefeld.de
- BiBoS Research Centre, Bielefeld, Germany
- CERFIM, Locarno, Switzerland
- Institute of Mathematics, NASU, Kiev, Ukraine

Summary. We give a review of recent results obtained by the authors on the existence, uniqueness and a priori estimates for Euclidean Gibbs measures corresponding to quantum anharmonic crystals. Especially we present a new method to prove existence and a priori estimates for Gibbs mesures on loop lattices, which is based on the alternative characterization of Gibbs measures in terms of their logarithmic derivatives through integration by parts formulas. This method allows us to get improvements of essentially all related existence results known so far in the literature. In particular, it applies to general (non necessary translation invariant) interactions of unbounded order and infinite range given by many-particle potentials of superquadratic growth. We also discuss different techniques for proving uniqueness of Euclidean Gibbs measures, including Dobrushin's criterion, correlation inequalities, exponential decay of correlations, as well as Poincaré and log-Sobolev inequalities for the corresponding Dirichlet operators on loop lattices. In the special case of ferromagnetic models, we present the strongest result of such a type saying that uniqueness occurs for sufficiently small values of the particle mass.

MATHEMATICAL SUBJECT CLASSIFICATION (2000):
Primary: 82B10; Secondary: 46G12, 60H30

1 Introduction

The aim of this paper is twofold: First, we give the reader an elementary introduction in the mathematical theory of *quantum lattice systems* (*QLS*, for short). In Statistical Physics they are commonly viewed as models for *quantum crystals* (see, e.g., [1], [30], [33]). Second, we present recent results on the existence, uniqueness and a priori estimates for the corresponding Euclidean Gibbs measures obtained by the authors in [2]–[13], and we demonstrate how methods of Stochastic Analysis can successfully be applied to this topic.

According to common knowledge, a mathematical description of equilibrium properties of quantum systems might be given in terms of their Gibbs states defined on proper algebras of observables (cf. [20]). Such an algebraic approach is especially applicable to spin models with finite dimensional physical Hilbert spaces associated with every single particle. But, unfortunately, in the realization of this general concept for the quantum lattice systems considered in our papers there occur important difficulties (see, e.g., the discussion in [3]). In order to overcome these difficulties we shall take the *Euclidean* (or *path space*) *approach,* which is conceptually analogous to the well-known Euclidean strategy in quantum field theory (see, e.g., [30], [33], [47]). This analogy was pointed out and first implemented to quantum lattice systems in [1]; for recent developments see the review articles [13], [3] and an extensive bibliography therein. Actually, the Euclidean approach remains so far the only method which allows to construct and study Gibbs states for infinite systems of quantum particles described by unbounded operators. Briefly speaking, it transforms the problem of giving a proper meaning to a *quantum Gibbs state* G_β into the problem of studying a certain *Euclidean Gibbs measure* μ on the *loop lattice* $\Omega := [C(S_\beta)]^{\mathbb{Z}^d}$ (cf. Sect. 3 below for rigorous definitions). Here $\beta := 1/T > 0$ is the inverse (absolute) temperature and $C(S_\beta)$ is the space of continuous functions (i.e., loops) on a circle $S_\beta \cong [0, \beta]$ of length β. For a more detailed discussion of the relations between quantum and Euclidean Gibbs states we refer to [3], [5] (for an explanation of main ideas see also Sect. 2 below).

As a consequence, various probabilistic techniques become available for investigating equilibrium properties of quantum infinite-particle systems (cf. the review on this account included in Sect. 5). But, as compared with classical lattice systems, the situation with Euclidean Gibbs measures is much more complicated, since now the spin (i.e., loop) spaces themselves are *infinite dimensional* and their topological features should be taken into account carefully. Also, as is typical for non-compact spin spaces, we have to restrict ourselves to the set $\mathcal{G}_t$ of *tempered* Gibbs measures μ, which we specify by some natural support conditions (cf. definitions (15) and (22) below).

Among possible applications of Stochastic Analysis in Quantum Statistical Physics, we especially present *a new method for proving existence and a priori estimates* for tempered Euclidean Gibbs measures, which is based on the alternative description of Gibbs measures in terms of integration by parts and was developed by the authors in [6]–[8], [13]. It allows us to obtain improvements and generalizations of essentially all corresponding existence results for Gibbs measures known so far in the literature. Moreover, this method seems to be quite universal for lattice models and gives additional structural inside.

The organization of this paper is as follows. Section 2 is devoted to general aspects of the theory of Euclidean Gibbs measures. Here we introduce the models of quantum lattice systems ("anharmonic crystals"), concentrating on the simplest case of the *QLS Model I* with *harmonic pair interaction* between *nearest neighbors* only. In Sect. 3 we recall details on the corresponding Gibbsian formalism for Euclidean Gibbs measures μ on the loop lattice Ω. In Subsect. 4.1 we formulate our main *Theorems 1–6* on the existence, uniqueness and a priori estimates for tempered Euclidean Gibbs measures $\mu \in \mathcal{G}_t$. In Subsect. 4.2 we discuss the above mentioned alternative description of $\mu \in \mathcal{G}_t$ in terms of their shift–Radon–Nikodym derivatives (cf. *Theorem 7*) and its infinitesimal form, i.e., in terms of their logarithmic derivatives via the integration by parts formulas (cf. *Theorem 8*). In Sect. 5 we outline some possible generalizations of our method to the *QLS Models II–IV* with *many-particle interaction* of possibly *infinite range*. In Sect. 6 we discuss fundamental problems and basic methods in the study of Euclidean Gibbs states (e.g., Dobrushin's existence and uniqueness criteria, correlation inequalities, exponential decay of correlations for $\mu \in \mathcal{G}_t$, Poincare and log-Sobolev inequalities for the corresponding Dirichlet operators $\mathbb{H}_\mu$), as well as compare our results with those previously obtained by other authors.

The results presented in the paper have been obtained within the DFG-Schwerpunktprogramm *"Interacting Stochastic Systems of High Complexity"*, Research Projects AL 214/17 *"Stochastic Differential Equations on Infinite Dimensional Manigfolds"* (duration 01.05.1997–30.04.1999) and RO 1195/5 *"Analysis of Gibbs Measures via Integration by Parts and Quasi-Invariance"* (duration 01.05.1999–30.04.2003). One basic idea our research is the systematical applications of the methods of *Infinite Dimensional Stochastic Analysis* to the study of the equilibrium properties of infinite particle systems in Statistical Mechnics, which entirely corresponds to the general aim of the whole Schwerpunktprogramm. More presicely, the research performed by the authors can be naturally placed to the following main topics of the Schwerpunktprogramm: *1. "Interacting Systems of Statistical Physics"* and *6. "Stochastic Analysis"*. Moreover, within the Schwerpunktprogramm there have been highly stimulating discussions (e.g., during our various meetings) with Professors A. Bovier, J.-D. Deuschel, H.-O. Georgii, F. Götze, and H. Spohn concerned with applications to Statistical Mechanics and Quantum Field Theory (cf. the corresponding contributions in this volume).

2 A Simple Model of Quantum Anharmonic Crystal

In order to fix the main ideas and make the reader more familiar with the topic, we start with the following simplest model of a quantum crystal, which was extensively studied in the literature.

Particular QLS Model I: Harmonic Pair Interaction. Let $\mathbb{Z}^d$ be the integer lattice in the Euclidean space $(\mathbb{R}^d, | \cdot |)$, $d \in \mathbb{N}$. We consider an infinite system of interacting quantum particles performing one-dimensional (i.e., polarized) oscillations with displacements $q_k \in \mathbb{R}$ around their equilibrium positions at points $k \in \mathbb{Z}^d$. Each particle individually is described by the quantum mechanical Hamiltonian

$$\mathbb{H}_k := -\frac{1}{2\mathfrak{m}} \frac{d^2}{dq_k^2} + \frac{a^2}{2} q_k^2 + V(q_k) \tag{1}$$

acting in the (physical) Hilbert state space $\mathcal{H}_k := L^2(\mathbb{R}, dq_k)$. Here $\mathfrak{m}$ $(= \mathfrak{m}_{ph}/\hbar^2) > 0$ is the (reduced) mass of the particles and $a^2 > 0$ is their rigidity w.r.t. the harmonic oscillations. Concerning the anharmonic self-interaction potential, we suppose that $V \in C^2(\mathbb{R})$, i.e., twice continuously differentiable, and, moreover, that it satisfies the following growth condition:

Assumption (V). *There exist some constants $P > 2$ and $K_V, C_V > 0$ such that for all $k \in \mathbb{Z}^d$ and $q \in \mathbb{R}$*

$$K_V^{-1}|q|^{P-l} - C_V \leq (\mathrm{sgn}q)^l \cdot V^{(l)}(q) \leq K_V|q|^{P-l} + C_V, \quad l = 0, 1, 2.$$

Next, we add the *harmonic nearest-neighbor interaction*

$$W(q_k, q_{k'}) := \frac{J}{2}(q_k - q_{k'})^2$$

with intensity $J > 0$, the sum being taken over all (unordered) pairs $\langle k, k' \rangle$ in $\mathbb{Z}^d$ such that $|k - k'| = 1$. The whole system is then described by a heuristic Hamiltonian of the form

$$\mathbb{H} := -\frac{1}{2\mathfrak{m}} \sum_{k \in \mathbb{Z}^d} \frac{d^2}{dq_k^2} + \frac{a^2}{2} \sum_{k \in \mathbb{Z}^d} q_k^2 + \sum_{k \in \mathbb{Z}^d} V_k(q_k) + \frac{J}{2} \sum_{\langle k, k' \rangle \subset \mathbb{Z}^d} (q_k - q_{k'})^2. \tag{2}$$

Actually, the infinite-volume Hamiltonian (2) cannot be defined directly as a mathematical object and is represented by the local (i.e., indexed by finite volumes $\Lambda \Subset \mathbb{Z}^d$) Hamiltonians

$$\mathbb{H}_\Lambda := \sum_{k \in \Lambda} \mathbb{H}_k + \frac{J}{2} \sum_{\langle k, k' \rangle \subset \Lambda} (q_k - q_{k'})^2 \tag{3}$$

(as self-adjoint and lower bounded Schrödinger operators) acting in the Hilbert spaces $\mathcal{H}_\Lambda := \otimes_{k \in \Lambda} \mathcal{H}_k$.

Lattice systems of the above type (as well as their generalizations discussed below) are commonly viewed in quantum statistical physics as mathematical models of a crystalline substance (for more physical background see, e.g., [3], [30], [33], [42]). The study of such systems is especially motivated by the reason, that they provide a mathematically rigorous and physically quite realistic description for the important phenomenon of phase transitions (i.e., non-uniqueness of Gibbs states). So, if the potential V has a double-well shape, in the large mass limit $\mathsf{m} \to \infty$ the QLS (2) may undergo (ferroelectic) structural phase transitions connected with the appearance of macroscopic displacements of particles for low temperatures $T < T_{cr}(\mathsf{m})$ (for the mathematical justification of this effect see, e.g., [3], [5], [35]).

Remark 1.

(i) Typical potentials satisfying Assumption (**V**) are polynomials of even degree and with a positive leading coefficient, i.e.,

$$V(q) := P(q) := \sum_{1 \le l \le 2n} b_l q^l \quad \text{with } b_{2n} > 0 \text{ and } n \ge 2. \tag{4}$$

In this case one speaks about so-called *ferromagnetic* $P(\phi)$*–models*, which also can naturally be looked upon as lattice discretizations of quantum $P(\phi)$–fields (cf. [33], [47]). Let us also mention a special choice in (4), when

$$P(q) := \sum_{0 \le l \le n} b_{2l} q^{2l} \quad \text{with } b_{2l} \ge 0 \text{ for all } 2 \le l \le n. \tag{5}$$

Since $b_2 \in \mathbb{R}$ can be a large negative number, such polynomials may have arbitrary deep double wells. The last systems are technically more suitable for the study of critical behaviour and the influence of quantum effects, since then one can use not only the FKG correlation inequalities, which are standard for ferromagnetic pair interactions, but also more advanced (e.g., GKS, Lebowitz) inequalities relying on the additional symmetry properties of the one-partical potential (5) (cf. [3] as well as Sect. 6 below).

(ii) All subsequent constructions remain true if one takes for $W(q_k, q_{k'})$ the general ferromagnetic interaction $U(q_k - q_{k'})$ given by a nonnegative convex function $U \in C^2(\mathbb{R} \to \mathbb{R})$ satisfying

$$0 \le \inf_{\mathbb{R}} U'' \le \sup_{\mathbb{R}} U'' < \infty, \quad 0 \le U(q) = U(-q), \ \forall q \in \mathbb{R}. \tag{6}$$

As was already mentioned in Sect. 1, we take the *Euclidean approach* based on a path space representation for the quantum Gibbs states corresponding to the quantum mechanical system (2). Here we only illustrate these deep relations between quantum states and measures on loop spaces following the initial paper [1]) (see also [3], [5], [6] for the extended and up-to-date presentation).

Let us fix some $\beta := 1/T > 0$ having the meaning of inverse (absolute) temperature. Due to Assumption (**V_i**), for each $k \in \mathbb{Z}^d$ the one-particle Hamiltonian $\mathbb{H}_k$ is a self-adjoint operator with trace class semigroup $e^{-\tau \mathbb{H}_k}$, $\tau \ge 0$. On the algebra $\mathcal{A}_k := \mathcal{L}(\mathcal{H}_k)$ of all bounded linear operators in $\mathcal{H}_k$, we may then define the authomorphism group $\alpha_{\theta,k}$, $\theta \in \mathbb{R}$, and the quantum Gibbs state $G_{\beta,k}$ acting respectively by

$$\alpha_{\theta,k}(B) := e^{i\theta \mathbb{H}_k} B e^{-i\theta \mathbb{H}_k}, \quad G_{\beta,k}(B) := Tr(Be^{-\beta \mathbb{H}_k})Tr(e^{-\beta \mathbb{H}_k}), \quad B \in \mathcal{A}_k.$$

For any finite set of multiplication operators $(B_i)_{i=1}^n \in L^\infty(\mathbb{R}^d)$ we next construct the so-called temperature *Euclidean Green functions*

$$\Gamma_{\beta,k}^{B_1,...,B_n}(\tau_1,...,\tau_n) := Tr_{\mathcal{H}_k}\left(\prod_{i=1}^n e^{-(\tau_{i+1}-\tau_i)\mathbb{H}_k} B_i\right)/Tr(e^{-\beta\mathbb{H}_k}), \quad (7)$$
$$0 \le \tau_1 \le ... \le \tau_n \le \tau_{n+1} := \tau_1 + \beta.$$

These functions have analytic continuations to the complex domain $0 < \operatorname{Re} z_1 < ... < \operatorname{Re} z_n < \beta$ with the boundary values at $z_i = -i\tau_i$:

$$\Gamma_{\beta,k}^{B_1,...,B_n}(-i\tau_1,...,-i\tau_n) = G_{\beta,k}\left(\prod_{i=1}^n \alpha_{\tau_i,k}(B_i)\right). \quad (8)$$

Moreover, it should be noted that relation (8) *uniquely* determines the Gibbs state $G_{\beta,k}$. A further *crucial* observation is that the Green functions (7) may be represented (by the Feynman–Kac formula) as the moments

$$\Gamma_{\beta,k}^{B_1,...,B_n}(\tau_1,...,\tau_n) = E_{\mu_{\beta,k}}\left(\prod_{i=1}^n B_i(\omega_k(\tau_i))\right) \quad (9)$$

of a certain probability measure μ_k on the loop space

$$C_\beta := \{\omega_k \in C([0,\beta] \to \mathbb{R}) \mid \omega_k(0) = \omega_k(\beta)\}. \quad (10)$$

More precisely,

$$d\mu_k(\omega_k) = \frac{1}{Z} E_\beta^{(x,x)}\left\{-\exp\int_0^\beta \left[\frac{a^2}{2}\omega_k^2(\tau) + V(\omega_k(\tau))\right]d\tau\right\}dx, \quad (11)$$

where Z is a normalization constant and $E_\beta^{(x,x)}$ is the conditional expectation, given that $\omega_k(0) = \omega_k(\beta) = x$, w.r.t. a Brownian bridge process of length β. So, we get a *one-to-one correspondence* between the quantum Gibbs state $G_{\beta,k}$, Euclidean Green functions (7) and the path measure μ_k. Respectively, for all local Hamiltonians $\mathbb{H}_\Lambda$ in volumes $\Lambda \Subset \mathbb{Z}^d$, relations similar to (7)–(9) are valid for the associated Gibbs states $G_{\beta,\Lambda}$ on the algebra $\mathcal{A}_\Lambda := \mathcal{L}(\mathcal{H}_\Lambda)$ and the measure μ_Λ on the loop space $[C_\beta]^\Lambda$. This gives a possible way to construct the limiting states when $\Lambda \nearrow \mathbb{Z}^d$, and hence motivates us to consider Gibbs measures μ on the "*temperature loop lattice*" $\Omega := [C_\beta]^{\mathbb{Z}^d}$. We stress that so far there is *no method at all within operator theory* which allows to prove convergence of local Gibbs states in our situation. What is important, is the fact that (analogously to the well-known Osterwalder-Schrader reconstruction theorem in Euclidean field theory, see e.g. [30], [33], [47]) from each such Gibbs measure μ it is possible to *reconstruct* the quantum Gibbs state G_β of the system (2). For the above reasons the measures μ will be called *Euclidean Gibbs states (in the temperature loop space representation)* for the quantum lattice system (2). We proceed with their rigorous definition in Sect. 3 below.

3 Definition of Euclidean Gibbs Measures

Here we briefly describe the corresponding *Euclidean Gibbsian formalism* just for the concrete class of quantum lattice systems (2); for a detailed exposition and an extensive bibliography we refer the reader to [3], [6].

Let $S_\beta \cong [0, \beta]$ be a circle of length β considered as a compact Riemannian manifold with Lebesgue measure $d\tau$ as a volume element and distance $\rho(\tau, \tau') := \min(|\tau - \tau'|, \beta - |\tau - \tau'|)$, $\tau, \tau' \in S_\beta$. For each $k \in \mathbb{Z}^d$, we denote by

$$L_\beta^r := L^r(S_\beta \to \mathbb{R}, d\tau), \quad r \geq 1,$$
$$C_\beta^\alpha := C^\alpha(S_\beta \to \mathbb{R}), \quad \alpha \geq 0,$$

the standard Banach spaces of all integrable resp. (Hölder) continuous functions (i.e., loops) $\omega_k : S_\beta \to \mathbb{R}$. In particular, C_β with sup-norm $|\cdot|_{C_\beta}$ will be treated as the *single spin space*, whereas $H := L_\beta^2$ with the inner product $(\cdot, \cdot)_H := |\cdot|_H^2$ as the Hilbert space tangent to C_β.

As the *configuration space* for the infinite volume system we define the space of all loop sequences over $\mathbb{Z}^d$

$$\Omega := [C_\beta]^{\mathbb{Z}^d} = \left\{ \omega = (\omega_k)_{k \in \mathbb{Z}^d} \,\middle|\, \omega : S_\beta \to \mathbb{R}^{\mathbb{Z}^d}, \ \omega_k \in C_\beta \right\}. \tag{12}$$

We endow Ω with the *product topology* (i.e., the weakest topology such that all finite volume projections

$$\Omega \ni \omega \mapsto \mathbb{P}_\Lambda \omega := \omega_\Lambda := (\omega_k)_{k \in \Lambda} \in C_\beta^\Lambda =: \Omega_\Lambda, \quad \Lambda \Subset \mathbb{Z}^d,$$

are continuous) and with the corresponding *Borel σ-algebra* $\mathcal{B}(\Omega)$ (which also coincides with the σ-algebra generated by all cylinder sets

$$\{\omega \in \Omega \mid \omega_\Lambda \in \Delta_\Lambda\}, \quad \Delta_\Lambda \in \mathcal{B}(\Omega_\Lambda), \ \Lambda \Subset \mathbb{Z}^d \,).$$

Let $\mathcal{M}(\Omega)$ denote the set of all probability measures on $(\Omega, \mathcal{B}(\Omega))$. Next, we define the subset of *(exponentially) tempered configurations*

$$\Omega_t := \left\{ \omega \in \Omega \,\middle|\, \forall \delta \in (0, 1) : \ ||\omega||_{-\delta} := \left[\sum_{k \in \mathbb{Z}^d} e^{-\delta |k|} |\omega_k|_{L_\beta^2}^2 \right]^{\frac{1}{2}} < \infty \right\}. \tag{13}$$

and respectively the subset of tempered measures supported by $\Omega_t \in \mathcal{B}(\Omega)$, i.e.,

$$\mathcal{M}_t := \{ \mu \in \mathcal{M}(\Omega) \mid \mu(\Omega_t) = 1 \}. \tag{14}$$

In the context below, Ω_t will be always considered as a locally convex Polish space with the topology induced by the system of seminorms $\left(||\omega||_{-\delta}, |\omega_k|_{C_\beta} \right)_{\delta > 0, \ k \in \mathbb{Z}^d}$.

Heuristically (cf. the discussion in Sect. 2 above), the Euclidean Gibbs measures μ we are interested in have the following representation

$$d\mu(\omega) := Z^{-1} \exp\left\{-\mathcal{I}(\omega)\right\} \prod_{k \in \mathbb{Z}^d} d\gamma_\beta(\omega_k), \qquad (15)$$

where Z is a normalization factor, γ_β is a centered Gaussian measure on $(C_\beta, \mathcal{B}(C_\beta))$ with correlation operator A_β^{-1}, and $A_\beta := -\mathfrak{m}\Delta_\beta + a^2\mathbf{1}$ is the shifted Laplace–Beltrami operator on the circle S_β. Respectively $\mathcal{I}$ is defined as the map

$$\Omega \ni \omega \longmapsto \mathcal{I}(\omega) := \int_{S_\beta} \left[\sum_{k \in \mathbb{Z}^d} V(\omega_k) + \sum_{<k,k'> \subset \mathbb{Z}^d} W(\omega_k, \omega_{k'})\right] d\tau, \qquad (16)$$

which can be viewed as a potential energy functional describing an interacting system of loops $\omega_k \in C_\beta$ indexed by $k \in \mathbb{Z}^d$. Of course it is impossible to use this presentation for μ literally, since the series in (16) do not converge in any sense. Relying on the *Dobrushin–Lanford–Ruelle (DLR) formalism* (cf. [23], [32]), a rigorous meaning can be given to the measures μ as random fields on $\mathbb{Z}^d$ with a prescribed family of *local specifications* $\{\pi_\Lambda\}_{\Lambda \in \mathbb{Z}^d}$ in the following way:

For every finite set $\Lambda \Subset \mathbb{Z}^d$, we define a probability kernel π_Λ on $(\Omega, \mathcal{B}(\Omega))$: for all $\Delta \in \mathcal{B}(\Omega)$ and $\xi \in \Omega$

$$\pi_\Lambda(\Delta|\xi) := Z_\Lambda^{-1}(\xi) \int_{\Omega_\Lambda} \exp\left\{-\mathcal{I}_\Lambda(\omega|\xi)\right\} \mathbf{1}_\Delta(\omega_\Lambda, \xi_{\Lambda^c}) \prod_{k \in \Lambda} d\gamma_\beta(\omega_k) \qquad (17)$$

(where $\mathbf{1}_\Delta$ denotes the indicator on Δ). Here $Z_\Lambda(\xi)$ is the normalization factor and

$$\mathcal{I}_\Lambda(\omega|\xi) := \int_{S_\beta} \left[\sum_{k \in \Lambda} V(\omega_k) + \sum_{<k,k'> \subset \Lambda} W(\omega_k, \omega_{k'}) + \sum_{k \in \Lambda, \, k' \in \Lambda^c} W(\omega_k, \xi_{k'})\right] d\tau$$
$$(18)$$

is the interaction in the volume Λ, subject to the boundary condition $\xi_{\Lambda^c} := (\xi_{k'})_{k' \in \Lambda^c}$ in the complement $\Lambda^c := \mathbb{Z}^d \setminus \Lambda$. Obviously, $\inf_{\omega \in \Omega} \mathcal{I}_\Lambda(\omega|\xi) > -\infty$ and the RHS in (18) makes sense for the potentials V, W we deal here with. Moreover,

$$\int_\Omega \exp\left\{\lambda|\omega_k|_{C_\beta^\alpha}\right\} d\pi_\Lambda(d\omega|\xi) < \infty, \quad \forall \lambda > 0, \ \alpha \in \alpha \in [0, 1/2) \qquad (19)$$

since the Gaussian measure γ_β has such exponential moments. An important point is the consistency property for $\{\pi_\Lambda\}_{\Lambda \in \mathbb{Z}^d}$: for all $\Lambda \subset \Lambda' \Subset \mathbb{Z}^d$, $\xi \in \Omega$ and $\Delta \in \mathcal{B}(\Omega)$

$$(\pi_{\Lambda'}\pi_\Lambda)(\Delta|\xi) := \int_\Omega \pi_{\Lambda'}(d\omega|\xi)\pi_\Lambda(\Delta|\omega) = \pi_{\Lambda'}(\Delta|\xi). \qquad (20)$$

Definition 1. *A probability measure μ on $(\Omega, \mathcal{B}(\Omega))$ is called Euclidean Gibbs measure for the specification $\Pi := \{\pi_\Lambda\}_{\Lambda \Subset \mathbb{Z}^d}$ (corresponding to the quantum lattice system (2) at inverse temperature $\beta > 0$) if it satisfies the DLR equilibrium equations: for all $\Lambda \Subset \mathbb{Z}^d$ and $\Delta \in \mathcal{B}(\Omega)$*

$$\mu\pi_\Lambda(\Delta) := \int_\Omega \mu(d\omega)\pi_\Lambda(\Delta|\omega) = \mu(\Delta). \qquad (21)$$

Fixing $\beta > 0$, let $\mathcal{G}$ denote the set of all such measures μ. We shall be mainly concerned with the subset $\mathcal{G}_t$ of *tempered* Gibbs measures supported by Ω_t, i.e.,

$$\mathcal{G}_t := \mathcal{G} \cap \mathcal{M}_t = \{\mu \in \mathcal{G} \mid \mu(\Omega_t) = 1\}. \qquad (22)$$

Remark 2.
(i) It is worthwhile to compare our results on quantum systems with the analogous classical ones. The large-mass limit $\mathrm{m} \to \infty$ (or $\hbar \to 0$) of model (2) gives us an infinite system of interacting classical particles moving in the external field V. Such system is described by the potential energy functional

$$\mathbb{H}_{cl}(q) = \frac{a^2}{2} \sum_{k \in \mathbb{Z}^d} q_k^2 + \sum_{k \in \mathbb{Z}^d} V(q_k) + \sum_{<k,k'>\subset \mathbb{Z}^d} W(q_k, q_{k'}) \qquad (23)$$

on the configuration space $\Omega_{cl} := \mathbb{R}^{\mathbb{Z}^d} \ni \{q_k\}_{k \in \mathbb{Z}^d} := q$ (cf. [3]). Again, the formal Hamiltonian (23) does not make sense itself and is represented by the local Hamiltonians

$$\mathbb{H}_{cl,\Lambda}(q|y) := \frac{a^2}{2} \sum_{k \in \Lambda} q_k^2 + \sum_{k \in \Lambda} V(q_k) + \sum_{<k,k'>\subset \Lambda} W(q_k, q_{k'}) + \sum_{k \in \Lambda,\ k' \in \Lambda^c} W(q_k, y_{k'}) \qquad (24)$$

in the volumes $\Lambda \Subset \mathbb{Z}^d$ given the boundary conditions $y \in \Omega_{cl}$. The corresponding Gibbs states $\mu \in \mathcal{G}_{cl}$ at inverse temperature $\beta > 0$ are defined as probability measures on Ω_{cl} satisfying the DLR equations $\mu\pi_\Lambda = \mu$, $\Lambda \Subset \mathbb{Z}^d$, with the family of local specifications

$$\pi_\Lambda(\Delta|y) := Z_\Lambda^{-1}(y) \int_{\mathbb{R}^\Lambda} \exp\{-\beta\mathbb{H}_{cl,\Lambda}(q|y)\} \mathbf{1}_\Delta(q_\Lambda, y_{\Lambda^c}) \prod_{k \in \Lambda} dq_k,$$

$$\Delta \in \mathcal{B}(\Omega_{cl}), \quad y \in \Omega_{cl}. \qquad (25)$$

Starting from the pioneering papers [15], [21], [40], [45], such unbounded spin systems have been under intensive investigation in classical statistical mechanics (for recent developments see, e.g., [13], [17], [50]). However, it should be noted that the results obtained in those papers principally do not apply in the more complex situation of quantum systems with infinite dimensional spin spaces as considered here.

(ii) The necessity of restricting to configurations $\omega \in \Omega_t$ will appear, once we proceed to getting uniform moment estimates on Gibbs measures (cf. Theorem 2 below). On the other hand, our definition of temperedness (as well as its modification to the classical systems (23) with $|q_k|$ substituting $|\omega_k|_{L^2_\beta}$) is less restrictive (and simpler) than those usually used in the literature (for comparison, see e.g. [15], [21]). So, obviously, $\Omega_t \supseteq \Omega_{(s)t}$ resp. $\mathcal{G}_t \supseteq \mathcal{G}_{(s)t}$, where the subsets of all (*"slowly increasing"*) *tempered* configurations resp. measures are defined by

$$\Omega_{\bullet s \bullet t} := \left\{ \omega \in \Omega \ \middle| \ \exists p = p(\omega) > 0 : \ ||\omega||_{\bullet \ p} := \left[\sum_{k \bullet \ \mathbb{Z}^d} (1 + |k|)^{\bullet \ \bullet p} |\omega_k|_{L^2_\beta}^{\bullet} \right]^{\frac{1}{2}} < \infty \right\},$$

$$\mathcal{G}_{\bullet s \bullet t} := \{ \mu \in \mathcal{G} \ | \ \exists p = p(\mu) > 0 : \ ||\omega||_{\bullet \ p} < \infty \ \text{for} \ \mu\text{-a.e.} \ \omega \in \Omega \}. \tag{26}$$

In turn, $\mathcal{G}_{(s)t}$ contains the subset $\mathcal{G}_{(ss)t}$ of the so-called *Ruelle type "superstable"* Gibbs measures named after the earlier papers [40], [46] on the classical case. In the context of translation invariant quantum systems with many particle interactions at most of quadratic growth, such measures were introduced in [44] by the following support condition

$$\Omega_{(ss)t} := \left\{ \omega \in \Omega \ \middle| \ \sup_{N \in \mathbb{N}} \left[(1 + 2N)^{-d} \sum_{|k| \leq N} |\omega_k|_{L^2_\beta}^2 \right] < \infty \right\},$$

$$\mathcal{G}_{(ss)t} := \{ \mu \in \mathcal{G} \ | \ \mu(\Omega_{(ss)t}) = 1 \}. \tag{27}$$

4 Formulation of the Main Results

Now we are ready to present our main results for the Euclidean Gibbs measures concerning the following two directions:

– **existence, uniqueness, and a priori estimates for $\mu \in \mathcal{G}_t$;**
– **alternative description of $\mu \in \mathcal{G}_t$ in terms of their Radon–Nikodym derivatives and integration by parts formulas.**

For the sake of simplicity we confine ourselves to the concrete set up of the QLS Model I. We assume that all the conditions on the interaction potentials imposed in Sect. 2 are fulfiled without mentioning this again in the formulations of our statements. It is worth noting that, even for this mostly studied model, all the results presented here (as well as their trivial modifications for classical lattice systems like (23) performed in [13]) either are completely new or essentially improve previous ones obtained by other authors. How far they can be extended to more general interactions will be discussed in Sect. 4 below.

4.1 Existence, Uniqueness and a Priori Estimates for Euclidean Gibbs Measures

The following provides us with basic information for any further investigation of $\mu \in \mathcal{G}_t$:

Theorem 1. (Existence of Tempered Gibbs States, cf. [6]–[8])
(i) For all values of the mass $\mathfrak{m} > 0$ and the inverse temperature $\beta > 0$:

$$\mathcal{G}_t \neq \emptyset.$$

(ii) Moreover, (in all translation invariant systems treated below) there exists at least one translation invariant $\mu \in \mathcal{G}_{(ss)t} \subseteq \mathcal{G}_t$ satisfying the (even stronger than (27)) support condition for all $Q \geq 1$ and $\alpha \in [0, 1/2)$:

$$\sup_{N \in \mathbb{N}} \left\{ (1 + 2N)^{-d} \sum_{|k| \leq N} |\omega_k|_{C_\beta^\alpha}^Q \right\} < \infty \quad \text{for } \mu\text{-a.e. } \omega \in \Omega. \tag{28}$$

Theorem 2. (A Priori Estimates on Tempered Gibbs States, cf. [6]–[8]) *Every $\mu \in \mathcal{G}_t$ is supported by the set of Hölder loops $\bigcap_{\alpha \in [0,1/2)} [C_\beta^\alpha]^{\mathbb{Z}^d}$. Actually, for all $Q \geq 1$ and $\alpha \in [0, 1/2)$*

$$\sup_{\mu \in \mathcal{G}_t} \sup_{k \in \mathbb{Z}^d} \int_\Omega |\omega_k|_{C_\beta^\alpha}^Q d\mu(\omega) < \infty. \tag{29}$$

Corollary 1. *The set $\mathcal{G}_t$ is compact w.r.t. the topology of weak convergence of measures on any of spaces $[C_\beta^\alpha]^{\mathbb{Z}^d}$, $\alpha \in [0, 1/2)$, equipped by the system of seminorms $|\omega_k|_{C_\beta^\alpha}$, $k \in \mathbb{Z}^d$.*

In particular, by Prokhorov's tightness criterion, the existence result for $\mu \in \mathcal{G}_t$ immediately follows from the moment estimate (30) below, which holds for the family $\pi_\Lambda(d\omega|\xi = 0)$, $\Lambda \Subset \mathbb{Z}^d$, uniformly in volume. Besides, the a priori estimates for the probability kernels $\pi_\Lambda(d\omega|\xi)$ of the local specification $\Pi := \{\pi_\Lambda\}_{\Lambda \Subset \mathbb{Z}^d}$ subject to the fixed boundary condition $\xi \in \Omega_t$, stated in Theorems 3 and 4 below, are also of an independent interest and have various applications.

Theorem 3. (Moment Estimates Uniformly in Volume, cf. [8]) *Let us fix any boundary condition $\xi \in \Omega$. Then for all $\delta > 0$, $\alpha \in [0, 1/2)$ and $Q \geq 1$*

$$\sup_{k \in \mathbb{Z}^d} |\xi_k|_{C_\beta^\alpha} < \infty \implies \sup_{\Lambda \Subset \mathbb{Z}^d} \sup_{k \in \mathbb{Z}^d} \int_\Omega |\omega_k|_{C_\beta^\alpha}^Q \pi_\Lambda(d\omega|\xi) =: C_{Q,\xi} < \infty, \tag{30}$$

$$\xi \in \Omega_t \implies \sup_{\Lambda \Subset \mathbb{Z}^d} \sum_{k \in \Lambda} e^{-\delta|k|} \int_\Omega |\omega_k|_{C_\beta^\alpha}^Q \pi_\Lambda(d\omega|\xi) =: C'_{Q,\xi} < \infty. \tag{31}$$

Theorem 4. (Dobrushin Type Exponential Bound, cf. [8]) *For any given*
$\lambda, \sigma > 0$, *there exist a constant* $A > 0$ *and a matrix* $\mathbf{I} = (I_{k,j})_{k,j \in \mathbb{Z}^d}$ *with*
entries $I_{k,j} \geq 0$ *and with the bounded operator norms*

$$||\mathbf{I}||_\delta := \sup_{k \in \mathbb{Z}^d} \left\{ \sum_{k' \in \mathbb{Z}^d} I_{k,j} \exp \delta |k - j| \right\} < \infty, \quad \forall \delta > 0,$$

$$||\mathbf{I}||_0 = ||\mathbf{I}||_{l^1(\mathbb{Z}^d)} := \sup_{k \in \mathbb{Z}^d} \sum_{j \in \mathbb{Z}^d} I_{k,j} < \sigma, \tag{32}$$

such that for all $k \in \mathbb{Z}^d$ *and* $\xi \in \Omega_t$

$$\int_\Omega \exp \lambda |\omega_k|_{L^2_\beta} \, d\pi_{\{k\}}(d\omega|\xi) \leq \exp \lambda \left(A + \sum_{j \in \mathbb{Z}^d} I_{k,j} |\xi_j|_{L^2_\beta} \right). \tag{33}$$

Of course, we cannot provide here the full proofs of the theorems formulated above. But it should be mentioned in this respect that we propose a new method of proving existence and a priori estimates for Gibbs measures, which is based on their alternative characterization via integration by parts (cf. Theorem 8 in Subsect. 4.2). This method and hence the statements of Theorems 1–4 obtained by it easily extend to general many-particle interactions (cf. Sect. 5). In contrast, the uniqueness results quoted below essentially use the concrete structure of the one-particle and pair potentials $V(q_k)$ and $W(q_k, q_{k'}) := \frac{J}{2}(q_k - q_{k'})^2$.

Theorem 5. (Uniqueness of Tempered Gibbs States, cf. [11], [12]) *Suppose*
that the anharmonic self-interaction possesses a decomposition

$$V = V_0 + U,$$

where $V_0 \in C^2(\mathbb{R})$ *is a strictly convex function with polynomially bounded*
derivatives (i.e., satisfying Assumption (V_0)*) and* $U \in C_b(\mathbb{R})$ *is a bounded*
perturbation (describing the presence of possible wells). Denote

$$b := \inf_{\mathbb{R}} V_0'' > 0, \quad \delta(U) := \sup_{\mathbb{R}} U - \inf_{\mathbb{R}} U < \infty.$$

Then, for all values of the mass $\mathfrak{m} > 0$, *the set* $\mathcal{G}_t$ *consists of exactly one*
point, provided the following relation between the parameters is satisfied:

$$\frac{e^{\beta \, \delta(U)}}{2d + J^{-1}(a^2 + b^2)} < \frac{1}{2d}. \tag{34}$$

Theorem 6. (Uniqueness in $P(\phi)$-models by small mass, cf. [2]–[8]) *For the*
quantum lattice model (2) *with the polynomial interaction of the form* (5),
there exists $\mathfrak{m}_* > 0$ *such that, for all* $\mathfrak{m} \in (0, \mathfrak{m}_*)$ *and all temperatures* $\beta > 0$,
the set $\mathcal{G}_t$ *consists of exactly one point.*

A general presentation of the "*state of the art*" in the literature concerning these problems will be given in Sect. 6.

4.2 Flow and Integration by Parts Characterization of Euclidean Gibbs Measures

Here we briefly discuss the main ingredients of our new approach for proving existence and uniform a priori estimates for Euclidean Gibbs measures (cf. Theorems 1–4). A basic idea of our method is to use an *alternative characterization of Gibbs measures via their Radon–Nikodym derivatives or via integration by parts* (instead of the usual one in terms of local specifications through the Dobrushin–Lanford–Ruelle equations (21)). Such alternative descriptions of Gibbs measures have long been known for a number of specific models in statistical mechanics and field theory (see, e.g., [28]–[31], [36], [45]). But for the quantum lattice systems under consideration, a complete characterization of the measures $\mu \in \mathcal{G}_t$ in terms of their Radon–Nikodym derivatives has first been proved in [10], Theorem 4.6. Assuming that the interaction potentials are differentiable, it was further shown in [6]–[8] that this description of Gibbs measures is equivalent to their characterization as differentiable measures satisfying integration by parts formulas.

So, we start with the *flow* description of $\mu \in \mathcal{G}_t$ in terms of their Radon–Nikodym derivatives w.r.t. shift transformations of the configuration space Ω. Let us consider $\mathcal{H} := l^2(\mathbb{Z}^d \to L_\beta^2)$ with the scalar product $< \omega, \omega >_\mathcal{H} = \|\omega\|_\mathcal{H}^2 := \sum_{k \in \mathbb{Z}^d} |\omega_k|_{L_\beta^2}^2$ as the tangent Hilbert space to Ω. We fix an orthonormal basis in $\mathcal{H}$ consisting of the vectors $h_i := \{\delta_{k-j}\varphi_n\}_{j \in \mathbb{Z}^d}$ indexed by $i = (k, n) \in \mathbb{Z}^{d+1}$, where $\{\varphi_n\}_{n \in \mathbb{Z}} \subset C_\beta^\infty$ is the complete orthonormal system of eigenvectors of the operator A_β in $H := L_\beta^2$, i.e., $A_\beta \varphi_n = \lambda_n \varphi_n$ with $\lambda_n = (2\pi n/\beta)^2 \mathrm{m} + a^2$.

Theorem 7. (Flow Description of $\mu \in \mathcal{G}_t$, cf. [6]–[10]). *Let $\mathcal{M}_t^a$ denote the set of all probability measures $\mu \in \mathcal{M}_t$ which are quasi-invariant w.r.t. the shifts $\omega \longmapsto \omega + \theta h_i$, $\theta \in \mathbb{R}$, $i = (k, n) \in \mathbb{Z}^{d+1}$, with Radon–Nikodym derivatives*

$$a_{\theta h_i}(\omega) := \exp\left\{-\theta(A_\beta \varphi_n, \omega_k)_H - \frac{\theta^2}{2}(A_\beta \varphi_n, \varphi_n)_H\right\} \tag{35}$$

$$\times \exp \int_{S_\beta} \{V(\omega_k) - V(\omega_k + \theta \varphi_n)$$

$$+ \sum_{j: |j-k|=1} [W(\omega_k, \omega_j) - W(\omega_k + \theta \varphi_n, \omega_j)]\right\} d\tau.$$

Then $\mathcal{G}_t = \mathcal{M}_t^a$.

However, in applications it is more convenient to use not the flow characterization itself, but its infinitesimal form which we shall describe now. To this end we define functions (which will turn out to be the *partial logarithmic derivatives* of measures $\mu \in \mathcal{M}_t^a$ along directions h_i) for $i = (k, n) \in \mathbb{Z}^{d+1}$ by

$$b_{h_i}(\omega) := \frac{\partial}{\partial\theta} a_{\theta h_i}(\omega)\Big|_{\theta=0} = -(A_\beta\varphi_n, \omega_k)_H - (F_k^{V,W}(\omega), \varphi_n)_H, \quad \omega \in \Omega_t.$$

(36)

Here $F_k^{V,W} : \Omega \to C_\beta$ is the nonlinear Nemytskii-type operator acting by

$$F_k(\omega) := V'(\omega_k) + \sum_{j: \, |j-k|=1} \partial_q W_{\{k,j\}}(q,q')\big|_{q=\omega_k, \, q'=\omega_{k'}}.$$

(37)

For every $i = (k,n) \in \mathbb{Z}^{d+1}$, we denote by $C^1_{\mathrm{dec},i}(\Omega_t)$ the set of all functions $f : \Omega_t \to \mathbb{R}$ which are bounded and continuous together with their partial derivatives $\partial_{h_i} f$ in the direction h_i and, moreover, satisfy the extra decay condition

$$\sup_{\omega \in \Omega_t} \left| f(\omega) \left(1 + |\omega_k|_{L^1_\beta} + |F_k(\omega)|_{L^1_\beta}\right) \right| < \infty.$$

(38)

Of course, $fb_i \in L^\infty(\mu)$ for any $f \in C^1_{\mathrm{dec},i}(\Omega_t)$ and any $\mu \in \mathcal{M}_t$, even though we do not know a priori whether $b_{h_i} \in L^1(\mu)$. Since the interaction potentials are assumed to be differentiable, next we can show that the above flow characterization of $\mu \in \mathcal{G}_t$ is equivalent to their characterization as *differentiable measures* solving the *integration by parts* (for short, *IbP*) equations

$$\partial_{h_i}\mu(d\omega) = b_{h_i}(\omega)\mu(d\omega)$$

with the prescribed logarithmic derivatives b_{h_i}.

Theorem 8. ((IbP)-Description of $\mu \in \mathcal{G}_t$, cf. [6]–[8]). *Let $\mathcal{M}_t^b$ denote the set of all probability measures $\mu \in \mathcal{M}_t$ which satisfy the (IbP)-formula*

$$\int_\Omega \partial_{h_i} f(\omega) d\mu(\omega) = -\int_\Omega f(\omega) b_{h_i}(\omega) d\mu(\omega)$$

(39)

for all test functions $f \in C^1_{\mathrm{dec},i}(\Omega_t)$ and all directions h_i, $i = \mathbb{Z}^{d+1}$. Then $\mathcal{G}_t = \mathcal{M}_t^b$.

Remark 3. **(i)** Let us stress that the b_{h_i} just depend on the given potentials V and W, and hence are the same for all $\mu \in \mathcal{G}_t$ associated with the heuristic Hamiltonian (2). The measures given by the probability kernels $\pi_\Lambda(d\omega|\xi)$ of the local specification Π satisfy Theorems 7, 8, but only in directions h_i, $i = (k,n)$ with $k \in \Lambda \Subset \mathbb{Z}^d$. Since $a_{\theta h_i}$ and b_{h_i} are continuous locally bounded functions on Ω_t, the latter implies that every accumulation point of the family $\{\pi_\Lambda(d\omega|\xi) \,|\, \Lambda \Subset \mathbb{Z}^d, \xi \in \Omega_t\}$ is surely Gibbs.

(ii) In Stochastic Analysis, solutions μ to the (IbP)-formula (39) are also called *symmetrizing* measures. For further connections to reversible diffusion processes and Dirichlet operators in infinite dimensions we refer e.g. to [9], [10], [13], [19].

(iii) The key point of the proofs of Theorems 1–4 stated in Subsect. 4.1 is that (according to Theorem 6) $\mu \in \mathcal{G}_t$ resp. $\pi_\Lambda(d\omega|\xi)$ are viewed as the solutions of the infinite system (39) of first order PDE's. Due to the assumptions on the potentials V and W imposed above, the corresponding vector fields $b_{h_i}, i \in \mathbb{Z}^{d+1}$, possess certain *coercivity properties* w.r.t. the tangent space $\mathcal{H}$. This enables us to employ an analog of the *Lyapunov function method* well-known from finite dimensional PDE's to get uniform moment estimates (29–33). It should be noted that this approach has been first implemented in [13], however in the much simpler situation of the classical spin systems (23). Since the concrete technique used in those papers does not apply to loop spaces, in [6]–[8] we develop its proper (highly non-trivial) modification for the quantum case, which involves a *"single spin space analysis"* depending on the spectral properties of the elliptic operator A.

5 Possible Generalizations of QLS Model I

Here we briefly discuss how to modify the previous setting in order to include many-particle interaction potentials.

Particular QLS Model II: Pair Interaction of Superquadratic Growth. Let us first consider the following generalization of the QLS (2) described by a heuristic Hamiltonian of the form

$$\mathbb{H} := -\frac{1}{2m} \sum_{k \in \mathbb{Z}^d} \frac{d^2}{dq_k^2} + \frac{a^2}{2} \sum_{k \in \mathbb{Z}^d} q_k^2 + \sum_{k \in \mathbb{Z}^d} V(q_k) + \sum_{\langle k, k' \rangle \subset \mathbb{Z}^d} W(q_k, q_{k'}). \quad (40)$$

The one-particle potential $V \in C^2(\mathbb{R} \to \mathbb{R})$ satisfies the same Assumption ($\mathbf{V_0}$) as in Sect. 2, i.e., has asymptotic behaviour at infinity as a polynomial of order $P > 2$. Concerning the pair potential, we suppose that $W \in C^2(\mathbb{R}^2 \to \mathbb{R})$ has respectively at most polynomial growth of any order $R < P$:

Assumption ($\mathbf{W_0}$). *There exist constants $R \in [2, P)$ and $K_W, C_W > 0$ such that for all $q, q' \in \mathbb{R}$*

$$|\partial_q^{(l)} W(q, q')| \leq K_W \left(|q|^{R-l} + |q'|^{R-l} \right) + C_W, \quad l = 0, 1, 2.$$

Remark 4. A trivial example for pair potentials which satisfy ($\mathbf{W_0}$) are the polynomials $W(q, q') := \sum_{l=0}^{2r} (q - q')^l$ of even degree $2r < P$. In other words, our assumptions mean that the pair interaction is dominated by the single-particle one, which implies the so-called *lattice stabilization*. The case $P = R$ is also allowed, but it needs a more accurate analysis (cf. [7], [8]).

As compared with the initial QLS model (2), the only principal difference in dealing with its generalization (40) is that we should proper change the notion of temperedness. Now we define the subset of *tempered configurations* by

$$\Omega_t^R := \left\{ \omega \in \Omega \; \middle| \; \forall \delta \in (0,1): \; ||\omega||_{-\delta,R} := \left[\sum_{k \in \mathbb{Z}^d} e^{-\delta|k|} |\omega_k|_{L_\beta^R}^2 \right]^{\frac{1}{2}} < \infty \right\}.$$

(41)

Note that for $R = 2$ this is just the previous definition (14). Then all our main Theorems 1–6 presented in Sect. 4.1 *remain true*, as soon as in their formulation we substitute the single spin space L_β^2 by L_β^R and respectively specify the subset $\mathcal{G}_t$ of *tempered Gibbs measures* as those supported by $\Omega_t := \Omega_t^R$. Let us stress that (even in the case of translation invariant interactions we now deal with) we cannot guarantee that (outside the uniqueness regime) any tempered Gibbs measure will be invariant w.r.t. lattice translations. So, the above set $\mathcal{G}_t^R$ is the largest one so that for any of its points μ we are technically able to get moment estimates like (29) *uniformly* w.r.t. the lattice parameter $k \in \mathbb{Z}^d$.

Particular QLS Model III: Pair Interaction of Infinite Range. A further generalization of the QLS Models (2) and (40) concerns the case of *not necessarily translation-invariant* pair interaction of possibly *infinite range*. Namely, let us consider a model described by a heuristic Hamiltonian of the form

$$\mathbb{H} := -\frac{1}{2m} \sum_{k \in \mathbb{Z}^d} \frac{d^2}{dq_k^2} + \frac{a^2}{2} \sum_{k \in \mathbb{Z}^d} q_k^2 + \sum_{k \in \mathbb{Z}^d} V_k(q_k)$$
$$+ \sum_{\{k,k'\} \subset \mathbb{Z}^d} W_{\{k,k'\}}(q_k, q_{k'}).$$

(42)

The one-particle potentials $V_k \in C^2(\mathbb{R})$ satisfy the same Assumption $(\mathbf{V_0})$ as before, but with $P > 2$ and $K_V, C_V > 0$ which are *uniform* for all $k \in \mathbb{Z}^d$. The two-particle interactions (taken over all unordered pairs $\{k, k'\} \subset \mathbb{Z}^d$, $k \neq k'$) are given by symmetric functions $W_{\{k,k'\}} \in C^2(\mathbb{R}^2)$ satisfying the following growth condition:

Assumption $(\mathbf{W_0^*})$: *There exist some constants* $R \in [2, P)$ *and* $J_{k,k'} \geq 0$ *such that for all* $\{k, k'\} \subset \mathbb{Z}^d$ *and* $q, q' \in \mathbb{R}$:

$$|\partial_q^{(l)} W_{\{k,k'\}}(q, q')| \leq J_{k,k'} \left(1 + |q|^{R-l} + |q'|^{R-l} \right), \quad l = 0, 1, 2.$$

For the matrix $\mathbf{J} := \{J_{k,j}\}$ we can allow different rates of decay (for instance, polynomial or exponential), when the distance $|k - j|$ between the points of the lattice gets large:

Assumption $(\mathbf{J_0})$. *For all* $p \geq 0$ *or resp. some* $\delta > 0$

(i) $||\mathbf{J}||_p := \sup_{k \in \mathbb{Z}^d} \left\{ \sum_{j \in \mathbb{Z}^d \setminus \{k\}} J_{k,j}(1 + |k - j|)^p \right\} < \infty,$

(ii) $\quad ||\mathbf{J}||_\delta := \sup_{k \in \mathbb{Z}^d} \left\{ \sum_{j \in \mathbb{Z}^d \setminus \{k\}} J_{k,j} e^{\delta|k-j|} \right\} < \infty.$

Obviously, (ii) is stronger than (i). Again, we first need to choose the proper notion of the temperedness, which essentially depends on the decay rate of the pair interaction. A new issue caused by the *infinite range* of the interaction is that one also has to check that the probability kernels $\pi_\Lambda(d\omega|\xi)$ are well defined for all boundary conditions $\xi \in \Omega_t$. So, in view of Assumption $(\mathbf{J_0})(\mathbf{i})$, we define the subset $\Omega^R_{(s)t} \subset \Omega^R_t$ of (*"slowly increasing"*) tempered configurations by

$$\Omega^R_{(s)t} := \left\{ \omega \in \Omega \;\middle|\; \exists p = p(\omega) > 0 : \; ||\omega||_{-p,R} := \left[\sum_{k \in \mathbb{Z}^d} (1 + |k|)^{-2p} |\omega_k|^2_{L^R_\beta} \right]^{\frac{1}{2}} < \infty \right\}. \tag{43}$$

Respectively, we introduce the subset of tempered Gibbs measures

$$\mathcal{G}^R_{(s)t} := \{ \mu \in \mathcal{G} \;|\; \exists p = p(\mu) > d : \; ||\omega||_{-p,R} < \infty \; \forall \omega \in \Omega \; (\mathrm{mod}\,\mu) \}. \tag{44}$$

Then our main theorems about existence and a priori estimates for the tempered Euclidean Gibbs measures *remain true*, provided in their formulation one substitutes the single spin spaces L^2_β by L^R_β and, respectively, Ω_t by $\Omega^R_{(s)t}$ and $\mathcal{G}_t$ by $\mathcal{G}^R_{(s)t}$. In the formulation of Theorems 3 and 4 describing the properties of the probability kernels $\pi_\Lambda(d\omega|\xi)$ one also needs obvious changes (e.g., by claiming that $||\mathbf{I}||_p < \infty$ for some $p \geq 0$ instead of $||\mathbf{I}||_\delta < \infty$ for all $\delta > 0$ as before), which are discussed in [5], [6]. On the other hand, if we want to deal with the larger subset $\mathcal{G}^R_t \supset \mathcal{G}^R_{(s)t}$ and completely keep the previous setup of the QLS Model II, we should correspondingly impose the stronger Assumption $(\mathbf{J_0})(\mathbf{ii})$ on the decay of matrix $\mathbf{J}$.

General QLS Model IV: Many Particle Interactions of Unbounded Order and Infinite Range. Here we mean the systems described by a heuristic infinite dimensional Hamiltonian

$$\mathbb{H} = -\frac{1}{2m} \sum_{k \in \mathbb{Z}^d} \frac{d^2}{dq_k^2} + \frac{a^2}{2} \sum_{k \in \mathbb{Z}^d} q_k^2 + \sum_{k \in \mathbb{Z}^d} V_k(q_k)$$

$$+ \sum_{n=2}^{\infty} \sum_{\{k_1,...,k_n\} \subset \mathbb{Z}^d} W_{\{k_1,...,k_n\}}(q_{k_1},...,q_{k_n}), \tag{45}$$

where the n-particle interaction potentials (taken over all finite sets $\{k_1, ..., k_n\} \subset \mathbb{Z}^d$ consisting of $n \geq 2$ different points) are given by twice continuously differentiable *symmetric* functions $W_{\{k_1,...,k_n\}} \in C^2(\mathbb{R}^{dn})$. Again, the statements of Theorems 1–4 (in the same formulation as that for the QLS Model III) still hold, if one uses e.g. the following modification of $(\mathbf{W^*_0})$ and $(\mathbf{J_0})$:

Assumption (W). There exist $R \geq 2$, $I \geq 0$ and symmetric matrices $\{J_{k_1,...,k_n}\}_{(k_1,...,k_n) \in \mathbb{Z}^{nd}}$ with positive entries, such that for all $n \in \mathbb{N}$, $\{k_1, ..., k_n\} \subset \mathbb{Z}^d$ and $q_1, ..., q_n \in \mathbb{R}^d$

$$|\nabla^{(l)}_{q_1} W_{\{k_1,...,k_M\}}(q_1,...,q_n)|(1 + \sum_{m=1}^{n} |q_m|^l) \leq J_{k_1,...,k_n} \sum_{m=1}^{n} |q_m|^R + I, \quad l = 0, 1, 2.$$

Assumption (J). *For all $p \geq 0$ or (even stronger) for some $\delta > 0$*

(i) $\|\mathbf{J}\|_p := \sum_{n \cdot \cdot}^{\cdot} n^R \sup_{k_1 \cdot \mathbb{Z}^d} \left\{ \sum_{\cdot k_2, \ldots, k_n \cdot \cdot \mathbb{Z}^d} J_{k_1, \ldots, k_n} \left(1 + \sum_{m \cdot \cdot}^n |k_{\cdot} - k_m|^p \right) \right\} < \infty,$

(ii) $\|\mathbf{J}\|_\delta := \sum_{n \cdot \cdot}^{\cdot} n^R \sup_{k_1 \cdot \mathbb{Z}^d} \left\{ \sum_{\cdot k_2, \ldots, k_n \cdot \cdot \mathbb{Z}^d} J_{k_1, \ldots, k_n} \exp \left(\delta \sum_{m \cdot \cdot}^n |k_{\cdot} - k_m| \right) \right\} < \infty.$

It should be particularly emphasized that our technique based on the (IbP)-formula (39) obviously extends even to the above non-trivial interactions, which were not covered at all by any previous work (cf. the discussion in Sect. 6). All details on the general QLS Model IV may be found in [6], [8]. Of course, from the physical point of view, it is more realistic to consider systems of D-dimensional quantum oscillators on the lattice $\mathbb{Z}^d$ $(d, D \in \mathbb{N})$ with interaction potentials $V_k : \mathbb{R}^D \to \mathbb{R}$, $W_{\{k_1, \ldots, k_M\}} : (\mathbb{R}^D)^M \to \mathbb{R}$. Also in this case our method works. In this respect, we refer the interested reader to [7] where such a *multidimensional* version of the particular QLS Model II was treated.

6 Comments on Theorems 1–6

In order to give the reader a wider insight into the subject, we present here a systematic account of the *fundamental problems, basic methods, known results and possible nearest goals* in the study of Euclidean Gibbs measures on loop lattices.

I. Existence Problem. As is typical for systems with noncompact (in our case, infinite-dimensional) spin spaces, even the initial question of whether the set $\mathcal{G}_t$ is not empty (and hence the positive answer on it given by **Theorem 1** for the QLS Models I–IV) is far from trivial. A useful observation in this respect is that, under natural assumptions on the interaction, any accumulating point of the family π_Λ, $\Lambda \Subset \mathbb{Z}^d$, is certainly Gibbs. Depending on the specific class of quantum lattice models one deals with, the required convergence $\pi_{\Lambda^{(N)}} \to \mu$, $\Lambda^{(N)} \nearrow \mathbb{Z}^d$, and thus the existence of $\mu \in \mathcal{G}_t$, are proved by the following main methods listed below:

(i) General Dobrushin's Criterion for Existence of Gibbs Distributions [23]. The validity of the sufficient conditions of the *Dobrushin existence theorem* for some classical unbounded spin systems (23) has been verified, e.g., in [15], [21], [48] (however, under assumptions on the interaction potentials more restrictive than $(\mathbf{V_0})$ and $(\mathbf{W_0})$). Contrary to the classical case, the same problem for quantum lattice systems was not covered at all by any previous work. More precisely, in order to apply the Dobrushin criterion to quantum lattice systems, one should estimate in a proper way the expectations $E_{\pi_\Lambda(d\omega|\xi)}(F(\omega_k))$, $k \in \Lambda \Subset \mathbb{Z}^d$, of some *compact* function

$F : C_\beta \to \mathbb{R}_+ \cup \{\infty\}$. For this reason, the Dobrushin type moment estimate in the spin space L_β^R, which we prove in **Theorem 3** as an extension to the quantum case of the corresponding result in [15], is yet not enough for the existence criterion. Naturally one could try to take for F the norm-function in C_β^α with $0 \leq \alpha < 1/2$, except that so far no technical means were available to get such moment estimates in Hölder spaces starting from the DLR equations. A proper improvement of Theorem 4 will be the subject of a forthcoming paper of the authors.

(ii) Ruelle's Technique of Superstability Estimates (cf. the original papers [40], [46] and resp. [44] for its extension to the quantum case). This technique in particular requires that the interaction is *translation invariant* and the many-particle potentials have *at most quadratic growth* (i.e., $(\mathbf{W_0})$ holds with $R = 2$). As was shown in [44], for the subclass of boundary conditions $\xi \in \Omega_{(ss)t} \subset \Omega_{(s)t}$ (for instance, such that $\sup_{k \in \mathbb{Z}^d} |\xi_k|_{L_\beta^2} < \infty$) the family of probability kernels $\pi_\Lambda(d\omega|\xi)$, $\Lambda \Subset \mathbb{Z}^d$, $\Lambda \nearrow \mathbb{Z}^d$, is tight (in the sense of local weak convergence on Ω) and has at least one accumulation point μ from the subset of superstable Gibbs measures $\mathcal{G}_{(ss)t} \subset \mathcal{G}_t$ defined by (27). This technique also shows that any $\mu \in \mathcal{G}_{(ss)t}$ is a priori of sub-Gaussian growth.

(iii) Cluster Expansions is one of the most powerful methods for the study of Gibbs fields, but it works only in a *perturbative regime*, i.e., when an effective parameter of the interaction is small. In particular, various versions of this technique imply both existence and also uniqueness (but in some weaker than the DLR sense) of the associated infinite volume Gibbs distributions (see, e.g., [41], [42] and references therein).

(iv) Method of Correlations Inequalities involves more detailed information about the structure of the interaction (for instance, whether they are ferromagnetic or convex, cf. Remark 1). Starting from a number of correlations inequalities (such as *FKG, GKS, Lebowitz, Brascamp-Lieb* etc.) commonly known for classical lattice systems, by a lattice approximation technique (similar to the one used in Euclidean field theory) one can extend them to the quantum case (cf., e.g., [3]).

(v) Method of Reflection Positivity (as a part of **(iv)**) applies to *translation invariant* systems with *nearest-neighbours pair interactions* (i.e., when $V_k := V$, $W_{\{k,k'\}} := W$, and $W_{\{k,k'\}} = 0$ if $|k - k'| > 1$). The proper modification of this technique for the QLS (2) gives the existence of so-called *periodic* Gibbs states at least under Assumptions $(\mathbf{V_0})$, $(\mathbf{W_0})$ imposed in Sect. 3. Moreover, the reflection positivity method can also be used to study phase transitions in such models with the double-well anharmonicity V. This has been implemented under certain conditions (in the dimension $d \geq 3$ and for large enough $\beta, \mathfrak{m} >> 1$), e.g., in [3], [24], [35].

(vi) Method of Stochastic Dynamics (also referred to in quantum physics as "*stochastic quantization*"; see, e.g., [25], [13] and the related bibliography therein). In this method the Gibbs measures are directly treated as invariant (more precise, reversible) distributions for the so-called Glauber or Langevin stochastic dynamics. However, some additional technical assumptions are required on the interaction (among them *at most quadratic growth* of the pair potentials $W_{\{k,k'\}}(q,q')$) in order to ensure the solvability of the corresponding stochastic evolution equations in infinite dimensions (not to mention the extremely difficult ergodicity problem for them). This method has been first applied in [13] to prove existence of Euclidean Gibbs states for the particular QLS model (2).

II. A Priori Estimates for Measures in $\mathcal{G}_t$. **Theorem 2** above contributes to the fundamental problem of getting *uniform estimates* on correlation functionals of Gibbs measures in terms of parameters of the interaction. This problem was initially posed for classical lattice systems in [15], [21] and is closely related with the compactness of the set of tempered Gibbs states (cf. **Corollary** from Theorem 2); we refer also to [13] for a detailed discussion of the classical lattice case. There are very few results in the literature about a priori integrability properties of tempered Gibbs measures on loop or path spaces (see, for instance, [31], [43] for the case of Euclidean $P(\phi)_1$-fields and resp. [13] for the case of quantum anharmonic crystals). All of them are based on the method of stochastic dynamics just mentioned above. It is worth noting that the other methods listed under I (i)–(v) give also some estimates on limit points for π_Λ, $\Lambda \in \mathbb{Z}^d$, but not uniformly for all $\mu \in \mathcal{G}_t$. Besides, the finiteness of the moments of the measures $\mu \in \mathcal{G}_t$ is also useful for the study of Gibbs measures by means of the associated Dirichlet operators $\mathbb{H}_\mu$ in the spaces $L^p(\mu)$, $p \geq 1$, (this is known as the Holley–Stroock approach). In particular, by [9], [10] μ is an *extreme point* (or pure phase) in $\mathcal{G}_t$, if and only if the corresponding Markov semigroup $\exp(-t\mathbb{H}_\mu)$, $t \geq 0$, is *ergodic* in $L^2(\mu)$ (which extends the well known results in [36] related to the Ising model).

III. Uniqueness Problem. The validity of the sufficient conditions of the *Dobrushin uniqueness criterion* [Do70] for the QLS's (42) with *convex pair interactions of at most quadratic growth* has been first verified in [11], [12]. In doing so, the coefficients of Dobrushin's matrix were estimated by means of log-Sobolev inequalities proved on the single loop spaces L_β^2 and the uniqueness of $\mu \in \mathcal{G}_t$ was established for small values of the inverse temperature $\beta \in (0, \beta_0)$, but under conditions *independent of the particle mass* $\mathrm{m} > 0$ (and hence in the quasiclassical regime also). The exact statement for the particular QLS Model I is contained in our main **Theorem 5** above. For a *special class* of *ferromagnetic models* with the polynomial self-interaction (5), these results have been essentially improved in the recent series of papers [2]–[4]. The latter papers propose a new technique which combines the classical ideas of [15], [40] based on the use of FKG and GKS correlation inequalities with the spectral analysis of one-site oscillators (1) specific for the quantum case. The strongest result of such type, obtained in [4] and quoted here as **Theorem 6**, establishes the uniqueness of $\mu \in \mathcal{G}_t$ in the small-mass domain $\mathfrak{m} \in (0, \mathfrak{m}_0)$ *uniformly at all values of* $\beta > 0$. This provides a mathematical justification for the well-known physical phenomenon that structural phase transition for a given mass $\mathfrak{m} > 0$ can be suppressed not only by thermal fluctuations (i.e., high temperatures $\beta^{-1} > \beta_{cr}^{-1}$), but for the light particles (with $\mathfrak{m} < \mathfrak{m}_{cr}$) also by the quantum fluctuations (i.e., tunneling in a double-well potential) simultaneously at all temperatures $\beta > 0$. On the other hand, for small masses $\mathfrak{m} << 1$, the convergence of cluster expansions (independently of the boundary condition) has even been proved uniformly for all values of the temperature including the ground state case $\beta = \infty$, see [42]. However, in the case of unbounded spin systems such convergence of cluster expansions does not yet imply the DLR uniqueness. The uniqueness of $\mu \in \mathcal{G}_t$ in the QLS (42) with *superquadratic growth* of the many-particle interaction is a completely open problem, which will be the subject of our forthcoming paper. Another important and long-standing analytical problem is to find sufficient conditions for the uniqueness of symmetrizing measures in infinite dimensional spaces satisfying (IbP)-formulas like (39) with the given logarithmic derivatives b_{h_i} (for particular results on this topic see [19]).

Although our results are mainly concerned with the first three problems described above, for completeness of the exposition we also mention the following important directions:

IV. Decay of Correlations. First of all, the exponential decay of spin correlations for Gibbs measures is one of the standard applications of the *Dobrushin contraction technique* (cf., e.g., [23], [27]). In particular, it implies that for $\mu \in \mathcal{G}_t$ (which in this case is a priori unique)

$$Cov_\mu \left(f(\omega_k, \varphi)_{L_\beta^2}, \; g(\omega_{k'}, \varphi')_{L_\beta^2} \right) \leq$$
$$\mathcal{K} \exp(-\varepsilon|k - k'|)||f'||_{L^\infty}||g'||_{L^\infty}||\varphi||_{L_\beta^2}||\varphi'||_{L_\beta^2}$$

with some $K, \varepsilon > 0$ uniformly for all $k, k' \in \mathbb{Z}^d$, $\varphi, \varphi' \in L_\beta^2$ and $f, g \in C_b^1(\mathbb{R})$ (cf. [12]). Another general analytical approach to the decay of correlations is based on the *spectral gap estimates* for the corresponding Dirichlet operator $\mathbb{H}_\mu$ (see [17], [38], [50] for its realization for the classical spin systems (23). For quantum systems, however, the problem of getting the spectral gap estimates for $\mathbb{H}_\mu$ (or equivalently, for all $\mathbb{H}_{\mu_{\pi\Lambda(d\omega|\xi)}}$ uniformly w.r.t. volume and boundary conditions) has not yet been studied in the literature (except the trivial case of strictly convex interaction potentials) and will be the subject of our forthcoming paper. On the other hand, for the quantum ferromagnets with polynomial self-interactions like (5), the exponential decay as $|k - k'| \to \infty$ of the two-point correlations $Cov_{\mu_{\Lambda(d\omega|\xi)}}(\omega_k(\tau), \omega_{k'}(\tau'))$, uniformly in $\tau, \tau' \in S_\beta$, $\Lambda \Subset \mathbb{Z}^d$ and $\xi \in \Omega_t$, has been used in [4] as a crucial step for proving uniqueness for $\mu \in \mathcal{G}_t$. Moreover, for such quantum systems one expects the *complete equivalence* between the Dobrushin-Shlosman mixing conditions, exponential decay of correlations and Poincaré and log-Sobolev inequalities for the corresponding Dirichlet operators $\mathbb{H}_\mu$ (this equivalence has been first shown in [49] for lattice systems with compact spins and respectively extended in [50] to classical ferromagnetic systems with unbounded spins). Poincaré and log-Sobolev inequalities are commonly accepted in the literature as important tools to describe the links between the relaxation of stochastic dynamics and its equilibrium (e.g., Gibbs) measures (cf., e.g., [50], [13]).

V. Phase Transitions. There are basically *two general methods* for proving existence of phase transitions (i.e., non-uniqueness of $\mu \in \mathcal{G}_t$) for *low temperatures* β^{-1}, namely, the *reflection positivity* (for $d \geq 3$) and the *energy-entropy (Peierls-type) argument* (for $d \geq 2$). However, in practice their successful applications to quantum lattice systems have been limited so far to ferromagnetic $P(\phi)$-models (cf., e.g., [14], [24], [33], [35]). The first method (already mentioned in Item I.(v)) enables one to prove the positivity of *a long-range order parameter* $\lim_{|A| \to \infty} E_{\mu_{per,A}} \left[\sum_{k \in A} \omega_k(\tau) \right]^2 / |A|^2$ for large enough $\mathfrak{m} > \mathfrak{m}_0$ and $\beta > \beta_0(\mathfrak{m}_0)$ via the so-called infrared (Gaussian) bounds on two-point correlation functions $E_{\mu_{per,A}} \omega_k(\tau) \omega_{k'}(\tau)$ w.r.t. the local Gibbs measures $\mu_{\text{per},A}$ with periodic boundary conditions (cf. [24]). The second method has originally been discovered (as the so-called Peierls argument) for the Ising model and further developed to apply to rather general classical spin systems (now known as the Pirogov-Sinai contour method, cf. [48]). Its quantum modification was firstly implemented to the study of phase transition in the $(\varphi^4)_2$-model of Euclidean field theory (cf. [33]) and then in [14] to its lattice approximation (2) with $V(q_k) = \lambda(q_k^2 - \lambda^{-1})^2 + \lambda q_k^2/2$, where $\lambda > 0$ is the strength of the interaction. Following the idea of [33], [14], one defines a "*collective spin variable*" $\sigma_k := \text{sign} \int_{S_\beta} \omega_k(\tau) d\tau$ taking values ± 1 and a long-range parameter as the correlation function $< \sigma_k \sigma_{k'} > := \lim_{|A| \to \infty} E_{\mu_A^{\text{per}}} \sigma_k \sigma_{k'}$. Then, for any fixed λ, the existence of long-range behaviour, and hence phase transition, follows from the estimate $< \sigma_k \sigma_{k'} > \geq 1/2$ valid for large enough values of $\mathfrak{m}$ and β.

VI. Euclidean Ground States. Of special interest for quantum systems is the case of *zero absolute temperature*, i.e., $\beta = \infty$, which is technically more complicated and less studied in the literature. In particular, it involves an important problem of the operator realization of the formal Hamiltonian (2) in quantum mechanics (cf. [1], [16]). The corresponding Gibbs measures $\mu \in \mathcal{G}_{gr}$ on the "*path lattice*" $[C(\mathbb{R})]^{\mathbb{Z}^d}$, so-called *Euclidean ground states*, also allow the DLR-description, but through a family of local specifications $\pi_{I \times A}$ indexed by "time-space" windows $I \times A$ with $I \Subset \mathbb{R}$, $A \Subset \mathbb{Z}^d$, cf. [41]. A principal difference with the previous case $0 < \beta < \infty$ is that now there is not available any such (independent from boundary conditions ξ) reference measure so that all $\pi_{I \times \Lambda}(d\omega|\xi)$ are defined as its Gibbs modifications. So far, there are very few rigorous results about Gibbs measures on the path space $[C(\mathbb{R})]^{\mathbb{Z}^d}$, which all are mainly related to the existence problem. A recent progress in this direction was achieved in the series of papers [41], [42], [39], where the limit measures $\lim_{\Lambda \nearrow \mathbb{Z}^d} \lim_{I \nearrow \mathbb{R}} \pi_{I \times \Lambda} \in \mathcal{G}_{gr}$ for the $P(\phi)$–lattice models (2) have been constructed through cluster expansions w.r.t. the small mass parameter $\mathrm{m} \ll 1$. At the same time, for fixed $\Lambda \Subset \mathbb{Z}^d$, the corresponding unique Gibbs measures $\mu_{gr,\Lambda} := \lim_{I \nearrow \mathbb{R}} \pi_{I \times \Lambda}$ on the path space $[C(\mathbb{R})]^\Lambda$ are well-known as the $P(\phi)_1$–processes and can be looked upon as a special case of Euclidean field theory in space-dimension zero (cf. [43], [18]). Besides, the Gibbs measures on the path space $[C(\mathbb{R})]^{\mathbb{Z}^d}$ also appear in a natural way as weak solutions for SDE's in $\mathbb{Z}^d$ ([22], [41]). Within the Schwerpunktprogramm, related Gibbs measures on path spaces have detailed been studied by J. Loerinczi and H. Spohn (see their survey in this volume).

In this respect it should be mentioned that in the recent preprint [34] some (deterministic) version of integration by parts for local specifications has been used to prove existence of Gibbs measures relative to Brownian motion on the path space $C(\mathbb{R})$. Finally, let us note that our method based on the (IbP)-formula (39) can also be modified to apply to the case of zero absolute temperature, i.e., $\beta = \infty$, and corresponding symmetrizing measures on $[C(\mathbb{R})]^{\mathbb{Z}^d}$. This case is under present investigation.

VI. Random Lattice systems. At this stage the case of spin systems with *random interactions* as those studied within the Schwerpunktprogramm by A. Bovier and Ch. Külske (see their contribution to this volume) has not yet been considered. We think, however, that our method can also be applied to some of such situations, at least in a modified way.

Acknowledgements

Financial support by the DFG through the Schwerpunkt- programm "Interacting Stochastic Systems of High Complexity" (Research Projects AL 214/17-2 and RO 1195/5) and by the Lise-Meitner Habilitation Stipendium (for the third-named author) is gratefully acknowledged.

References

1. Albeverio, S., Høegh-Krohn, R.: Homogeneous random fields and quantum statistical mechanics. J. Funct. Anal., **19**, 241–272 (1975)
2. Albeverio, S., Kondratiev, Yu.G., Kozitzky, Yu.V., Röckner, M.: Uniqueness of Gibbs states of quantum lattices in small mass regime. Ann. Inst. H. Poincaré Probab. Statist., **37**, 43–69 (2001)
3. Albeverio, S., Kondratiev, Yu.G., Kozitzky, Yu.V., Röckner, M.: Euclidean Gibbs states of quantum lattice systems. Rev. Math. Phys., **14**, 1335–1401 (2002)

4. Albeverio, S., Kondratiev, Yu.G., Kozitzky, Yu.V., Röckner, M.: Small mass implies uniqueness of Gibbs states of a quantum crystal. To appear in Comm. Math. Phys. (2004)
5. Albeverio, S., Kondratiev, Yu.G., Kozitzky, Yu.V., Röckner, M.: Anharmonic quantum crystals: A functional integral approach. Monograph in preparation
6. Albeverio, S., Kondratiev, Yu.G., Pasurek (Tsikalenko), T., Röckner, M.: A priori estimates and existence for Euclidean Gibbs measures. BiBoS Preprint Nr. 02-06-089, Universität Bielefeld, 90 pages (2002)
7. Albeverio, S., Kondratiev, Yu.G., Pasurek, T., Röckner, M.: Euclidean Gibbs measures on loop lattices: existence and a priori estimates. Annals of Probab., **32**, 153–190 (2004)
8. Albeverio, S., Kondratiev, Yu.G., Pasurek, T., Röckner, M.: Euclidean Gibbs measures. To appear in Transactions of Moscow Math. Society
9. Albeverio, S., Kondratiev, Yu.G., Röckner, M.: Ergodicity of L^*-semigroups and extremality of Gibbs states. J. Funct. Anal., **144**, 394–423 (1997)
10. Albeverio, S., Kondratiev, Yu.G., Röckner, M.: Ergodicity of the stochastic dynamics of quasi-invariant measures and applications to Gibbs states. J. Funct. Anal. **149**, 415–469 (1997)
11. Albeverio, S., Kondratiev, Yu.G., Röckner, M., Tsikalenko, T.: Uniqueness of Gibbs states for quantum lattice systems. Prob. Theory Rel. Fields, **108**, 193–218 (1997)
12. Albeverio, S., Kondratiev, Yu.G., Röckner, M., Tsikalenko, T.: Dobrushin's uniqueness for quantum lattice systems with nonlocal interaction. Comm. Math. Phys., **189**, 621–630 (1997)
13. Albeverio, S., Kondratiev, Yu.G., Röckner, M., Tsikalenko, T.: Glauber dynamics for quantum lattice systems. Rev. Math. Phys., **13**, 51–124 (2001)
14. Albeverio, S., Kondratiev, A.Yu., Rebenko, A.L.: Peiers argument and long-range-behaviour in quantum lattice systems with unbounded spins. J. Stat. Phys. **92**, 1153–1172 (1998)
15. Bellissard, J., Høegh- Krohn, R.: Compactness and the maximal Gibbs states for random Gibbs fields on a lattice. Comm. Math. Phys., **84**, 297–327 (1982)
16. Berezansky, Yu.M., Kondratiev, Yu.G.: Spectral Methods in Infinite Dimensional Analysis. Kluwer Academic Publishes, Dordrecht Boston London (1993)
17. Bodineau, T., Helffer, B.: Correlations, spectral gap and log-Sobolev inequalities for unbounded spin systems. In: Differential Equations and Mathematical Physics (Birmingham, AL, 1999). AMS/IP Stud. Adv. Math., AMS, Providence, **16**,.51–66 (2000)
18. Betz, V., Lörinczi, J.: A Gibbsian description of $P(\phi)$. -process. Preprint (2000)
19. Bogachev, V.I., Röckner, M.: Elliptic equations for measures on infinite-dimensional spaces and applications. Prob. Theory Rel. Fields, **120**, 445–496 (2001)
20. Bratteli, O., Robinson, D.W.: Operator Algebras and Quantum Statistical Mechanics, I, II. Springer, Berlin Heidelberg New York (1981).
21. Cassandro, M., Olivieri, E., Pellegrinotti, A., Presutti, E.: Existence and uniqueness of DLR measures for unbounded spin systems. Z. Wahrsch. verw. Gebiete, **41**, 313–334 (1978)
22. Deuschel, J.-D.: Infinite-dimensional diffusion processes as Gibbs measures on $C[0,1]^{Z^d}$. Probab. Th. Rel. Fields, **76**, 325–340 (1987)

23. Dobrushin, R.L.: Prescribing a system of random variables by conditional distributions. Theory Prob. Appl., **15**, 458–486 (1970)
24. Driesler, W., Landau, L., Perez, J.F.: Estimates of critical length and critical temperatures for classical and quantum lattice systems. J. Stat. Phys., **20**, 123–162 (1979)
25. Da Prato, G., Zabczyk, J.: Ergodicity for Infinite-Dimensional Systems. Cambridge University Press (1996).
26. Doss, H., Royer, G.: Processus de diffusion associe aux mesures de Gibbs sur $\mathbb{R}^{\mathbb{Z}^d}$. Z. Wahrsch. verw. Geb., **46**, 106–124, (1979)
27. Föllmer, H.: A covariance estimate for Gibbs measures. J. Funct. Anal , **46**, 387–395 (1982)
28. Föllmer, H.: Random Fields and Diffusion Processes. In: Lect. Notes Math., **1362**, 101–204. Springer, Berlin Heidelberg New York (1988)
29. Fritz, J.: Stationary measures of stochastic gradient dynamics, infinite lattice models. Z. Wahrsch. verw. Gebiete, **59**, 479–490 (1982)
30. Fröhlich, J.: Schwinger functions and their generating functionals, II. Advances in Math., **33**, 119–180 (1977)
31. Funaki, T.: The reversible measures of multi-dimensional Ginzburg–Landau type continuum model. Osaka J. Math., **28**, 462–494 (1991)
32. Georgii, H.-O.: Gibbs Measures and Phase Transitions. Studies in Mathematics, **9**. Walter de Gruyter, Berlin New York (1988).
33. Glimm, J., Jaffe, A.: Quantum Physics. A Functional Integral Point of View. Springer, Berlin Heidelberg New York (1981).
34. Hariya, Y.: A new approach to constructing Gibbs measures on $C(\mathbb{R}; \mathbb{R}^d)$ – an application for hard-wall Gibbs measures on $C(\mathbb{R}, \mathbb{R})$. Preprint (2001)
35. Helffer, B.: Splitting in large dimensions and infrared estimates. II. Moment inequalities. J. Math. Phys., **39**, 760–776 (1998)
36. Holley, R., Stroock, D.: L.-theory for the stochastic Ising model. Z. Wahrsch. verw. Gebiete, **35**, 87–101 (1976)
37. Klein, A., Landau, L.: Stochastic processes associated with KMS states. J. Funct. Anal., **42**, 368–428 (1981)
38. Ledoux, M.: Logarithmic Sobolev inequalities for unbounded spin systems revisited. In: Séminaire de Probabilités, **XXXV**, 167–194. Lecture Notes Math., **1755** (2001).
39. Lőrinczi, J., Minlos, R.A.: Gibbs measures for Brownian paths under the effect of an external and a small pair potential. J. Stat. Phys., **105**, 607–649 (2001)
40. Lebowitz, J.L., Presutti, E.: Statistical mechanics of systems of unbounded spins. Comm. Math. Phys., **50**, 195–218 (1976)
41. Minlos, R.A., Roelly, S., Zessin, H.: Gibbs states on space-time. Potent. Anal., **13**, 367–408 (2000)
42. Minlos, R.A., Verbeure, A., Zagrebnov, V.A.: A quantum crystal model in the light mass limit: Gibbs state. Rev. Math. Phys., **12**, 981–1032 (2000)
43. Osada, H., Spohn, H.: Gibbs measures relative to Brownian motion. Ann. Prob., **27**, 1183–1207 (1999)
44. Park, Y.M., Yoo, H.J.: A characterization of Gibbs states of lattice boson systems. J. Stat. Phys., **75**, 215–239 (1994)
45. Royer, G.: Étude des champs Euclidiens sur un resau $\mathbb{Z}^\gamma$. J. Math. Pures et Appl., **56**, 455–478 (1977)

46. Ruelle, D.: Statistical Mechanics. Rigorous Results. Benjamin, New York Amsterdam (1969)
47. Simon, B.: The $P(\varphi)$. Euclidean Field Theory. Princeton University Press, Princeton New York (1974)
48. Sinai, Ya.: Theory of Phase Transitions. Rigorous Results. Pergamon Press, Oxford (1982)
49. Stroock, D.W., Zegarlinski, B.: The logarithmic Sobolev inequality for continuous spin systems on a lattice. J. Funct. Anal., **104**, 299–326 (1982)
50. Yoshida, N.: The equivalence of the log-Sobolev inequality and a mixing condition for unbounded spin systems on the lattice. Ann. Inst. H. Poincare Prob. Statist., **37**, 223–243 (2001)

Some Jump Processes
in Quantum Field Theory

Roderich Tumulka[1] and Hans-Otto Georgii[2]

. Dipartimento di Fisica and INFN sezione di Genova, Università di Genova
  Via Dodecaneso 33, 16146 Genova, Italy
  tumulka@mathematik.uni-muenchen.de
. Mathematisches Institut, Ludwig-Maximilians-Universität
  Theresienstr. 39, 80333 München, Germany
  georgii@mathematik.uni-muenchen.de

Summary. A jump process for the positions of interacting quantum particles on a lattice, with time-dependent transition rates governed by the state vector, was first considered by J.S. Bell. We review this process and its continuum variants involving "minimal" jump rates, describing particles as they get created, move, and get annihilated. In particular, we sketch a recent proof of global existence of Bell's process. As an outlook, we suggest how methods of this proof could be applied to similar global existence questions, and underline the particular usefulness of minimal jump rates on manifolds with boundaries.

MSC (2000). 60J75, 81T25

1 Introduction

This contribution deals with Markov jump processes Q_t describing the positional time evolution of finitely many interacting quantum particles. These processes are characterized by a specific form of time-dependent jump rates induced by the Schrödinger equation for the quantum state vector Ψ_t of the underlying quantum field theory (QFT). As a typical example, suppose that the particles live in the physical three-space $\mathbb{R}^3$. The process Q_t then takes values in the space of all finite subsets of $\mathbb{R}^3$, the configuration space of a variable number of identical particles (corresponding to the Fock space of QFT), and the jumps of Q_t describe the creation or annihilation of particles; between these stochastic jumps, Q_t evolves deterministically according to some ordinary differential equation governed by Ψ_t. Alternatively, one may think of quantum particles on a lattice; the jumps of Q_t then record all changes of the particle configuration. We will portray several processes of this type, present some common principles, and in particular discuss some results of three recent papers [GT03, DGTZ03c, DGTZ04], the work on which was supported by the DFG Priority Program.

As the state vector Ψ_t determining the jump rates follows the time-dependent Schrödinger equation, the jump rates themselves are explicitly time-dependent, so that the processes Q_t considered here do not admit an invariant measure. However, the jump rates are designed in such a way that Q_t does admit an *equivariant measure*, namely the quantum distribution $|\Psi_t|^2$, which means that Q_t has distribution $|\Psi_t|^2$ at any time t. This is the key feature justifying the particular form of the jump rates, and on the other hand the main fact on which one can build the analysis of these processes. So, the issue here is not the analysis of distributional properties of a given process, but the converse: the equivariant distribution is given, and the objective is to prove the existence of the associated process, and to check that it really does have the equivariant distribution. In [GT03], we carried out this program for the case of a discrete configuration space, including in particular a lattice model proposed by J.S. Bell [Bel86]; the main arguments will be sketched in Sect. 4.

From the probabilistic viewpoint, one has to overcome two difficulties. First, the transition rates exhibit singularities, in that they become ill-defined at certain time and space points. One has to show that the process avoids these singularities. The second (and more important) task is to rule out the possibility of explosion, i.e., the accumulation of infinitely many jumps in finite time. Due to the unbounded growth of the rates near the singularities, the standard methods fail, and one has to use the particular relation between transition rates and equivariant distribution.

Besides the results on the discrete case mentioned above, we will also describe the continuum analogues of Bell's process investigated in [DGTZ03a, DGTZ03c, DGTZ04]; as a special case these include Bohmian mechanics [Boh52, Bel66, BDDGZ95, Dür01] for the appropriate Hamiltonian with a conserved number of particles. On the basis of what we learned from our existence proof for Bell's model, we also propose here some new methods for proving the existence of Bohmian mechanics.

Let us now discuss how the models considered here relate to the topics of other research in the DFG Priority Program and other articles in this volume. First, the existence problem for a model of quantum field theory is also the subject of the contribution of S. Albeverio, Y. Kondratiev, M. Röckner, and T. Pasurek. The issue there is the existence, and uniqueness, of Euclidean Gibbs measures for infinitely many interacting quantum spins. These concern an equilibrium setting, and time appears only via path integration to make the connection with the quantum states. The difficulty there is the infinite number of spins, requiring particular assumptions on the interaction. By way of contrast, the models considered here involve only finitely many particles, but in a nonequilibrium situation, and we do not need any particular assumptions on the interaction.

From the methodological side, there is a closer connection to the research in the Priority Program dealing with population biology, in particular that by R. Höpfner and E. Löcherbach (not included in this volume). The similarities concern the creation and annihilation of particles in the vicinity of other particles, and the necessity of proving non-explosion. Also, the space-dependence of the reproduction rates in Höpfner and Löcherbach's article implies non-exponential life-times of individuals, just as the time-dependent jump rates considered here imply non-exponential interjump times. However, in our case the paths between the jumps are smooth and deterministic.

This note is organized as follows. In Sect. 2 we derive the fundamental formula (8) for minimal jump rates, defining the jump process associated with a certain type of Hamiltonian. This involves consideration of the equivariant probability distribution (4) and probability current (6) provided by quantum theory. In Sect. 3 we explain the connection with Bohmian mechanics and with Bell's model. We also describe the process for a concrete example QFT, introduced in [DGTZ03a]. In Sect. 4 we sketch the global existence proof for the discrete case, including Bell's model, that we developed in [GT03]. In Sect. 5 we point out how the methods of [GT03] could be adapted to other global existence problems. In Sect. 6 we indicate some perspectives for future research concerning a process for quantum theory on a manifold with boundaries, and the special role the minimal jump rate (8) plays for this process.

2 Jump Rates Induced by Schrödinger Equations

We now introduce the class of jump processes we are concerned with, starting with a general framework. For our purposes, a quantum theory is abstractly given by a Hilbert space $\mathscr{H}$ containing the state vectors, a one-parameter group U_t of unitary operators on $\mathscr{H}$ defining the time evolution

$$\Psi_t = U_t \Psi_0 \tag{1}$$

of the state vector, and a measurable space $(\mathcal{Q}, \mathscr{F})$ of configurations describing the locations of particles. $\mathcal{Q}$ is tied to $\mathscr{H}$ via a projection-valued measure (PVM) $P(\mathrm{d}q)$ on $\mathcal{Q}$ acting on $\mathscr{H}$, i.e., a mapping from the σ-algebra $\mathscr{F}$ to the family of projection operators on $\mathscr{H}$ that is, like a measure, countably additive (in the sense of the weak operator topology) and normalized, in that $P(\mathcal{Q}) = I$, the identity operator on $\mathscr{H}$. If $\mathscr{H} = L^2(\mathcal{Q}, \mathbb{C}^k)$ with respect to some measure on $\mathcal{Q}$ then $\mathscr{H}$ is equipped with a natural PVM, namely $P(B)$ being multiplication by the indicator function of the set B. In nonrelativistic quantum mechanics, another way of saying this is that P is the PVM corresponding to the joint spectral decomposition of all position operators.

By Stone's theorem, the unitary operators U_t are of the form

$$U_t = \mathrm{e}^{-iHt/\hbar} \tag{2}$$

with H a self-adjoint operator on $\mathscr{H}$, called the Hamiltonian. Equations (1) and (2) together correspond to the formal Schrödinger equation

$$i\hbar \frac{d\Psi_t}{dt} = H\Psi_t. \qquad (3)$$

We will show how this Schrödinger equation, together with the PVM P, gives rise to a natural Markov process on $\mathcal{Q}$.

In this section we focus on the case in which this Markov process is a pure jump process. (Roughly speaking, this will require that the Hamiltonian is an integral operator in the "position representation" defined by P; differential operators will be considered in Sect. 3.) So we ask: *Is there any distinguished Markovian jump process (Q_t) on $\mathcal{Q}$ describing the evolution of the particle configuration, and what are its transition rates?* To answer this question we note that the quantum theoretical probability distribution of the configuration at time t is given by

$$\pi_t(\,\cdot\,) = \langle \Psi_t | P(\,\cdot\,)\Psi_t \rangle. \qquad (4)$$

(We generally assume that $\|\Psi_0\| = 1$.) It is therefore natural to stipulate that π_t is *equivariant* for (Q_t), meaning that Q_t has distribution π_t at every time t. Can one choose some (time-dependent) transition rates (σ_t) for (Q_t) to satisfy this requirement of equivariance? Yes indeed, in view of (3) the time derivative of π_t is given by

$$\dot\pi_t(B) = \tfrac{2}{\hbar} \operatorname{Im} \langle \Psi_t | P(B)H\Psi_t \rangle = \int J_t(B, dq'), \qquad (5)$$

where

$$J_t(B, B') = \tfrac{2}{\hbar} \operatorname{Im} \langle \Psi_t | P(B)HP(B')\Psi_t \rangle \qquad (6)$$

is the quantum theoretical current between two sets $B, B' \in \mathscr{F}$. On the other hand, suppose (Q_t) is a pure jump process on $\mathcal{Q}$ jumping at time t with rate $\sigma_t(B, q')$ from $q' \in \mathcal{Q}$ to some configuration in $B \in \mathscr{F}$. Then its distribution $\rho_t = \mathbb{P} \circ Q_t^{-1}$ evolves according to the equation

$$\dot\rho_t(B) = \int_{\mathcal{Q}} \sigma_t(B, q)\,\rho_t(dq) - \int_B \sigma_t(\mathcal{Q}, q)\,\rho_t(dq). \qquad (7)$$

To satisfy the condition of equivariance we need to find jump rates σ_t such that the right-hand sides of the evolution equations (5) and (7) coincide whenever $\rho_t = \pi_t$. We see that this is the case when σ_t is given by the Radon–Nikodym derivative

$$\sigma_t(dq, q') = \frac{J_t^+(dq, dq')}{\pi_t(dq')} = \frac{\left[\tfrac{2}{\hbar}\operatorname{Im} \langle \Psi_t | P(dq)HP(dq')\Psi_t \rangle\right]^+}{\langle \Psi_t | P(dq')\Psi_t \rangle} \qquad (8)$$

of the positive part J_t^+ of J_t in its second variable q', provided this makes sense. For, the antisymmetry of J_t then implies that

$$\sigma_t(dq, q')\pi_t(dq') - \sigma_t(dq', q)\pi_t(dq) = J_t(dq, dq').$$

To make formula (8) meaningful one needs some assumptions which roughly require that H is an integral operator in the position representation given by P, and $(\mathcal{Q}, \mathcal{F})$ is standard Borel. This is discussed in detail in [DGTZ03c], where (8) was written down for the first time in this generality; special cases had been used before in [Bel86, Sud87, DGTZ03a]. For the precise conditions we refer to Theorem 1 (Sect. 4.1) and Corollaries 1–3 (Sect. 4.2) of [DGTZ03c]. Under these conditions, formula (8) can (and has to) be read as follows: A priori, J_t is a signed bi-measure on $\mathcal{F} \times \mathcal{F}$ (a measure in each of the two variables q, q'). This has to (and then can) be extended to a signed measure on the product σ-algebra $\mathcal{F} \otimes \mathcal{F}$. The positive part in (8) is then to be understood in the sense of the Hahn–Jordan decomposition of this extended measure. Next one notes that, for each $B \in \mathcal{F}$, $J_t(B, \cdot) \ll \pi_t$ because $P(B')\Psi_t = 0$ whenever $\pi_t(B') = 0$. One can thus form the Radon–Nikodym derivatives $\sigma_t(B, \cdot) = J_t^+(B, \mathrm{d}q')/\pi_t(\mathrm{d}q')$, which finally have to be chosen in such a way that σ_t becomes a measure kernel.

According to our derivation above, the transition rates (8) have been chosen to satisfy the requirement of equivariance. There was, however, still some freedom of choice. The particular rate (8) is singled out by the following additional facts.

1. Suppose there exists a jump process (Q_t) on $\mathcal{Q}$ with rates (8). As is evident from the arguments above, the net probability current of (Q_t) between two sets $B, B' \in \mathcal{F}$,

$$j_t(B, B') = \lim_{\varepsilon \searrow 0} \frac{1}{\varepsilon} \Big(\mathbb{P}(Q_t \in B', Q_{t+\varepsilon} \in B) - \mathbb{P}(Q_t \in B, Q_{t+\varepsilon} \in B') \Big),$$

then coincides with the quantum theoretical current J_t defined by (6). Conversely, if $(\tilde{Q}_t)$ is any pure jump process having initial distribution π_0 at time 0, some jump rates $\tilde{\sigma}_t$ and probability current $\tilde{j}_t = J_t$, it turns out that necessarily $\tilde{\sigma}_t(\mathrm{d}q, q') \geq \sigma_t(\mathrm{d}q, q')$ [RS90, BD99, DGTZ03c]. This follows from the minimality of the Hahn–Jordan decomposition. The rates (8) are therefore called the *minimal jump rates*, and a process with rates (8) is distinguished among all processes with the right probability current by having the least frequent jumps, or the smallest amount of randomness.

2. Always one of the transitions $q' \to q$ or $q \to q'$ is forbidden. More precisely, for every time t there exists a set $S_t^- \in \mathcal{F} \otimes \mathcal{F}$ which, together with its transposition S_t^+, covers $\mathcal{Q} \times \mathcal{Q}$ (except possibly the diagonal), and such that

$$\sigma_t(\{q : (q, q') \in S_t^-\}, q') = 0 \quad \text{for } \pi_t\text{-almost-every } q'.$$

Indeed, by the anti-symmetry of J_t, its positive and negative part J_t^+ and J_t^- admit supports S_t^+ and S_t^- that are transpositions of each other, whence the result follows.

Put more simply, the mechanism is this: When the current $J_t(dq, dq')$ is positive, meaning that there should be a net flow from dq' to dq, then $\sigma_t(dq, q') > 0$ and $\sigma_t(dq', q) = 0$, i.e., only jumps from q' to q are allowed; the converse holds in the case of a negative current. Under all rates with $j_t = J_t$, the minimal rates (8) are characterized by this property.

3 Bohmian Mechanics and Bell-Type QFT

In this section we discuss three particular instances in which jump rates of the form (8) play a significant role.

A. Bohmian mechanics as continuum limit of jump processes. Consider nonrelativistic quantum mechanics for N particles: the configuration space is $Q = \mathbb{R}^{3N}$, the Hilbert space $\mathscr{H} = L^2(\mathbb{R}^{3N}, \mathbb{C}^k)$ and the Hamiltonian

$$H = -\sum_{i=1}^{N} \frac{\hbar^2}{2m_i} \Delta_i + V(x_1, \ldots, x_N) \tag{9}$$

with Δ_i the Laplacian acting on the variable x_i, m_i the mass of the i-th particle, and V the potential function (possibly having values in the Hermitian $k \times k$ matrices). We obtain a Markov process on the configuration space in the following way: first discretize space, i.e., replace $\mathbb{R}^3$ by a lattice $\Lambda = \varepsilon \mathbb{Z}^3$ and the Laplacian Δ_i by the corresponding lattice Laplacian Δ_i^ε. We then can consider the jump process Q_t^ε on Λ^N with rates (8). As the lattice shrinks, $\varepsilon \to 0$, one obtains [Sud87, Vin93] in the limit the deterministic process Q_t satisfying the ordinary differential equation

$$\frac{dQ_{t,i}}{dt} = \frac{\hbar}{m_i} \text{Im} \frac{\Psi_t^* \nabla_i \Psi_t}{\Psi_t^* \Psi_t}(Q_{t,1}, \ldots, Q_{t,N}). \tag{10}$$

Here $Q_{t,i}$ is the i-th component of Q_t, i.e., the position of the i-th particle, Ψ_t obeys the Schrödinger equation (3) with Hamiltonian (9), and $\Phi_1^* \Phi_2$ is the inner product in $\mathbb{C}^k$. The process (10) is known as *Bohmian mechanics* [Boh52, Bel66, BDDGZ95, Dür01]. For a suitable other choice of jump rates [Vin93], also satisfying $j_t = J_t$ but greater than minimal, one obtains in the continuum limit $\varepsilon \to 0$ the diffusion process introduced by E. Nelson and known as *stochastic mechanics* [Nel85, Gol87].

What makes Bohmian mechanics (or, for that matter, stochastic mechanics) particularly interesting to quantum physicists is that in a Bohmian universe – one in which the particles move according to (10) and the initial configuration is chosen according to the $|\Psi|^2$ distribution – the inhabitants find all their observations in agreement with the probabilistic predictions of quantum mechanics – in sharp contrast with the traditional belief that it be impossible to explain the probabilities of quantum mechanics by any theory describing events objectively taking place in the outside world.

B. Bell's jump process for lattice QFT. The study of jump processes with rates (8) has been inspired by Bohmian mechanics, in particular by the wish for a theory similar to Bohmian mechanics covering quantum field theory. The first work in this direction was Bell's seminal paper [Bel86]. For simplicity, Bell replaces physical 3-space by a lattice Λ and considers a QFT on that lattice. A configuration is specified in his model by the number of fermions $q(x)$ at every lattice site x. Thus, with the notation $\mathbb{Z}_+ = \{0, 1, 2, \ldots\}$, the configuration space is

$$\mathcal{Q} = \Gamma(\Lambda) := \Big\{ q \in \mathbb{Z}_+^\Lambda : \sum_{x \in \Lambda} q(x) < \infty \Big\},$$

the space of all configurations of a variable (but finite) number of identical particles on the lattice. (In fact, he imposes a bound on the total number of particles and assumes that Λ is finite, but this is not necessary.) The PVM $P(\cdot)$ that he suggests arises from the joint spectral decomposition of the fermion number operators $N(x)$ for every lattice site, i.e., $P(q) := P(\{q\})$ is the projection to the joint eigenspace of the (commuting) operators $N(x)$ for the eigenvalues $q(x)$. The jump rate Bell uses is the appropriate special case of (8): the rate of jumping from q' to q is

$$\sigma_t(q, q') = \frac{\big[\frac{2}{\hbar} \mathrm{Im} \, \langle \Psi_t | P(q) H P(q') \Psi_t \rangle\big]^+}{\langle \Psi_t | P(q') \Psi_t \rangle}. \tag{11}$$

For studies of Bell's process we refer to [Sud87, Vin93, BD99, Col03a, Col03b, DGTZ03c, GT03], and for some numerical simulations and applications to [DR03, Den03]. We return to it in more detail in the next section.

C. Bohmian mechanics with variable number of particles. A third example of a process for a QFT was considered in [DGTZ03a]. It arose from an attempt to include particle creation and annihilation into Bohmian mechanics by simply introducing the possibility that world lines of particles can begin and end. That is, the aim is to provide a generalization of the Bohmian motion (10) to a configuration space of a variable number of particles. Here we describe this model in a simplified version. For the numerous similarities between our model process and Bell's discrete process, we called it a "Bell-type QFT." In [DGTZ03c, DGTZ04], methods are developed for obtaining a canonical Bell-type process for more or less any regularized QFT.

A configuration of finitely many identical particles can be described by a finite counting measure on $\mathbb{R}^3$. Since the coincidence configurations, those in which there are two or more particles at the same location, form a subset of codimension 3, they are basically irrelevant, and it will be convenient to exclude them from the configuration space. What remains, as the space of "simple configurations", is the set of all finite subsets of $\mathbb{R}^3$,

$$\Gamma_{\neq}(\mathbb{R}^3) = \{q \subset \mathbb{R}^3 : \#q < \infty\}.$$

Under the physical conditions prevailing in everyday life, the most frequent type of particle creation and annihilation is the emission and absorption of photons by electrons. This can be described in a model QFT as follows. Particles (photons) move in a Bohmian way and can be emitted and absorbed by another kind of particles (electrons). For simplicity, we will assume here that the electrons remain at fixed locations, given by a finite set $\eta \subset \mathbb{R}^3$; the case of moving electrons is described in [DGTZ03a]. The configuration space is thus the space of photon configurations, $\mathcal{Q} = \Gamma_{\neq}(\mathbb{R}^3)$, and Ψ_t a square-integrable complex-valued function on $\mathcal{Q}$; the Hilbert space $\mathscr{H}$ of these functions is known as the bosonic Fock space arising from $L^2(\mathbb{R}^3)$.

The Markov process Q_t in $\mathcal{Q}$ has piecewise smooth paths. It obeys the deterministic motion (10), interrupted by stochastic jumps. The process is piecewise deterministic in the sense that, conditional on the times of two subsequent jumps and the destination of the first, the path in between these jumps is deterministic. The jumps correspond to creation or annihilation of a photon near some point of η; in particular, every jump changes the number of photons by one. The process is thus a special kind of a spatial birth-and-death process with moving individuals [Pre76].

The deterministic motion, during which the number of photons is kept constant, is defined by (10); for simplicity, we deviate a little from the physical facts and assume that the "photons" have a positive mass m_{ph}. The rate for the transition $q \to q \cup x := q \cup \{x\}$, i.e., the creation of a new photon at the location $x \in \mathbb{R}^3 \setminus q$, has density (with respect to Lebesgue measure $\mathrm{d}x$)

$$\sigma_t(q \cup x, q) = \frac{\left[\frac{2}{\hbar} \operatorname{Im} \Psi_t^*(q \cup x) \, (\#q + 1)^{1/2} \sum_{y \in \eta} \varphi(x - y) \, \Psi_t(q) \right]^+}{\Psi_t^*(q) \, \Psi_t(q)}$$

where $\varphi : \mathbb{R}^3 \to \mathbb{R}$ is a fixed function, a spherically symmetric, square-integrable potential supported by the ball of radius $\delta > 0$. Likewise, for any $x \in q$ the rate for the transition $q \to q \setminus x := q \setminus \{x\}$, i.e., annihilation of the photon at x, is

$$\sigma_t(q \setminus x, q) = \frac{\left[\frac{2}{\hbar} \operatorname{Im} \Psi_t^*(q \setminus x) \, (\#q)^{-1/2} \sum_{y \in \eta} \varphi(x - y) \, \Psi_t(q) \right]^+}{\Psi_t^*(q) \, \Psi_t(q)}$$

These rates, together with vanishing rate for any other transition, are in fact a special case of (8), for a suitable integral operator H_I in place of H. For the definition of H_I and the derivation of the above expressions from (8) we refer to Sect. 3.12 of [DGTZ03c].

Now, H_I is not the Hamiltonian of the relevant QFT, but its *interaction part*; i.e., the complete Hamiltonian is $H = H_0 + H_I$, where H_0, the *free Hamiltonian*, is given by

$$H_0 \Psi(q) = -\sum_{i=1}^{\#q} \frac{\hbar^2}{2 m_{\mathrm{ph}}} \Delta_i \Psi(q).$$

Observe that there is a correspondence between the splitting $H = H_0 + H_I$ and the two constituents of the process, the motion (10) and the jump rates just given. Deterministic motion corresponds to H_0 while the jumps correspond to H_I. Indeed, the *minimal process*, the one arising as a limiting case from jump processes with minimal rates (8), associated with H_0 alone is the continuous motion (10) while the minimal process associated with H_I is the pure jump process with the above rates.

This is an instance of the general rule of *process additivity*: If the minimal process associated with H_1 has generator $\mathscr{L}_{1,\Psi_t}$ and the one associated with H_2 has generator $\mathscr{L}_{2,\Psi_t}$, then the minimal process associated with $H_1 + H_2$ has generator $\mathscr{L}_{1,\Psi_t} + \mathscr{L}_{2,\Psi_t}$, provided that the (formal) integral kernels of H_1 and H_2 have disjoint supports. The sum of the generators of a deterministic flow and of a pure jump process generates the piecewise deterministic process that follows the flow between stochastic jumps. In QFT, it is a typical situation that $H = H_0 + H_I$ where H_0 is a differential operator associated with continuous motion while H_I is an integral operator (often linking different particle numbers) associated with jumps.

4 Global Existence of Bell's Jump Process

In this section we deal with Bell's jump process introduced as model B in the last section. As we have shown in [GT03], this process exists globally in time. In fact, for our proof it is not relevant whether the configuration corresponds to the fermion number operators. We only need that $\mathcal{Q}$ is any countable set, and $P(\,\cdot\,)$ a PVM on $\mathcal{Q}$ acting on $\mathscr{H}$. In fact we can allow that $P(\,\cdot\,)$ is a positive-operator-valued measure (POVM), a concept somewhat weaker than a PVM.[1] Here is our result.

Theorem 1. *Let $\mathscr{H}$ be a Hilbert space, H a self-adjoint operator on $\mathscr{H}$, $\mathcal{Q}$ a countable set, and $P(\,\cdot\,)$ a POVM on $\mathcal{Q}$ acting on $\mathscr{H}$. For every initial state vector Ψ_0 with $\|\Psi_0\| = 1$ satisfying*

$$\Psi_t \in \mathrm{domain}(H) \quad \forall t \in \mathbb{R}, \tag{12a}$$

$$P(q)\Psi_t \in \mathrm{domain}(H) \quad \forall t \in \mathbb{R}, q \in \mathcal{Q}, \tag{12b}$$

$$\int_{t_1}^{t_2} dt \sum_{q,q' \in \mathcal{Q}} \left| \langle \Psi_t | P(q) H P(q') \Psi_t \rangle \right| < \infty \quad \forall t_1, t_2 \in \mathbb{R} \text{ with } t_1 < t_2, \tag{12c}$$

there exists a (right-continuous) Markovian pure jump process $(Q_t)_{t \in \mathbb{R}}$ on $\mathcal{Q}$ with transition rates (11) such that, for every t, Q_t has distribution $\pi_t = \langle \Psi_t | P(\,\cdot\,)\Psi_t \rangle$. This process is unique in distribution.

---

[*] That is, P takes values in the positive (bounded, self-adjoint) operators on $\mathscr{H}$ (instead of the projection operators as a PVM) and shares the countable additivity and normalization of a PVM.

Some comments are necessary. First of all, how could the process fail to exist globally in time? Two kinds of catastrophes could occur. On the one hand, the jump rate (11) is singular at the nodes of Ψ (i.e., at such q and t for which $\langle \Psi_t | P(q)\Psi_t \rangle = 0$). While Q_t is sitting on a configuration q it might become a node, and then the process would not know how to proceed. It turns out that this problem does not arise because, with probability one, there is no t at which Q_t is a node. This is because the increase of the rates close to the nodes has the positive effect of forcing the process to jump away before the singularity time is reached.

The second kind of possible catastrophe would be an explosion, i.e., an accumulation of infinitely many jumps in finite time. The main task is to show that this does not occur, with probability one. The standard criteria for non-explosion of pure jump processes are confined to transition rates that are homogeneous in time, relying heavily on the fact that the holding times are then exponentially distributed and independent conditionally on the positions; see, e.g., Sect. 2.7 of [Nor97]. This conditional independence, however, fails to hold in the case of time-dependent jump rates, and the singularities of Bell's transition rates do not allow any a priori bounds as they were used, e.g., in [Pre76, RL53] to exclude explosion. The only thing one knows is that the process is designed to have the prescribed quantum distribution (4) at fixed (deterministic) times, and it is this fact we will exploit. We will sketch our main arguments below.

Steps towards a proof of global existence of Bell's process have also been made by G. Bacciagaluppi [Bac96, BD99]; his approach is, however, very different from ours.

Concerning the technical assumptions (12) on H, P, and Ψ_0 we note the following. For fixed H and P, the conditions (12) define a set of "good" initial state vectors Ψ_0 for which we can prove global existence; this set is obviously invariant under the time evolution (1). In fact, when H is a Hilbert–Schmidt operator (i.e., $\operatorname{tr} H^2 < \infty$), the conditions (12) are satisfied for *all* POVMs P and *all* $\Psi_0 \in \mathscr{H}$; this is also true when H is bounded and Q is finite. (Usually, the Hamiltonian of a lattice QFT is not Hilbert–Schmidt but can, at least, be approximated by Hilbert–Schmidt operators. And it is not unusual in quantum field theory that Hamiltonians need to be "cut off" in one way or another to make them treatable, or well-defined at all.) Condition (12b) ensures that the expression $P(q)HP(q')\Psi_t$ can be formed, and thus that (11) is well-defined whenever q' is not a node.

The main construction is obvious. Starting from any fixed initial time t_0, the process Q_t can be expressed for $t > t_0$ in terms of T_k and X_k, the time and the destination of the k-th jump after time t_0, via

$$Q_t = X_k \quad \text{if } T_k \leq t < T_{k+1}$$

with $T_0 := t_0$ and $X_0 = Q_{t_0}$. The variables T_k and X_k are defined by their conditional distributions:

$$\mathbb{P}(T_{k+1} \in dt, X_{k+1} = q | T_0, X_0, \ldots, T_k, X_k) =$$

$$1_{\{T_k < t\}}\, \sigma_t(q, X_k)\, \exp\Big(-\int_{T_k}^{t} \sigma_s(\mathcal{Q}, X_k)\, ds\Big) dt,$$

where the role of the "failure rate function" is played by

$$\sigma_t(\mathcal{Q}, q) = \sum_{q' \in \mathcal{Q}} \sigma_t(q', q),$$

the total jump rate, to whatever destination, at q.

Here is the reason why the process cannot run into a node. By definition, the conditional probability of remaining at q until at least time t_2, given that $Q_{t_1} = q$, is

$$\exp\Big(-\int_{t_1}^{t_2} \sigma_t(\mathcal{Q}, q)\, dt\Big).$$

We want to show that this probability vanishes whenever q is a node at t_2 but not at any t with $t_1 \leq t < t_2$. Ignoring some technical subtleties, this can be derived by the following simple calculation. Since a sum of positive parts exceeds the positive part of the sum, we conclude from (11) that

$$\sigma_t(\mathcal{Q}, q') = \sum_{q \in \mathcal{Q}} \frac{[\frac{2}{\hbar} \operatorname{Im} \langle \Psi_t | P(q) H P(q') \Psi_t \rangle]^+}{\langle \Psi_t | P(q') \Psi_t \rangle} \geq \frac{[\frac{2}{\hbar} \operatorname{Im} \langle \Psi_t | H P(q') \Psi_t \rangle]^+}{\langle \Psi_t | P(q') \Psi_t \rangle}.$$

Omitting the positive part and using (5) we find

$$\sigma_t(\mathcal{Q}, q') \geq \frac{-(d/dt)\langle \Psi_t | P(q') \Psi_t \rangle}{\langle \Psi_t | P(q') \Psi_t \rangle} = -\frac{d}{dt} \log \langle \Psi_t | P(q') \Psi_t \rangle$$

at every t with $\langle \Psi_t | P(q') \Psi_t \rangle > 0$. Hence, by the fundamental theorem of calculus,

$$\int_{t_1}^{t_2} \sigma_t(\mathcal{Q}, q)\, dt \geq -\log \langle \Psi_{t_2} | P(q) \Psi_{t_2} \rangle + \log \langle \Psi_{t_1} | P(q) \Psi_{t_1} \rangle = \infty,$$

since q is a node at t_2 (so that the first term is infinite) but not at t_1 (so that the second term is finite).

Another part of the proof we would like to sketch here is the core of the argument why the jump times cannot accumulate. As a convenient notation, we introduce an additional "cemetery" configuration ∞ and set $Q_t := \infty$ for all t after the explosion time $\sup_k T_k$. Let $S(t_1, t_2)$ be the number of jumps that the process performs in the time interval $[t_1, t_2]$. The random variable $S(t_1, t_2)$ is either a nonnegative integer or infinite. Our assumption (12c) implies in fact that $S(t_1, t_2)$ has finite expectation, for all $t_1 < t_2$, and thus is finite almost surely. To see this we use the equation

$$\mathbb{E}\, S(t_1, t_2) = \int_{t_1}^{t_2} \sum_{q,q' \in \mathcal{Q}} \sigma_t(q, q')\, \rho_t(q')\, \mathrm{d}t, \tag{13}$$

with $\rho_t(q') = \mathbb{P}(Q_t = q')$. Intuitively, equation (13) can be understood as follows. $\sigma_t(q, q')\, \rho_t(q')\, \mathrm{d}t$ is the probability of a jump from q' to q in the infinitesimal time interval $[t, t + \mathrm{d}t]$. Summing over q and q', we obtain (the expected total jump rate and thus) the total probability of a jump during $[t, t + \mathrm{d}t]$. Integrating over t, we obtain the expected number of jumps. The point of equation (13) is that its right-hand side involves exclusively the one-time quantities σ_t and ρ_t. Now, one can deduce from the definition of the process that

$$\rho_t(q) \leq \langle \Psi_t | P(q) \Psi_t \rangle, \tag{14}$$

where any case of strict inequality would have to go along with a positive probability $\mathbb{P}(Q_t = \infty)$ of accumulation before t. Combining (13) and (14), we obtain

$$\begin{aligned}
\mathbb{E}\, S(t_1, t_2) &\leq \int_{t_1}^{t_2} \sum_{q,q' \in \mathcal{Q}} \sigma_t(q, q')\, \langle \Psi_t | P(q') \Psi_t \rangle\, \mathrm{d}t \\
&= \int_{t_1}^{t_2} \sum_{q,q' \in \mathcal{Q}} \left[\tfrac{2}{\hbar} \operatorname{Im} \langle \Psi_t | P(q) H P(q') \Psi_t \rangle \right]^+ \mathrm{d}t \\
&\leq \tfrac{2}{\hbar} \int_{t_1}^{t_2} \sum_{q,q' \in \mathcal{Q}} \left| \langle \Psi_t | P(q) H P(q') \Psi_t \rangle \right| \mathrm{d}t < \infty
\end{aligned}$$

by assumption (12c) of Theorem 1. Indeed, this reasoning more or less dictates the assumption (12c).

5 Other Global Existence Questions

Variants of the reasoning in the previous paragraph could be used in the future in other global existence proofs. Here is an example concerning the *global existence of Bohmian mechanics* (with fixed number of particles). This was first proved in [BDGPZ95] under suitable assumptions on the potential V and the initial wavefunction Ψ_0. One way in which a solution Q_t of (10) may fail to exist globally is by escaping to infinity (i.e., leaving every bounded set in $\mathbb{R}^{3N}$) in finite time. To control this behavior, we suggest to consider an analogue of $S(t_1, t_2)$: Let $D(t_1, t_2)$ be the Euclidean distance in $\mathbb{R}^{3N}$ traveled by Q_t between t_1 and t_2. Then

$$\mathbb{E}\, D(t_1, t_2) = \int_{t_1}^{t_2} \mathrm{d}t \int_{\mathbb{R}^{3N}} \mathrm{d}q\, |v_t(q)|\, \rho_t(q), \tag{15}$$

where v_t is the Bohmian velocity vector field on $\mathbb{R}^{3N}$, with i-th component given by the right-hand side of (10), and the expectation is taken over the randomness coming from the initial configuration. If this expectation can be shown to be finite, $D(t_1, t_2)$ must be almost surely finite. Using $\rho_t(q) \leq |\Psi_t(q)|^2$, the analogue of (14), we see that the escape to infinity is almost surely excluded provided that

$$\int_{t_1}^{t_2} dt \int_{\mathbb{R}^{3N}} dq \, |\Psi_t^*(q) \, \nabla \Psi_t(q)| < \infty \quad \text{whenever } t_1 < t_2.$$

This is a condition analogous to (12c); it is almost equivalent to the assumption A4 of [BDGPZ95]. The proof there, however, is different, using skillful estimates of the probability flux across suitable surfaces in configuration space-time $\mathbb{R}^{3N} \times \mathbb{R}$ surrounding the "bad" points (nodes, infinity, points where Ψ is not differentiable). The above argument based on (15) might contribute to a simpler global existence proof [TT04].

The global existence question is still open for the *Bohm–Dirac law of motion* [Boh53, BH93], a version of Bohmian mechanics suitable for wavefunctions ψ obeying the Dirac equation. The Dirac equation is a relativistic version of the Schrödinger equation and reads

$$i\hbar \frac{\partial \psi}{\partial t} = -\sum_{i=1}^{N} \left(i c \hbar \alpha_i \cdot \nabla_i \psi + \beta_i m_i c^2 \psi\right) \tag{16}$$

where ψ_t is a function $\mathbb{R}^{3N} \to (\mathbb{C}^4)^{\otimes N}$, c is the speed of light, m_i the mass of the i-th particle, α_i the vector of Dirac alpha matrices acting on the i-th spin index of the wavefunction, and β_i the Dirac beta matrix acting on the i-th index. The Bohm–Dirac equation of motion reads

$$\frac{dQ_{t,i}}{dt} = c \frac{\psi_t^* \alpha_i \psi_t}{\psi_t^* \psi_t}(Q_{t,1}, \ldots, Q_{t,N}) \tag{17}$$

where $\phi^* \psi$ is the inner product in Dirac spin-space. Since these velocities are bounded by the speed of light, the question of escape to infinity does not arise here. But the question of running into a node does because, like the minimal jump rate (8), the velocity formula (17) is ill-defined at nodes.

This question can be treated in a way analogous to the previous arguments based on (13) and (15). To this end, let $L(t_1, t_2)$ be the total variation, between t_1 and t_2, of $\log |\psi_t(Q_t)|^2$; in other words,

$$L(t_1, t_2) = \int_{t_1}^{t_2} dt \left| \frac{d}{dt} \log |\psi_t(Q_t)|^2 \right|.$$

It takes values in $[0, \infty]$ and is infinite if the trajectory Q_t runs into a node between t_1 and t_2. This must be a null event if $L(t_1, t_2)$ has finite expectation; for the latter we have the formula

$$\mathbb{E} \, L(t_1, t_2) = \int_{t_1}^{t_2} dt \int_{\mathbb{R}^{3N}} dq \left| \left(\frac{\partial}{\partial t} + v_t(q) \cdot \nabla \right) \log |\psi_t(q)|^2 \right| \rho_t(q),$$

where v_t is the vector field on $\mathbb{R}^{3N}$ whose i-th component is the right-hand side of (17). Using $\rho_t(q) \leq |\psi_t(q)|^2$ we obtain

$$\mathbb{E}\, L(t_1, t_2) \leq \int_{t_1}^{t_2} dt \int_{\mathbb{R}^{3N}} dq \left| \left(\frac{\partial}{\partial t} + v_t(q) \cdot \nabla \right) |\psi_t(q)|^2 \right|.$$

As the velocities are bounded, the last expression is at most

$$\int_{t_1}^{t_2} dt \int_{\mathbb{R}^{3N}} dq \left(\left| \frac{\partial}{\partial t} |\psi_t(q)|^2 \right| + c \left| \nabla |\psi_t(q)|^2 \right| \right).$$

Inserting (16) and using that the alpha and beta matrices have norm 1 we see that this in turn is not larger than

$$\int_{t_1}^{t_2} dt \int_{\mathbb{R}^{3N}} dq \left(2c \left| \psi_t^* \nabla \psi_t \right| + \frac{2}{\hbar} \left(\sum_i m_i c^2 \right) \psi_t^* \psi_t + 2c \left| \psi_t^* \nabla \psi_t \right| \right).$$

Since $\|\psi_t\| = 1$, the last integral coincides with

$$\frac{2}{\hbar} \left(\sum_i m_i c^2 \right) (t_2 - t_1) + 4c \int_{t_1}^{t_2} dt \int_{\mathbb{R}^{3N}} dq \left| \psi_t^* \nabla \psi_t \right|.$$

The question remains under which conditions on ψ_0 the last term is finite for arbitrary $t_1 < t_2$. This is presumably the case when ψ_0 lies in Schwartz space (containing all smooth functions f such that f and all its derivatives decay, at infinity, faster than $|q|^{-n}$ for any $n > 0$). To work out the proof remains for future research [TT04].

While global existence of the Bohm–Dirac trajectories can presumably be proved with the same methods as used in [BDGPZ95] for (10) (estimating the flux across suitable surfaces surrounding the bad points), such a proof requires a lot of effort. It seems that the argument just sketched would be much easier and more elegant.

Another global existence question that is still open is that concerning the process defined in [DGTZ03a] and described above in Part C of Sect. 3, and for similar processes, on a configuration space that is a disjoint union of manifolds, following deterministic trajectories interrupted by stochastic jumps.

6 Deterministic Jumps and Boundaries in Configuration Space

In this last section we describe another application of minimal jump rates, one that has not yet been discussed in the literature and that raises some questions for further research. Suppose that the configuration space $\mathcal{Q}$ is a

Riemannian manifold with boundaries, or more generally the disjoint union of (at most countably many) Riemannian manifolds with boundaries. We write $Q = \partial Q \cup Q^\circ$ where ∂Q denotes the boundary and Q° the interior.

We develop below an analogue of Bohmian mechanics on Q, consisting of smooth motion interrupted by jumps from the boundary to the interior or vice versa. The jumps from ∂Q to Q° are deterministic and occur whenever the process hits the boundary. The jumps from Q° to ∂Q are stochastic, and their rates are fully determined by requiring that (i) the process is Markovian and equivariant, and (ii) the construction is invariant under time reversal, in that the processes associated with Ψ_t resp. Ψ^*_{-t} are reverse to each other, in distribution. These rates are, in fact, another instance of minimal jump rates.

Configuration spaces with boundaries arise from QFT if a particular kind of "ultraviolet cutoff" is applied, which can be regarded as corresponding to smearing out the charge of an electron over a sphere rather than a ball. Here is an example. Consider again electrons and photons, with the electrons fixed at locations given by the finite set $\eta \subset \mathbb{R}^3$, and suppose that photons cannot get closer than a fixed (small) distance $\delta > 0$ to an electron, as they get absorbed when they reach that distance. Thus, the available configuration space is

$$Q = \{q \in \Gamma_{\neq}(\mathbb{R}^3) : d(q, \eta) \geq \delta\}, \tag{18}$$

where $d(q, \eta)$ is the Euclidean distance of the finite sets q and η. The space Q is a countable disjoint union of Riemannian manifolds with boundary,

$$Q = \bigcup_{n=0}^{\infty} \{q \in Q : \#q = n\}.$$

Its interior is $Q^\circ = \{q \in \Gamma_{\neq}(\mathbb{R}^3) : d(q, \eta) > \delta\}$, and its boundary $\partial Q = \{q \in \Gamma_{\neq}(\mathbb{R}^3) : d(q, \eta) = \delta\}$.

For this or any other configuration space with boundaries, the law of motion

$$\frac{dQ_t}{dt} = v_t(Q_t) = \frac{\hbar}{m} \operatorname{Im} \frac{\Psi^*_t \nabla \Psi_t}{\Psi^*_t \Psi_t}(Q_t) \tag{19}$$

on Q° must be completed by specifying what should happen to the process at the time τ when it reaches the boundary. (No specification is needed, however, for what should happen when two photons collide, as this has probability zero ever to occur.) The specification we consider here is a deterministic jump law

$$Q_{\tau+} = f(Q_{\tau-})$$

for a fixed mapping $f : \partial Q \to Q^\circ$. In our example (18), the obvious choice of f is

$$f(q) = \{x \in q : d(x, \eta) > \delta\},$$

which means that all photons having reached the critical distance δ to some electron disappear.

Since we want the theory to be reversible, we must also allow for spontaneous jumps from interior points to boundary points. Since we want the process to be an equivariant Markov process, the rate for a jump from $q' \in \mathcal{Q}^\circ$ to a surface element $dq \subseteq \partial\mathcal{Q}$ must be, as one can derive,

$$\sigma_t(dq, q') = \frac{[n(q) \cdot v_t(q) \, |\Psi_t(q)|^2]^+}{|\Psi_t(q')|^2} \, \nu(dq, q'), \tag{20}$$

where $n(q)$ is the inward unit normal vector to the boundary at $q \in \partial\mathcal{Q}$, the dot $\cdot$ denotes the Riemannian inner product, and $\nu(dq, q')$ is the measure-valued function defined in terms of the Riemannian volume measure μ on $\mathcal{Q}$ and the Riemannian surface area measure λ on $\partial\mathcal{Q}$ by

$$\nu(B, q') = \frac{\lambda(B \cap f^{-1}(dq'))}{\mu(dq')},$$

with $B \subseteq \partial\mathcal{Q}$, and the right-hand side denoting a Radon–Nikodym derivative (the existence of which we presuppose). The measure $\nu(\,\cdot\,, q')$ is concentrated on the subset $f^{-1}(q')$ of the boundary for almost every q'. For a probability distribution on $\mathcal{Q}$ having a density function ρ with respect to μ, one obtains the following probability transport equation at $q' \in \mathcal{Q}^\circ$:

$$\frac{\partial \rho_t}{\partial t}(q') = -\nabla \cdot (\rho_t v_t)(q') - \sigma_t(\partial\mathcal{Q}, q') \, \rho_t(q') + \int_{\partial\mathcal{Q}} \nu(dq, q') \big[n(q) \cdot v_t(q) \, \rho_t(q) \big]^-. \tag{21}$$

For equivariance we need that (21), with $|\Psi_t|^2$ in place of ρ_t, has the structure of the transport equation for $|\Psi_t|^2$ that follows from the Schrödinger equation (3),

$$\frac{\partial |\Psi_t(q')|^2}{\partial t} = \frac{2}{\hbar} \, \mathrm{Im} \, \Psi_t^*(q') \, (H\Psi_t)(q'). \tag{22}$$

This is not automatically the case, but it follows from (and therefore suggests) the following boundary condition relating $\Psi_t|_{\partial\mathcal{Q}}$ to $\Psi_t|_{\mathcal{Q}^\circ}$: for all $q \in \partial\mathcal{Q}$,

$$n(q) \cdot \nabla\Psi_t(q) = \gamma(q) \, \Psi_t(f(q)) \tag{23}$$

where $\gamma(q)$ is any complex number.[2] This condition prescribes the normal derivative of the wavefunction on the boundary. Some boundary condition would be needed anyway to define the evolution of the wavefunction, i.e., to select a self-adjoint extension of the Laplacian; whether (23) actually suffices for this, we have to leave to future research. Note that (23) is a linear condition and thus defines a subspace of the Hilbert space $L^2(\mathcal{Q})$. From (23) and (19), one obtains equivariance with respect to the formal Hamiltonian $H = -\frac{\hbar^2}{2m}\Delta + H_I$, where

---

$\cdot$ More generally, if Ψ takes values not in $\mathbb{C}$ but in a higherdimensional complex vector space, $\gamma(q)$ would be a $\mathbb{C}$-linear mapping from the value space at $f(q)$ to the value space at q.

$$\langle \Phi | H_I \Psi \rangle = \frac{\hbar^2}{2m} \int_Q \mu(dq') \int_{\partial Q} \nu(dq, q') \, \Phi^*(q') \, \gamma^*(q) \, \Psi(q)$$

$$+ \frac{\hbar^2}{2m} \int_Q \mu(dq) \int_{\partial Q} \nu(dq', q) \, \Phi^*(q') \, \gamma(q') \, \Psi(q).$$

Furthermore, the jump rate (20) is indeed the minimal jump rate (8) associated with H_I, thanks to (23).

Let us emphasize the following aspects. First, starting from the picture of a piecewise deterministic process that jumps whenever it hits the boundary, we arrived with remarkable ease at the probability transport equation (21) and thus at the boundary condition (23). Second, we have *derived* what the interaction Hamiltonian H_I is; once the destination mapping f and the corresponding coefficients γ had been selected, there was no further freedom. Third, it turned out that the minimal jump rate is the only *possible* jump rate in this setting. Its very minimality plays a crucial role: a jump to a boundary point q at which the velocity field is pointing *towards* the boundary, $n(q) \cdot v_t(q) < 0$, would not allow any continuation of the process since there is no trajectory starting from q. The problem is absent if the velocity at q is pointing *away* from the boundary, $n(q) \cdot v_t(q) > 0$. (We are leaving out the case $n(q) \cdot v_t(q) = 0$.) On the other hand, jumps from q to $f(q)$ cannot occur when $v_t(q)$ is pointing away from the boundary since in that case there is no trajectory arriving at q. Thus, the jumps must be such that at each time t, one of the transitions $q \to f(q)$ or $f(q) \to q$ is forbidden, and the decision is made by the sign of $n(q) \cdot v_t(q)$.

References

[Bac96]   Bacciagaluppi, G.: Topics in the Modal Interpretation of Quantum Mechanics. Ph. D. thesis, University of Cambridge (1996)

[BD99]   Bacciagaluppi, G., Dickson, M.: Dynamics for modal interpretations. Found. Phys., **29**, 1165–1201 (1999)

[Bel66]   Bell, J.S.: On the problem of hidden variables in quantum mechanics. Rev. Mod. Phys., **38**, 447–452 (1966). Reprinted in: Bell, J.S.: Speakable and unspeakable in quantum mechanics. Cambridge University Press, Cambridge (1987), p. 1.

[Bel86]   Bell, J.S.: Beables for quantum field theory. Phys. Rep., **137**, 49–54 (1986). Reprinted in: Bell, J.S.: Speakable and unspeakable in quantum mechanics. Cambridge University Press, Cambridge (1987), p. 173. Also reprinted in: Peat, F.D., Hiley, B.J. (eds) Quantum Implications: Essays in Honour of David Bohm. Routledge, London (1987), p. 227. Also reprinted in: Bell, M., Gottfried, K., Veltman, M. (eds) John S. Bell on the Foundations of Quantum Mechanics. World Scientific Publishing (2001), chap. 17.

[BDGPZ95]   Berndl, K., Dürr, D., Goldstein, S., Peruzzi, G., Zanghì, N.: On the global existence of Bohmian mechanics. Commun. Math. Phys., **173**, 647–673 (1995). quant-ph/9503013

[BDDGZ95]   Berndl, K., Daumer, M., Dürr, D., Goldstein, S., Zanghì, N.: A Survey
            of Bohmian Mechanics. Il Nuovo Cimento B, **110**, 737–750 (1995).
            quant-ph/9504010
[Boh52]   Bohm, D.: A Suggested Interpretation of the Quantum Theory in
            Terms of "Hidden" Variables, I and II. Phys. Rev., **85**, 166–193 (1952)
[Boh53]   Bohm, D.: Comments on an Article of Takabayasi concerning the
            Formulation of Quantum Mechanics with Classical Pictures. Progr.
            Theoret. Phys., **9**, 273–287 (1953)
[BH93]   Bohm, D., Hiley, B.J.: The Undivided Universe: An Ontological In-
            terpretation of Quantum Theory. Routledge, London (1993)
[Col03a]   Colin, S.: The continuum limit of the Bell model. quant-ph/0301119
[Col03b]   Colin, S.: A deterministic Bell model. Phys. Lett. A, **317**, 349–358
            (2003). quant-ph/0310055
[DR03]   Dennis, E., Rabitz, H.: Bell trajectories for revealing quantum control
            mechanisms. Phys. Rev. A, **67**, 033401 (2003). quant-ph/0208109
[Den03]   Dennis, E.: Purifying Quantum States: Quantum and Classical Algo-
            rithms. Ph.D. thesis, University of California, Santa Barbara (2003)
[Dür01]   Dürr, D.: Bohmsche Mechanik als Grundlage der Quantenmechanik.
            Springer, Berlin (2001)
[DGTZ03a]   Dürr, D., Goldstein, S., Tumulka, R., Zanghì, N.: Trajectories and
            Particle Creation and Annihilation in Quantum Field Theory. J. Phys.
            A: Math. Gen., **36**, 4143–4149 (2003). quant-ph/0208072
[DGTZ03b]   Dürr, D., Goldstein, S., Tumulka, R., Zanghì, N.: Bohmian Mechanics
            and Quantum Field Theory. To appear in Phys. Rev. Lett. (2004).
            quant-ph/0303156
[DGTZ03c]   Dürr, D., Goldstein, S., Tumulka, R., Zanghì, N.: Quantum Hamil-
            tonians and Stochastic Jumps. To appear in Commun. Math. Phys.
            (2004). quant-ph/0303056
[DGTZ04]   Dürr, D., Goldstein, S., Tumulka, R., Zanghi, N.: Bell-Type Quantum
            Field Theories. quant-ph/0407116
[GT03]   Georgii, H.-O., Tumulka, R.: Global Existence of Bell's Time-
            Inhomogeneous Jump Process for Lattice Quantum Field Theory. To
            appear in Markov Proc. Rel. Fields (2004). math.PR/0312294 and
            mp_arc 04-11
[Gol87]   Goldstein, S.: Stochastic Mechanics and Quantum Theory. J. Statist.
            Phys., **47**, 645–667 (1987)
[Nel85]   Nelson, E.: Quantum Fluctuations. Princeton University Press,
            Princeton (1985)
[Nor97]   Norris, J. R.: Markov chains. Cambridge University Press, Cambridge
            (1997)
[Pre76]   Preston, C. J.: Spatial birth-and-death processes. Bull. Inst. Internat.
            Statist., **46**, no. 2, 371–391, 405–408 (1975)
[RL53]   Reuter, G. E. H., Ledermann, W.: On the differential equations for the
            transition probabilities of Markov processes with enumerably many
            states. Proc. Cambridge Philos. Soc., **49**, 247–262 (1953)
[RS90]   Roy, S.M., Singh, V.: Generalized beable quantum field theory. Phys.
            Lett. B, **234**, 117–120 (1990)
[Sud87]   Sudbery, A.: Objective interpretations of quantum mechanics and the
            possibility of a deterministic limit. J. Phys. A: Math. Gen., **20**, 1743–
            1750 (1987)

[TT04]   Teufel, S., Tumulka, R.: Simple Proof for Global Existence of
            Bohmian Trajectories. In preparation.
[Vin93]   Vink, J.C.: Quantum mechanics in terms of discrete beables. Phys.
            Rev. A, **48**, 1808–1818 (1993)

Gibbs Measures on Brownian Paths: Theory and Applications

Volker Betz*, József Lőrinczi, and Herbert Spohn

Zentrum Mathematik, Technische Universität München
Boltzmannstr. 3, 85747 Garching bei München, Germany

Summary. We review our investigations on Gibbs measures relative to Brownian motion, in particular the existence of such measures and their path properties, uniqueness, resp. non-uniqueness. For the case when the energy only depends on increments, we present a functional central limit theorem. We also explain connections with other work and state open problems of interest.

1 Introduction

The probability measures studied in Statistical Mechanics have the generic structure

$$\frac{1}{Z} \exp[-\mathcal{E}] \quad \times \quad a\ priori \text{ measure} \tag{1}$$

The *a priori* measure is explicit and simple. The energy function $\mathcal{E}$ is defined on the same space as the *a priori* measure and the partition function Z makes (1) a probability measure. Of course, it is understood that $\mathcal{E}$ has a natural structure, as dictated by concrete applications.

One much studied class of examples is that of lattice spin systems with finite state space S. Then the *a priori* measure is the product over the lattice points of the counting measure on S. The energy function typically has the form

$$\mathcal{E}_\Lambda = (k_B T)^{-1} \sum_{x,y \in \Lambda} U(\sigma_x, \sigma_y, |x-y|) \tag{2}$$

where $U : S^2 \times \mathbb{R}^+ \to \mathbb{R}$ is a pair potential, and $\sigma_x \in S$ is the value of the spin at site x of the finite subset Λ of the lattice. The inverse temperature $(k_B T)^{-1}$ appears as a strength factor multiplying the energy.

The specific expression of the measures as formally given by (1) is actually firmly grounded in the experience of rigorous statistical mechanics. At least in the context of lattice spin systems with compact state space the so emerging *Gibbs measures* prove to provide a proper mathematical description of thermodynamic equilibrium states and thus they play a fundamental

* Institute for Biomathematics and Biometry, GSF Forschungszentrum, Postfach 1129, D-85758 Neuherberg, Germany

role in the theory of phase transitions. In more specific cases, such as the Potts model, these measures make a strong link between locality properties and memory effects (Markov random fields), variational principles involving the minimization of free energy so that states appear as tangent functionals (large deviation theory), and the understanding in terms of percolation properties of how macroscopic long range order builds up from small scale events governed by chance (stochastic geometry). Although as soon as we leave the class of discrete models these relationships are not as clear any longer, these signposts pinpoint a programme of a general theory of Gibbs measures from which one can take an inspiration. In this paper we present the first steps in developing a theory of Gibbs measures on path space.

We will study the case where the *a priori* measure is Brownian motion in $\mathbb{R}^d$. Let us denote by $t \to X_t \in \mathbb{R}^d$ a Brownian path and by $\mathcal{W}$ the Wiener measure. Since $t \in \mathbb{R}$, in the parlance of Statistical Mechanics our model is one-dimensional with d components. The finite box Λ corresponds to the time interval $[-T, T]$. $\mathcal{W}$ has then to be supplied with appropriate boundary conditions. For example one could pin the path at both endpoints, $X_{-T} = 0$, $X_T = 0$, in which case $\mathcal{W}$ would turn into a Brownian bridge. The simplest energy function is given through an "on site" potential $V : \mathbb{R}^d \to \mathbb{R}$ and takes the form

$$\mathcal{E}_{1,T} = \int_{-T}^{T} V(X_t) dt. \tag{3}$$

The analogue of the pair interaction energy (2) transcribes as

$$\mathcal{E}_{2,T} = \int_{-T}^{T} \int_{-T}^{T} W(X_t, X_s, t - s) dt ds \tag{4}$$

with $W : \mathbb{R}^{2d} \times \mathbb{R} \to \mathbb{R}$, $W(x, x', t) = W(x', x, t)$, and $\int |W(x, x', t)| dt < \infty$. In Statistical Mechanics energies are proportional to the volume, i.e. proportional to T in our case. Clearly, in spirit this is satisfied by both energies (3) and (4). With these preparations a Gibbs measure on path space reads as

$$\frac{1}{Z(T)} \exp\left[-\mathcal{E}_{1,T}(X) - \mathcal{E}_{2,T}(X) \right] \delta(X_{-T}) \delta(X_T) d\mathcal{W}(X). \tag{5}$$

Of course, there is considerable freedom in how to pick the energy function. (3) and (4) come up naturally from applications. A further set of examples is obtained by replacing in (3), (4) the Riemann integrals by stochastic integrals as

$$\widetilde{\mathcal{E}}_{1,T}(X) = \int_{-T}^{T} a(X_t) dX_t, \quad \widetilde{\mathcal{E}}_{2,T}(X) = \int_{-T}^{T} \int_{-T}^{T} dX_s W(X_t, X_s, t-s) dX_t, \tag{6}$$

with $a(x)$ a vector field and $W(x, x', t)$ a $d \times d$ matrix. Since our own work is centered more around (5), we will concentrate exclusively on this case.

Our plan is first to explore the probabilistic structure. In the final chapter we list various applications for which measures of the form (5) with specific choices of $\mathcal{E}_1$ and $\mathcal{E}_2$ appear. From there it will also be apparent that each application poses specific questions not covered by general theory.

Broadly speaking, given the measure in (5) there are two limiting procedures of interest.

i) Short distance (ultraviolet) limit. The box $[-T,T]$ is fixed and the interaction is singular on the diagonal. The prototype are polymer measures, where self-crossings are penalized by the energy

$$\mathcal{E}_{T,\text{poly}}(X) = \int_{-T}^{T} \int_{-T}^{T} \delta_n(X_t - X_s)dtds. \tag{7}$$

Here $\delta_n \geq 0$ with support in a ball of radius $1/n$ centered at the origin. One goal is then to prove that the Gibbs measure in (5) with the energy (7) has a limit as $n \to \infty$. Problems of these type also come up in proving renormalizability of quantum field theories. They have been studied in considerable detail. We refer to [20, 35, 36, 40, 6] and references therein. A more detailed discussion is outside the scope of the present review and we will always assume that W is locally bounded.

ii) Large distance (infinite volume) limit. The goal is to show that the measure in (5) has a limit as $T \to \infty$. The limit measure has then conditional expectations à la Dobrushin, Lanford, and Ruelle. As standard in the theory of Gibbs measures, the issue divides into the existence of a limit measure and the dependence of the limit measure on the choice of boundary conditions.

The infinite volume limit will be discussed in Section 2. A prerequisite is the case $W \equiv 0$, which leads to the theory of $P(\phi)_1$-processes, i.e. reversible diffusion processes with constant diffusion, which will be taken up in Section 2.1. If the interaction W is weak, one expects that the qualitive properties of the stationary $P(\phi)_1$-process remain intact. Technically, a cluster expansion will be used to establish such a result. The basic set-up will be explained in Section 2.2. It differs from the more convential cluster expansions because the *a priori* measure is not a product measure and the configurations are segments of Brownian paths rather than the better understood $\mathbb{R}$- or $\mathbb{Z}$-valued spins. To prove existence of the limit measure with no restriction on the interaction strength requires other methods. One possibility is domination and monotonicity [32]. In Section 2.3 we explain a more general scheme, which relies on having an essentially bounded interaction energy between the path $\{X_t, t \leq 0\}$ and the path $\{X_t, t \geq 0\}$. Under such a condition we prove that the Gibbs measure is unique, i.e. independent of the choice of boundary conditions within a reasonable class. To have non-uniqueness, the interaction energy must increase at least as $\log T$, or equivalently $W(x, x', t)$ has to decay at least as slow as $|t|^{-2}$ for large $|t|$. In Section 2.4 we discuss a specific example, for which it can be shown that the limit measure depends on the choice of the boundary conditions.

Another case of interest is the energy

$$\mathcal{E}(X) = \int_{-T}^{T} \int_{-T}^{T} W(X_t - X_s, t - s) dt ds \tag{8}$$

with $\int |t| W(x,t) dt < \infty$, hence zero external potential V. As the expression shows, the energy depends only on path increments. Thus one expects that, under the Gibbs measure (5) for $\mathcal{E}_{1,T} + \mathcal{E}_{2,T}$ replaced by $\mathcal{E}$, X_t behaves like Brownian motion with some effective diffusion coefficient. For example, if X is pinned as $X_{-T} = 0 = X_T$, then $\mathbb{E}_T(X_0^2) \simeq T$, at large T. The $T \to \infty$ limit of the measure (5) will not exist and the more sensible project is to prove an invariance principle under suitable rescaling. This will be explained in Section 3. Finally, in Section 4 we discuss some specific applications.

At this point we would like to take the opportunity to thank the organizers of the SPP 1033 "Interacting Stochastic Systems of High Complexity" for their initiative. The Schwerpunkt turned out to be a successful enterprise for joint research in the applied areas of probability theory.

2 Gibbs Measures

2.1 The Case of External Potential

First we outline a method on how to represent $P(\phi)_1$-processes (i.e., Brownian motion in the presence of an external potential) in terms of Gibbs measures. Since B_t, the outcomes of Brownian motion, are correlated for different values of t, Wiener measure carries some dependence and is not as simple as a product measure. However, by its Markovianness and since this property survives under the potentials we consider, $P(\phi)_1$-processes are tractable to a fair extent, which is a first step toward understanding more complicated cases, such as (4) when also a pair interaction in present. For early results we refer to [37, 38], for details of Gibbsian description as well as proofs and a discussion of the related literature see [3]; the arguments used here are largely based on a spectral theoretic analysis.

Denote $V^+ = \sup\{0, V\}$ and $V^- = \inf\{-V, 0\}$. Two classes of external potential $V : \mathbb{R}^d \to \mathbb{R}$ will be considered:

(V1) *Kato-class.*   Here $V^- \in K_d$ and $V^+ \in K_d^{\text{loc}}$, with

$$K_1 = \{V : \sup_{x \in \mathbb{R}} \int_{|x-y| \leq 1} |V(y)| \, dy < \infty\},$$

$$K_d = \{V : \lim_{r \to 0} \sup_{x \in \mathbb{R}^d} \int_{|x-y| \leq r} |V(y)| \, q(|x-y|) \, dy = 0\} \quad \text{if} \quad d \geq 3,$$

with $q(x) = -\log|x|$ for $d = 2$, and $q(x) = 1/|x|^{d-2}$ for $d \geq 3$, and the local Kato-class

$$K_d^{\mathrm{loc}} = \{f : f1_A \in K_d \text{ for each compact } A \subset \mathbb{R}^d\}. \tag{1}$$

(V2) *Confining potentials.* V is bounded from below and continuous, moreover $V(x) = a|x|^{2s} + o(|x|^{2s})$, with some $s > 1$ and $a > 0$.

Examples of Kato-class potentials include smooth functions bounded from below, but also some local (e.g. Coulomb) singularities are allowed. In particular, (V2) is a specific case of (V1). The sets K_d can also be characterized in terms of Wiener integrals.

For V having either of the regularity properties above define the Schrödinger operator $H = -1/2\Delta + V(x)$ on $L^2(\mathbb{R}^d, dx)$ as a sum of quadratic forms (V is regarded as a multiplication operator). Then $C_0^\infty(\mathbb{R}^d)$ is a form core on which H is essentially self-adjoint and bounded from below. If the bottom of the spectrum E_0 of H is a simple eigenvalue, then the corresponding eigenfunction ψ_0 (ground state) is strictly positive. The semigroup e^{-tH}, $t \geq 0$, exists on $L^2(\mathbb{R}^d, dx)$, and it is an integral operator with positive, continuous, uniformly bounded kernel $G_t(x, y)$. For (V2)-type potentials the semigroup is moreover intrinsically ultracontractive. That is, with the probability measure $d\nu = \psi_0^2 dx$ on $\mathbb{R}^d$, and isometry $j : L^2(\mathbb{R}^d, d\nu) \to L^2(\mathbb{R}^d, dx)$, $f \mapsto \psi_0 f$, the operator

$$
H_\nu f = (j^{-1}(H - E_0)j)f = \frac{1}{\psi_0}(H - E_0)(\psi_0 f)
$$
$$
= -\frac{1}{2}\Delta f - (\nabla \ln \psi_0, \nabla f)_{\mathbb{R}^d}, \tag{2}
$$

with $\mathrm{Dom}\, H_\nu = j^{-1}(\mathrm{Dom}\, H)$, defines a semigroup e^{-tH_ν} for all $f \in L^2(\mathbb{R}^d, d\nu)$ and $t \geq 0$. Intrinsic ultracontractivity of e^{-tH} means that e^{-tH_ν} is ultracontractive, i.e. it maps $L^2(\mathbb{R}^d, d\nu)$ into $L^\infty(\mathbb{R}^d, d\nu)$ continuously, or equivalently, $\|e^{-tH_\nu}\|_{2,\infty} < \infty, \forall t \geq 0$.

Choose now H to be a Schrödinger operator such that its ground state ψ_0 exists. For convenience and without loss we shift the potential by E_0 so that the bottom of the spectrum of H is 0. For $t_1 < \ldots < t_n \in \mathbb{R}$, $f_1, \ldots, f_n \in L^2(\mathbb{R}^d, dx) \cap L^\infty(\mathbb{R}^d, dx)$, the $P(\phi)_1$-*process* associated with H is the unique probability measure P on path space $C(\mathbb{R}, \mathbb{R}^d)$ defined by

$$
\int f_1(X_{t_1}) \ldots f_n(X_{t_n}) dP(X)
$$
$$
= (\psi_0 f_1, e^{-(t_2 - t_1)H} f_2 \ldots e^{-(t_n - t_{n-1})H} f_n \psi_0)_{L^2(\mathbb{R}^d, dx)}. \tag{3}
$$

P is indeed a probability measure as $e^{-tH}\psi_0 = \psi_0$ and $\|\psi_0\|_2 = 1$. A $P(\phi)_1$-process is a reversible stationary Markov process with stationary measure $d\nu$ and generator H_ν, and it has almost surely continuous paths. It is moreover the stationary solution of the stochastic differential equation (Itô-diffusion)

$$dX_t = (\nabla \ln \psi_0)(X_t)\, dt + dB_t,$$

where B_t denotes Brownian motion on $\mathbb{R}^d$.

Processes of this type can be given a Gibbsian description. We emphasize that since in the present stage there are no useful relationships available with variational principles etc as discussed in the Introduction, here the basic fact is that there is at all a probability measure associated with the scalar product in (3), a consequence of the Riesz-representation theorem, while its Gibbsianness comes second to it. That we are able to identify this measure as a Gibbs measure leads however to further insight.

Denote $\mathcal{X} = C(\mathbb{R}, \mathbb{R}^d)$, the space of continuous functions from $\mathbb{R}$ to $\mathbb{R}^d$, and its σ-field $\mathcal{A} = \sigma(\pi_t : t \in \mathbb{R})$ generated by the point evaluations $\pi_t : \mathcal{X} \to \mathbb{R}^d$, $X \mapsto \pi_t(X) = X_t$. These will be the configuration space and σ-field for the Gibbs measure, respectively. For $[-T, T] \subset \mathbb{R}$ we denote by $\mathcal{A}_T$ the σ-field $\sigma(\pi_t : t \in [-T, T]) \subset \mathcal{A}$; also, we put $[-T, T]^c = \mathbb{R} \smallsetminus [-T, T]$.

Write as before $\mathcal{W}$ for Wiener measure, and $\mathcal{W}^{\xi, \eta}_{[-T,T]}$ for the Wiener measure conditional on starting in ξ at time $-T$ and ending in η at time T. This Brownian bridge can be extended to a measure on $\mathcal{X}$ by picking an $Y \in \mathcal{X}$ and putting $\mathcal{W}^Y_T = \mathcal{W}^{Y_{-T}, Y_T}_{[-T,T]} \otimes \delta^Y_{[-T,T]^c}$, with Dirac measure on $C([-T,T]^c, \mathbb{R}^d)$ concentrated on $Y|_{[-T,T]^c}$. $\mathcal{W}^Y_T$ is thus a finite measure on $(\mathcal{X}, \mathcal{A})$; it will serve as reference measure for the Gibbs measure to be constructed.

Take any $A \in \mathcal{A}$ and consider

$$dP_T(A|Y) = \frac{1}{Z_T(Y)} 1_A(X) e^{-\int_{-T}^{T} V(X_s)\, ds}\, d\mathcal{W}^Y_T(X), \qquad (4)$$

where

$$Z_T(Y) = \int e^{-\int_{-T}^{T} V(X_s)\, ds}\, d\mathcal{W}^Y_T(X) \qquad (5)$$

is the partition function turning P_T into a probability measure.

Definition 2.1. *Let $\mathcal{X}^* \subset \mathcal{X}$. A probability measure $\mathcal{P}$ on $(\mathcal{X}, \mathcal{A})$ is called a Gibbs measure for potential V and reference measure $\mathcal{W}^Y$, if for every bounded interval $[-T, T] \subset \mathbb{R}$*

1. *$\mathcal{P}|_{\mathcal{A}_T} \ll \mathcal{W}|_{\mathcal{A}_T}$,*
2. *for every $A \in \mathcal{A}$ the function $Y \mapsto \mathcal{P}_T(A|Y)$ given by the right hand side of (4) is a regular version of the conditional probability $\mathcal{P}(A|\mathcal{A}_{[-T,T]^c})$.*

A probability measure $\mathcal{P}_T$ on $([-T, T], \mathcal{A}_T)$ is called a finite time interval Gibbs measure for V and reference measure W^Y_T if for every bounded interval $[-S, S] \subset [-T, T]$ the function $Y \mapsto \mathcal{P}_S(A|Y)$ as above is a regular version of the conditional probability $\mathcal{P}_S(A|\mathcal{A}_{[-S,S]^c})$. Furthermore, a Gibbs measure $\mathcal{P}$ is said to be supported by $\mathcal{X}^*$ whenever $\mathcal{P}(\mathcal{X}^*) = 1$.

This definition rests on the DLR conception of Gibbs measure. In this sense we then have

Theorem 2.2. *A $P(\phi)_1$-measure P corresponding to potential V is a Gibbs measure with respect to V and Wiener measure.*

On the other hand, that $P_T(\cdot|Y)$ are a family of finite time interval Gibbs measures indexed by bounded intervals can be seen in a straightforward way. It can be proven by a monotone class argument that (infinite time interval) Gibbs measures on path space can be obtained by limits of finite interval Gibbs measures, similarly to the case known from lattice spin models. In this limiting procedure thus one must have a control of boundary conditions.

A Gibbs measure associated with a $P(\phi)_1$-process need not be unique. This non-uniqueness appears as a dependence of the Gibbs measure on the boundary conditions. An example showing this is the Ornstein-Uhlenbeck process, in which case uncountably many Gibbs measures can occur for the same potential. This is related with the rate how the boundary paths increase, or in other words, how fast for each T the boundary path on $[-T,T]^c$ has to "forget" that it was free Brownian motion before stepping in $[-T,T]$ where it must "steady down" to conform with the correct distribution prescribed by (4). A condition for uniqueness of the Gibbs measure is provided by the following theorem.

Theorem 2.3. *Let H be a Schrödinger operator for a Kato-class potential V such that the spectral gap Λ of H is strictly positive, and let ψ_0 be its ground state. Put*

$$\mathcal{X}^* = \{X \in \mathcal{X} : \lim_{|t|\to\infty} \frac{e^{-\Lambda|t|}}{\psi_0(X_t)} = 0\}. \tag{6}$$

Then the $P(\phi)_1$-measure P corresponding to V is the unique Gibbs measure for V supported by $\mathcal{X}^*$. If, furthermore, V is a $(V2)$-type confining potential, then P is the unique Gibbs measure supported on the entire $\mathcal{X}$.

The first part of the statement results from an argument using direct estimates, the second relies on ultracontractivity.

By restricting to (V2)-type potentials and making use of the fact that for this class ψ_0 is bounded both from below and above by $C\exp(-\theta|x|^{s+1})$, with suitable constants $C, \theta > 0$ for the two bounds respectively, we obtain from Theorem 2.3 that those paths are typical for the $P(\phi)_1$-measure that grow asymptotically like $t^{1/(s+1)}$.

2.2 Weak Pair Potential: Cluster Expansion

Next we turn to discussing whether Gibbs measures can be defined also for Brownian motion subjected to both an external and a pair interaction potential. Such a process is not Markovian and therefore not accessible to spectral analysis. Instead, we will develop a cluster expansion; for details and proofs see [26].

We use the same set-up as before. The pair interaction potential is a measurable function $W : \mathbb{R}^d \times \mathbb{R}^d \times \mathbb{R} \to \mathbb{R}$ with the (inessential) symmetry properties $W(\cdot, \cdot, t - s) = W(\cdot, \cdot, |t - s|)$, $W(x, y, \cdot) = W(y, x, \cdot)$, $x, y \in \mathbb{R}^d$, $s, t \in \mathbb{R}$, and satisfying either of the following regularity conditions:

(W1) There is $R > 0$ and $\alpha > 2$ such that

$$|W(x, y, t - s)| \leq R \frac{|x|^2 + |y|^2}{1 + |t - s|^\alpha} \tag{7}$$

for every $x, y \in \mathbb{R}^d$ and $t, s \in \mathbb{R}$.

(W2) There is $R > 0$ and $\alpha > 1$ such that

$$|W(x, y, t - s)| \leq \frac{R}{1 + |t - s|^\alpha} \tag{8}$$

for every $s, t \in \mathbb{R}$ and $x, y \in \mathbb{R}^d$.

For $[-T, T] \subset \mathbb{R}$ write

$$W_{[-T,T]}(X|Y) = W_{[-T,T]}(X) + W^Y_{[-T,T]}(X) \tag{9}$$

for the "total energy" associated with configuration $X \in \mathcal{X}_{[-T,T]}$ given the boundary configuration $Y = Y^- \cup Y^+$, with $Y^- \in \mathcal{X}_{(-\infty,-T]}$ resp. $Y^+ \in \mathcal{X}_{[T,\infty)}$. Term by term,

$$W_{[-T,T]}(X) = \int_{-T}^{T} \int_{-T}^{T} W(X_t, X_s, s - t) ds dt \tag{10}$$

is the "internal energy" associated with the path inside $[-T, T]$, and

$$W^Y_{[-T,T]}(X) = 2 \int_{-\infty}^{-T} dt \int_{-T}^{T} ds\, W(Y_t^-, X_s, t - s) +$$
$$+ 2 \int_{T}^{\infty} dt \int_{-T}^{T} ds\, W(Y_t^+, X_s, t - s) \tag{11}$$

is the "interaction energy" between X and the boundary path Y. We calibrate the interaction energy such that $W^0_{[-T,T]}(X) = 0$. As before, $P_T(\,\cdot\,|Y) = P_{[-T,T]}(\,\cdot\,|Y^-_{-T}, Y^+_T)$ is the conditional distribution of the reference measure for the given boundary condition Y (which by Markovianness obviously depends only on the positions attained at the ends of the interval). It is readily checked that $\mu_{[-T,T]}(\,\cdot\,|Y)$, with $Y \in C([-T,T]^c, \mathbb{R}^d)$, is a family of finite time interval Gibbs measures. We also allow $\lambda \in \mathbb{R}$, a parameter which can be interpreted as the strength of the coupling of the pair interaction to the Brownian paths.

Consider now a $P(\phi)_1$-process with stationary measure $d\nu = \psi_0^2 dx$ and transition probability density

$$g_t(x|y) = \frac{\psi_0(x) G_t(x, y)}{\psi_0(y) e^{-E_0 t}}. \tag{12}$$

Denote again the probability distribution of this process by P, and by P_T its restriction to the field $\mathcal{A}_{[-T,T]}$. We take this as reference measure in constructing the finite time interval Gibbs measures on $\mathcal{X}_{[-T,T]}$ for the pair potentials above:

$$d\mu_T(A|Y) = \frac{1}{\mathcal{Z}_T(Y)} 1_A(X) e^{-\lambda W_{[-T,T]}(X|Y)} dP_T(X|Y), \qquad (13)$$

for any $A \in \mathcal{A}$ and boundary condition Y. Here we speak about Gibbs measure μ in the same sense as in Definition 2.1, now for potential W and reference measure P. The partition function is

$$\mathcal{Z}_T(Y) = \int e^{-\lambda W_{[-T,T]}(X|Y)} dP_T(X|Y). \qquad (14)$$

As said before, Gibbs measures can be obtained as limits over finite time interval Gibbs measures. Thus it is of interest whether the sequence of Gibbs measures μ_T has a limit as $T \to \infty$; if it does then it provides a Gibbs measure on the full path space as soon as also condition (i) of Definition 2.1 is met.

Theorem 2.4. *Suppose V and W satisfy assumptions (V2), respectively either (W1) or (W2). Take any unbounded increasing sequence $T^{(n)}$ of positive real numbers, and suppose $0 < |\lambda| \leq \lambda^*$ with λ^* small enough. Then the local weak limit $\lim_{n\to\infty} \mu_{T^{(n)}} = \mu$ exists and is a Gibbs probability measure on $(\mathcal{X}, \mathcal{A})$ with respect to W and reference measure P. Moreover, μ does not depend on the choice of sequence $T^{(n)}$.*

In order to prove this convergence we use a cluster expansion controlled by the small parameter λ. Next we sketch the cluster representation of the partition function (14) and outline the main steps of the proof. For simplicity we start with free boundary conditions, i.e. $Y = 0$ in (13).

Take a division of $[-T,T]$ into disjoint intervals $\tau_k = (t_k, t_{k+1})$, $k = 0, ..., N-1$, with $t_0 = -T$ and $t_N = T$, each of length b, i.e. fix $b = 2T/N$; for convenience we choose N to be an even number so that the origin is endpoint to some intervals. We break up a path X into pieces X_{τ_k} by restricting it to τ_k. The total energy contribution of the pair interaction then becomes

$$W_T := \int_{-T}^{T} \int_{-T}^{T} W(X_t, X_s, s-t) ds dt = \sum_{0 \leq i < j \leq N-1} W_{\tau_i, \tau_j} \qquad (15)$$

where with the notation $\mathcal{J}_{ij} = \int_{\tau_i} dt \int_{\tau_j} W(X_s, X_t, s-t) ds$ we have

$$W_{\tau_i, \tau_j} = \begin{cases} \mathcal{J}_{ij} + \mathcal{J}_{ji} & \text{if } |i-j| \geq 2 \\ \frac{1}{2}(\mathcal{J}_{ii} + \mathcal{J}_{jj}) + \mathcal{J}_{ij} + \mathcal{J}_{ji} & \text{if } |i-j| = 1, \text{ and } i \neq 0, j \neq N-1 \\ \mathcal{J}_{ij} + \mathcal{J}_{ji} + \frac{1}{2}\mathcal{J}_{00} & \text{if } i = 0 \text{ and } j = 1 \\ \mathcal{J}_{ij} + \mathcal{J}_{ji} + \frac{1}{2}\mathcal{J}_{N-1\,N-1} & \text{if } i = N-1 \text{ and } j = N-2 \end{cases}$$

(For keeping the notation simple we do not make explicit the X dependence of these objects.) By using (15) we obtain

$$e^{-\lambda W_T} = \prod_{0 \le i < j \le N-1} (e^{-\lambda W_{\tau_i, \tau_j}} + 1 - 1) = 1 + \sum_{\mathcal{R} \neq \emptyset} \prod_{(\tau_i, \tau_j) \in \mathcal{R}} (e^{-\lambda W_{\tau_i, \tau_j}} - 1). \quad (17)$$

Here the summation is performed over all nonempty sets of different pairs of intervals, i.e. $\mathcal{R} = \{(\tau_i, \tau_j) : (\tau_i, \tau_j) \neq (\tau_{i'}, \tau_{j'})$ whenever $(i, j) \neq (i', j')\}$.

In order to keep this and the forthcoming summations in hand we need a few more notations. Two distinct pairs of intervals (τ_i, τ_j) and $(\tau_{i'}, \tau_{j'})$ will be called *directly connected* and denoted $(\tau_i, \tau_j) \sim (\tau_{i'}, \tau_{j'})$ if one interval of the pair (τ_i, τ_j) coincides with one interval of the pair $(\tau_{i'}, \tau_{j'})$. A set of connected pairs of intervals is a collection $\{(\tau_{i_1}, \tau_{j_1}), ..., (\tau_{i_n}, \tau_{j_n})\}$ in which each pair of intervals is connected to another through a sequence of directly connected pairs, i.e., for any $(\tau_i, \tau_j) \neq (\tau_{i'}, \tau_{j'})$ there exists $\{(\tau_{k_1}, \tau_{l_1}), ..., (\tau_{k_m}, \tau_{l_m})\}$ such that $(\tau_i, \tau_j) \sim (\tau_{k_1}, \tau_{l_1}) \sim ... \sim (\tau_{k_m}, \tau_{l_m}) \sim (\tau_{i'}, \tau_{j'})$. A maximal set of connected pairs of intervals is called a *contour* and denoted by γ. We denote by $\bar{\gamma}$ the set of all intervals that are elements of the pairs of intervals belonging to contour γ, and by γ^* the set of time-points of intervals appearing in $\bar{\gamma}$. Clearly, $\mathcal{R}$ can be decomposed into maximal connected components, i.e. contours: $\mathcal{R} = \{\gamma_1, ..., \gamma_r\}$ with $\bar{\gamma}_i \cap \bar{\gamma}_j = \emptyset$, $i \neq j$; $i, j = 1, ..., r$.

The sum in (17) is then further expanded as

$$\sum_{\mathcal{R} \neq \emptyset} \prod_{(\tau_i, \tau_j) \in \mathcal{R}} (e^{-\lambda W_{\tau_i, \tau_j}} - 1) = \sum_{r \ge 1} \sum_{\{\gamma_1, ..., \gamma_r\}} \prod_{k=1}^{r} \prod_{(\tau_i, \tau_j) \in \gamma_k} (e^{-\lambda W_{\tau_i, \tau_j}} - 1) \quad (18)$$

where now summation goes over collections $\{\gamma_1, ..., \gamma_r\}$ of contours such that $\bar{\gamma}_k \cap \bar{\gamma}_{k'} = \emptyset$ unless $k = k'$.

A collection of consecutive intervals $\{\tau_j, \tau_{j+1}..., \tau_{j+k}\}$, $j \ge 0$, $j+k \le N-1$ is called a *chain*. As in the case of contours, $\bar{\varrho}$ and ϱ^* mean the set of intervals belonging to the chain ϱ and the set of time-points in ϱ, respectively. We call two contours γ_1, γ_2 disjoint if they have no intervals in common, i.e. $\bar{\gamma}_1 \cap \bar{\gamma}_2 = \emptyset$. Two chains ϱ_1, ϱ_2 are called disjoint if they have no common time-points, i.e. $\varrho_1^* \cap \varrho_2^* = \emptyset$. Take now a non-ordered set of disjoint contours and disjoint chains, $\Gamma = \{\gamma_1, ..., \gamma_r; \varrho_1, ..., \varrho_s\}$, with some $r \ge 1$ and $s \ge 0$. Note that such contours and chains may have common time-points. We use the notation $\Gamma^* = (\cup_i \gamma_i^*) \cup (\cup_j \varrho_j^*)$ for the set of all time-points appearing as beginnings or ends of intervals belonging to some contour or chain in Γ. Also, we put $\bar{\Gamma} = (\cup_i \bar{\gamma}_i) \cup (\cup_j \bar{\varrho}_j)$ for the set of intervals appearing in Γ through entering some contours or chains. Denote by $\partial^- \varrho$ resp. $\partial^+ \varrho$ the leftmost resp. rightmost time-points belonging to ϱ. Γ is called a *cluster* if $\{\gamma_1^*, ..., \gamma_r^*; \varrho_1^*, ..., \varrho_s^*\}$ is a connected collection of sets and for every $\varrho \in \Gamma$ we have that $\partial^- \varrho, \partial^+ \varrho \in \cup_{j=1}^{r} \gamma_j^*$. This means that in a cluster chains have no loose ends.

Next we fix the positions of path X at the time-points of the division, i.e. we put $X_{t_k} = x_k$, for all $k = 0, ..., N$, with $-T = t_0 < t_1 < ... < t_N = T$.

The distribution of path X in interval $[-T, T]$ conditional on the positions attained at the fixed times is

$$dP_T(X_{\tau_0}, \ldots, X_{\tau_{N-1}} | X_{t_0} = x_0, \ldots, X_{t_N} = x_N) = \prod_{k=0}^{N-1} dP_{\tau_k}(X_{\tau_k} | x_k, x_{k+1}). \quad (19)$$

We use the shorthand at the right hand side for the corresponding conditional probabilities for easing the notation. Let $p_{t_0,\ldots,t_N}(x_0, \ldots, x_N)$ be the density with respect to $\prod_{k=0}^{N} d\nu_k(x_k)$ of the joint distribution of positions of path X recorded at the time-points $t_0, \ldots, t_N$. Here $d\nu_k$ denotes a copy of $d\nu$ for each $k = 0, \ldots, N$. By Markovianness it then follows that

$$p_{t_0,\ldots,t_N}(x_0, \ldots, x_N) = \prod_{k=0}^{N-1} g_b(x_{k+1} | x_k) = \prod_{k=0}^{N-1} (g_b(x_{k+1} | x_k) - 1 + 1)$$

$$= 1 + \sum_{\mathcal{S}} \prod_{k:\tau_k \in \mathcal{S}} (g_b(x_{k+1} | x_k) - 1).$$

The summation is extended over all nonempty sets $\mathcal{S} = \{\tau_k = (t_k, t_{k+1})\}$ of different pairs of consecutive time-points. In a similar way as before the latter formula can be recast in the form

$$\sum_{\mathcal{S}} \prod_{k:\tau_k \in \mathcal{S}} (g_b(x_{k+1} | x_k) - 1) = \sum_{s \geq 1} \sum_{\{\varrho_1,\ldots,\varrho_s\}} \prod_{j=1}^{s} \prod_{k:\tau_k \in \varrho_j} (g_b(x_{k+1} | x_k) - 1). \quad (20)$$

Here $\{\varrho_1, \ldots, \varrho_s\}$ is a collection of disjoint chains, and this formula explains the way we defined them before.

For every cluster $\Gamma = \{\gamma_1, \ldots, \gamma_r; \varrho_1, \ldots, \varrho_s\}$ define the function

$$\kappa_\Gamma = \prod_{l=1}^{r} \prod_{(\tau_i,\tau_j) \in \gamma_l} (e^{-\lambda W_{\tau_i \tau_j}} - 1) \prod_{m=1}^{s} \prod_{k:\tau_k \in \varrho_m} (g_b(x_{k+1} | x_k) - 1). \quad (21)$$

Also, introduce the auxiliary probability measure on $\mathcal{X}_T$

$$d\mathcal{P}_T(X) = \prod_{k=0}^{N-1} dP_{\tau_k}(X_{\tau_k} | x_k, x_{k+1}) \prod_{k=0}^{N} d\nu_k(x_k) \quad (22)$$

and look at

$$K_\Gamma = \mathbb{E}_{\mathcal{P}_T}[\kappa_\Gamma]. \quad (23)$$

Note that $\int (g_b(x_{k+1}|x_k) - 1) d\nu(x_{k+1}) = \int (g_b(x_{k+1}|x_k) - 1) d\nu(x_k) = 0$. This is the reason why from a cluster we rule out chains having loose ends; for any such chain $\mathbb{E}_{\mathcal{P}_T}[\kappa_\Gamma] = 0$.

By putting (18), (19), (20), (21) and (23) together we obtain the cluster representation of the partition function $\mathcal{Z}_T$:

Proposition 2.5. *For every $T > 0$*

$$\mathcal{Z}_T = 1 + \sum_{n \geq 1} \sum_{\{\Gamma_1, \ldots, \Gamma_n\}} \prod_{l=1}^{n} K_{\Gamma_l}. \tag{24}$$

Here the summation is performed over all sets of clusters $\{\Gamma_1, \ldots, \Gamma_n\} \neq \emptyset$ for which $\Gamma_i^* \cap \Gamma_j^* = \emptyset$ whenever $i \neq j$.

As soon as the cluster representation of $\mathcal{Z}_T$ is established, the existence of the weak limit measure $\mu = \lim_{T \to \infty} \mu_T$ follows by the cluster estimates below and the general arguments presented in e.g. [29], Chapter 3.

We conclude the presentation of the expansion by briefly explaining the two crucial cluster estimates. The first one is given by

Proposition 2.6. *For every cluster Γ we have the bound*

$$|K_\Gamma| \leq \prod_{\varrho \in \Gamma} (c_1 |\lambda|^{1/3})^{|\bar{\varrho}|} \prod_{\gamma \in \Gamma} \prod_{(\tau_i, \tau_j) \in \gamma} \frac{c_2 |\lambda|^{1/3}}{(|i - j - 1|b)^\delta + 1} \tag{25}$$

with $|\bar{\varrho}|$ denoting the number of intervals contained in ϱ, constants $c_1, c_2 > 0$ and exponent $\delta > 1$.

In estimate (25) the factor accounting for the contribution of chains comes from the uniform upper bound $Ce^{-\Lambda b}$ on $|g_b(x|x') - 1|$ (see second factor in (21)), where Λ is the spectral gap of the Schrödinger operator of the underlying $P(\phi)_1$-process, and $C > 0$. This bound, in its turn, is a consequence of the intrinsic ultracontractivity of e^{-tH}, compare Section 2.1. The factor accounting for the contribution of contours comes from an estimate using a generalized variant of the Hölder inequality applied to the products over $e^{-\lambda W_{\tau_i \tau_j}} - 1$ (see first factor in (21)). b is finally chosen in such a combination with λ and Λ that the expression (25) results.

The second fundamental estimate ensuring the convergence of the cluster expansion is

Proposition 2.7. *There is a constant $c > 0$, independent of λ, and a number $0 < \eta(\lambda) < 1$ with $\eta \to 0$ as $\lambda \to 0$, such that*

$$\sum_{\substack{\Gamma : \Gamma^* \ni 0 \\ |\bar{\Gamma}| = n}} |K_\Gamma| \leq c\, \eta^n. \tag{26}$$

with $|\bar{\Gamma}|$ denoting the number of intervals contained in Γ through some contour or chain.

This estimate follows through a procedure of translating the summation in the left hand side of (26) into a combinatorial problem and resumming over (and counting of) first graphs and then trees. The contours are assigned vertices and they are linked into graphs according to the rules connecting them up into clusters.

So far we assumed free boundary conditions. By an extension of the argument sketched above also other boundary conditions can be taken into account, picked from $\mathcal{X}^*$, the subset provided by Theorem 2.3. Then an important question is how the limiting measure depends on the boundary conditions. Uniqueness (in DLR-sense) means that for any increasing sequence of real numbers $\{T_n\}$ and any corresponding sequence $\{Y_n\} \subset \mathcal{X}^*_{[-T_n,T_n]^c}$, $\lim_{n\to\infty} \mathbb{E}_{\mu_{T_n}}[F_B|Y_n] = \mathbb{E}_\mu[F_B]$, for every bounded $B \subset \mathbb{R}$, and each bounded measurable function F_B.

Theorem 2.8. *Suppose V is of class (V2) and W satisfies (W2). Then we have the following cases:*

1. *If $\alpha > 2$, then whenever the Gibbs measure μ exists, it is unique in DLR sense.*
2. *If $\alpha > 1$, then for sufficiently small $|\lambda|$ the limiting Gibbs measure μ is unique in DLR sense whenever the reference measure is unique.*

If $\alpha > 2$, $|W_T^Y(X)|$ (given by (11)) is uniformly bounded in T, and in paths X and Y. This implies that only one Gibbs measure can exist, and the argument requires no restriction on the values of λ. For $1 < \alpha \leq 2$ this uniform boundedness does not hold any longer and we once again take recourse to cluster expansion.

Having a Gibbs measure at hand, an important aspect in its understanding is to see what a typical configuration looks like under it. This is answered by

Theorem 2.9. *Under the same conditions as in the previous theorem, with μ-probability 1 we have*

$$|X_t| \leq C \left(\log(|t| + 1)\right)^{1/(s+1)} + Q(\{X\}) \tag{27}$$

with a suitable number $C > 0$ and a functional Q, independent of t.

The strategy of proving Theorem 2.9 goes by boosting the typical behaviour of the reference process explained above to the level of the Gibbs measure. First it is shown that for any $a > 0$

$$P\left(\{X \in \mathcal{X} : \max_{0 \leq t \leq 1} |X_t| \geq a\}\right) \leq C\, e^{-\theta a^{s+1}} \tag{28}$$

with appropriate $C, \theta > 0$. This can be proven by using Varadhan's Lemma taken together with the upper bound $\exp(-\theta|x|^{s+1})$ for ψ_0 (the ground state of the Schrödinger operator generating the reference process). Then Theorem 2.9 comes about by proving that also $C' > 0$ and $\theta' > 0$ can be found such that

$$\mu\left(\{X \in \mathcal{X} : \max_{0 \leq t \leq 1} |X_t| \geq a\}\right) \leq C' e^{-\theta' a^{s+1}}. \tag{29}$$

The proof requires once again the use of cluster expansion.

Finally, we list some additional properties of Gibbs measures for (W2)-type pair potentials, useful in various contexts. This case in particular covers Nelson's scalar field model, see Section 4 below.

Theorem 2.10. *Let μ be a Gibbs measure for W satisfying (W2). Suppose V is of (V2)-type and $|\lambda|$ is small enough. Then the following hold:*

1. *[Invariance properties] μ is invariant with respect to time shift and time reflection:*

$$\mu \circ \tau_t = \mu, \quad \forall t \in \mathbb{R}, \quad \text{where } (\tau_s X)_t = X_{s+t},$$
$$\mu \circ \vartheta = \mu, \quad \text{where } (\vartheta X)_t = X_{-t}.$$

2. *[Single time distributions] The distributions φ_T under μ_T of positions x at time $t = 0$ are equivalent to ν, i.e. there exist $C_1, C_2 \in \mathbb{R}$, independent of T and x such that*

$$C_1 \leq \frac{d\varphi_T}{d\nu}(x) \leq C_2 \tag{30}$$

for every $x \in \mathbb{R}^d$ and $T > 0$. Moreover $\lim_{T \to \infty}(d\varphi_T/d\nu)(x) = (d\varphi/d\nu)(x)$ exists pointwise.

3. *[Single time conditional distributions] The conditional distributions $\mu_T(\cdot \,|X_0 = x)$ converge locally weakly to $\mu(\cdot \,|X_0 = x)$, for all $x \in \mathbb{R}^d$.*

4. *[Mixing properties] For any bounded functions F, G on $\mathbb{R}^d$ we have on the covariance the estimate*

$$|\text{cov}_\mu (F_s; G_t)| \leq \text{const } \frac{\sup |F_s| \sup |G_t|}{1 + |t - s|^\beta} \tag{31}$$

where $\beta > 0$, $F_s := F(X_s)$, $G_t := G(X_t)$, and the constant prefactor is independent of s, t and F, G.

2.3 Existence for Pair Potential of Arbitrary Strength

The main restriction in the previous section was that the pair potential W had to carry a small prefactor λ. This restriction is inherent in the cluster expansion. An alternative route to the existence of Gibbs measures are compactness arguments; the main tool is the concept of uniform domination [21], which in our context reads as follows:

Definition 2.11. *Let P, $(\mu_T)_{T \geq 0}$ be probability measures on $C(\mathbb{R}, \mathbb{R}^d)$. We say that the family $(\mu_T)_{T \geq 0}$ is locally uniformly dominated by P if the following holds true: For each $\varepsilon > 0$ and $S > 0$ there exists $\delta > 0$ such that $P(A) < \delta$ implies $\limsup_{T \to \infty} \mu_T(A) < \varepsilon$ uniformly in sets A depending on $X_s, |s| < S$, $(X_s)_{s \in \mathbb{R}} \in C(\mathbb{R}, \mathbb{R}^d)$.*

The important fact is that each family $(\mu_T)_{T \geq 0}$ of probability measures that is locally uniformly dominated by a probability measure P has at least one cluster point as $T \to \infty$ in the topology of local convergence. In order to apply this to Gibbs measures we adopt the general set-up from the previous section. As a first assumption on the potentials we need

(A1) V is Kato-class, i.e. satisfies (V1) from Section 2.1. Moreover, the Schrödinger operator H corresponding to V has a unique, square-integrable ground state ψ_0.

(A2) W is extensive, i.e. there exists $C_\infty > 0$ such that

$$\int_{-\infty}^{\infty} \sup_{x,y \in \mathbb{R}^d} |W(x, y, |s|)| \, ds < C_\infty. \tag{32}$$

Comparing with the previous section, we find that (W2) implies $(A2)$.

Let P be the measure of the stationary $P(\phi)_1$-process as given in Section 2.1, and let $W_{[-T,T]}$ be given by (10). We will use finite time interval Gibbs measures with free boundary conditions as approximants for our infinite time interval Gibbs measures, i.e. we put

$$d\mu_T = \frac{1}{Z_T} e^{-W_{[-T,T]}(X)} \, dP.$$

Using the concept of local uniform domination, it is now possible to prove

Proposition 2.12. *[2] Assume (A1) and (A2). Suppose that for each $\varepsilon > 0$ there exists $R > 0$ such that*

$$\mu_T(|X_0| > R) < \varepsilon \tag{33}$$

uniformly in $T > 0$. Then there exists an (infinite time interval) Gibbs measure for the potentials V and W and the reference measure $\mathcal{W}$ (Wiener measure).

We have thus reduced the problem to proving (33). For this we need some further assumptions.

(A1') In addition to (A1) suppose $\psi_0 \in L^1(\mathbb{R}^d)$.

Condition (A1') is not very restrictive; in many cases ψ_0 decays exponentially at infinity. The additional condition on W will be more restrictive and requires some preparations to formulate. Let $C^{(0)}(\mathbb{R}, \mathbb{R}^d)$ denote the space of functions which are continuous with the possible exception of the origin but have left and right hand side limits there. For $\tau > 0$ we define the map

$$\theta_\tau : C(\mathbb{R}, \mathbb{R}^d) \to C^{(0)}(\mathbb{R}, \mathbb{R}^d), \quad (\theta_\tau X)_t = \begin{cases} X_{t+\tau} & \text{if } t \geq 0, \\ X_{t-\tau} & \text{if } t < 0. \end{cases} \tag{34}$$

With $E_0 = \inf \text{Spec}(H)$ as before, and H the Schrödinger operator corresponding to the $P(\phi)_1$-process P, put

$$\alpha = \liminf_{|x| \to \infty} V(x) - E_0 \le \infty. \tag{35}$$

Our assumption on W now reads

(A2') In addition to (A2), we assume that there exist $D \ge 0$ and $0 \le C < \alpha$ such that

$$-W_{[-T,T]}(X) \le -W_{[-T,T]}(\theta_\tau X) + C\tau + D \tag{36}$$

for all $T, \tau > 0$ and all $X \in C(\mathbb{R}, \mathbb{R}^d)$.

In words, (A2') means that we can control, uniformly in T, the change of energy induced by cutting out a piece of the path X around $t = 0$ and gluing the remaining pieces together again. If we have finite interaction energy between the positive and the negative half-line, i.e.

$$\sup_{X \in C(\mathbb{R}, \mathbb{R}^d)} \left| \int_{-\infty}^0 ds \int_0^\infty dt\, W(X_t, X_s, |t - s|) \right| < \infty, \tag{37}$$

then (A2') holds with $C = 0$. In particular, (37) holds when W fulfills (W2) with $\alpha > 2$. (37) is, however, not necessary for (A2'), and part of the interest in condition (A2') is that it also covers cases where (37) is not met. Some sufficient conditions for (A2') are given in [1].

Theorem 2.13. *Assume (A1') and (A2'). Then (33) holds, and consequently an infinite volume Gibbs measure μ for the potentials V and W, and reference measure $\mathcal{W}$ exists.*

The theorem above does not make any statement about uniqueness. However, in conjunction with (2) of Theorem 2.8 it leads to

Corollary 2.14. *Provided (W2) with $\alpha > 2$ holds, and V satisfies (A1') and (A2'), a unique Gibbs measure exists supported by $\mathcal{X}$.*

Hariya [22] arrives at a similar result under different hypotheses.

The proof of (33) relies on the equality

$$\mu_T(|X_0| > R) = \frac{1}{Z_T} \int_{|y| > R} \psi_0^2(y) \mathbb{E}_P\left[e^{-W_{[-T,T]}} \Big| X_0 = y \right] dy. \tag{38}$$

We first prove

$$\frac{1}{Z_T} \mathbb{E}_P\left[e^{-W_{[-T,T]}} \Big| X_0 = y \right] \le \frac{\text{const}}{\psi_0(y)} \tag{39}$$

and then use (A1') in order to obtain (33). To get an idea about the proof of the latter inequality, note that (39) involves expectation with respect to a Markov process conditioned at its 'midpoint' $t = 0$. For making use of the strong Markov property of P, we flip the negative time axis to the right and obtain a Markov process with a doubled state space $\mathbb{R}^{2d}$, now conditioned on its starting point. Now we start the new process in $y \in \mathbb{R}^{2d}$ and stop it when it reaches the ball B_r around zero with radius r. By the properties of the $P(\phi)_1$-process, the stopping time τ_r the process needs to reach B_r is exponentially integrable. More explicitly, $\mathbb{E}_P^x[\exp(\beta\tau_r)] < \infty$ if $\beta < \alpha$, and the expectation value grows with the starting point x like $1/\psi_0(x)$ as $x \to \infty$. Condition (A2') is now tailor-made to ensure that the energy $\tilde{W}_{[-T,T]}$ acquired by a (flipped) path X on its way down to the B_r is no larger than $\exp(C\tau_r + D)$. Together with the strong Markov property and some technical estimates, this yields (33).

2.4 Phase Transition

In one-dimensional statistical mechanical sytems the entropy increases as $\log T$. To have a phase transition the interaction energy for the paths $\{X_t, \ -T \le t \le 0\}$ and $\{X_t, \ 0 \le t \le T\}$ must be at least comparable. Transcribed to the Gibbs measures under study this means

$$W(x,x',t) \cong |t|^{-\gamma} \quad \text{for large } |t| \tag{40}$$

with $1 < \gamma \le 2$. The lower bound on γ is needed for having the energy extensive. To carry out a proof more specific assumptions will be needed. We set $d = 1$. For the external potential we choose a double well potential of the form

$$V(x) = \beta(x^4 - x^2), \quad \beta > 0. \tag{41}$$

In fact, as long as $V(x) = V(-x)$, a general class of double well type potentials can be handled. The pair interaction is quadratic,

$$W(x,x',t) = \alpha\rho(t)\frac{1}{2}(x - x')^2, \alpha > 0, \quad \rho(t) = (1 + |t|)^{-\gamma}. \tag{42}$$

Since we rely on comparison inequalities, the interaction needs to be quadratic, at least at the present stage of understanding. Thus the only non-Gaussian piece of the Gibbs measure is $\exp[-\beta \int_{-T}^{T}(X_t)^4 dt]$. Let $\langle \ \cdot \ \rangle_{b,T}$ be the expectation of the Gibbs measure for the potentials V and W from (41), (42), with the pinned boundary conditions $X_{-T} = b = X_T$, $b \in \mathbb{R}$. Then, for $b > 0$, $\langle X_0 \rangle_{b,T} \ge 0$ and $\langle X_0 \rangle_{b,T}$ is decreasing in T. Hence the limit

$$\lim_{T\to\infty} \langle X_0 \rangle_{b,T} = \langle X_0 \rangle_{b,\infty} \tag{43}$$

exists.

Theorem 2.15. Let V, W be as in (41), (42) and fix $1 < \gamma \le 2$. If $b > 0$, then there exist $\alpha, \beta, m^* > 0$ such that

$$\langle X_0 \rangle_{b,\infty} \ge m^*. \tag{44}$$

By symmetry, $\langle X_0 \rangle_{-b,\infty} = -\langle X_0 \rangle_{b,\infty}$. Thus there must be at least two distinct extreme Gibbs measures for the same interaction. Most likely there are no others, but this problem has not been approached yet.

The strategy of proof is to reduce the bound in (44) to a corresponding one for a one-dimensional Ising spin system with long-range ferromagnetic pair interaction, for which the famous proofs of Dyson [13] and of Fröhlich and Spencer [19] on the existence of long range order are available. The reduction is based on ferromagnetic type inequalities. With the block variables

$$\phi_j = \frac{1}{\delta} \int_{(j-\frac{1}{2})\delta}^{(j+\frac{1}{2})\delta} X_t dt \,, \quad j \in \mathbb{Z} \,, \tag{45}$$

by Griffiths II we obtain that $\langle X_0 \rangle_{b,\infty} \geq c_G \langle \phi_0 \rangle_{b,\infty}^c$, where $\langle \ \rangle^c$ is a Gibbs measure over $\mathbb{Z}$ with long range interaction for the continuous spin variables ϕ_j, and $c_G > 0$. Secondly, the Wells inequality, see [38] in the case of stochastic processes, [8] implies that $\langle \phi_0 \rangle_{b,\infty}^c \geq c_W \langle \sigma_0 \rangle_{+,\infty}$ with $c_W > 0$. Here $\langle \ \rangle_{+,\infty}$ is an Ising spin system, $\sigma_0 = \pm$, with ferromagnetic interaction which decays as $|i - j|^\gamma$ for large $|i - j|$ and $+$ boundary conditions. The complete proof is given in [32], where also explicit bounds for the phase diagram are discussed.

3 A Central Limit Theorem

In this section we study the case where $V = 0$, i.e. we consider

$$\mu_T = \frac{1}{\mathcal{Z}_T} \exp\left(-\int_{-T}^{T} \int_{-T}^{T} W(X_t - X_s, t - s)\, dt\, ds \right) \mathcal{W}_T^0 \,. \tag{1}$$

Here, $\mathcal{W}_T^0$ is two-sided Brownian motion in $[-T, T]$ pinned at 0 at $t = 0$. The interaction depends only on the increments $X_t - X_s$. Provided W has a decent decay in the t-variable, one would thus expect a functional central limit theorem to hold, i.e. after rescaling the path measure μ_T should look like Brownian motion with some effective diffusion matrix D. Such a general result is not available. In case $t \mapsto W(\cdot, t)$ decays exponentially, one can use Dobrushin's theory of one-dimesional spin systems [10, 11] to establish exponential mixing of the increment process [39]. This implies the central limit theorem for X_t properly rescaled. Our approach is less restrictive in terms of decay conditions, but assumes W to be of the special form

$$W(x,t) = -\frac{1}{2} \int |\widehat{\rho}(k)|^2 e^{ik \cdot x} e^{-\omega(k)|t|} \frac{1}{2\omega(k)} dk \tag{2}$$

with

$$\omega(k) \geq 0, \quad \omega(k) = \omega(-k), \quad \text{and} \quad \widehat{\rho}(k) = \widehat{\rho}(-k)^* . \tag{3}$$

In addition, we assume

$$\int |\widehat{\rho}(k)|^2 (\omega^{-1} + \omega^{-2} + \omega^{-3}) dk < \infty. \tag{4}$$

(4) is in fact a (mild) decay condition. For example, if $d = 3$, $\omega(k) = |k|$ and $\widehat{\rho}$ is compactly supported, then the most stringent condition is $\int |\widehat{\rho}|^2 \omega^{-3} d^3 k < \infty$, which corresponds to a decay of W as

$$|W(x,t)| \le c(1 + |t|^{3+\delta})^{-1},$$

for some $\delta > 0$. The above choice of parameters represents a physically relevant model, see (iv) of Section 4.

Theorem 3.1. *Define μ_T as in (1) with W given by (2).*

(i): *μ_T converges to a measure μ as $T \to \infty$ in the topology of local convergence.*

(ii): *The stochastic process X_t, $t \ge 0$, induced by μ satisfies a functional central limit theorem*

$$\lim_{\varepsilon \to 0} \sqrt{\varepsilon} X_{t/\varepsilon} = \sqrt{D} B_t$$

in distribution, where $0 \le D \le 1$ as a $d \times d$ matrix, and B_t is standard Brownian motion.

(iii): *In addition to (3),(4) suppose*

$$\int |\widehat{\rho}(k)|^2 |k|^2 \left(\omega^{-2} + \omega^{-4} \right) dk < \infty. \tag{5}$$

Then $D > 0$.

In the remainder of this section, we will give an outline of the proof of Theorem 3.1. A full account is [5]. We will do the proof in three steps.

(1) We use the special form (2) of W in order to linearize the interaction in (1) by introducing an auxiliary Gaussian process. As a result, we will prove (i) above, and the stochastic process X_t under μ is driven by a reversible Markov process η_t.

(2) In the so obtained representation, we use the by now well-established technique of Kipnis and Varadhan [24]; we write X_t as the sum of a martingale and an additive functional of η_t. X_t is then the sum of two martingales and a negligible process, and the martingale central limit theorem applies, proving (ii).

(3) In order to show that the diffusion is nondegenrate, we rely on an idea of Brascamp, Lebowitz and Lieb [7], which in the present context has been employed before [39].

To carry out step (1), let $\mathcal{K}_0$ be the real Hilbert space obtained by completing the subspace of $L^2(\mathbb{R}^d)$ on which the inner product given by

$$\langle a, b \rangle_{\mathcal{K}_0} = \int \widehat{a}(k) \frac{1}{2\omega(k)} \widehat{b}(k)^* \, dk \qquad (6)$$

is finite. Let $\mathcal{G}$ be the path measure of the infinite dimensional Ornstein-Uhlenbeck process with mean 0 and covariance

$$\mathbb{E}_{\mathcal{G}}[\phi_s(a)\phi_t(b)] = \int \widehat{a}(k) \frac{1}{2\omega(k)} e^{-|t-s|\omega(k)} \widehat{b}(k)^* \, dk \qquad (a, b \in \mathcal{K}_0).$$

There exists a Hilbert space $\mathcal{K} \supset \mathcal{K}_0$ such that $\mathcal{G}$ is a reversible Gaussian Markov process with values in $\mathcal{K}$ and continuous paths. The reversible measure G is the Gaussian measure on $\mathcal{K} \ni \phi$ with mean zero and covariance

$$\mathbb{E}_{\mathsf{G}}[\phi(a)\phi(b)] = \langle a, b \rangle_{\mathcal{K}_0}.$$

For $x \in \mathbb{R}^d$, let τ_x be the shift by x on $\mathcal{K}$, i.e. $(\tau_x \phi)(a) = \phi(\tau_x a)$ and $\tau_x a(y) = a(y - x)$. More generally, for $f \in L^2(\mathsf{G})$, $(\tau_x f)(\phi) = f(\tau_x \phi)$

For $T > 0$ we put

$$\mathcal{P}_T = \frac{1}{\mathcal{Z}_T} \exp\left(-\int_{-T}^{T} \tau_{X_s} \phi_s(\rho) \, ds\right) \mathcal{W}^0 \otimes \mathcal{G}. \qquad (7)$$

With $\mathcal{P}_T$ we achieved our first goal, the linearization of the interaction: Indeed, for functions F depending on x only,

$$\mathbb{E}_{\mathcal{P}_T}[F] = \mathbb{E}_{\mu_T}[F],$$

as can be seen by carrying out the Gaussian integration. $\mathcal{P}_T$ is the measure of a Markov process, more specifically a $P(\phi)_1$-process with state space $\mathbb{R}^d \times \mathcal{K}$. The role of the Schrödinger operator is now played by

$$Hf(x, \phi) = -\frac{1}{2}\Delta f(x, \phi) + H_{\mathsf{f}} f(x, \phi) + V_\rho(x, \phi) f(x, \phi), \qquad (8)$$

where H_{f} is the generator of $\mathcal{G}$ and $V_\rho(x, \phi) = \tau_x \phi(\rho)$. The semigroup Π_T generated by H is strongly continuous on $C_0(\mathbb{R}^d, L^2(\mathsf{G}))$. More importantly, it is also strongly continuous on the Hilbert space $\mathcal{T}$ of functions that are invariant under shift over the x-variable. Explicitly, $\mathcal{T}$ is the image of $L^2(\mathsf{G})$ under the operator

$$U : L^2(\mathsf{G}) \to C(\mathbb{R}, L^2(\mathsf{G})), \quad Uf(x, \phi) = \tau_x f(\phi),$$

equipped with the scalar product

$$\langle f, g \rangle_{\mathcal{T}} = \mathbb{E}_{\mathsf{G}}[(U^{-1}f)(U^{-1}g)^*] = \langle U^{-1}f, U^{-1}g \rangle_{L^2(\mathsf{G})}. \qquad (9)$$

H is self-adjoint on $\mathcal{T}$, and (4) implies

$$\|\Pi_T 1\|_{\mathcal{T}}^2 \leq C \langle 1, \Pi_T 1 \rangle_{\mathcal{T}}^2.$$

Now from spectral theory we obtain

Theorem 3.2. *The infimum E_0 of the spectrum of H acting in $\mathcal{T}$ is an eigenvalue of multiplicity one. The corresponding eigenfunction $\Psi \in \mathcal{T}$ can be chosen strictly positive.*

An alternative proof of Theorem 3.2, using a completely different method, can be found in [18].

It is now easy to identify the infinite volume limit of the families $\mathcal{P}_T$ and $\mathcal{N}_T$. Let $\mathcal{P}$ be the probability measure on paths $(X_t, \phi_t)_{t \in \mathbb{R}}$ determined by

$$\mathbb{E}_{\mathcal{P}}(f) = e^{2TE_0} \mathbb{E}_{\mathcal{W}^0 \otimes \mathcal{G}} \left[\Psi(X_{-T}, \phi_{-T}) e^{-\int_{-T}^{T} \tau_{X_s} \phi_s(\rho)\, ds} \Psi(X_T, \phi_T) f \right] \quad (10)$$

for functions f depending only on X_t, ϕ_t with $|t| < T$. Above, $\mathcal{W}^0$ is the measure of two-sided Brownian motion or, equivalently, Wiener measure conditioned on $X_0 = 0$. Let μ be the measure $\mathcal{P}$ when applied to functions of x only. Then $\mathcal{P}$ is the measure of a Markov process with generator L acting as

$$Lf = -\frac{1}{\Psi}(H - E_0)(\Psi f). \quad (11)$$

$\mathcal{P}_T \to \mathcal{P}$ in the topology of local convergence, and by integrating out the Gaussian field, $\mu_T \to \mu$. The $\mathcal{K}$-valued process $\eta_t = \tau_{X_t} \phi_t$ is reversible with reversible measure $(U^{-1}\Psi)^2 G$, and its generator is unitarily equivalent to L.

Let $\gamma \in \mathbb{R}^d$ be fixed, and $h_\gamma(x) = \gamma \cdot x$. Then $L(h_\gamma) = j(\eta)$ with

$$j = U^{-1}(\gamma \cdot \nabla_x \ln \Psi) \in L^2(G). \quad (12)$$

Since the result of the generator L of process $\mathcal{P}$ applied to $\gamma \cdot x$ is a function of η, only η_t influences the behavior of $\gamma \cdot X_t$, i.e. X_t is driven by η_t. Step one is completed.

Next we write

$$\gamma \cdot X_t = M_t + \int_0^t Lh_\gamma(X_s, \phi_s))\, ds \quad (13)$$

with

$$M_t = \gamma \cdot X_t - \int_0^t Lh_\gamma(X_s, \phi_s)\, ds = \gamma \cdot X_t - \int_0^t j(\eta_s)\, ds.$$

Then M_t is a martingale with stationary increments and quadratic variation $|\gamma|^2 t$, and

$$\int_0^t Lh_\gamma(X_s, \phi_s))\, ds = \int_0^t j(\eta_s)\, ds$$

is an additive functional of η_t satisfying the assumptions of [24]. It is thus the sum of a martingale N_t with stationary increments and a negligible process. Now the martingale central limit theorem proves Theorem 3.1 (ii) and finishes step 2.

In principle, it could happen that M_t and N_t are strongly dependent and cancel each other. Then the diffusion matrix D would be zero and X_t would behave subdiffusively. We already know the central limit theorem holds with diffusion matrix $D \geq 0$. Thus it is enough to investigate

$$\lim_{t \to \infty} \frac{1}{t} \mathbb{E}_\mu[(\gamma \cdot X_t)^2] = \langle \gamma, D\gamma \rangle_{\mathbb{R}^d} . \tag{14}$$

It turns out that

$$\langle \gamma, D\gamma \rangle_{\mathbb{R}^d} = |\gamma|^2 - 2 \langle \gamma \cdot \nabla_x \Psi, (H - E_0)^{-1} \gamma \cdot \nabla_x \Psi \rangle_{\mathcal{T}} . \tag{15}$$

The standard technique is to turn (15) into a variational problem and find a reasonably explicit lower bound to the variational functional. We did not succeed in carrying out the second step of this procedure. Instead, we show directly that

$$\mathbb{E}_\mu[(\gamma \cdot X_t)^2] \geq c|\gamma|^2|t| \tag{16}$$

for some $c > 0$, by using ideas from Brascamp *et al* [7] originally developed to study fluctuations for anharmonic lattices. Together with (14) this immediately shows $D \geq c$.

4 Applications and Open Problems

The scheme outlined so far is a probabilistically natural way of constructing through the limit $T \to \infty$ stationary stochastic processes with continuous sample paths. Moreover, specific choices of V and W correspond to particular applications on which there is already a large body of literature using a variety of methods. Very roughly, and as far as we are aware of, the applications originate from three distinct corners of low energy physics.

i) Self-avoiding random walks. Polymers with interaction due to excluded volume is an important statistical mechanics topic, in particular because of the connections with critical phenomena [14]. It is tempting to model the free polymer as Brownian motion and the excluded volume through an interaction of the form (7). Note, however, that by the nature of the interaction there is no decay in t. In particular the energy is not extensive. Thus, while the energy depends only on the increments, for large T the statistical properties of the self-avoiding polymer are qualitatively different from a free random walk. One conjecture is that the self-similar scaling theory is obtained from the ultraviolet limit. So far most of the mathematical effort went into constructing the limit measure [41, 6]. But it is not obvious how to extract scale invariant properties from this measure. In fact, self-similarity is now established through lace expansion and other methods [30]. The link between the two approaches remains unexplored.

ii) Statistical hydrodynamics. There is general agreement that fully developed turbulence should be described by a suitable measure over divergence free vorticity fields. One attempt to write down such a measure is to assume that the velocity field $w(x) = \nabla \wedge u(x)$ is concentrated along Brownian curves $X_t \in \mathbb{R}^3$ [9]. Under the Eulerian incompressible flow, the kinetic energy $\frac{1}{2} \int u(x)^2 d^3x$ is conserved. Thus it seems natural to use it as energy in the Gibbs measure. This yields the formal expression

$$\mathcal{E}(X) = \int_{-T}^{T} \int_{-T}^{T} \frac{1}{|X_t - X_s|} dX_t \cdot dX_s \,. \tag{1}$$

In order to have $\exp[-\mathcal{E}(X)]$ as a well-defined random variable, [17] required the condition that the Coulomb potential in (1) is smoothened such that it has a finite electrostatic energy.

Our own investigations mostly draw on applications in quantum mechanics. Since upon Wick rotation the free Schrödinger equation turns into the diffusion equation, Brownian motion as *a priori* measure is in fact forced by the problem. Several interesting cases can be distinguished.

iii) Electron coupled to the quantized radiation field. Upon Wick rotation the free Maxwell field is isomorphic to a stationary infinite-dimensional Ornstein-Uhlenbeck process, see Section 3, for the transverse vector potential $A(x,t)$. It has the covariance

$$\mathbb{E}\left[A_\alpha(x,t)A_\beta(x',t')\right] = \int d^3k \frac{1}{2\omega(k)} e^{-\omega(k)|t-s|} e^{ik\cdot(x-x')}(\delta_{\alpha\beta} - |k|^{-2}k_\alpha k_\beta)$$
$$= W_{\alpha\beta}(x - x', t - s) \tag{2}$$

$\alpha, \beta = 1, 2, 3$. The dispersion relation of the Maxwell field is

$$\omega(k) = |k| \,. \tag{3}$$

Within the Euclidean framework, the electron is governed by the Hamiltonian

$$H = \frac{1}{2}\left(-i\nabla_x - eA(x,t)\right)^2, \tag{4}$$

on ignoring the electron spin. The units are such that $\hbar = 1$, $c = 1$, mass of electron $m = 1$; e is the charge of the electron expressing the strength of coupling to the Maxwell field. We use the Feynman-Kac-Ito formula for the propagator for H [38]. Then the joint X_t and $A(x,t)$ path measure is given by

$$\exp\left(-ie \int_{-T}^{T} A(X_t, t) \cdot dX_t\right) \mathcal{W}(X) \otimes \mathcal{G}, \tag{5}$$

where $\mathcal{G}$ is the Gaussian measure of the A-field with covariance (2). Note that $\nabla \cdot A(x,t) = 0$ almost surely. Since the exponent is linear in A, the averaging over the Ornstein-Uhlenbeck process can be done explicitly. This results in a finite volume Gibbs measure with energy

$$\mathcal{E}(X) = \frac{1}{2} \int_{-T}^{T} \int_{-T}^{T} dX_t \cdot W(X_t - X_s, t - s) dX_s. \qquad (6)$$

This is of the form (6) and should be read as a double Ito stochastic integral.

W is singular on the diagonal, roughly $W_{\alpha\beta}(x, t) = \delta_{\alpha\beta}(x^2 + t^2)^{-1}$. Thus it is necessary to smear out the charge distribution which leads to the regularized version

$$W^{\rho}_{\alpha\beta}(x, t) = \int d^3 k |\widehat{\rho}(k)|^2 \frac{1}{2\omega(k)} e^{-\omega(k)|t|} e^{ik \cdot x} (\delta_{\alpha\beta} - |k|^{-2} k_\alpha k_\beta). \qquad (7)$$

Here $\widehat{\rho}$ is rotation invariant, decays rapidly for large $|k|$, and $\widehat{\rho}(0) = (2\pi)^{-3/2}$ by charge normalization. A problem which appears to be very challenging, is to establish that, for fixed T and $X_{-T} = 0 = X_T$, the Gibbs measure for the energy (6) is well defined. In other words, with a smoothening as in (7) we would like to study the sequence of Gibbs measures as $\widehat{\rho}(k) \to (2\pi)^{-3/2}$ pointwise (ultraviolet or point charge limit). In favorable cases the existence of the limit can be shown by suitable centering and by possibly adding other counter terms. Such a procedure seems unlikely to work in the present context. Thus the ultraviolet limit has to be linked with a change of the diffusion coefficient D of the underlying Wiener process $\mathcal{W}$ (= mass renormalization). We expect $D \to \infty$ in this limit.

iv) Quantum particle coupled to a scalar Bose field. This model was studied by Nelson [31] in the context of energy renormalization. The Bose field translates to the scalar field $\phi(x, t)$, which again is an infinite-dimensional Ornstein-Uhlenbeck process this time with covariance

$$\mathbb{E}[\phi(x, t)\phi(x', t')] = \int d^3 k |\widehat{\rho}(k)|^2 \frac{1}{2\omega(k)} e^{-\omega(k)|t-t'|} e^{ik \cdot (x-x')}$$

$$= W(x - x', t - t'). \qquad (8)$$

The quantum particle "sees" ϕ as a fluctuating electrostatic potential. Thus the Hamiltonian becomes $H = -\frac{1}{2}\Delta + e\phi(x, t)$. Then through the Feynman-Kac formula the path measure, jointly for X_t and $\phi(x, t)$, is given by

$$\exp\left(-e \int_{-T}^{T} \phi(X_t, t) dt\right) \mathcal{W}^0(X) \otimes \mathcal{G}, \qquad (9)$$

which has the structure of a $P(\phi)_1$-process, since the *a priori* measure is Markovian and the energy is local in time. The only difference to our discussion in Section 2.1 is that $\mathbb{R}^d$ is replaced by the state space $\mathbb{R}^d \times \mathcal{K}$, compare with the discussion preceding (8).

The exponent in (9) is linear in ϕ. Thus we can perform the integration over ϕ resulting in the following path measure for X,

$$\frac{1}{Z(T)} \exp\left[-\int_{-T}^{T} V(X_t) dt + \frac{e^2}{2} \int_{-T}^{T} \int_{-T}^{T} W(X_t - X_s, t - s) dt ds\right] \mathcal{W}^0, \quad (10)$$

where we added an external potential V. Thus the Nelson model naturally yields Gibbs measures of the form studied in Sections 2 and 3. In fact, the Nelson model was our source of motivation for studying Gibbs measures over Brownian motion. The existence of the infinite volume limit can be deduced from (10) which requires that

$$\int dk |\widehat{\rho}(k)|^2 (\omega(k)^{-3} + \omega(k)^{-1}) < \infty. \tag{11}$$

We can also use the cluster expansion which holds provided e^2 is sufficiently small and

$$\int dk |\widehat{\rho}(k)|^2 (\omega(k)^{-1} + \omega(k)^{-2-\delta}) < \infty \tag{12}$$

for some $\delta > 0$. Since it is possible to express the ground state of the Nelson model directly in terms of data of these Gibbs measures, given the existence of the infinite time interval measures we have a useful tool at hand for studying qualitative properties of the ground state. We refer to [4] for details.

The Nelson model, in the case of massless bosons $\omega(k) = |k|$, is both ultraviolet and infrared divergent. The ultraviolet divergence is mild and can be handled by energy renormalization. This is the content of the famous work [31], which uses exclusively functional analytic methods. Somewhat surprisingly, no one has succeeded in a proper transcription of Nelson's results into the framework of path measures. The infrared divergence translates into a somewhat unexpected feature of the joint $(X_t, \phi(x,t))$ process. From (12) and suitable conditions on V, we infer that the infinite volume Gibbs measure exists. However, the limiting procedure changes the situation seen by the a priori measure dramatically. For instance, the $t = 0$ joint distribution is not absolutely continuous with respect to the $t = 0$ projection of the a priori distribution. One way to cope is to introduce a suitable shifted Gaussian measure which takes on the role of a new a priori measure making the model infrared regular. We refer for more details to [27, 28].

v) The polaron. Physically the polaron is an electron coupled to the optical mode of an ionic crystal. It can be viewed as a particular case of the Nelson model with the choice $\omega(k) = \omega_0$ and $\widehat{\rho}(k) = |k|^{-1}$. Then

$$W(x,t) = -\frac{\alpha}{|x|} e^{-\omega_0 |t|}. \tag{13}$$

Here $\alpha > 0$ and subsumes all dimensional coupling coefficients. The ground state energy of the polaron is defined through

$$E_g(\alpha) = -\lim_{T \to \infty} \frac{1}{T} \log Z(T, \alpha). \tag{14}$$

For small α one can use perturbation theory in α. For large α Pekar [33] developed an approximate strong coupling theory. Thus the challenge was to have reliable predictions at moderate values of α, which turned out to be difficult. Feynman [15, 16] had the insight from functional integration and used a quadratic functional as upper variational bound. Optimizing the quadratic form yields $E_g(\alpha)$ roughly 2% away from Pekar's result and even better at smaller values when compared with machine computations. The strong coupling (Pekar) limit of the ground state energy has been established by Donsker and Varadhan [12] using functional integration, and by Lieb and Thomas [25] using functional analytic methods.

A long standing open problem is to obtain a corresponding result for the effective mass $m(\alpha)$. In fact, as shown in [39], $m(\alpha) = D(\alpha)^{-1}$ with $D(\alpha)$ the diffusion coefficient in Section 3 with the specific choice (13) for W. On heuristic grounds one can guess the behavior of $D(\alpha)$ for large α and relate it to Pekar's variational problem [39]. A proof is missing with the exception of [34] in the simplification where Brownian motion on $\mathbb{R}^3$ is replaced by Brownian motion on the circle.

References

1. Betz, V.: Existence of Gibbs measures relative to Brownian motion. *Markov Proc. Rel. Fields* **9** (2003), 85-102
2. Betz, V.: *Gibbs measures relative to Brownian motion and Nelson's model*, PhD thesis, TU München, 2002
3. Betz, V. and Lőrinczi, J.: Uniqueness of Gibbs measures relative to Brownian motion, *Ann. I. H. Poincaré* PR**39** (2003), 877-889
4. Betz, V., Hiroshima, F., Lőrinczi, J., Minlos, R.A. and Spohn, H.: *Ground state properties of the Nelson Hamiltonian - A Gibbs measure-based approach*, Rev. Math. Phys **14** (2002), 173-198
5. Betz, V. and Spohn, H.: A central limit theorem for Gibbs measures relative to Brownian motion, submitted for publication, 2003
6. Bolthausen, E.: On the construction of the three dimensional polymer measure, *Probab. Theory Rel. Fields* **97** (1993), 81-101
7. Brascamp, H.J., Lebowitz, J.L. and Lieb, E.: The Statistical Mechanics of anharmonic lattices, in *Proceedings of the 40th session of the International Statistics Institute*, Warsaw, 1975, Vol. 9, pp. 1-11
8. Bricmont, J., Lebowitz, J.L. and Pfister, Ch.-E.: On the equivalence of boundary conditions, *J. Stat. Phys.* **21** (1979), 573-582
9. Chorin, A.: *Vorticity and Turbulence*, Springer, 1994
10. Dobrushin, R.L.: Analyticity of correlation functions in one-dimensional classical systems with slowly decreasing potentials, *Commun. Math. Phys.* **32** (1973), 269-289
11. Dobrushin, R.L.: Analyticity of correlation functions for one-dimensional classical systems with power-law decay of the potential, *Math. USSR Sbornik* **23** (1973), 13-44
12. Donsker, M.D. and Varadhan, S.R.S.: Asymptotic for the polaron, *Commun. Pure Appl. Math.* **36** (1983), 505-528
13. Dyson, F.J.: Existence of a phase transition in a one-dimensional Ising ferromagnet, *Commun. Math. Phys.* **12** (1969), 91-107

14. Fernández, R., Fröhlich, J. and Sokal, A.D.: *Random Walks, Critical Phenomena, and Triviality in Quantum Field Theory*, Springer, 1992
15. Feynman, R.P.: Slow electrons in a polar crystal, *Phys. Rev.* **97** (1955), 660-665
16. Feynman, R.P. and Hibbs, A.: *Quantum Mechanics and Path Integrals*, McGraw-Hill, New York, 1965
17. Flandoli, F.: On a probabilistic description of small scale structures in 3D fluids, *Ann. I.H. Poincaré* PR**38** (2002), 207-228
18. Fröhlich, J.: Existence of dressed one electron states in a class of persistent models, *Fortschr. Phys.* **22** (1974), 159-198
19. Fröhlich, J. and Spencer, T.: The phase transition in the one-dimensional Ising model with $1/r^*$ interaction energy, *Commun. Math. Phys.* **84** (1982), 87-101
20. Le Gall, J.F.: Sur le temps local d'intersection du mouvement Brownien plan, et la méthode de renormalisation de Varadhan, *Séminaire des Probabilités XIX, 1983/84*, Lecture Notes in Mathematics **1123**, Springer, 1985, pp. 314-331
21. Georgii, H.-O.: *Gibbs Measures and Phase Transitions*, Berlin, New York: de Gruyter, 1988
22. Hariya, Y.: A new approach to construct Gibbs measures on $C(\mathbb{R}, \mathbb{R}^d)$, preprint, 2001
23. Hariya, Y. and Osada, H.: Diffusion processes on path spaces with interactions, *Rev. Math. Phys.* **13** (2001), 199-220
24. Kipnis, C. and Varadhan, S.R.S.: Central limit theorem for additive functionals of reversible Markov processes and applications to simple excursions, *Commun. Math. Phys.* **104** (1986), 1-19
25. Lieb, E. and Thomas, L.: Exact ground state energy of the strong-coupling polaron, *Commun. Math. Phys.* **183** (1997), 511-519
26. Lőrinczi, J. and Minlos, R.A.: Gibbs measures for Brownian paths under the effect of an external and a small pair potential, *J. Stat. Phys.* **105** (2001), 605-647
27. Lőrinczi, J., Minlos, R.A. and Spohn, H.: The infrared behaviour in Nelson's model of a quantum particle coupled to a massless scalar field, *Ann. Henri Poincaré* **3** (2002), 1-28
28. Lőrinczi, J., Minlos, R.A. and Spohn, H.: Infrared regular representation of the three dimensional massless Nelson model, *Lett. Math. Phys.* **59** (2002), 189-198
29. Malyshev, V.A. and Minlos, R.A.: *Gibbs Random Fields*, Kluwer Academic Publishers, 1991
30. Madras N. and Slade, G.: *The Self-Avoiding Walk*, Birkhäuser, 1996
31. Nelson, E.: Interaction of nonrelativistic particles with a quantized scalar field, *J. Math. Phys.* **5** (1964), 1990-1997
32. Osada, H. and Spohn, H.: Gibbs measures relative to Brownian motion, *Ann. Probab.* **27** (1999), 1183-1207
33. Pekar, S.I.: *Untersuchungen zur Elektronentheorie der Kristalle*, Akademie Verlag, Berlin, 1954
34. Petermann, M.: PhD Thesis, University of Zürich, 2001
35. Rosen, J.: A local time approach to the self-intersections of Brownian paths in space, *Commun. Math. Phys.* **88** (1983), 327-338
36. Rosen, J.: A representation for the intersection local time of Brownian motion in space, *Ann. Probab.* **13** (1985), 145-153
37. Simon, B.: *Functional Integration and Quantum Physics*, Academic Press, 1979
38. Simon, B.: Schrödinger semigroups, *Bull. AMS* **7** (1982), 447-526

39. Spohn, H.: Effective mass of the polaron: a functional integral approach, *Ann. Phys.* **175** (1987), 278-318
40. Varadhan, S.R.S.: in *Local Quantum Theory*, R. Jost (ed.), *Enrico Fermi School*, Academic Press, New York, 1969
41. Westwater, J.: On Edwards model for long polymer chains, *Commun. Math. Phys.* **72** (1980), 131-174

Spectral Theory
for Nonstationary Random Potentials

Stefan Böcker[1], Werner Kirsch[1], and Peter Stollmann[2]

Institut für Mathematik, Ruhr-Universität Bochum
Institut für Mathematik, TU Chemnitz

1 Introduction: Leaving Stationarity

One-particle Schrödinger operators with random potentials are used to model quantum aspects of disordered electronic systems like unordered alloys, amorphous solids or liquids.

Starting from periodic potentials to describe perfect crystals, one is interested in the spectral properties of these operators. Periodic Schrödinger operators usually have absolutely continuous spectrum, which is connected physically to good conductance properties. The spectrum consists of bands, that is intervals, where one finds (absolutely continuous) spectrum. Between the bands there are so called forbidden zones without spectrum.

Anderson, Mott and Twose [4, 84] discovered – based on physical reasoning – that disordered systems should have a tendency to "localized states" in certain regimes of the energy spectrum, which reflects bad conductance properties of the solid in this energy regime.

In recent years there has been considerable progress concerning mathematically rigorous results on this phenomenon of **localization**. We refer to the bibliography where we chose some classics, some recent articles as well as books on the subject. However, all these results provide only one part of the picture that is accepted since the ground breaking work [4, 84] by Anderson, Mott and Twose. The effect of metal insulator transition is supposed to depend upon the dimension and the general picture is as follows: Once translated into the language of spectral theory there is a transition from a **localized phase** that exhibits pure point spectrum (= only bound states = no transport) to a **delocalized phase** with absolutely continuous spectrum (= scattering states = transport). What has been proven so far is the occurence of the former phase, well known under the name of localization. The missing part, delocalization, has not been settled for genuine random models.

There is need for an immediate disclaimer or, put differently, for an explanation of what we mean by "genuine".

An instance where a metal insulator transition has been verified rigorously is supplied by the **almost Mathieu operator**, a model with modest disorder for which the parameter that triggers the transition is the strength of the coupling. As references let us mention [7, 42, 59, 60, 61, 78] where the reader can find more about the literature on this true evergreen. Quite recently it has attracted a lot of interest especially among harmonic analysts; see [8, 9, 10, 11, 12, 13, 14, 15, 16, 46, 87]

The underlying Hilbert space is $l^2(\mathbb{Z})$. Consider parameters $\alpha, \lambda, \theta \in \mathbb{R}$ and define the selfadjoint, bounded operator $h_{\alpha,\lambda,\theta}$ by

$$(h_{\alpha,\lambda,\theta}u)(n) = u(n+1) + u(n-1) + \lambda \cos(2\pi(\alpha n + \theta))u(n),$$

for $u = (u(n))_{n \in \mathbb{Z}} \in l^2(\mathbb{Z})$.

Note that this operator is a discrete Schrödinger operator with a potential term with the coupling constant λ in front and the discrete analog of the Laplacian. For irrational α the potential term is an almost periodic function on $\mathbb{Z}$.

Basically, there is a metal insulator transition at the critical value 2 for the coupling constant λ. Since these operators are very close to periodic ones one can fairly label them as poorly disordered. Moreover, the proof of delocalization boils down to the proof of localization for a "dual operator" that happens to have the same form. In this sense, the almost Mathieu operator is not a genuine random model.

A second instance, where a delocalized phase is proven to exist is the Bethe lattice. See Klein's article [63].

Quite recently, an order parameter has been introduced by Germinet and Klein to characterize the range of energies where a multiscale scenario provides a proof of a localized regime, [44]. In their work the important parameter is the energy.

However, as we already pointed out above, for genuine random models, there is no rigorous proof of the existence of a transition or even of the appearance of spectral components other than pure point, so far. This is a quite strange situation: the unperturbed problem exhibits extended states and purely a.c. spectrum but for the perturbed one can prove the opposite spectral behavior only.

In this survey we are dealing with models that are not stationary in the sense that the influence of the random potential is not uniform in space. The precise meaning of this admittedly vague description differs from case to case but will be clear for each of them.

2 Sparse Random Potentials

The term sparse potentials is mostly known for potentials that have been introduced in the 1970's by Pearson [86] to construct Schrödinger operators on the line with singular continuous spectrum. To use similar geometries to obtain a metal insulator transition can be traced back to Molchanov, Molchanov and Vainberg [82, 83] and Krishna [75, 76], see also [65, 67, 77]. Let us introduce three models for operators with sparse random potentials in $L^2(\mathbb{R}^d)$ taken from [51].

Model I:
$$H(\omega) = -\Delta + V_\omega,$$

where
$$V_\omega(x) = \sum_{k \in \mathbb{Z}^m} \xi_k(\omega) f(x - k),$$

$f \leq 0$ is a compactly supported single site potential and the ξ_k are independent Bernoulli variables with $p_k := \mathbb{P}\{\xi_k = 1\}$.

Model II:
$$H(\omega) = -\Delta + V_\omega,$$

where
$$V_\omega(x) = \sum_{k \in \mathbb{Z}^m} q_k(\omega)\xi_k(\omega) f(x - k),$$

with f and ξ as above and independent identically distributed random variables q_k.

Model III:
$$H(\omega) = -\Delta + V_\omega,$$

where
$$V_\omega(x) = \sum_{k \in \mathbb{Z}^m} a_k q_k(\omega) f(x - k),$$

with q_i as above and a deterministic sequence a_i decaying fast enough at infinity (see [67] for a discrete analog of this model).

For the first two models it was proven in [51] that $[0, \infty)$ belongs to the absolutely continuous spectrum as long as p_k decays fast enough. To understand this appearance of a metallic regime, we recall the following facts from scattering theory:

We write $-\Delta = H_0$ so that the operators we are interested in can be written as $H = H_0 + V$. By $\sigma_{ac}(H)$ we denote the absolutely continuous spectrum, related to delocalized states.

Theorem 1. *(Cooks criterion)*
If for some $T_0 > 0$ and all ϕ in a dense set

$$\int_{T_0}^{\infty} \|V e^{-itH_0}\phi\| dt < \infty \qquad (*)$$

then $\Omega_- := \lim_{t \to \infty} e^{itH} e^{-itH_0}$ exists and, consequently, $[0, \infty) \subset \sigma_{ac}(H)$, i.e., there are scattering states for H and any nonnegative energy.

The typical application rests on the fact that $(*)$ is satisfied if

$$|V(x)| \leq C(1 + |x|)^{-(1+\epsilon)}, \qquad (**)$$

a condition that obviously fails to hold for almost every V_ω provided the p_k are not summable. However, in [51] the following result has been proved:

Theorem 2. *Assume that*

$$W(x) := \left(\mathbb{E}(V_\omega(x)^2)\right)^{\frac{1}{2}} \le C(1 + |x|)^{-(1+\epsilon)}.$$

Then V_ω satisfies Cook's criterion for a.e. ω.

The proof is short. So we reproduce it here.

Proof.

$$\mathbb{E}\left(\int_{T_0}^{\infty} \|V_\omega e^{-itH_0}\phi\|dt\right)$$

$$= \int_{T_0}^{\infty} \mathbb{E}\left(\int V_\omega(x)^2|e^{-itH_0}\phi(x)|^2 dx\right)^{\frac{1}{2}} dt$$

$$= \int_{T_0}^{\infty} \left(\mathbb{E}\int V_\omega(x)^2|e^{-itH_0}\phi(x)|^2 dx\right)^{\frac{1}{2}} dt$$

$$\le \int_{T_0}^{\infty} \left(\int \mathbb{E}(V_\omega(x)^2)|e^{-itH_0}\phi(x)|^2 dx]\right)^{\frac{1}{2}} dt$$

$$= \int_{T_0}^{\infty} \|W(x)e^{-itH_0}\phi\|dt$$

$\square$

One can apply this result if the p_k decay fast enough to guarantee sufficient decay of $W(x)$. For **Model I** and **Model II** operators (see [51]) with $p_k \sim k^{-\alpha}$ and f with compact support this condition reads $2 < \alpha$.

On the other hand one wants to have that $\sum_k p_k = \infty$, since otherwise V_ω has compact support a.s. by the Borel-Cantelli Lemma. Thus one gets both essential spectrum below zero and absolutely continuous spectrum above zero if

$$2 < \alpha < d.$$

For fixed $d \ge 3$ and $p_k \sim k^{-\alpha}$ one can moreover control the essential spectrum below 0 as done in [51]: the negative essential spectrum consists of a sequence of energies that can at most accumulate at 0.

To control the essential spectrum below zero, it is useful to introduce the model operators $H_f := H_0 + f$ and $H_n := H_0 + f(x - x_n)$ for a sequence of points $\{x_m\}_{m\in\mathbb{N}}$.

Theorem 3. *(Klaus) Let $\mathcal{E}$ denote the set of energies*

$$\mathcal{E} = \{E < 0 \mid \text{there exists a sequence } n_j, \text{ and energies}$$
$$E_{n_j} \in \sigma(H_{n_j}) \text{ with } E_{n_j} \to E\}.$$

Then $\sigma_{ess}(H) = \mathcal{E} \cup [0, \infty)$.

Using the sets

$$A_k(\Lambda) := \{\omega \mid \exists \text{ at least } k \text{ distinct points } n_l \in \Lambda \text{ with } \xi_{n_l} = 1\},$$

together with the bound

$$\mathbb{P}[A_k(\Lambda)] \leq \left(\sum_{i \in \Lambda} p_i\right)^k \leq |\Lambda|^{k-1} \sum_{i \in \Lambda} p_n^k,$$

one gets by a Borel-Cantelli argument, that the event

$$A_k := \bigcup_{L=1}^{\infty} A_k(\Lambda_L)$$

has zero probability. Calling a finite subset F of $\mathbb{Z}^d$ essential for H_ω if

$$\mathbb{P}\left[F + n \subset \Xi \text{ for infinitely many } n\right] = 1$$

with $\Xi := \{n \mid \xi_n = 1\}$ and

$$H_F := H_0 + \sum_{i \in F} f(\cdot - i),$$

from Klaus' theorem one gets the following.

Theorem 4. Let $\mathcal{E} = \{E_n(F) \in H_F \mid F \text{ is essential for } H_\omega\}$. If $\sum p_n^k < \infty$ for some k, then

$$\sigma_{ess}(H_\omega) = \mathcal{E} \cup [0, \infty) \quad \mathbb{P}\text{-almost surely}$$

Therefore, there exists essential spectrum below zero, if $\sum p_i = \infty$, and it is pure point, if $\sum p_i^k < \infty$ for some k.

For **Model III** operators the distribution $\mathbb{P}_0$ of the i.i.d. random variables q_k has a strong influence on the spectral behaviour in the following sense: If $\mathbb{P}_0$ has a bounded support, every realization of **Model III** is decaying at infinity and there is no hope to find essential spectrum below 0. However, if $\mathbb{P}_0$ does not have compact support, the potential $V_\omega(x)$ may admit a sequence $x_i \to \infty$ such that $V_\omega(x_i) \to -\infty$, thus allowing for essential, in fact, dense pure point spectrum below zero.

For details and more on sparse potentials, especially for models for which the negative spectrum has a richer structure and contains intervals, we refer the reader to [51].

Remarks 5. In [18] absence of (absolutely) continuous spectrum outside the spectrum of the unperturbed operator for certain random sparse models is proved reminiscent of Model I above and Model II from [51] but considerably more general. This is based on the techniques from [54, 95, 96].

3 Sparse Random Potentials
and the Integrated Density of States

A useful object in studying random Schrödinger operators is the (integrated) density of states $N(H_\omega, E)$. For ergodic random Schrödinger operators it can be defined as the thermodynamic limit of the eigenvalue counting function up to energy E, that is

$$N(H_\omega, E) := \lim_{L \to \infty} \frac{1}{|\Lambda_L|} N(H_{\omega,L}^D, E)$$

and

$$N(H_{\omega,L}^D, E) := \#\{\lambda_i \le E \mid \lambda_i \in \sigma(H_{\omega,L}^D)\},$$

where $H_{\omega,L}^D$ denotes the operator H_ω restricted to a box Λ_L of side length L with Dirichlet boundary conditions. Under appropriate conditions on the random potential this limit exists and is non-random. Also this limit tends to be independent of the boundary conditions used in the definition, that is, using Neumann boundary conditions on the boxes Λ_L gives the same limit.

It is not immediately clear, how to define an analogous object for sparse random potentials, because on the one hand there is nothing like ergodicity around them, on the other hand one has to guess the right normalisation substituting the volume $|\Lambda_L|$.

At least for sparse random potentials with $\sum p_n = \infty$, but $\sum p_n^2 < \infty$ it is possible to define a modified integrated density of states for energies $E < 0$ as

$$N(H_\omega, E) := \lim_{L \to \infty} \frac{1}{P_L} N(H_{\omega,L}^D, E)$$

with

$$P_L := \sum_{i \in \Lambda_L} p_i.$$

For **Model I** operators this gives

Theorem 6. *Let V_ω be a sparse random potential as in **Model I** with $\sum p_n = \infty$ and $\sum p_n^2 < \infty$. Then the limits*

$$N(H_\omega, E) := \lim_{L \to \infty} \frac{1}{P_L} N(H_{\omega,L}^D, E)$$

and

$$\tilde{N}(H_\omega, E) := \lim_{L \to \infty} \frac{1}{P_L} N(H_{\omega,L}^N, E)$$

exist for $E < 0$ and are non-random. Moreover

$$\tilde{N}(H_\omega, E) = N(H_\omega, E) = N(H_0 + f, E)$$

at every point of continuity of $N(H_0 + f, E)$ with $E < 0$.

An analogues proof works for **Model II** operators and gives:

Theorem 7. *Let V_ω be a sparse random potential as in **Model II** with $\sum p_n = \infty$ and $\sum p_n^2 < \infty$. Then the limits*

$$N(H_\omega, E) := \lim_{L \to \infty} \frac{1}{P_L} N(H_{\omega,L}^D, E)$$

and

$$\tilde{N}(H_\omega, E) := \lim_{L \to \infty} \frac{1}{P_L} N(H_{\omega,L}^N, E)$$

exist for $E < 0$ and are non-random. Moreover

$$\tilde{N}(H_\omega, E) = N(H_\omega, E) = \mathbb{E}\left[N(H_0 + q_0(\omega)f, E)\right]$$

at every point of continuity of $\mathbb{E}\left[N(H_0 + q_0(\omega)f, E)\right]$ with $E < 0$.

Remarks 8. Since $\sum p_n^2 < \infty$, by a Borel-Cantelli argument the essential set for H_ω consists of only one point. So the theorem tells us, that the modified integrated density of states "sees" exactly the essential spectrum below zero which is located at the eigenvalues of the model operator $H_0 + f$.

Since these operators are not ergodic in any sense, one has to look for a substitute for the ergodic theorems used to proof the existence and non-randomness of the integrated density of states. In [6] a strong law of large numbers has been proven to cover also these cases of sparse random potentials.

For details on the integrated density of states for sparse random potentials see [6].

4 Random Surface Models

Consider the following self-adjoint random operator in $L^2(\mathbb{R}^d)$ or $\ell^2(\mathbb{Z}^d)$, $\mathbb{R}^d = \mathbb{R}^m \times \mathbb{R}^{d-m}$:

$$H(\omega) = -\Delta + V_\omega,$$

where

$$V_\omega(x) = \sum_{k \in \mathbb{Z}^m} q_k(\omega) f(x - (k, 0)),$$

the q_k are i.i.d. random variables and $f \geq 0$ is a single site potential that satisfy certain technical assumptions. This leads to the following geometry characterizing random surface models. Sometimes the upper half plane is considered only.

There is a lot of literature, mostly on the discrete case, using a decomposition into a bulk and a surface term see [5, 19, 22, 48, 52, 53, 56, 55, 57, 58].

The moral of the story is the appearance of a metal insulator transition at the edges of the unperturbed operator. We now concentrate on the continuum case, where we only know of [17, 51] as references. The existence of an a.c. component is proven in [51]. In the following, we present the result from [17], giving strong dynamical localization. Similar but somewhat different results have been announced in [51]. As discussed there, an additional Dirichlet boundary condition "stabilizes" the spectrum so that the appearance of negative spectrum requires a certain strength of the random perturbation. Therefore, proving localization at negative energies is easier (compared to the case without Dirichlet boundary conditions) since one is automatically dealing with a "large coupling" regime.

It is not hard to see that

$$\sigma(H(\omega)) = [E_0, \infty) \text{ where } E_0 = \inf \sigma(-\Delta + q_{\min} \cdot f^{\mathrm{per}}),$$

and

$$f^{\mathrm{per}} = \sum_{k \in \mathbb{Z}^m} f(x - (k, 0))$$

denotes the periodic continuation of f along the surface. Near the bottom of the spectrum E_0 one expects localization, i.e. suppression of transport as is typical for insulators. For nonnegative energies one expects extended states. To stress the existence of a metallic phase let us cite Theorem 4.3 of [51]

Theorem 9. *Let $H(\omega)$ be as below. Then we have, for every $\omega \in \Omega$: $[0, \infty) \subset \sigma_{\mathrm{ac}}(H(\omega))$.*

The idea of the *Proof* is that a wave packet with velocity pointing away from the surface will escape the influence of the surface potential and is asymptotically free. The rigorous implementation of this idea uses Enss' technique from scattering theory.

The model. 1. $0 < m < d$ and points in $\mathbb{R}^d = \mathbb{R}^m \times \mathbb{R}^{d-m}$ are written as pairs, if convenient;

2. The single site potential $f \geq 0$, $f \in L^p(\mathbb{R}^d)$ where $p \geq 2$ if $d \leq 3$ and $p > d/2$ if $d > 3$, and $f \geq \sigma > 0$ on some open set $U \neq \emptyset$ for some $\sigma > 0$.

3. The q_k are i.i.d. random variables distributed with respect to a probability measure μ on $\mathbb{R}$, such that $\operatorname{supp} \mu = [q_{\min}, 0]$ with $q_{\min} < 0$.

We will sometimes need further assumptions on the single site distribution μ:

4. μ is *Hölder continuous*, i.e. there are constants $C, \alpha > 0$ such that

$$\mu[a, b] \leq C(b - a)^{\alpha} \text{ for } q_{\min} \leq a \leq b \leq 0.$$

5. *Disorder assumption*: there exist $C, \tau > 0$ such that

$$\mu[q_{\min}, q_{\min} + \varepsilon] \leq C \cdot \varepsilon^{\tau} \text{ for } \varepsilon > 0.$$

What follows is the main result of [17].

Theorem 10. *Let $H(\omega)$ be as above with $\tau > d/2$ and assume that $E_0 < 0$.*

(a) There exists an $\varepsilon > 0$ such that in $[E_0, E_0 + \varepsilon]$ the spectrum of $H(\omega)$ is pure point for almost every $\omega \in \Omega$, with exponentially decaying eigenfunctions.

(b) Assume that $p < 2(2\tau - m)$. Then there exists an $\varepsilon > 0$ such that in $[E_0, E_0 + \varepsilon] = I$ we have strong dynamical localization in the sense that for every compact set $K \subset \mathbb{R}^d$:

$$\mathbb{E}\{\sup_{t>0} \||X|^p e^{-itH(\omega)} P_I(H(\omega)) \chi_K\|\} < \infty$$

A consequence is pure point spectrum in the interval $[E_0, E_0 + \varepsilon] = I$. Together with the previous result on extended states we get the picture from Figure 4 that still leaves open some important questions.

5 The Density of Surface States

For random surface models it is also possible to define an appropriate analogue of the density of states, the density of surface states (see [33, 73, 74]).

Since one has to get rid of the bulk spectrum to recover properties of the surface potential (see [33]), one defines the density of surface states as the following limit.

$$\nu^S[f] := \lim_{L \to \infty} \frac{1}{L^m} \operatorname{tr}\{\chi_L (f(H_\omega) - f(H_0))\}$$

for sufficiently smooth functions f, where χ_L is the characteristic function of a cube Λ_L. In [33] it was proven, that one needs $f \in C_0^3(\mathbb{R})$ at most. [73, 74] showed, that one needs $f \in C_0^1(\mathbb{R})$, but they used sign definite surface potentials.

This definition doesn't make use of boundary conditions, because one might guess, that the introduction of boundary conditions may have an effect on the density of surface states since they are at least of the same order as the perturbation by the potential.

So it may be surprising that it is also possible to define the density of surface states as the limit

$$\nu^{S,D}[f] := \lim_{L \to \infty} \frac{1}{L^m} \operatorname{tr}\{f(H_{\omega,L}^D) - f(H_0^D)\}$$

with Dirichlet boundary conditions or as the limit

$$\nu^{S,N}[f] := \lim_{L \to \infty} \frac{1}{L^m} \operatorname{tr}\{f(H_{\omega,L}^N) - f(H_0^N)\}$$

with Neumann boundary conditions. In [6] it has been proved, that for a wider class of random surface potentials (that allows also for sign indefinite random potentials) all these definitions give the same limit. This limit is also non-random and gives a distribution of order 1 (in fact it is the distributional derivative of a signed measure).

To prove this we strongly used Feynman-Kac representations of the Laplace transforms of the finite volume measures and showed that the difference of these Laplace transforms converge to 0 as $L \to \infty$. Together with the regularity of the density of surface states this gives also their independence of boundary conditions.

Using this Laplace transforms it was also possible to calculate the asymptotic behaviour of the integrated density of surface states for random Gaussian surface potentials (for details see [6]).

6 Lifshitz Tails & Localization

Below the energy $E = 0$, the bottom of the spectrum of the unperturbed (free) operator the density of surface states is easily seen to be a positive Borel measure. Hence, we may speak of its distribution function

$$N_S(E) = \nu^s \left((-\infty, E) \right) \quad for \quad E < 0.$$

Very recently, it was proved in [66] that $N_S(E)$ shows a very characteristic decay at E_0, the bottom of the spectrum. This behavior, known as Lifshitz behavior in the case of stationary random potentials, is given by:

$$N_S(E) \sim e^{-c(E-E_0)^{-m/2}} \qquad (*)$$

as $E \searrow E_0$. This result is then used in [66] to prove Anderson localization for surface models without assuming the disorder assumption (5). [66] modify the proof in [17] replacing (5) by estimates that follow from $(*)$.

References

1. M.Aizenman and S.A.Molchanov: Localization at large disorder and at extreme energies: An elementary derivation. *Commun. Math. Phys.*, **157**, 245–278 (1993)
2. M.Aizenman and G.M.Graf: Localization bounds for an electron gas. *J. Phys. A: Math. Gen.* **31**, 6783 – 6806 (1998)
3. M.Aizenman, A. Elgart, S. Naboko, J.H. Schenker, G. Stolz: Moment Analysis for Localization in Random Schroedinger Operators, `eprint`, `arXiv`, `math-ph/0308023`
4. P.W. Anderson: Absence of diffusion in certain random lattices. *Phys. Rev.* **109**, 1492–1505 (1958)
5. Bentosela, F.; Briet, Ph.; Pastur, L.: On the spectral and wave propagation properties of the surface Maryland model. *J. Math. Phys.* **44** (2003), no. 1, 1–35.

6. Böcker, St.: Zur integrierten Zustandsdichte von Schrödingeroperatoren mit zufälligen, inhomogenen Potentialen. *Doctoral thesis* Ruhr-Universität Bochum (2003).

7. J. Bourgain, M. Goldstein: On nonperturbative localization with quasiperiodic potential. *Ann. of Math.*, **152**, 835–879 (2000).

8. J. Bourgain, On the spectrum of lattice Schrödinger operators with deterministic potential. II. Dedicated to the memory of Tom Wolff. *J. Anal. Math.* **88** (2002), 221–254;

9. J. Bourgain, M. Goldstein and W. Schlag, Anderson localization for Schrödinger operators on Z^{*} with quasi-periodic potential. *Acta Math.* **188** (2002), no. 1, 41–86;

10. J. Bourgain, New results on the spectrum of lattice Schrödinger operators and applications. In *Mathematical results in quantum mechanics (Taxco, 2001)*, 27–38, Contemp. Math., 307, Amer. Math. Soc., Providence, RI, 2002;

11. J. Bourgain, On the spectrum of lattice Schrödinger operators with deterministic potential. Dedicated to the memory of Thomas H. Wolff. *J. Anal. Math.* **87** (2002), 37–75;

12. J. Bourgain, Exposants de Lyapounov pour opérateurs de Schrödinger discrètes quasi-périodiques. *C. R. Math. Acad. Sci. Paris* **335** (2002), no. 6, 529–531;

13. J. Bourgain and S. Jitomirskaya, Continuity of the Lyapunov exponent for quasiperiodic operators with analytic potential. Dedicated to David Ruelle and Yasha Sinai on the occasion of their 65th birthdays. *J. Statist. Phys.* **108** (2002), no. 5-6, 1203–1218

14. J. Bourgain and S. Jitomirskaya, Absolutely continuous spectrum for 1D quasiperiodic operators. *Invent. Math.* **148** (2002), no. 3, 453–463;

15. J. Bourgain, On random Schrödinger operators on Z^{*}. *Discrete Contin. Dyn. Syst.* **8** (2002), no. 1, 1–15;

16. J. Bourgain, M. Goldstein and W. Schlag, Anderson localization for Schrödinger operators on Z with potentials given by the skew-shift. *Comm. Math. Phys.* **220** (2001), no. 3, 583–621.

17. A. Boutet de Monvel, P. Stollmann: Dynamical localization for continuum random surface models, *Arch. Math.*, **80**, 87–97 (2003)

18. A. Boutet de Monvel, P. Stollmann and G. Stolz: Absence of (absolutely) continuous spectrum for certain nonstationary random models. In preparation

19. A.Boutet de Monvel, A. Surkova: Localisation des états de surface pour une classe d'opérateurs de Schrödinger discrets à potentiels de surface quasi-périodiques, *Helv. Phys. Acta* **71**(5), 459–490 (1998).

20. R.Carmona and J.Lacroix: *Spectral Theory of Random Schrödinger Operators*. Birkhäuser Boston, 1990

21. R.Carmona and A.Klein and F.Martinelli: Anderson Localization for Bernoulli and Other Singular Potentials. *Commun. Math. Phys.*, **108**, 41–66 (1987)

22. Chahrour, A. and Sahbani, J.: On the spectral and scattering theory of the Schrödinger operator with surface potential. *Rev. Math. Phys.* **12** (2000), no. 4, 561–573

23. J.M.Combes and P.D.Hislop: Localization for some continuous, random Hamiltonians in d-dimensions, *J. Funct. Anal.*, **124**, 149–180 (1994)

24. J.M.Combes, and P.D.Hislop: Landau Hamiltonians with random potentials: Localization and density of states. *Commun. Math. Phys.*, **177**, 603–630 (1996)

25. D.Damanik and P.Stollmann: Multi-scale analysis implies strong dynamical localization. GAFA, *Geom. funct. anal.* **11**, 11–29 (2001)

26. R.del Rio, S.Jitomirskaya, Y.Last and B.Simon: What is localization? *Phys. Rev. Letters* **75**, 117–119 (1995)

27. R.del Rio, S.Jitomirskaya, Y.Last and B.Simon: Operators with singular continuous spectrum IV: Hausdorff dimensions, rank one perturbations and localization. *J. d'Analyse Math.* **69**, 153–200 (1996)

28. F.Delyon, Y.Levy and B.Souillard: Anderson localization for multidimensional systems at large disorder or low energy. *Commun. Math. Phys.* **100**, 463–470 (1985)

29. F.Delyon, Y.Levy and B.Souillard: Approach à la Borland to multidimensional localisation. *Phys. Rev. Lett.* **55**, 618–621 (1985)

30. T.C.Dorlas, N.Macris and J.V.Pulé: Localization in a single-band approximation to random Schrödinger operators in a magnetic field. *Helv. Phys. Acta*, **68**, 329–364 (1995)

31. H.von Dreifus: On the effects of randomness in ferromagnetic models and Schrödinger operators. NYU, Ph. D. Thesis (1987)

32. H.von Dreifus and A.Klein: A new proof of localization in the Anderson tight binding model. *Commun. Math. Phys.* **124**, 285–299 (1989)

33. H. Englisch, W. Kirsch, M. Schröder, B. Simon: Random Hamiltonians Ergodic in All But One Direction. *Comm. Math. Phys.* **128** (1990)

34. Faris, W.G. Random waves and localization. *Notices Amer. Math. Soc.* **42** (1995), no. 8, 848–853.

35. A.Figotin and A.Klein: Localization of classical waves I: Acoustic waves. *Commun. Math. Phys.* **180**, 439–482 (1996)

36. W.Fischer, H.Leschke and P.Müller: Spectral localization by Gaussian random potentials in multi-dimensional continuous space. *J. Stat. Phys.* **101**, 935–985 (2000)

37. J.Fröhlich: Les Houches lecture. In: *Critical Phenomena, Random Systems, Gauge Theories,* K. Osterwalder and R. Stora, eds. North-Holland, 1984

38. J.Fröhlich, F.Martinelli, E.Scoppola and T.Spencer: Constructive Proof of Localization in the Anderson Tight Binding Model. *Commun. Math. Phys.* **101**, 21–46 (1985)

39. J.Fröhlich and T.Spencer: Absence of Diffusion in the Anderson Tight Binding Model for Large Disorder or Low Energy. *Commun. Math. Phys.* **88**, 151–184 (1983)

40. F.Germinet: Dynamical Localization II with an Application to the Almost Mathieu Operator. *J. Statist. Phys.* **98** (2000), no. 5-6, 1135–1148.

41. F.Germinet and S.de Bievre: Dynamical Localization for Discrete and Continuous Random Schrödinger operators. *Commun. Math. Phys.* **194**, 323–341 (1998)

42. F.Germinet and S.Jitomirskaya: Strong dynamical localization for the almost Mathieu model. *Rev. Math. Phys.* **13** (2001), no. 6, 755–765

43. F.Germinet and A.Klein: Bootstrap multiscale analysis and localization in random media. *Comm. Math. Phys.* **222** (2001), no. 2, 415–448;

44. F.Germinet and H.Klein: A characterization of the Anderson metal-insulator transport transition, submitted.

45. I.Ya.Goldsheidt, S.A.Molchanov and L.A.Pastur: Typical one-dimensional Schrödinger operator has pure point spectrum, *Funktsional. Anal. i Prilozhen.* **11** (1), 1–10 (1977); Engl. transl. in *Functional Anal. Appl.* **11** (1977)

46. Goldstein, M. Schlag, W.: Hölder continuity of the integrated density of states for quasi-periodic Schrödinger equations and averages of shifts of subharmonic functions. *Ann. of Math.* (2) **154** (2001), no. 1, 155–203.

47. G.M.Graf: Anderson localization and the space-time characteristic of continuum states. *J. Stat. Phys.* **75**, 337–343 (1994)

48. V. Grinshpun: Localization for random potentials supported on a subspace, *Lett. Math. Phys.* **34**(2), 103–117 (1995).

49. H.Holden and F.Martinelli: On the absence of diffusion for a Schrödinger operator on $L^\cdot(R^\nu)$ with a random potential. *Commun. Math. Phys.* **93**, 197–217 (1984)

50. D.Hundertmark: On the time-dependent approach to Anderson localization. *Math. Nachrichten*, **214**, 25–38 (2000)

51. D. Hundertmark, W. Kirsch: Spectral theory of sparse potentials, in "Stochastic processes, physics and geometry: new interplays, I (Leipzig, 1999)," Amer. Math. Soc., Providence, RI, 2000, pp. 213–238.

52. V. Jakšić, Y. Last: Corrugated surfaces and a.c. spectrum, *Rev. Math. Phys.* **12**(11), 1465–1503 (2000).

53. V. Jakšić, Y. Last: Spectral structure of Anderson type hamiltonians, *Inv. Math.*(2000) **141** no. 3, 561–577

54. I. McGillivray, P. Stollmann and G. Stolz: Absence of absolutely continuous spectra for multidimensional Schrödinger operators with high barriers. *Bull. London Math. Soc.* **27** (1995), no. 2, 162–168.

55. V. Jakšić, S. Molchanov: On the surface spectrum in dimension two, *Helv. Phys. Acta* **71**(6), 629–657 (1998).

56. V. Jakšić, S. Molchanov: On the spectrum of the surface Maryland model, *Lett. Math. Phys.* **45**(3), 189–193 (1998).

57. V. Jakšić, S. Molchanov: Localization of surface spectra, *Comm. Math. Phys.* **208**(1), 153–172 (1999).

58. V. Jakšić, S. Molchanov, L. Pastur: On the propagation properties of surface waves, in "Wave propagation in complex media (Minneapolis, MN, 1994)," IMA Math. Appl., Vol. **96**, Springer, New York, 1998, pp. 143–154.

59. S.Jitomirskaya: Almost everything about the almost Mathieu operator II. In: *Proc. of the XIth International Congress of Mathematical Physics, Paris 1994*, Int. Press, 366–372 (1995)

60. S.Jitomirskaya and Y.Last: Anderson Localization for the Almost Mathieu Operator III. Semi-Uniform Localization, Continuity of Gaps and Measure of the Spectrum. *Commun. Math. Phys.* **195**, 1–14 (1998)

61. S.Jitomirskaya: Metal-Insulator transition for the Almost Mathieu operator. *Annals of Math.*, **150**, 1159–1175 (1999)

62. S.John: The Localization of Light and Other Classical Waves in Disordered Media. *Comm. Cond. Mat. Phys.* **14**(4), 193–230 (1988)

63. A. Klein: Extended states in the Anderson model on the Bethe lattice. *Adv. Math.* **133** (1998), no. 1, 163–184;

64. W.Kirsch: Wegner estimates and Anderson localization for alloy-type potentials. *Math. Z.* **221**, 507–512 (1996)

65. W.Kirsch: Scattering theory for sparse random potentials. *Random Oper. Stochastic Equations* **10** (2002), no. 4, 329–334.

66. W. Kirsch, S. Warzel: Lifshitz tails and Anderson localization for surface random potentials, in preparation.

67. W.Kirsch, M.Krishna and J.Obermeit: Anderson model with decaying randomness: Mobility edge, *Math. Z.*, **235**, 421–433 (2000)

68. W. Kirsch, P. Stollmann, G. Stolz: Localization for random perturbations of periodic Schrödinger operators, *Random Oper. Stochastic Equations* **6**(3), 241–268 (1998).

69. W. Kirsch, P. Stollmann, G. Stolz: Anderson localization for random Schrödinger operators with long range Interactions, *Comm. Math. Phys.* **195**(3), 495–507 (1998).

70. F.Kleespies and P.Stollmann: Localization and Lifshitz tails for random quantum waveguides, *Rev. Math. Phys.* **12**, No. 10, 1345–1365 (2000)

71. F.Klopp: Localization for semiclassical Schrödinger operators II: The random displacement model. *Helv. Phys. Acta* **66**, 810–841 (1993)

72. F.Klopp: Localization for some continuous random Schrödinger operators. *Commun. Math. Phys.* **167**, 553 – 569 (1995)

73. V. Kostrykin, R. Schrader: The density of states and the spectral shift density of random Schrdinger operators. *Rev. Math. Phys.* **12** (2000)

74. V. Kostrykin, R. Schrader: Regularity of the surface density of states. *J. Func. Anal.* **187** (2001)

75. M. Krishna: Anderson model with decaying randomness: Existence of extended states. *Proc. Indian Acad. Sci. Math. Sci.* **100**, no. 3, 285–294

76. M. Krishna: Absolutely continuous spectrum for sparse potentials. *Proc. Indian Acad. Sci. Math. Sci.* **103** (1993), no. 3, 333–339

77. Krishna, M.; Sinha, K. B. Spectra of Anderson type models with decaying randomness. *Proc. Indian Acad. Sci. Math. Sci.* **111** (2001), no. 2, 179–201

78. Y.Last: Almost everything about the almost Mathieu operator I. In: *Proc. of the XIth International Congress of Mathematical Physics,* Paris 1994, Int. Press, 373–382 (1995)

79. F.Martinelli and H.Holden: On absence of diffusion near the bottom of the spectrum for a random Schrödinger operator on $L^{\cdot}(\mathbb{R}^{\nu})$. *Commun. Math. Phys.* **93**, 197–217 (1984)

80. F.Martinelli and E.Scoppola: Introduction to the mathematical theory of Anderson localization. *Rivista Nuovo Cimento* **10**, 10 (1987)

81. F.Martinelli and E.Scoppola: Remark on the absence of the absolutely continuous spectrum for d-dimensional Schrödinger operator with random potential for large disorder or low energy. *Commun. Math. Phys.* **97**, 465–471 (1985), Erratum ibid. **98**, 579 (1985)

82. Molchanov, S. Multiscattering on sparse bumps. Advances in differential equations and mathematical physics (Atlanta, GA, 1997), 157–181, Contemp. Math., 217, Amer. Math. Soc., Providence, RI, 1998.

83. Molchanov, S.; Vainberg, B.: Multiscattering by sparse scatterers. Mathematical and numerical aspects of wave propagation (Santiago de Compostela, 2000), 518–522, SIAM, Philadelphia, PA, 2000.

84. N.F.Mott and W.D.Twose: The theory of impurity conduction. *Adv. Phys.* **10**, 107–163 (1961)

85. L.A.Pastur and A.Figotin: *Spectra of Random and Almost-Periodic Operators.* Springer-Verlag, Berlin, 1992

86. D. Pearson: Singular Continuous Measures in Scattering Theory. *Commun. Math. Phys.* **60**, 13–36 (1976)

87. Schlag, W.; Shubin, C.; Wolff, T.: Frequency concentration and location lengths for the Anderson model at small disorders. Dedicated to the memory of Tom Wolff. *J. Anal. Math.* 88 (2002), 173–220.

88. C.Shubin, R.Vakilian and T.Wolff: Some harmonic analysis questions suggested by Anderson–Bernoulli models. *GAFA* **8**, 932–964 (1998)

89. B.Simon and T.Wolff: Singular continuous spectrum under rank one perturbations and localization for random Hamiltonians. *Comm. Pure Appl. Math.* **39**, 75–90 (1986)

90. R.Sims and G.Stolz: Localization in one dimensional random media: a scattering theoretic approach. *Comm. Math. Phys.* **213**, 575–597 (2000)

91. T.Spencer: The Schrödinger equation with a random potential – a mathematical review. In: *Critical Phenomena, Random Systems, Gauge Theories*, K. Osterwalder and R. Stora, eds. North Holland 1984

92. T.Spencer: Localization for random and quasi-periodic potentials. *J. Stat. Phys.* **51**, 1009 (1988)

93. P. Stollmann: *Caught by disorder – bound states in random media*, Progress in Math. Phys., vol. 20, Birkhäuser, Boston, 2001

94. P. Stollmann: Wegner estimates and localization for continuum Anderson models with some singular distributions, *Arch. Math.* **75**(4), 307–311 (2000)

95. P. Stollmann and G. Stolz: Singular spectrum for multidimensional operators with potential barriers. *J. Operator Theory* **32**, 91–109 (1994)

96. G. Stolz: Localization for Schrödinger operators with effective barriers. *J. Funct. Anal.* **146** (1997), no. 2, 416–429.

97. G.Stolz: Localization for random Schrödinger operators with Poisson potential. *Ann. Inst. Henri Poincaré* **63**, 297–314 (1995)

98. Thouless, D. J.: Introduction to disordered systems. Phénomènes critiques, systèmes aléatoires, théories de jauge, Part I, II (Les Houches, 1984), 681–722, North-Holland, Amsterdam, 1986.

99. I.Veselic: Localisation for random perturbations of periodic Schrödinger operators with regular Floquet eigenvalues. *Ann. Henri Poincare* **3** (2002), no. 2, 389–409.

100. W.M.Wang: Microlocalization, Percolation, and Anderson Localization for the Magnetic Schrödinger operator with a Random Potential. *J. Funct. Anal.* **146**, 1–26 (1997)

101. H.Zenk: Anderson localization for a multidimensional model including long range potentials and displacements. *Rev. Math. Phys.* **14** (2002), no. 3, 273–302

A Survey of Rigorous Results on Random Schrödinger Operators for Amorphous Solids[*]

Hajo Leschke[1], Peter Müller[2,3], and Simone Warzel[4,5]

· Institut für Theoretische Physik, Universität Erlangen-Nürnberg
  Staudtstraße 7, D–91058 Erlangen, Germany
  hajo.leschke@physik.uni-erlangen.de
· Institut für Theoretische Physik, Georg-August-Universität Göttingen
  Tammannstraße 1, D–37077 Göttingen, Germany
  peter.mueller@physik.uni-goettingen.de
· Present address: Fakultät und Institut für Mathematik
  Ruhr-Universität Bochum
  Universitätsstraße 150, D–44780 Bochum, Germany
  peter.mueller@mathphys.rub.de
· Institut für Theoretische Physik, Universität Erlangen-Nürnberg
  Staudtstraße 7, D–91058 Erlangen, Germany
  simone.warzel@physik.uni-erlangen.de
· Present address: Princeton University
  Jadwin Hall, Princeton, NJ 08544, USA
  swarzel@princeton.edu

Summary. Electronic properties of amorphous or non-crystalline disordered solids are often modelled by one-particle Schrödinger operators with random potentials which are ergodic with respect to the full group of Euclidean translations. We give a short, reasonably self-contained survey of rigorous results on such operators, where we allow for the presence of a constant magnetic field. We compile robust properties of the integrated density of states like its self-averaging, uniqueness and leading high-energy growth. Results on its leading low-energy fall-off, that is, on its Lifshits tail, are then discussed in case of Gaussian and non-negative Poissonian random potentials. In the Gaussian case with a continuous and non-negative covariance function we point out that the integrated density of states is locally Lipschitz continuous and present explicit upper bounds on its derivative, the density of states. Available results on Anderson localization concern the almost-sure pure-point nature of the low-energy spectrum in case of certain Gaussian random potentials for arbitrary space dimension. Moreover, under slightly stronger conditions all absolute spatial moments of an initially localized wave packet in the pure-point spectral subspace remain almost surely finite for all times. In case of one dimension and a Poissonian random potential with repulsive impurities of finite range, it is known that the whole energy spectrum is almost surely only pure point.

---

[*] A shorter and less up-to-date version appeared in: Markov Process. Relat Fields, **9**, 729–760 (2003)

1 Introduction

Over the last three decades a considerable amount of rigorous results on random Schrödinger operators have been achieved by many researchers. Good general overviews of such results can be found in the review articles [127, 104, 73, 115, 133, 140] and the monographs [33, 29, 116, 131]. We also recommend these sources for the general background in the field.

Most works concern Schrödinger operators with random potentials that possess an underlying lattice structure even if they are defined on continuous space. Such operators are suitable to model *random alloys*. As against that, the present survey aims to collect rigorous results on one-particle Schrödinger operators with (and only with) *truly continuum* random potentials modelling *amorphous solids*. For the sake of simplicity we will refrain from stating these results under the weakest assumptions available for their validity, but only provide sufficient conditions which are easy to check. For weaker assumptions and related slightly stronger results the interested reader is referred to the original cited works where also the corresponding proofs can be found, which will be omitted here. As far as results by the present authors are mentioned, they have been obtained in collaboration with Jean-Marie Barbaroux, Markus Böhm, Kurt Broderix (1962–2000), Werner Fischer, Nils Heldt, Dirk Hundertmark, Thomas Hupfer and/or Werner Kirsch.

1.1 Motivation and Models

Almost half a century after Anderson's pioneering paper [9], one-particle Schrödinger operators with random potentials continue to play a prominent rôle for understanding electronic properties of disordered solids. While perfect solids or crystals are characterized by the periodic arrangement of identical atoms (or ions) on the sites of a lattice, disordered solids lack any kind of long-range order, but may exhibit some vestige of short-range order. In some disordered solids, like in random alloys, an underlying lattice structure may still exist so that on average the solid remains homogeneous with respect to lattice translations. Accordingly, the random potentials employed to model such solids should be lattice-homogeneous, or even lattice-ergodic, in order to also take into account that different parts of the solid are practically decoupled at large separation [21, 99]. In extremely disordered materials, like liquids, glasses or amorphous solids [149, 81, 59, 86], an underlying lattice structure is no longer available and the corresponding random potentials should even be ergodic with respect to the group of *all* Euclidean translations, not only lattice ones.

The present survey is concerned with one-particle Schrödinger operators with random potentials modelling amorphous solids. More generally, we will consider a single quantum particle in d-dimensional Euclidean configuration space $\mathbb{R}^d$, $d \in \mathbb{N}$, subject to a random potential in the sense of

Definition 1.1. *By a* random potential *we mean a* random scalar field *V : $\Omega \times \mathbb{R}^d \to \mathbb{R}$, $x \mapsto V^{(\omega)}(x)$ on a complete probability space $(\Omega, \mathcal{A}, \mathbb{P})$, such that V is jointly measurable with respect to the product of the sigma-algebra $\mathcal{A}$ of (event) subsets of Ω and the sigma-algebra $\mathcal{B}(\mathbb{R}^d)$ of Borel sets in $\mathbb{R}^d$. We further suppose that V is $\mathbb{R}^d$-ergodic and that it has one of the following two properties:*

(V−)   *V has a finite pth absolute moment, that is, $\mathbb{E}\big[\,|V(0)|^p\,\big] < \infty$ with some real $p > \max\{3, d+1\}$.*

(V+)   *V is non-negative and has a finite pth moment, that is, $\mathbb{E}\big[V(0)^p\big] < \infty$ with $p = 2$ if $d \in \{1, 2, 3\}$ or with some real $p > d/2$ if $d \geq 4$.*

Here $\mathbb{E}[\cdot] := \int_\Omega \mathbb{P}(\mathrm{d}\omega)\,(\cdot)$ denotes the expectation functional *(in other words: ensemble averaging) induced by the probability measure $\mathbb{P}$.*

The precise definition of $\mathbb{R}^d$-*ergodicity* of V requires [88, 73] the existence of a group $\{T_x\}_{x \in \mathbb{R}^d}$ of probability-preserving transformations on $(\Omega, \mathcal{A}, \mathbb{P})$ such that (i) the group is ergodic in the sense that every event $\mathcal{E} \in \mathcal{A}$ which is invariant under the whole group is either almost impossible or almost sure, $\mathbb{P}(\mathcal{E}) \in \{0, 1\}$, and that (ii) V is $\mathbb{R}^d$-*homogeneous* in the sense that $V^{(T_x \omega)}(y) = V^{(\omega)}(y - x)$ for (Lebesgue-) almost all x, $y \in \mathbb{R}^d$ and all $\omega \in \Omega$. Roughly speaking, V is $\mathbb{R}^d$-ergodic if its fluctuations in different regions become sufficiently fast decorrelated with increasing distance between the regions.

The quantum particle subject to the random potential may also be exposed to a *constant magnetic field* characterized by a skew-symmetric $d \times d$-matrix with real entries $B_{ij} = -B_{ji}$, where $i, j \in \{1, \ldots, d\}$. The components of the corresponding vector potential $A : \mathbb{R}^d \to \mathbb{R}^d$ in the Poincaré gauge are defined by $A_j(x) := \frac{1}{2}\sum_{i=1}^d x_i B_{ij}$ for all $x = (x_1, \ldots, x_d) \in \mathbb{R}^d$.

Choosing physical units such that the mass and electric charge of the particle as well as Planck's constant (divided by 2π) are all equal to one, the Schrödinger operator for the quantum particle subject to a *realization* $V^{(\omega)} : \mathbb{R}^d \to \mathbb{R}$, $x \mapsto V^{(\omega)}(x)$ of a random potential V and a constant magnetic field is informally given by the differential expression

$$H\big(A, V^{(\omega)}\big) := \frac{1}{2}\sum_{j=1}^d \left(\mathrm{i}\,\frac{\partial}{\partial x_j} + A_j\right)^2 + V^{(\omega)}, \qquad (1.1)$$

where $\mathrm{i} = \sqrt{-1}$ is the imaginary unit. According to the basic postulates of quantum mechanics, this expression has to be well defined as a self-adjoint operator acting on a dense domain in $\mathrm{L}^2(\mathbb{R}^d)$, the Hilbert space of all complex-valued, Lebesgue square-integrable functions on $\mathbb{R}^d$, which is equipped with the usual scalar product $\langle \varphi, \psi \rangle := \int_{\mathbb{R}^d} \mathrm{d}x\,\varphi(x)^* \psi(x)$, for φ, $\psi \in \mathrm{L}^2(\mathbb{R}^d)$. In fact, our assumptions on V guarantee the existence of some subset $\Omega_0 \in \mathcal{A}$ of Ω with full probability, $\mathbb{P}(\Omega_0) = 1$, such that for every $\omega \in \Omega_0$ the right-hand side of (1.1) is essentially self-adjoint on the dense subspace $\mathcal{C}_0^\infty(\mathbb{R}^d) \subset \mathrm{L}^2(\mathbb{R}^d)$

of all complex-valued, arbitrarily often differentiable functions with compact supports, see for example [76, 73, 29, 51]. This justifies

Definition 1.2. *By a random Schrödinger operator $H(A, V)$ with a random potential V and a constant magnetic field, corresponding to the vector potential A, we mean the family $\Omega_0 \ni \omega \mapsto H(A, V^{(\omega)})$ of Schrödinger operators given by (1.1).*

In this survey we basically focus on two examples of random potentials in the sense of Definition 1.1, namely Gaussian and non-negative Poissonian ones. Both are rather popular in the physics literature [150, 21, 123, 99, 71, 44], see also [87] and references therein.

Definition 1.3. *By a Gaussian random potential we mean a Gaussian random field [1, 100] which is $\mathbb{R}^d$-ergodic. It has zero mean, $\mathbb{E}[V(0)] = 0$, and its covariance function $C : \mathbb{R}^d \to \mathbb{R}$, $x \mapsto C(x) := \mathbb{E}[V(x)V(0)]$ is continuous at the origin where it obeys $0 < C(0) < \infty$.*

The covariance function C of a Gaussian random potential is bounded and uniformly continuous on $\mathbb{R}^d$ by definition. Consequently, [48, Thm. 3.2.2] implies the existence of a separable version of this field which is jointly measurable. Referring to a Gaussian random potential, we will tacitly assume that only this version will be dealt with. By the Bochner-Khinchin theorem there is a one-to-one correspondence between Gaussian random potentials and finite positive (and even) Borel measures on $\mathbb{R}^d$. A simple sufficient criterion ensuring $\mathbb{R}^d$-ergodicity is the mixing condition $\lim_{|x| \to \infty} C(x) = 0$. Furthermore, the explicit formula

$$
\begin{aligned}
&\mathbb{E}\big[|V(0)|^p\big] \\
&= (2\pi C(0))^{-1/2} \int_{\mathbb{R}} dv \, e^{-v^2/2C(0)} \, |v|^p \\
&= \Gamma\left(\frac{p+1}{2}\right) \frac{[2C(0)]^{p/2}}{\pi^{1/2}},
\end{aligned}
\tag{1.2}
$$

where Γ stands for Euler's gamma function [60], shows that a Gaussian random potential has property (V–) and is therefore a random potential in the sense of Definition 1.1.

The second example of a random potential considered subsequently is a non-negative Poissonian one, which we define as follows.

Definition 1.4. *By a non-negative Poissonian random potential with single-impurity potential $U \geq 0$ and mean concentration $\varrho > 0$ we mean a random field with realizations given by*

$$
V^{(\omega)}(x) = \int_{\mathbb{R}^d} \mu_\varrho^{(\omega)}(dy) \, U(x - y).
\tag{1.3}
$$

Here μ_ϱ denotes the Poissonian (random) measure on $\mathbb{R}^d$ with intensity parameter $\varrho > 0$ and the (shape) function $U : \mathbb{R}^d \to [0, \infty[$ is supposed to be non-negative and strictly positive on some non-empty open set in $\mathbb{R}^d$. Moreover, we assume that U satisfies the Birman-Solomyak condition $\sum_{m \in \mathbb{Z}^d} \left(\int_{\Lambda_m} dx \, |U(x)|^p \right)^{1/p} < \infty$ with $p = 2$ if $d \in \{1, 2, 3\}$ or with some real $p > d/2$ if $d \geq 4$. Here Λ_m denotes the unit cell of the d-dimensional simple cubic lattice $\mathbb{Z}^d \subset \mathbb{R}^d$ which is centred at the site $m \in \mathbb{Z}^d$.

Since the Poissonian measure is a random Borel measure which is only pure point and positive integer-valued, each realization of a non-negative Poissonian random potential is informally given by

$$V^{(\omega)}(x) = \sum_j U\left(x - p_j^{(\omega)}\right). \tag{1.4}$$

It can therefore be interpreted as the potential generated by immobile *impurities*, located at $\{p_j^{(\omega)}\} \subset \mathbb{R}^d$, each of which is characterized by the same repulsive potential U. The random variable $\mu_\varrho(\Lambda)$ then equals the number of impurities in the bounded Borel set $\Lambda \subset \mathbb{R}^d$. It is Poissonian distributed according to

$$\mathbb{P}\left(\left\{ \omega \in \Omega \,\middle|\, \mu_\varrho^{(\omega)}(\Lambda) = n \right\}\right) = \frac{(\varrho|\Lambda|)^n}{n!} e^{-\varrho|\Lambda|}, \qquad n \in \mathbb{N} \cup \{0\}, \tag{1.5}$$

where $|\Lambda| := \int_\Lambda dx$ denotes the (Lebesgue-) volume of Λ, so that ϱ is indeed the mean (spatial) concentration of impurities. Employing for example [70, Lemma 3.10], one makes sure that a non-negative Poissonian random potential satisfies property (V+). Since it is also $\mathbb{R}^d$-ergodic due to $U \in L^2(\mathbb{R}^d)$, it is therefore a random potential in the sense of Definition 1.1.

1.2 Interesting Quantities and Basic Questions

A quantity of primary interest in the theory and applications of random Schrödinger operators is the (specific) integrated density of states N. Its knowledge allows one to compute the specific free energy of the corresponding non-interacting many-particle system in the thermodynamic limit, see for example (1.10) below. It also enters formulae for transport coefficients.

For its definition we first introduce $\Theta(E - H(A, V^{(\omega)}))$, the spectral projection operator of $H(A, V^{(\omega)})$ associated with the open half-line $]-\infty, E[\subset \mathbb{R}$ up to a given energy $E \in \mathbb{R}$. This notation complies with the functional calculus for self-adjoint operators in that $\Theta : \mathbb{R} \to \{0, 1\}$ stands for Heaviside's left-continuous unit-step function. For a random potential in the sense of Definition 1.1, this spectral projection possesses a complex-valued integral kernel (in other words: position representation) $\Theta(E - H(A, V^{(\omega)}))(x, y)$, which is a jointly continuous function of $x, y \in \mathbb{R}^d$ and a $\mathbb{P}$-integrable function of $\omega \in \Omega_0$, see [124, 27, 28]. This justifies

Definition 1.5. *The* integrated density of states *is the function* $N : E \mapsto N(E)$ *defined through the expectation value*

$$N(E) := \mathbb{E}\left[\Theta\big(E - H(A,V)\big)(x,x)\right]$$
$$= \int_\Omega \mathbb{P}(\mathrm{d}\omega)\, \Theta\left(E - H(A,V^{(\omega)})\right)(x,x). \tag{1.6}$$

Thanks to the unitary invariance of the kinetic-energy operator $H(A,0)$ under so-called magnetic translations [148] and to the $\mathbb{R}^d$-homogeneity of V, N is independent of the chosen $x \in \mathbb{R}^d$. Moreover, N is non-negative, non-decreasing and left-continuous.

There are some other universally valid properties of the integrated density of states N which do not depend on the specific choice of the random potential. For example, for theoretical and experimental reasons it is a comforting fact to learn that N, which is defined above as an ensemble average involving the infinite-volume random operator $H(A,V)$, may be viewed as a spatial average for a given typical realization $V^{(\omega)}$ of V. This property, often dubbed *self-averaging* [99], and the arising *uniqueness* problem will be made more precise in Section 2 below. Basically, self-averaging is a consequence of the assumed ergodicity of the random potential. The latter is also responsible for the almost-sure non-randomness of the spectrum of $H(A,V)$ and of its spectral components in the Lebesgue decomposition [114, 74]. By (1.6), the location of the almost-sure spectrum of $H(A,V)$, as a closed subset of the real line $\mathbb{R}$, coincides with the *set of growth points* $\{E \in \mathbb{R} \,|\, N(E) < N(E+\varepsilon) \text{ for all } \varepsilon > 0\}$ of N. But the location of the spectral components (in other words: the nature of the spectrum) cannot be inferred from N alone. For rather simple exceptions see the paragraph below (1.9). In any case, the spectrum depends on the choice of V. For example, due to the unboundedness of the negative fluctuations of a Gaussian random potential V, the almost-sure spectrum of the corresponding $H(A,V)$ coincides with the whole real line. In case of a non-negative Poissonian potential the spectrum of $H(A,V)$ is almost surely equal to the half-line starting at the ground-state energy of $H(A,0)$, that is, at the infimum of its spectrum.

Another universally valid property of N is its leading high-energy growth which is given by

$$N(E) \sim \frac{1}{\Gamma(1+d/2)}\left(\frac{E}{2\pi}\right)^{d/2} \qquad (E \to \infty). \tag{1.7}$$

see [73, 29, 116, 105, 138]. Here we use the notation $f(E) \sim g(E)$ as $E \to E'$ to indicate the asymptotic equivalence in the sense that $\lim_{E \to E'} f(E)/g(E) = 1$ for some $E' \in [-\infty, \infty]$. Remarkably, the asymptotics (1.7) is purely classical in the sense that $N(E) \sim N_c(E)$ as $E \to \infty$, where

$$N_c(E) := \frac{1}{|\Lambda|}\, \mathbb{E}\left[\int_{\Lambda \times \mathbb{R}^d} \frac{\mathrm{d}x\,\mathrm{d}k}{(2\pi)^d}\; \Theta\left(E - \tfrac{1}{2}\,|k - A(x)|^2 - V(x)\right)\right]$$

$$= \frac{1}{\Gamma(1 + d/2)}\, \mathbb{E}\left[\left(\frac{E - V(0)}{2\pi}\right)^{d/2}\; \Theta\big(E - V(0)\big)\right] \qquad (1.8)$$

defines the *(quasi-) classical integrated density of states*, see also [72]. In accordance with a theorem of Bohr and van Leeuwen on the non-existence of diamagnetism in classical physics [106], the integration with respect to the canonical momentum $k \in \mathbb{R}^d$ shows that N_c does not depend on A and hence not on the magnetic field (B_{ij}). Furthermore, the $\mathbb{R}^d$-homogeneity of V ensures that N_c is independent of the chosen bounded Borel set $\Lambda \subset \mathbb{R}^d$ of strictly positive volume $|\Lambda|$. The asymptotics (1.7) does not even depend on the random potential. Rather it is consistent with a famous result of Weyl [146].

In contrast to the high-energy growth (1.7), the low-energy fall-off of N near the almost-sure ground-state energy of $H(A, V)$ is not universal, more complicated and in general harder to obtain. It typically stems from exponentially rare low-energy fluctuations of the random potential, which compete with the quantum fluctuations related to the kinetic energy. As a result, N exhibits a much faster low-energy fall-off in comparison to the non-random case $V = 0$. This is commonly referred to as a *Lifshits tail* in honour of the theoretical physicist I. M. Lifshits (1917–1982) who was the first to develop a quantitative theory [96, 97] in case $A = 0$. Lifshits' arguments can be summarized in terms of the so-called *optimal-fluctuation ideology* [62, 151, 98] (see also [123, 99, 116]) according to which the low-energy fall-off of N near the almost-sure ground-state energy $E_0 \in [-\infty, \infty[$ of $H(A, V)$ is (universally) given by the formula

$$\log N(E) \sim \inf_{\tau > 0}\left(\tau E + \sup_{\substack{\psi \in \mathcal{C}_0^\infty(\mathbb{R}^d) \\ \langle \psi, \psi \rangle = 1}} \log \mathbb{E}\big[e^{-\tau \langle \psi,\, H(A,V)\psi \rangle}\big]\right) \qquad (E \downarrow E_0). \ (1.9)$$

Here and in the following we suppress suitable constants ensuring dimensionless arguments of the natural logarithm because they become irrelevant in the limit. Actually all Lifshits tails we know of, in particular the ones caused by Gaussian and Poissonian random potentials presented in Sections 3 and 4 below, are consistent with (1.9).

As a left-continuous (unbounded) distribution function, N has at most countably many discontinuity points. For one space dimension, N is known to be even globally continuous [114]. If this is also true in the multi-dimensional case (of a continuum and $A = 0$), is an open problem. Using (1.6) it is not hard to show (for example by [70, Cor. 3.7] and [116, Thm. 2.12]) that N is discontinuous at a given energy $E \in \mathbb{R}$ if and only if E is almost surely an infinitely degenerate eigenvalue of $H(A, V)$. In the other extreme case in which $H(A, V)$ has almost surely only absolutely continuous spectrum in a given bounded energy interval $I \subset \mathbb{R}$, it follows directly from (1.6) that N is absolutely continuous on I. This means that the *density of states* $D : E \mapsto D(E) := \mathrm{d}N(E)/\mathrm{d}E$ exists as a non-negative Lebesgue-integrable function on I. In any case, under more specific assumptions on V regularity of N beyond absolute continuity is expected on the whole almost-sure energy spectrum, even in regimes with (dense) pure-point spectrum. For example, in case of a Gaussian random potential with a non-negative covariance function, N turns out to be (at least) locally Lipschitz continuous, so that D exists on the whole real line and is locally bounded, see Theorem 3.3 below. The regularity of N is physically relevant, for instance, for the basic thermal-equilibrium properties of the corresponding macroscopic system of non-interacting (spinless) fermions in a random medium. These properties are determined by the *specific free energy*

$$F(T, \bar{n}) := \sup_{\mu \in \mathbb{R}} \left[\mu \bar{n} - T \int_{\mathbb{R}} \mathrm{d}N(E) \log \left(1 + \mathrm{e}^{(\mu - E)/T} \right) \right] \qquad (1.10)$$

as a function of the absolute temperature $T > 0$ (multiplied by Boltzmann's constant), the spatial concentration $\bar{n} > 0$ of the fermions and (possibly) the magnetic field. In fact, Sommerfeld's ubiquitous asymptotic low-temperature expansion of F presupposes a sufficiently smooth integrated density of states [126], [89, Chap. 4]. In particular, one then has $F(T, \bar{n}) \sim \int_{-\infty}^{\mu_0} \mathrm{d}E \, D(E) E - T^2 D(\mu_0) \pi^2 / 6$ as $T \downarrow 0$, where the *Fermi energy* $\mu_0 \in \mathbb{R}$ is the solution of the equation $N(\mu_0) = \bar{n}$. Given (1.7), a sufficient condition for the existence of the Lebesgue-Stieltjes integral in (1.10) is the finiteness $\mathbb{E} \left[\exp \left(-\tau V(0) \right) \right] < \infty$ for all $\tau \in [0, \infty[$. In fact, this condition implies the (quasi-) classical estimate $N(E) \leq (2\pi\tau)^{-d/2} e^{\tau E} \mathbb{E} \left[\exp \left(-\tau V(0) \right) \right]$ for all $E \in \mathbb{R}$ with arbitrary $\tau \in$ $]0, \infty[$ by [116, Thm. 9.1] and the diamagnetic inequality, see also [28].

   In perfect solids or crystals the (generalized) one-electron energy eigenfunctions are given by Bloch-Floquet functions on $\mathbb{R}^d$, which are (time-independent) plane waves modulated by lattice-periodic functions [81, 90]. Accordingly these eigenfunctions are delocalized over the whole solid, hence not square-integrable, and the whole energy spectrum is only absolutely continuous. That the spectrum is absolutely continuous even in the low-energy regime, may be viewed as a consequence of the tunnelling effect. According to classical mechanics the electron would be localized in one of the identical atomic potential wells making up the crystal. Since even small differences in the potential wells may suppress "quantum coherence" and hence tunnelling, localized states given by square-integrable energy eigenfunctions associated with pure-point spectrum should emerge in disordered solids at least at low energies. In particular, this should be true for amorphous solids. As we will see in Theorem 3.7 and Theorem 3.9, localization at low energies can indeed be proven in case of certain Gaussian random potentials for arbitrary $d \in \mathbb{N}$. For Poissonian random potentials a proof of localization is so far only available for $d = 1$, see Theorem 4.9.

By their very nature, localized states are not capable of contributing to macroscopic charge transport. Following Mott [108], a *mobility edge* is expected to occur at a certain (non-random) energy in an amorphous solid for $d \geq 3$ which separates localized states in the Lifshits-tail regime from delocalized ones at higher energies. At zero temperature, the passing of the Fermi energy through a mobility edge from delocalized to localized states results in a metal-insulator transition known as the Anderson transition. Likewise, the phenomenon of localization by disorder is called *Anderson localization* [9], see also [87] and references therein. In one space dimension, it is well established that even small disorder leads to localization at all energies under rather general circumstances [33, 29, 116, 133] so that there is no mobility edge in $d = 1$.

In order to better understand the suppression of charge transport in disordered solids by Anderson localization, both spectral and dynamical criteria for localization are commonly studied. *Spectral localization* means that there is only (dense) pure-point spectrum in certain energy regimes. In addition, the corresponding eigenfunctions are often required to decay not slower than exponentially at infinity instead of being merely square-integrable. One criterion for *dynamical localization* is a sufficiently slow spatial spreading of initially localized wave functions which evolve under the unitary time evolution in the spectral subspace $\chi_I(H(A, V^{(\omega)}))\mathrm{L}^2(\mathbb{R}^d)$ corresponding to a certain (Borel) energy regime $I \subset \mathbb{R}$. More precisely, one requires the finiteness

$$\sup_{t \in \mathbb{R}} \int_{\mathbb{R}^d} \mathrm{d}x \, \left| \psi_{t,I}^{(\omega)}(x) \right|^2 |x|^q < \infty \qquad (1.11)$$

of the qth absolute spatial moment of $\psi_{t,I}^{(\omega)} := \mathrm{e}^{-\mathrm{i}tH(A,V^{(\omega)})}\chi_I(H(A, V^{(\omega)}))\psi_0$ for all times $t \in \mathbb{R}$, all $\psi_0 \in \mathcal{C}_0^\infty(\mathbb{R}^d)$ and, for example, a real $q \geq 2$. Usually (1.11) is demanded either for $\mathbb{P}$-almost all $\omega \in \Omega$ or even upon integration over ω with respect to $\mathbb{P}$, see [55, 15, 34, 56] and [131, Sec. 3.4]. While this type of dynamical localization implies spectral localization in I by the RAGE theorem [8, Sec. 5.1] (see also [33], [131, Sec. 4.1.5], [145]), the converse is not generally true, since pure-point spectrum may occur together with a sub-ballistic long-time behaviour [119]. Even more physical is another criterion for dynamical localization, namely the vanishing of the direct-current conductivity of the corresponding non-interacting fermion system at zero temperature. The interrelations between the different localization criteria are complicated and not yet understood in sufficient generality; in this context we recommend [125, 120, 94, 14, 80, 136].

1.3 Random Landau Hamiltonian and Its Single-Band Approximation

In recent decades the physics of (quasi-) two-dimensional electronic structures has attracted considerable attention [10, 83, 91, 129]. Some of the occurring phenomena, like the integer quantum Hall effect, are believed to be microscopically explainable in terms of a system of non-interacting electrically charged fermions in the Euclidean plane $\mathbb{R}^2$ subject to a perpendicular constant magnetic field of strength $B := B_{12} > 0$ and a random potential [71, 147]. The underlying one-particle Schrödinger operator is known as the *random Landau Hamiltonian* which acts on the Hilbert space $L^2(\mathbb{R}^2)$. Apart from numerous theoretical and numerical studies in the physics literature, there are nowadays quite a lot of rigorous results available for this and related models, see for example [92, 102, 16, 32, 13, 141, 3, 40, 12, 45, 58] and references therein.

The kinetic-energy part of the random Landau Hamiltonian is the well understood *Landau Hamiltonian*

$$H(A,0) = \frac{1}{2}\left[\left(\mathrm{i}\frac{\partial}{\partial x_1} - \frac{B}{2}x_2\right)^2 + \left(\mathrm{i}\frac{\partial}{\partial x_2} + \frac{B}{2}x_1\right)^2\right]$$

$$= B\sum_{l=0}^{\infty}\left(l+\frac{1}{2}\right)P_l. \tag{1.12}$$

Its spectral resolution expressed by the second equality dates back to Fock [52] and Landau [93]. The energy eigenvalue $(l + 1/2)B$ is called the lth *Landau level* and the corresponding orthogonal eigenprojection P_l is an integral operator with continuous complex-valued kernel

$$P_l(x,y) := \frac{B}{2\pi}\exp\left[\mathrm{i}\frac{B}{2}(x_2y_1 - x_1y_2) - \frac{B}{4}|x-y|^2\right]\mathrm{L}_l\left(\frac{B}{2}|x-y|^2\right) \tag{1.13}$$

given in terms of the lth Laguerre polynomial $\mathrm{L}_l(\xi) := \frac{1}{l!}\mathrm{e}^\xi\frac{d^l}{d\xi^l}(\xi^l\mathrm{e}^{-\xi})$, $\xi \geq 0$ [60]. The diagonal $P_l(x,x) = B/2\pi$ is naturally interpreted as the specific degeneracy of the lth Landau level.

Accordingly, the corresponding integrated density of states is the well-known discontinuous staircase function

$$N(E) = \frac{B}{2\pi}\sum_{l=0}^{\infty}\Theta\big(E - (l+1/2)\,B\big), \qquad V = 0. \tag{1.14}$$

Informally, the associated density of states $\mathrm{d}N(E)/\mathrm{d}E$ is a series of Dirac delta functions supported at the Landau levels. By adding a random potential to (1.12) the corresponding peaks are expected to be smeared out. In fact, in case of Gaussian random potentials with a non-negative covariance function they are smeared out completely, see Theorem 3.3 in case $d = 2$ and Figure 1 below.

In the limit of a strong magnetic field the spacing B of successive Landau levels approaches infinity and the so-called *magnetic length* $1/\sqrt{B}$ tends to zero. Therefore the effect of so-called level mixing should be negligible if either the strength of the random potential V is small compared to the level spacing or if the (smallest) correlation length of V is much larger than the magnetic length. In both cases a reasonable approximation to N should then read

$$N(E) \approx \frac{B}{2\pi} \sum_{l=0}^{\infty} R_l\big(E - (l+1/2)\, B\big). \tag{1.15}$$

Here the lth term of the infinite series stems from the probability distribution function R_l on $\mathbb{R}$ defined by

$$R_l(E) := \frac{2\pi}{B}\, \mathbb{E}\left[\big(P_l \,\Theta\, (E - P_l V P_l)\, P_l\big)(x,x)\right], \qquad x \in \mathbb{R}^2. \tag{1.16}$$

We refer to it as the (centred and normalized) lth *restricted integrated density of states*. It describes the broadening of the lth Landau level by the random potential to the lth *Landau band*, when considered in isolation. Assuming (without loss of generality) that the mean of the $\mathbb{R}^2$-homogeneous random potential V is zero, $\mathbb{E}\left[V(0)\right] = 0$, the variance of the energy in the lth Landau band is given by [24]

$$\sigma_l^2 := \int_{\mathbb{R}} \mathrm{d}R_l(E)\, E^2 = \frac{2\pi}{B} \int_{\mathbb{R}^2} \mathrm{d}x\; |P_l(0,x)|^2\, C(x) \tag{1.17}$$

in terms of the covariance function $C(x) := \mathbb{E}\left[V(x)V(0)\right]$ of V. The standard deviation $\sigma_l := \sqrt{\sigma_l^2}$ is physically interpreted as the (effective) *width* of the lth Landau band. The exact formula (1.17) first appeared in approximations to R_l, like the so-called self-consistent Born approximation, see [91] and references therein. In case of the *Gaussian covariance function* (with $x \in \mathbb{R}^2$)

$$C(x) = C(0) \exp\left(-\frac{|x|^2}{2\lambda^2}\right), \qquad C(0) > 0, \quad \lambda > 0, \tag{1.18}$$

the lth band width can be calculated exactly and explicitly as a function of the squared length ratio $B\lambda^2$. The result is

$$\sigma_l^2 = C(0)\, \frac{B\lambda^2}{B\lambda^2 + 1} \left(\frac{B\lambda^2 - 1}{B\lambda^2 + 1}\right)^l \mathrm{P}_l\left(\frac{(B\lambda^2)^2 + 1}{(B\lambda^2)^2 - 1}\right) =: w_l\,(B\lambda^2) > 0, \tag{1.19}$$

where $\mathrm{P}_l(\xi) := \frac{1}{l! 2^l} \frac{d^l}{d\xi^l}\left(\xi^2 - 1\right)^l$, $\xi \in \mathbb{R}$, is the lth Legendre polynomial [60].

Neglecting effects of level mixing by only dealing with the sequence of restricted random operators $\{P_l V P_l\}_l$ is a simplifying approximation which is often made. The interest in these operators relates to spectral localization [63, 37, 38, 39, 117, 118] and to properties of their (restricted) integrated density of states R_l, see for example [144, 22, 82, 18, 17, 11, 122, 24, 20, 68]. ¿From the physical point of view most interesting is the restriction to the lowest Landau band ($l = 0$). For strong enough magnetic fields all fermions may be accommodated in the lowest band without conflicting with Pauli's exclusion principle, since the specific degeneracy increases with the magnetic field. The contribution of $2\pi R_0(\mu_0 - B/2)/B$ to the sum of the series in (1.15) at the Fermi energy μ_0 should then already be a good approximation to $N(\mu_0)$, since the effects of higher Landau bands are negligible if B is large compared to the strength $\sqrt{C(0)}$ of the random potential. For Gaussian and non-negative Poissonian random potentials, rigorous statements in support of this heuristics can be found in [102, 26, 142].

Remark 1.6. A sufficient condition [37, 25] for the almost-sure self-adjointness of the restricted random operator $P_l V P_l$ on $P_l L^2(\mathbb{R}^2)$ is the following growth limitation for the even moments of the random potential,

$$\mathbb{E}\left[|V(0)|^{2n}\right] \leq (2n)! \, M^{2n} \qquad (1.20)$$

for all $n \in \mathbb{N}$ with some constant $M < \infty$. While (1.20) is satisfied for all Gaussian random potentials with $M = \sqrt{C(0)}$ (see (1.2)), its validity for non-negative Poissonian random potentials is ensured [25] by the additional (Lebesgue-essential) boundedness of the single-impurity potential, $U \in L^\infty(\mathbb{R}^2)$. Moreover, R_l in (1.16) is well defined, because the integral kernel $\left(P_l \Theta(E - P_l V^{(\omega)} P_l) P_l\right)(x, y)$ of the spectral projection $P_l \Theta(E - P_l V^{(\omega)} P_l) P_l$ is jointly continuous in x, $y \in \mathbb{R}^2$ and integrable with respect to $\mathbb{P}$ as a function of $\omega \in \Omega$. Thanks to magnetic translation invariance R_l is independent of the chosen $x \in \mathbb{R}^2$.

2 Self-averaging and Uniqueness of the Integrated Density of States

Since spatially separated, large parts of a macroscopic sample of an amorphous solid become decoupled rather fast with increasing distance, they effectively correspond to different realizations of the ergodic random potential modelling the solid. As a consequence, it should make no difference whether the integrated density of states N is defined as an ensemble average or as a spatial average for a given typical realization.

To specify the notion of a spatial average associated with N, one first has to consider a restriction of the (infinite-volume) random Schrödinger operator $H(A, V)$ to a bounded open cube $\Lambda \subset \mathbb{R}^d$. The resulting *finite-volume random Schrödinger operator* $H_{\Lambda,X}(A, V)$ is rendered almost surely self-adjoint on the Hilbert space $L^2(\Lambda)$ by imposing, for example, Dirichlet, $X = D$, or Neumann, $X = N$, boundary conditions on the wave functions in its domain of definition, see for example [73, 29, 69]. Since the spectrum of $H_{\Lambda,X}(A, V)$ almost surely consists only of isolated eigenvalues of finite multiplicity, the *finite-volume integrated density of states* $N_{\Lambda,X}^{(\omega)}$ is well defined as the (specific) eigenvalue counting-function

$$N_{A,X}^{(\omega)}(E) := \frac{1}{|A|} \left\{ \begin{array}{l} \text{number of eigenvalues of } H_{A,X}(A, V^{(\omega)}), \text{ counting} \\ \text{multiplicity, which are strictly smaller than } E \in \mathbb{R} \end{array} \right\}$$

$$= \frac{1}{|A|} \int_A \mathrm{d}x \, \Theta\big(E - H_{A,X}(A, V^{(\omega)})\big)(x,x) \tag{2.1}$$

for both boundary conditions $X = D$ and $X = N$, and $\mathbb{P}$-almost all $\omega \in \Omega$. The next theorem shows that in the infinite-volume limit $N_{A,X}^{(\omega)}$ coincides with the above-defined ensemble average (1.6), and therefore becomes independent of X and $\mathbb{P}$-almost all ω.

Theorem 2.1. *Let V be a random potential in the sense of Definition 1.1. Moreover, let $A \subset \mathbb{R}^d$ stand for bounded open cubes centred at the origin. Then there is a set $\Omega_0 \in \mathcal{A}$ of full probability, $\mathbb{P}(\Omega_0) = 1$, such that*

$$N(E) = \lim_{A \uparrow \mathbb{R}^d} N_{A,X}^{(\omega)}(E) \tag{2.2}$$

holds for both boundary conditions $X = D$ and $X = N$, all $\omega \in \Omega_0$ and all energies $E \in \mathbb{R}$ except for the (at most countably many) discontinuity points of N.

By a suitable ergodic theorem [88] the *existence* and almost-sure *non-randomness* of both infinite-volume limits in (2.2) are basically due to the assumed $\mathbb{R}^d$-ergodicity of V. Under slightly different hypotheses the actual proof was outlined in [105]. It uses functional-analytic arguments first presented in [75] for the case $A = 0$. A different approach to the existence of these limits for $A \neq 0$, using Feynman-Kac(-Itô) functional-integral representations of Schrödinger semigroups [124, 27], can be found in [138, 26]. It dates back to [111, 110] for the case $A = 0$ and, to our knowledge, works straightforwardly in the case $A \neq 0$ for $X = D$ only. For $A \neq 0$ *uniqueness* of the infinite-volume limit in (2.2), that is, its independence of the boundary condition X (previously claimed without proof in [105]) follows from [109] if the random potential V is bounded and from [35, 65] if V is bounded from below. For random potentials V yielding Schrödinger operators $H(A, V)$ which are almost surely unbounded from below, the proof of (2.2) can be found in [70, 28], see also [116, Thm. 5.20] for $A = 0$.

Remark 2.2. Similar as in equilibrium statistical-mechanics [121] there are more general sequences of regions expanding to $\mathbb{R}^d$ than concentric open cubes A for which (2.2) is true, see for example [116, p. 105], [29, p. 304] or [35]. Moreover, the convergence (2.2) holds for any other boundary condition X for which the self-adjoint operator $H_{A,X}(A, V^{(\omega)})$ obeys the inequalities $H_{A,N}(A, V^{(\omega)}) \leq H_{A,X}(A, V^{(\omega)}) \leq H_{A,D}(A, V^{(\omega)})$ in the sense of (sesquilinear) forms. The case of those mixed (in other words: Robin) boundary conditions, which cannot be sandwiched between Dirichlet and Neumann boundary conditions, is treated in [107].

3 Results in Case of Gaussian Random Potentials

This section compiles rigorous results on Lifshits tails, the density of states and spectral as well as dynamical localization in case of Gaussian random potentials. The corresponding theorems are formulated under increasingly stronger conditions on the covariance function.

3.1 Lifshits Tails

Since Gaussian random potentials V have unbounded negative fluctuations, it is not surprising that the leading low-energy fall-off of the integrated density of states is also Gaussian, even in the presence of a magnetic field. In particular, this type of fall-off ensures that $H(A, V)$, although the latter is almost surely unbounded from below, may serve as the one-particle Schrödinger operator of a macroscopic system of non-interacting fermions in a random medium with well-defined specific free energy (1.10) and related thermodynamic quantities.

Theorem 3.1. *Let V be a Gaussian random potential with covariance function C. Then the leading low-energy fall-off of the integrated density of states N is Gaussian in the sense that*

$$\log N(E) \sim -\frac{E^2}{2C(0)} \qquad (E \to -\infty). \qquad (3.1)$$

Theorem 3.1 dates back to Pastur [112, 113], see also [110, 75], and [23, 26, 105, 138] for the magnetic case, where the last two references even allow the presence of certain *random* magnetic fields. The by now standard way of proving (3.1), already used by Pastur, is in the spirit of (1.9) (setting there $E_0 = -\infty$). One first determines the leading asymptotic behaviour of the Laplace-Stieltjes transform, $\widetilde{N}(\tau) := \int_{\mathbb{R}} dN(E) \exp(-\tau E)$, $\tau > 0$, of N as $\tau \to \infty$ and then applies an appropriate Tauberian theorem [19].

The Lifshits tail (3.1) in case of Gaussian random potentials is highly universal. It only depends on the single-site variance $C(0) = \mathbb{E}\left[V(0)^2\right] > 0$, but not on further details of the covariance function, the space dimension or the magnetic field. As physical heuristics and formula (1.9) already suggest, the (non-negative) kinetic-energy operator $H(A, 0)$ becomes irrelevant at extremely negative energies and the tail (3.1) is purely classical in the sense that $\log N(E) \sim \log N_c(E)$ as $E \to -\infty$.

In contrast to the universal classical Lifshits tail (3.1) of N, the analogous tail of the restricted integrated density of states R_l exhibits non-universal quantum behaviour in that it depends on the magnetic field and on details of the covariance function.

Theorem 3.2 ([24]). *Let $d = 2$ and $B > 0$. Suppose that V is a Gaussian random potential with covariance function C. Moreover, let $\sigma_l^2 > 0$, see*

(1.17). *Then the leading low-energy fall-off of the restricted integrated density of states R_l is Gaussian in the sense that*

$$\log R_l(E) \sim -\frac{E^2}{2\Gamma_l^2} \qquad (E \to -\infty). \qquad (3.2)$$

Here the fall-off energy $\Gamma_l > 0$ is given by a solution of the maximization problem

$$\Gamma_l^2 := \sup_{\substack{\varphi \in P_l L^2(\mathbb{R}^2) \\ \langle \varphi, \varphi \rangle = 1}} \int_{\mathbb{R}^2} \mathrm{d}x \int_{\mathbb{R}^2} \mathrm{d}y \, |\varphi(x)|^2 \, C(x - y) \, |\varphi(y)|^2 . \qquad (3.3)$$

A proof of Theorem 3.2 follows the lines of reasoning in [24], which amounts to establish the appropriate version of (1.9). The symmetry $R_l(E) = 1 - R_l(-E)$, for all $E \in \mathbb{R}$, then immediately gives the high-energy growth $\log (1 - R_l(E)) \sim -E^2/(2\Gamma_l^2)$ as $E \to \infty$. For the Gaussian covariance (1.18) a maximizer in (3.3) is given by $\varphi(x) = \sqrt{B/(l! \, 2\pi)} \left[\sqrt{B/2} \, (x_1 - ix_2) \right]^l \exp(-B \, |x|^2/4)$ and the squared fall-off energy is explicitly found to be

$$\Gamma_l^2 = \left[B\lambda^2/(B\lambda^2 + 1) \right] w_l(B\lambda^2 + 1), \qquad (3.4)$$

see [24] and (1.19), and also [11] for $l = 0$. For a comparison of the fall-off energies in (3.1) and (3.2), we offer the chain of inequalities $\sigma_l^4/C(0) \leq \Gamma_l^2 \leq \sigma_l^2 \leq C(0)$ which is actually valid [24] for the covariance function of a general $\mathbb{R}^2$-homogeneous random potential, not only of a Gaussian one.

3.2 Existence of the Density of States

The continuity and non-negativity of the covariance function of a Gaussian random potential already imply that the corresponding integrated density of states N is locally Lipschitz continuous, equivalently, that N is absolutely continuous on any bounded interval and its (Lebesgue-) derivative $D(E) = \mathrm{d}N(E)/\mathrm{d}E$, the *density of states*, is locally bounded.

Theorem 3.3 ([50, 69]). *Let V be a Gaussian random potential with non-negative covariance function C. Then the integrated density of states N is locally Lipschitz continuous and the inequality*

$$\frac{\mathrm{d}N(E)}{\mathrm{d}E} \leq W(E) \qquad (3.5)$$

holds for Lebesgue-almost all energies $E \in \mathbb{R}$ with some non-negative $W \in L^\infty_{\mathrm{loc}}(\mathbb{R})$, which is independent of the magnetic field.

A simple, but not optimal choice for the Lipschitz constant is given by

$$W(E) = \left(r^{-1} + (2\pi\tau)^{-1/2} \right)^d \frac{\exp\{\tau E + \tau^2 C(0) \left[1 - \varkappa_r^2/2 \right]\}}{\varkappa_r \sqrt{2\pi C(0)}}. \qquad (3.6)$$

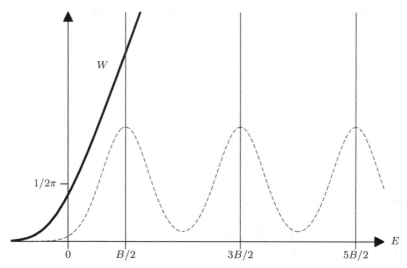

Fig. 1. Plot of an upper bound $W(E)$ on $D(E)$ as a function of the energy E. Here $D(E) = \mathrm{d}N(E)/\mathrm{d}E$ is the density of states of the Landau Hamiltonian with a Gaussian random potential with Gaussian covariance function (1.18). The dashed line is a plot of the graph of an approximation to $D(E)$. When zooming in, its bell-shaped parts near the Landau-level energies exhibit asymmetries. The exact $D(E)$ is unknown. Vertical lines indicate the delta peaks which reflect the non-existence of the density of states without random potential, $V = 0$. The single-step function $\Theta(E)/2\pi$ (not shown) is the free density of states characterized by $V = 0$ and $B = 0$, confer (1.8)

Here $r, \tau \in {]0, \infty[}$ are arbitrary except that $\varkappa_r := \inf_{|x| \leq r/\sqrt{d}} C(x)/C(0) > 0$ must be strictly positive. By the assumed continuity of the covariance function, the latter condition is fulfilled at least for all sufficiently small r. Figure 1, which is taken from [69], contains the graph of W (after a certain numerical minimization) for the case of the Gaussian covariance function (1.18) and $d = 2$. The upper bound reveals that the density of states has no infinities for arbitrarily weak disorder, that is, for arbitrarily small $C(0) > 0$.

Theorem 3.3 relies on a so-called *Wegner estimate* [143], which bounds the average number of eigenvalues of the finite-volume random Schrödinger operator $H_{\Lambda,\mathrm{X}}(A, V)$ in a given energy interval $[E_1, E_2[\subset \mathbb{R}$ from above. More precisely, under the assumptions of Theorem 3.3 one has [50, 69] the inequality

$$\mathbb{E}\left[N_{\Lambda,\mathrm{X}}(E_2) - N_{\Lambda,\mathrm{X}}(E_1)\right] \leq W(E) \left|E_2 - E_1\right| \tag{3.7}$$

for both boundary conditions $\mathrm{X} = \mathrm{D}$ and $\mathrm{X} = \mathrm{N}$, and all $E \geq E_2$, provided the bounded open cube $\Lambda \subset \mathbb{R}^d$ is not too small. The proof of (3.7) uses the one-parameter decomposition $V^{(\omega)}(x) =: V_0^{(\omega)}(x) + V^{(\omega)}(0) C(x)/C(0)$ of the Gaussian random potential V with covariance function C. By this definition V_0 is a non-homogeneous Gaussian random field which is stochastically independent of the Gaussian random variable $V(0)$. A second essential ingredient of the proof is the abstract one-parameter spectral-averaging estimate of Combes and Hislop [31, Cor. 4.2] (for an equivalent formulation see [68, Lemma 1]). Finally, the independence of W of the magnetic field arises from the diamagnetic inequality for Neumann partition functions [69, App. A], see also [65].

The analogue of Theorem 3.3 for the single-band approximation of the random Landau Hamiltonian yields the existence and boundedness of the probability density $\mathrm{d}R_l(E)/\mathrm{d}E$, the *restricted density of states*.

Theorem 3.4 ([68]). *Let $d = 2$ and $B > 0$. Suppose that V is a Gaussian random potential with non-negative covariance function C. Then the restricted integrated density of states R_l is Lipschitz continuous and the inequality*

$$\frac{\mathrm{d}R_l(E)}{\mathrm{d}E} \leq \frac{1}{\sqrt{2\pi \Gamma_l^2}} \tag{3.8}$$

holds for Lebesgue-almost all energies $E \in \mathbb{R}$, where the constant Γ_l is the fall-off energy of the lth Landau band, see (3.3). If additionally C is circularly symmetric, one also has

$$\frac{\mathrm{d}R_l(E)}{\mathrm{d}E} \leq \frac{C(0)}{\sigma_l^2} \frac{1}{\sqrt{2\pi C(0)}} \exp\left(-\frac{E^2}{2C(0)}\right), \tag{3.9}$$

where σ_l is the width of the lth Landau band, see (1.17).

In the physics literature [151, 44, 134] one often considers the limit of a *delta-correlated* Gaussian random potential informally characterized by $C(x) = u^2 \delta(x)$ with some constant $u > 0$. It emerges from the Gaussian random potential with the Gaussian covariance function (1.18) by choosing $C(0) = u^2(2\pi\lambda^2)^{-d/2}$ and letting $\lambda \downarrow 0$. In particular, several simplifications occur in this limit for the random Landau Hamiltonian ($d = 2$) in its single-band approximation. For example, the band width becomes independent of the Landau-level index, $\sigma_l^2 = \sigma_0^2 = u^2 B/(2\pi)$, and the squared fall-off energy (3.4) takes the form $\Gamma_l^2 = \left(u^2 B/4\pi\right)(2l)!/(l!\,2^l)^2$. More interestingly, there is an explicit expression for $\mathrm{d}R_0(E)/\mathrm{d}E$ in this limit and also for $\mathrm{d}R_l(E)/\mathrm{d}E$, if the subsequent high Landau-level limit $l \to \infty$ is performed. The first result is due to Wegner [144] and reads

$$\frac{\mathrm{d}R_0(E)}{\mathrm{d}E} = \frac{2}{\pi^{3/2}\sigma_0} \frac{\exp(\eta^2)}{1 + \left[2\pi^{-1/2} \int_0^\eta \mathrm{d}\xi \, \exp(\xi^2)\right]^2}, \qquad \eta := \frac{E}{\sigma_0}, \tag{3.10}$$

see also [22, 82, 102]. As for the second result, it is known [18, 122] that $\mathrm{d}R_l(E)/\mathrm{d}E$ becomes semi-elliptic,

$$\lim_{l \to \infty} \frac{\mathrm{d}R_l(E)}{\mathrm{d}E} = \frac{1}{2\pi\sigma_0} \Theta\left(4 - \eta^2\right) \sqrt{4 - \eta^2}. \tag{3.11}$$

While the right-hand side of (3.8) remains finite in the delta-correlated limit, it diverges asymptotically as $l^{1/4}/(\pi^{1/4}\sigma_0)$ in the additional limit $l \to \infty$, which is in accordance with [17].

The existence of a bounded restricted density of states $dR_0(E)/dE$ in the delta-correlated limit of a Gaussian random potential stands in contrast to certain situations with random *point* impurities. In all of the considered cases [22, 39, 117, 46, 118], it has been shown that the restricted random operator $P_0 V P_0$ has almost surely an infinitely degenerate eigenvalue, if the magnetic-field strength B is sufficiently large; see also (4.6) below. Part of these results have been unified in [30].

3.3 Spectral and Dynamical Localization

Since even weak disorder should be able to suppress quantum-mechanical tunnelling at sufficiently negative energies, bound states should emerge in such energy (or disorder) regimes in case of Gaussian random potentials. Therefore, the content of the subsequent theorems is often taken for granted in the physics literature. To our knowledge however, Theorem 3.7 and Theorem 3.9 contain the first rigorous results on localization by an $\mathbb{R}^d$-ergodic random potential in the multi-dimensional case $d \geq 2$. For their precise formulations we first need to define a property the covariance function C may have or not.

Definition 3.5. *A covariance function $C : \mathbb{R}^d \to \mathbb{R}$ of a (Gaussian) random potential is said to have* property (R) *if it admits the representation $C(x) = \int_{\mathbb{R}^d} dy\, \gamma(x+y)\,\gamma(y)$ for all $x \in \mathbb{R}^d$ with some non-zero $\gamma \in L^2(\mathbb{R}^d)$ which is both*

(i) *Hölder continuous of order α, that is, there exist constants $\alpha \in]0,1]$ and $a > 0$ such that $|\gamma(x+y) - \gamma(x)| \leq a|y|^\alpha$ for all $x \in \mathbb{R}^d$ and all y in some neighbourhood of the origin $0 \in \mathbb{R}^d$, and*

(ii) *non-negative and has sufficiently rapid decay at infinity in the sense that $0 \leq \gamma(x) \leq \gamma_0 (1 + |x|)^{-\beta}$ for Lebesgue-almost all $x \in \mathbb{R}^d$ with some constant $\gamma_0 > 0$ and some exponent $\beta > 13\,d/2 + 1$.*

Remarks 3.6. (i) In accordance with the Wiener-Khinchin criterion, the representation in Definition 3.5 (without requiring (i) and (ii)) is equivalent to the assumption that the (non-negative) spectral measure of V is absolutely continuous with respect to Lebesgue measure. A prominent example having property (R) is the Gaussian covariance function (1.18) with arbitrary correlation length $\lambda > 0$ for which $\gamma(x) = \left(2/(\pi\lambda^2)\right)^{d/4} \sqrt{C(0)} \exp\left(-|x|^2/\lambda^2\right)$.

(ii) Property (R) in particular ensures the non-negativity of C. It also implies [48] the $\mathbb{P}$-almost sure continuity, and hence local boundedness, of the realizations of V. The decay of γ at infinity yields the mixing property of V, hence ergodicity. In fact, property (R) requires a compromise between local dependence and global independence of V. From a physical point of view, both requirements are plausible: The effective one-particle potential V should be smooth due to screening. By the same token it is expected that impurities only weakly influence each other over long distances.

Theorem 3.7 ([51]). *Let V be a Gaussian random potential with a covariance function having property (R) in the sense of Definition 3.5. Then*

(i) *for every coupling constant $\zeta > 0$ there exists an energy $E_{pp} < 0$ such that the spectrum of $H(A, \zeta V^{(\omega)})$ is only pure point in the half-line $] - \infty, E_{pp}]$ for $\mathbb{P}$-almost all $\omega \in \Omega$.*
(ii) *for every energy $E < 0$ there exists a coupling constant $\zeta_{pp} > 0$ such that for every $\zeta \in]0, \zeta_{pp}]$ there exists $\Omega_\zeta \in \mathcal{A}$ with $\mathbb{P}(\Omega_\zeta) = 1$ and the spectrum of $H(A, \zeta V^{(\omega)})$ is only pure point in $] - \infty, E]$ for all $\omega \in \Omega_\zeta$.*

Given the Wegner estimate (3.7), the proof of Theorem 3.7 is based on a so-called *multi-scale analysis* in the spirit of the fundamental work of Fröhlich and Spencer [54]. They gave the first rigorous result on localization in case of a multi-dimensional lattice model, the original one due to Anderson [9]. The multi-scale analysis invokes elements from Kolmogorov-Arnold-Moser theory for coping with small denominators in order to bound resolvents of finite-volume random Schrödinger operators with high probabilities. Its applicability to the present situation requires additional assumptions on the covariance function beyond those needed for the Wegner estimate. The technical realization of the proof of Theorem 3.7 is patterned after the "fixed-energy" analysis of von Dreifus and Klein [42] in order to handle the long-ranged correlations of the fluctuations of Gaussian random potentials. In addition to that, different random potentials are used on different length scales. The idea behind this is to replace the given long-ranged correlated random potential V on the length scale ℓ_n by the element V_n of a sequence $\{V_n\}_{n\in\mathbb{N}}$ of (Gaussian) random potentials such that   (i) $\{V_n\}$ converges to V in a suitable sense as $n \to \infty$ and that   (ii) each V_n has finite-ranged correlations, but with a spatial extent that grows with increasing n. For the adaptation to the continuous space $\mathbb{R}^d$ the proof follows the lines of Combes and Hislop [31] and Figotin and Klein [49].

Remarks 3.8.   (i) The assumptions of Theorem 3.7 also allow for so-called deterministic random potentials, as is illustrated by the Gaussian covariance (1.18) for one space dimension.
(ii) It would be interesting to see whether the proof of Theorem 3.7 can be simplified by adapting the continuum-extension [7] of the fractional-moment approach by Aizenman and Molchanov [4, 2, 3, 5, 6] to (spectral) localization, which was originally developed for the lattice $\mathbb{Z}^d$.

As a consequence of the unbounded negative fluctuations of Gaussian random potentials the authors of [51] were only able to prove algebraic instead of exponential decay of the eigenfunctions corresponding to the pure-point spectrum. This technical problem should be mastered by extending either the "variable-energy" multi-scale analysis [128, 41, 78] or the powerful bootstrap programme of Germinet and Klein [56, 58] to certain Schrödinger operators which are almost surely unbounded from below. Indeed, by doing the latter, Ueki [139] succeeded in showing exponential localization for certain Gaussian random potentials provided their covariance function is compactly supported. Along the same lines, he obtained first results on dynamical localization. The following theorem may be deduced by specializing his results.

Theorem 3.9 (cf. [139]). *Let V be a Gaussian random potential with covariance function C having property* (R) *in the sense of Definition 3.5. Additionally assume that C is compactly supported. Then there exists an energy $E_{pp} < 0$ such that*

(i) *for $\mathbb{P}$-almost all $\omega \in \Omega$ the spectrum of $H(A, V^{(\omega)})$ in the half-line $] - \infty, E_{pp}]$ is only pure point with exponentially localized eigenfunctions.*

(ii) *for every $E \in] - \infty, E_{pp}[$ there exists $\epsilon > 0$ such that strong dynamical localization holds in the energy interval $I :=]E - \epsilon, E + \epsilon[$ in the sense that*

$$\mathbb{E}\left[\sup_{t \in \mathbb{R}} \int_{\mathbb{R}^d} \mathrm{d}x \, \left|\psi_{t,I}(x)\right|^2 |x|^q \right] < \infty \qquad (3.12)$$

for all $q \geq 0$ and all $\psi_0 \in C_0^\infty(\mathbb{R}^d)$. [Recall the definition of $\psi_{t,I}^{(\omega)}$ below (1.11).]

Remark 3.10. As is shown in [139] the assumptions of Theorem 3.9 even imply dynamical localization in the (strong) Hilbert-Schmidt sense. In case of $\mathbb{Z}^d$-ergodic random potentials dynamical localization has been proven in various forms [54, 103, 55, 3, 15, 34, 131, 56, 57] under practically no further assumptions than required to prove spectral localization. It would be desirable to prove some sort of dynamical localization also in the situation of Theorem 3.7.

4 Results in Case of Poissonian Random Potentials

In comparison to Gaussian random potentials, less is known about regularity properties of the integrated density of states N and localization in case of Poissonian random potentials for arbitrary space dimension. Most results concern the Lifshits tail of N.

4.1 Lifshits Tails

In contrast to the case of Gaussian random potentials, there is a huge multiformity of Lifshits tails in the Poissonian case.

For vanishing magnetic field and non-negative Poissonian random potentials with rapidly decaying single-impurity potentials, the leading low-energy fall-off of N was first identified by Lifshits [96, 97]. Using the strategy already described after Theorem 3.1, a rigorous proof [113, 110] relies on Donsker and Varadhan's celebrated large-deviation result [36] on the long-time asymptotics of certain Wiener integrals. We summarize the rigorous version of Lifshits' result together with the corresponding one of Pastur [113] for slowly decaying single-impurity potentials in

Theorem 4.1 ([36, 113, 110]). *Let V be a non-negative Poissonian random potential with mean concentration $\varrho > 0$. Furthermore, assume that the single-impurity potential $U \geq 0$ has one of the following two decays at infinity:*

(D1) *U is compactly supported or has a decay more rapid than algebraic with exponent $d + 2$ in the sense that $\limsup_{|x| \to \infty} |x|^{d+2} U(x) = 0$,*

(D2) *U has (definite) algebraic decay with exponent $\alpha \in\ ]d, d + 2[$ in the sense that $\lim_{|x| \to \infty} |x|^\alpha U(x) = g$ with some constant $g > 0$.*

Moreover, assume that the magnetic field vanishes, $(B_{ij}) = 0$. Then the leading low-energy fall-off of the integrated density of states reads

$$\log N(E) \sim \begin{cases} -\varrho \left(\dfrac{\kappa_d}{2E} \right)^{d/2} & \text{for the decay (D1)} \\[2ex] -C_d(\alpha, \varrho) \left(\dfrac{g}{E} \right)^{d/(\alpha-d)} & \text{for the decay (D2)} \end{cases} \qquad (E \downarrow 0). \quad (4.1)$$

Here κ_d is the lowest eigenvalue of the negative Laplacian $2H(0,0)$, when Dirichlet restricted to a ball in $\mathbb{R}^d$ of unit volume. Moreover, we have introduced the constant $C_d(\alpha, \varrho) := \frac{\alpha-d}{d} \left[\varrho \frac{d}{\alpha} \frac{\pi^{d/2}}{\Gamma(1+d/2)} \Gamma\left(\frac{\alpha-d}{\alpha} \right) \right]^{\alpha/(\alpha-d)}.$

Remarks 4.2. (i) One has, for example, $\kappa_1 = \pi^2$, $\kappa_2 = \pi a_0^2$, where $a_0 = 2.4048\ldots$ is the smallest positive zero of the zeroth Bessel function of the first kind [60], and $\kappa_3 = \pi^2(4\pi/3)^{2/3}$.

(ii) Convincing arguments for the validity of Lifshits' result (4.1) for the decay (D1) were also given in [53, 101]. An alternative (rigorous) proof of the underlying long-time asymptotics is due to Sznitman who invented a coarse-graining scheme called the *method of enlargement of obstacles* [135]. More elementary proofs which rely on Dirichlet-Neumann bracketing were found in [77, 130, 79], but for the price of obtaining only the so-called *Lifshits exponent* (here: $d/2$) and not the other constants in (4.1) for the decay (D1). Ref. [79] also generalizes the Lifshits exponent $d/(\alpha - d)$ to an anisotropic version of the decay (D2).

(iii) As an aside, we note that (4.1) for the decay (D1) with $d = 1$ remains valid in the limiting case of Poissonian point impurities, $U(x) = u_0 \delta(x)$, $u_0 > 0$; see [43, 61, 84] and [116, Thm. 6.7].

For (non-negative) single-impurity potentials U with rapid decay (D1), the Lifshits tail is insensitive to the details of the decay of U and is dominated by the quantum kinetic energy. It has therefore a quantum character. Over against this, if $U \geq 0$ has the slow decay (D2), the Lifshits tail sensitively depends on the details of this decay. It is classical in character in that $\log N(E) \sim \log N_c(E)$ as $E \downarrow 0$. Therefore, the character of the tail (4.1) changes from purely quantum to purely classical, when the decay changes from (D1) to (D2). The Lifshits tail for the borderline case of algebraic decay with exponent $\alpha = d + 2$ seems to be open. In view of (1.9) we have the following

Conjecture 4.3. If $\lim_{|x| \to \infty} |x|^{d+2} U(x) = g$ with some constant $g > 0$, then

$$\log N(E) \sim -\left[\varrho^{\frac{2}{d+2}} \left(\frac{\kappa_d}{2E}\right)^{\frac{d}{d+2}} + [C_d(d+2, \varrho)]^{\frac{2}{d+2}} \left(\frac{g}{E}\right)^{\frac{d}{d+2}} \right]^{\frac{d+2}{2}} \quad (E \downarrow 0).$$
(4.2)

This tail exhibits a mixed quantum/classical character.

A similar transition from a purely quantum to a purely classical Lifshits tail can be observed in case of the random Landau Hamiltonian with non-negative Poissonian potential. However, since the Landau Hamiltonian possesses ground-state (wave) functions with Gaussian decay, the borderline decay of U turns out to be Gaussian and not algebraic (with exponent $d + 2 = 4$).

Theorem 4.4 ([25, 46, 66, 67, 47]). *Let $d = 2$ and $B > 0$. Suppose that V is a non-negative Poissonian random potential with mean concentration $\varrho > 0$. Furthermore, assume that the single-impurity potential $U \geq 0$ has one of the following three decays at infinity:*

- (D3) *U is compactly supported or has super-Gaussian decay in the sense that $\limsup_{|x| \to \infty} |x|^{-2} \log U(x) = -\infty$.*

- (D4) *U has (definite) Gaussian decay in the sense that $\limsup_{|x| \to \infty} |x|^{-2} \log U(x) = -\lambda^{-2}$ with some length $\lambda > 0$.*

- (D5) *U has sub-Gaussian decay in the sense that $\limsup_{|x| \to \infty} |x|^{-2} \log U(x) = 0$. Moreover, the decay of U is integrable and regular in the sense of [67, Def. 3.5].*

Then the leading low-energy fall-off of the integrated density of states reads

$$\log N\left(\frac{B}{2} + E\right) \sim \begin{cases} \dfrac{2}{B} \pi \varrho \log E & \text{for the decay (D3)} \\[2mm] \left(\dfrac{2}{B} + \lambda^2\right) \pi \varrho \log E & \text{for the decay (D4)} \\[2mm] \log N_c(E) & \text{for the decay (D5)} \end{cases} \quad (E \downarrow 0).$$
(4.3)

Remarks 4.5. (i) For super-Gaussian decay (D3) and Gaussian decay (D4) the integrated density of states has a power-law fall-off (on a logarithmic scale). The corresponding exponent $2\pi\varrho/B$ in (4.3) for the decay (D3) is just the mean number of impurities in a disc of radius $\sqrt{2/B}$. Two important examples for (D5) are an algebraic decay with exponent $\alpha \in$ $]2,\infty[$ (see Theorem 4.1) and a stretched-Gaussian decay in the sense that $\lim_{|x|\to\infty} |x|^{-\beta} \log U(x) = -\lambda^{-\beta}$ with some exponent $\beta \in]0,2[$ and some decay length $\lambda > 0$. For stretched-Gaussian decay the explicit form of (4.3) reads [66]

$$\log N(B/2 + E) \sim -\lambda^2 \pi\varrho |\log E|^{2/\beta} \qquad (E \downarrow 0). \qquad (4.4)$$

For an algebraic decay the explicit form of (4.3) coincides with the right-hand side of (4.1) for the decay (D2) with $d = 2$, even if $\alpha \geq 4 (= d+2)$ [25]. Other examples for (D5) causing more exotic fall-offs can be found in [67].

(ii) The hard part of the proof of Theorem 4.4 deals with the compactly supported case of (D3). It is due to Erdős who developed a version [46] of the method of enlargement of obstacles [135] which takes into account the presence of a magnetic field. With this method he also confirmed [47] the mixed quantum/classical character of the Lifshits tail in case of the Gaussian decay (D4), which was conjectured in [66].

We have seen that the Gaussian Lifshits tail of the integrated density of states N of the random Landau Hamiltonian with a Gaussian random potential is slower than that of R_0, the integrated density of states of its lowest band (since $\Gamma_0^2 < C(0)$ unless $C(x) = C(0)$ for all $x \in \mathbb{R}^2$, which however is ruled out by ergodicity). Not unexpectedly, this ceases to be so in case of a Poissonian random potential with a non-negative single-impurity potential.

Theorem 4.6 ([142]). *Assume the situation of Theorem 4.4. Furthermore, let the single-impurity potential be locally bounded, $U \in \mathrm{L}^\infty_{\mathrm{loc}}(\mathbb{R}^2)$. Then the leading low-energy fall-off of the restricted integrated density of states R_0 to zero near zero energy coincides with that of N near $B/2$, that is*

$$\log R_0(E) \sim \log N\left(\frac{B}{2} + E\right) \qquad (E \downarrow 0). \qquad (4.5)$$

The proof of Theorem 4.6 follows from Theorem 4.4 and a two-sided estimate on the Laplace-Stieltjes transform $\widetilde{R}_0$ of R_0. For the lower estimate see [25, Eq. (3.7)]. The upper estimate, $\widetilde{R}_0(\tau) \leq e^{\tau B/2} \widetilde{N}(\tau)\, 2\pi/B$, stems from the Jensen-Peierls inequality applied to the right-hand side of [66, Eq. (3.7)]. For the case of an algebraic decay, (4.5) has been given already in [25], and for compactly supported U implicitly in [46].

Remark 4.7. Some results on Lifshits tails of higher Landau bands ($l \geq 1$) are available in [25, 66]. The leading high-energy growth of R_l (for any $l \geq 0$) corresponding to a non-negative single-impurity potential U coincides with the leading low-energy fall-off of the lth restricted integrated density of states corresponding to $-U$, because $P_l - P_l \Theta(E - P_l V P_l)P_l = P_l \Theta(P_l V P_l - E)P_l$.

For all three types of decay (D3)–(D5) considered in Theorems 4.4 and 4.6, N is continuous at $B/2$ and hence R_0 is continuous at zero energy. This stands in contrast to the case of Poissonian *point*-impurities, $U(x) = u_0 \, \delta(x)$, $u_0 > 0$, for which Brézin, Gross and Itzykson [22] managed to calculate R_0 exactly and explicitly by using (non-rigorous) supersymmetric functional integration, see also [82]. For the leading low-energy fall-off they obtained

$$\lim_{E \downarrow 0} R_0(E) = 1 - \frac{2\pi\varrho}{B} \qquad \text{if} \qquad \frac{2\pi\varrho}{B} < 1, \qquad (4.6)$$

$$\lim_{E \downarrow 0} |\log E| \, R_0(E) = 1 \qquad \text{if} \qquad \frac{2\pi\varrho}{B} = 1, \qquad (4.7)$$

$$\lim_{E \downarrow 0} \frac{\log R_0(E)}{|\log E|} = 1 - \frac{2\pi\varrho}{B} \qquad \text{if} \qquad \frac{2\pi\varrho}{B} > 1. \qquad (4.8)$$

Erdős [46] has given a rigorous proof that the size of the jump of R_0 at zero energy in case $2\pi\varrho/B < 1$ is not smaller than the right-hand side of (4.6).

We close this subsection with two complementary remarks.

Remarks 4.8.   (i)   First rigorous results on Lifshits tails in case of three space dimensions $d = 3$ and a constant magnetic field of strength $B > 0$ are available. For super-Gaussian decay (D3), Gaussian decay (D4) and stretched-Gaussian decay the so-called Lifshits exponent is identified [142] to coincide in all three cases with the corresponding one for $d = 1$,

$$-\lim_{E \downarrow 0} \frac{\log \left| \log N(B/2 + E) \right|}{\log E} = \frac{1}{2}. \qquad (4.9)$$

This may be ascribed to the effective zero-field motion of the particle parallel to the magnetic field, which dominates the low-energy asymptotics. Actually, in [142] somewhat more detailed information on the fall-offs depending on the actual decay can be found. For example, for algebraic decay (D2) with $d = 3$, that is, with exponent $\alpha \in]3, 5[$, the tail coincides with the corresponding one for $B = 0$ (see Theorem 4.1) and has therefore a classical character [64]. For heuristic explanations of (4.3) and their (conjectured) analogues for $d = 3$, see [95].

(ii)   For vanishing magnetic field, Lifshits tails have been investigated also for non-positive single-impurity potentials, $U \leq 0$. Depending on whether U is singular or not, the corresponding tail exhibits a quantum or classical character. For details, see [113, 99, 116, 85]. The results are again consistent with formula (1.9) (setting there $E_0 = -\infty$).

4.2 Existence of the Density of States and Spectral Localization

There are only a few rigorous results on these issues for Poissonian random potentials. For a special class of non-negative single-impurity potentials U, Tip [137] has proven that the integrated density of states N is absolutely continuous at sufficiently high energies. The only localization result is due to Stolz [132]. It concerns the case of one space dimension.

Theorem 4.9 ([132]). *Let $d = 1$ and let V be a non-negative Poissonian random potential. Moreover, let the single-impurity potential $U \geq 0$ be compactly supported. Then the almost-sure spectrum $[0, \infty[$ of $H(0, V)$ is only pure point with exponentially localized eigenfunctions.*

The proof builds on techniques which are available only for one dimension and are nicely summarized and discussed in the recent survey [133].

Remark 4.10. To our knowledge, the only other rigorous works [104, 31] which, among the rest, deal with localization proofs for Poissonian random potentials in multi-dimensional situations have to assume an additional randomness of the impurity coupling-strengths.

5 Some Open Problems

While most rigorous works on random Schrödinger operators concern $\mathbb{Z}^d$-ergodic random potentials, the present survey has focused on $\mathbb{R}^d$-ergodic ones. More precisely, for Gaussian and Poissonian random potentials, rigorous results have been presented on the integrated density of states and Anderson localization. In this context, a lot of issues, which are interesting from the (theoretical-) physics point of view, are still unsolved.

One major open problem concerns a proof of Anderson localization in case of (non-negative) Poissonian random potentials for arbitrary space dimension. Another problem is to isolate the weakest possible conditions for an $\mathbb{R}^d$-ergodic random potential which imply continuity of the corresponding integrated density of states if $d \geq 2$ and $A = 0$. In particular, one may ask: Is $\mathbb{R}^d$-ergodicity already enough?

Most striking is definitely the fact that there is not a single non-zero $\mathbb{R}^d$- or $\mathbb{Z}^d$-ergodic random potential for which the existence of an absolutely continuous component in the energy spectrum has been proven, that is, spectral delocalization in certain energy regimes. After all, physical intuition, approximate calculations and numerical studies suggest the occurrence of a mobility edge if $d \geq 3$. If $d = 2$ and $A = 0$ it is not yet rigorously settled whether the whole energy spectrum is almost surely only pure point or not.

Last but not least, one can hardly claim to utterly understand electronic properties of disordered solids without having a more solid foundation of their transport theory.

Acknowledgement Our thanks go to Alexandra Weichlein for helpful remarks. This work was partially supported by the Deutsche Forschungsgemeinschaft (DFG) under grant no. Le 330/12. The latter is a project within the DFG Priority Programme SPP 1033 "Interagierende stochastische Systeme von hoher Komplexität". Peter Müller acknowledges partial financial support of the DFG under grant no. Zi 209/6-1 and SFB 602.

References

1. Adler, R.J.: The geometry of random fields. Wiley, Chichester (1981) [A major revision, co-authored by J. Taylor, will appear with Birkhäuser, Boston]
2. Aizenman, M.: Localization at weak disorder: some elementary bounds. Rev. Math. Phys., **6**, 1163–1182 (1994)
3. Aizenman, M., Graf, G.M.: Localization bounds for an electron gas. J. Phys. A, **31**, 6783–6806 (1998)
4. Aizenman, M., Molchanov, S.: Localization at large disorder and at extreme energies: an elementary derivation. Commun. Math. Phys., **157**, 245–278 (1993)
5. Aizenman, M., Schenker, J.H., Friedrich, R.M., Hundertmark, D.: Constructive fractional-moment criteria for localization in random operators. Physica A, **279**, 369–377 (2000)
6. Aizenman, M., Schenker, J.H., Friedrich, R.M., Hundertmark, D.: Finite-volume criteria for Anderson localization. Commun. Math. Phys., **224**, 219–253 (2001)
7. Aizenman, M., Elgart, A., Naboko, S., Schenker, J.H., Stolz, G.: Moment analysis for localization in random Schrödinger operators. Preprint math-ph/0308023
8. Amrein, W.O.: Non-relativistic quantum dynamics. Reidel, Dordrecht (1981)
9. Anderson, P.W.: Absence of diffusion in certain random lattices. Phys. Rev., **109**, 1492–1505 (1958)
10. Ando, T., Fowler, A.B., Stern, F.: Electronic properties of two-dimensional systems. Rev. Mod. Phys., **54**, 437–672 (1982)
11. Apel, W.: Asymptotic density of states for a $2D$ electron system in a strong magnetic field. J. Phys. C, **20**, L577–L581 (1987)
12. Avron, J.E., Sadun, L.: Fredholm indices and the phase diagram of quantum Hall systems. J. Math. Phys., **42**, 1–14 (2001)
13. Barbaroux, J.-M., Combes, J.M., Hislop, P.D.: Landau Hamiltonians with unbounded random potentials. Lett. Math. Phys., **40**, 335–369 (1997)
14. Barbaroux, J.M., Combes, J.M., Montcho, R.: Remarks on the relation between quantum dynamics and fractal spectra. J. Math. Anal. Appl., **213**, 698–722 (1997)
15. Barbaroux, J.M., Fischer, W., Müller, P.: Dynamical properties of random Schrödinger operators. Preprint math-ph/9907002
16. Bellissard, J., van Elst, A., Schulz-Baldes, H.: The non-commutative geometry of the quantum Hall effect. J. Math. Phys., **35**, 5373–5451 (1994)
17. Benedict, K.A.: The fate of the Lifshitz tails of high Landau levels. Nucl. Phys. B, **280**, 549–560 (1987)
18. Benedict, K.A., Chalker, J.T.: An exactly solvable model of the disordered two-dimensional electron gas in a strong magnetic field. J. Phys. C, **19**, 3587–3604 (1986)

19. Bingham, N.H., Goldie, C.M., Teugels, J.L.: Regular variation. Paperback edition with additions. Cambridge UP, Cambridge (1989)
20. Böhm, M., Broderix, K., Leschke, H.: Broadening of the lowest Landau level by a Gaussian random potential with an arbitrary correlation length: an efficient continued–fraction approach. Z. Physik B, **104**, 111–115 (1997)
21. Bonch-Bruevich, V.L., Enderlein, R., Esser, B., Keiper, R., Mironov, A.G., Zvyagin, I.P.: Elektronentheorie ungeordneter Halbleiter. VEB Deutscher Verlag der Wissenschaften, Berlin (1984) [in German. Russian original: Nauka, Moscow (1981)]
22. Brézin, E., Gross, D.J., Itzykson, C.: Density of states in the presence of a strong magnetic field and random impurities. Nucl. Phys. B, **235**, 24–44 (1984)
23. Broderix, K., Heldt, N., Leschke, H.: Partition function and the density of states for an electron in the plane subjected to a random potential and a magnetic field. Phys. Rev. B, **40**, 7479–7486 (1989)
24. Broderix, K., Heldt, N., Leschke, H.: Exact results on Landau-level broadening. J. Phys. A: Math. Gen., **24**, L825–L831 (1991)
25. Broderix, K., Hundertmark, D., Kirsch, W., Leschke, H.: The fate of Lifshits tails in magnetic fields. J. Stat. Phys., **80**, 1–22 (1995)
26. Broderix, K., Hundertmark, D., Leschke, H.: Self-averaging, decomposition and asymptotic properties of the density of states for random Schrödinger operators with constant magnetic field. In: Grabert, H., Inomata, A., Schulman, L.S., Weiss, U. (eds) Path integrals from meV to MeV: Tutzing '92. World Scientific, Singapore (1993), pp. 98–107
27. Broderix, K., Hundertmark, D., Leschke, H.: Continuity properties of Schrödinger semigroups with magnetic fields. Rev. Math. Phys., **12**, 181–225 (2000)
28. Broderix, K., Leschke, H., Müller, P.: Continuous integral kernels for unbounded Schrödinger semigroups and their spectral projections. J. Funct. Anal., **212**, 287–323 (2004)
29. Carmona, R., Lacroix, J.: Spectral theory of random Schrödinger operators. Birkhäuser, Boston (1990)
30. Chistyakov, G., Lyubarskii, Yu., Pastur, L.: On completeness of random exponentials in the Bargmann-Fock space. J. Math. Phys., **42**, 3754–3768 (2001)
31. Combes, J.-M., Hislop, P.D.: Localization for some continuous, random Hamiltonians in d-dimensions. J. Funct. Anal., **124**, 149–180 (1994)
32. Combes, J.M., Hislop, P.D.: Landau Hamiltonians with random potentials: Localization and the density of states. Commun. Math. Phys., **177**, 603–629 (1996)
33. Cycon, H.L., Froese, R.G., Kirsch, W., Simon, B.: Schrödinger operators. Springer, Berlin (1987)
34. Damanik, D., Stollmann, P.: Multi-scale analysis implies strong dynamical localization. Geom. Funct. Anal., **11**, 11–29 (2001)
35. Doi, S., Iwatsuka, A., Mine, T.: The uniqueness of the integrated density of states for the Schrödinger operators with magnetic fields. Math. Z., **237**, 335–371 (2001)
36. Donsker, M.D., Varadhan, S.R.S.: Asymptotics of the Wiener sausage. Commun. Pure Appl. Math., **28**, 525–565 (1975); Errata: *ibid.*, 677–678
37. Dorlas, T.C., Macris, N., Pulé, J.V.: Localisation in a single-band approximation to random Schrödinger operators in a magnetic field. Helv. Phys. Acta, **68**, 329–364 (1995)

38. Dorlas, T.C., Macris, N., Pulé, J.V.: Localization in single Landau bands. J. Math. Phys., **37**, 1574–1595 (1996)

39. Dorlas, T.C., Macris, N., Pulé, J.V.: The nature of the spectrum for a Landau Hamiltonian with delta impurities. J. Stat. Phys., **87**, 847–875 (1997)

40. Dorlas, T.C., Macris, N., Pulé, J.V.: Characterization of the spectrum of the Landau Hamiltonian with delta impurities. Commun. Math. Phys., **204**, 367–396 (1999)

41. von Dreifus, H., Klein, A.: A new proof of localization in the Anderson tight binding model. Commun. Math. Phys., **124**, 285–299 (1989)

42. von Dreifus, H., Klein, A.: Localization for random Schrödinger operators with correlated potentials. Commun. Math. Phys., **140**, 133–147 (1991)

43. Eggarter, T.P.: Some exact results on electron energy levels in certain one-dimensional random potentials. Phys. Rev. B, **5**, 3863–3865 (1972)

44. Efetov, K.: Supersymmetry in disorder and chaos. Cambridge UP, Cambridge (1997)

45. Elbau, P., Graf, G.M.: Equality of bulk and edge Hall conductance revisited. Commun. Math. Phys., **229**, 415–432 (2002)

46. Erdős, L.: Lifschitz tail in a magnetic field: the non-classical regime. Probab. Theory Relat. Fields, **112**, 321–371 (1998)

47. Erdős, L.: Lifshitz tail in a magnetic field: coexistence of classical and quantum behavior in the borderline case. Probab. Theory Relat. Fields, **121**, 219–236 (2001)

48. Fernique, X.M.: Regularité des trajectoires des fonctions aléatoires Gaussiennes. In: Hennequin, P.-L. (ed) Ecole d'Eté de Probabilités de Saint-Flour IV - 1974. Lecture Notes in Mathematics **480**. Springer, Berlin (1975), pp. 1–96 [in French]

49. Figotin, A., Klein, A.: Localization of classical waves I: acoustic waves. Commun. Math. Phys., **180**, 439–482 (1996)

50. Fischer, W., Hupfer, T., Leschke, H., Müller, P.: Existence of the density of states for multi-dimensional continuum Schrödinger operators with Gaussian random potentials. Commun. Math. Phys., **190**, 133–141 (1997)

51. Fischer, W., Leschke, H., Müller, P.: Spectral localization by Gaussian random potentials in multi-dimensional continuous space. J. Stat. Phys., **101**, 935–985 (2000)

52. Fock, V.: Bemerkung zur Quantelung des harmonischen Oszillators im Magnetfeld. Z. Physik, **47**, 446–448 (1928) [in German]

53. Friedberg, R., Luttinger, J.M.: Density of electronic levels in disordered systems. Phys. Rev. B, **12**, 4460–4474 (1975)

54. Fröhlich, J., Spencer, T.: Absence of diffusion in the Anderson tight binding model for large disorder or low energy. Commun. Math. Phys., **88**, 151–184 (1983)

55. Germinet, F., De Bièvre, S.: Dynamical localization for discrete and continuous random Schrödinger operators. Commun. Math. Phys., **194**, 323–341 (1998)

56. Germinet, F., Klein, A.: Bootstrap multiscale analysis and localization in random media. Commun. Math. Phys., **222**, 415–448 (2001)

57. Germinet, F., Klein, A.: A characterization of the Anderson metal-insulator transport transition. Preprint mp_arc 02-19, to appear in Duke Math. J.

58. Germinet, F., Klein, A.: Explicit finite volume criteria for localization in continuous random media and applications. Geom. Funct. Anal., **13**, 1201–1238 (2003)
59. Gersten, J.I., Smith, F.W.: The physics and chemistry of materials. Wiley, New York (2001)
60. Gradshteyn, I.S., Ryzhik, I.M.: Table of integrals, series, and products. Corrected and enlarged edition. Academic, San Diego (1980)
61. Gredeskul, S.A., Pastur, L.A.: Behavior of the density of states in one-dimensional disordered systems near the edges of the spectrum. Theor. Math. Phys., **23**, 404-409 (1975) [Russian original: Teor. Mat. Fiz., **23**, 132-139 (1975)]
62. Halperin, B.I., Lax, M.: Impurity-band tails in the high-density limit. I. Minimum counting methods. Phys. Rev., **148**, 722–740 (1966)
63. Huckestein, B.: Scaling theory of the integer quantum Hall effect. Rev. Mod. Phys., **67**, 357–396 (1995)
64. Hundertmark, D., Kirsch, W., Warzel, S.: Classical magnetic Lifshits tails in three space dimensions: impurity potentials with slow anisotropic decay. Markov Process. Relat. Fields, **9**, 651–660 (2003)
65. Hundertmark, D., Simon, B.: A diamagnetic inequality for semigroup differences. J. Reine Angew. Math., **571**, 107–130 (2004)
66. Hupfer, T., Leschke, H., Warzel, S.: Poissonian obstacles with Gaussian walls discriminate between classical and quantum Lifshits tailing in magnetic fields. J. Stat. Phys., **97**, 725–750 (1999)
67. Hupfer, T., Leschke, H., Warzel, S.: The multiformity of Lifshits tails caused by random Landau Hamiltonians with repulsive impurity potentials of different decay at infinity. AMS/IP Stud. Adv. Math., **16**, 233–247 (2000)
68. Hupfer, T., Leschke, H., Warzel, S.: Upper bounds on the density of states of single Landau levels broadened by Gaussian random potentials. J. Math. Phys., **42**, 5626–5641 (2001)
69. Hupfer, T., Leschke, H., Müller, P., Warzel, S.: The absolute continuity of the integrated density of states for magnetic Schrödinger operators with certain unbounded random potentials. Commun. Math. Phys., **221**, 229–254 (2001)
70. Hupfer, T., Leschke, H., Müller, P., Warzel, S.: Existence and uniqueness of the integrated density of states for Schrödinger operators with magnetic fields and unbounded random potentials. Rev. Math. Phys., **13**, 1547–1581 (2001)
71. Janßen, M., Viehweger, O., Fastenrath, U., Hajdu, J.: Introduction to the theory of the integer quantum Hall effect. VCH, Weinheim (1994)
72. Kane, E.O.: Thomas-Fermi approach to impure semiconductor band structure. Phys. Rev., **131**, 79–88 (1963)
73. Kirsch, W.: Random Schrödinger operators: a course. In: Holden, H., Jensen, A. (eds) Schrödinger operators. Lecture Notes in Physics **345**. Springer, Berlin (1989), pp. 264–370
74. Kirsch, W., Martinelli, F.: On the ergodic properties of the spectrum of general random operators. J. Reine Angew. Math., **334**, 141–156 (1982)
75. Kirsch, W., Martinelli, F.: On the density of states of Schrödinger operators with a random potential. J. Phys. A: Math. Gen., **15**, 2139–2156 (1982)
76. Kirsch, W., Martinelli, F.: On the essential selfadjointness of stochastic Schrödinger operators. Duke Math. J., **50**, 1255–1260 (1983)

148   Hajo Leschke, Peter Müller, and Simone Warzel

77. Kirsch, W., Martinelli, F.: Large deviations and Lifshitz singularity of the
    integrated density of states of random Hamiltonians. Commun. Math. Phys.,
    89, 27–40 (1983)
78. Kirsch, W., Stollmann, P., Stolz, G.: Anderson localization for random Schrö-
    dinger operators with long range interactions. Commun. Math. Phys., **195**,
    495–507 (1998)
79. Kirsch, W., Warzel, S.: Lifshits tails caused by anisotropic decay: the emer-
    gence of a quantum-classical regime. Preprint math-ph/0310033
80. Kiselev, A., Last, Y.: Solutions, spectrum, and dynamics for Schrödinger
    operators on infinite domains. Duke Math. J., **102**, 125–150 (2000)
81. Kittel, C.: Introduction to solid-state physics. 7th edition. Wiley, New York
    (1996)
82. Klein, A., Perez, J.F.: On the density of states for random potentials in the
    presence of a uniform magnetic field. Nucl. Phys. B, **251**, 199–211 (1985)
83. von Klitzing, K.: The quantized Hall effect. Rev. Mod. Phys., **58**, 519–531
    (1986)
84. Kotani, S.: On asymptotic behaviour of the spectra of a one-dimensional
    Hamiltonian with a certain random coefficient. Publ. Res. Inst. Math. Sci.,
    Kyoto Univ., **12**, 447–492 (1976)
85. Klopp, F., Pastur, L.: Lifshitz tails for random Schrödinger operators with
    negative singular Poisson potential. Commun. Math. Phys., **206**, 57–103
    (1999)
86. Kovalenko, N.P., Krasny, Y.P., Krey, U.: Physics of amorphous metals. Wiley
    VCH, Berlin (2001)
87. Kramer, B., MacKinnon, A.: Localization: theory and experiment. Rep. Prog.
    Phys., **56**, 1469–1564 (1993)
88. Krengel, U.: Ergodic theorems. de Gruyter, Berlin (1985)
89. Kubo, R.: Statistical mechanics. North Holland, Amsterdam (1965)
90. Kuchment, P.: Floquet theory for partial differential equations. Birkhäuser,
    Boston (1993)
91. Kukushkin, I.V., Meshkov, S.V., Timofeev, V.B.: Two-dimensional electron
    density of states in a transverse magnetic field. Sov. Phys. Usp., **31**, 511–534
    (1988) [Russian original: Usp. Fiz. Nauk, **155**, 219–264 (1988)]
92. Kunz, H.: The quantum Hall effect for electrons in a random potential. Com-
    mun. Math. Phys., **112**, 121–145 (1987)
93. Landau, L.: Diamagnetismus der Metalle. Z. Physik, **64**, 629–637 (1930) [in
    German]
94. Last, Y.: Quantum dynamics and decompositions of singular continuous spec-
    tra. J. Funct. Anal., **142**, 406–445 (1996)
95. Leschke, H., Warzel, S.: Quantum-classical transitions in Lifshitz tails with
    magnetic fields. Phys. Rev. Lett., **92**, 086402:1–4 (2004). For a slightly longer
    version see the preprint cond-mat/0310389
96. Lifshitz, I.M.: Structure of the energy spectrum of the impurity bands in
    disordered solid solutions. Sov. Phys. JETP, **17**, 1159–1170 (1963) [Russian
    original: Zh. Eksp. Teor. Fiz., **44**, 1723–1741 (1963)]
97. Lifshitz, I.M.: The energy spectrum of disordered systems. Adv. Phys., **13**,
    483–536 (1964); Energy spectrum structure and quantum states of disordered
    condensed systems. Sov. Phys. Usp., **7**, 549–573 (1965) [Russian original: Usp.
    Fiz. Nauk., **83**, 617–663 (1964)]

98. Lifshitz, I.M.: Theory of fluctuating levels in disordered systems. Sov. Phys. JETP, **26**, 462–479 (1968) [Russian original: Zh. Eksp. Teor. Fiz., **53**, 743–758 (1967)]

99. Lifshits, I.M., Gredeskul, S.A., Pastur, L.A.: Introduction to the theory of disordered systems. Wiley, New York (1988) [Russian original: Nauka, Moscow (1982)]

100. Lifshits, M.A.: Gaussian random functions. Kluwer, Dordrecht (1995)

101. Luttinger, J.M.: New variational method with applications to disordered systems. Phys. Rev. Lett., **37**, 609–612 (1976)

102. Macris, N., Pulé, J.V.: Density of states of random Schrödinger operators with uniform magnetic field. Lett. Math. Phys., **24**, 307–321 (1992)

103. Martinelli, F., Holden, H.: On absence of diffusion near the bottom of the spectrum for a random Schrödinger operator on $L^\cdot(\mathbb{R}^\nu)$. Commun. Math. Phys., **93**, 197–217 (1984)

104. Martinelli, F., Scoppola, E.: Introduction to the mathematical theory of Anderson localization. Rivista del Nuovo Cimento, **10** (10), 1–90 (1987)

105. Matsumoto, H.: On the integrated density of states for the Schrödinger operators with certain random electromagnetic potentials. J. Math. Soc. Japan, **45**, 197–214 (1993)

106. Mattis, D. C.: The theory of magnetism I. Corrected 2nd printing. Springer, Berlin (1988)

107. Mine, T.: The uniqueness of the integrated density of states for the Schrödinger operators for the Robin boundary conditions. Publ. Res. Inst. Math. Sci., Kyoto Univ., **38**, 355-385 (2002)

108. Mott, N.F.: Electrons in disordered structures. Adv. Phys., **16**, 49–144 (1967)

109. Nakamura, S.: A remark on the Dirichlet-Neumann decoupling and the integrated density of states. J. Funct. Anal., **179**, 136–152 (2001)

110. Nakao, S.: On the spectral distribution of the Schrödinger operator with random potential. Japan. J. Math., **3**, 111–139 (1977)

111. Pastur, L.A.: On the Schrödinger equation with a random potential. Theor. Math. Phys., **6**, 299-306 (1971) [Russian original: Teor. Mat. Fiz., **6**, 415–424 (1971)]

112. Pastur, L.A.: On the distribution of the eigenvalues of the Schrödinger equation with a random potential. Funct. Anal. Appl., **6**, 163–165 (1972) [Russian original: Funkts. Anal. Prilozh., **6**, 93–94 (1972)]

113. Pastur, L.A.: Behavior of some Wiener integrals as $t \to \infty$ and the density of states of Schrödinger equations with random potential. Theor. Math. Phys., **32**, 615–620 (1977) [Russian original: Teor. Mat. Fiz., **32**, 88–95 (1977)]

114. Pastur, L.A.: Spectral properties of disordered systems in the one-body approximation. Commun. Math. Phys., **75**, 179–196 (1980)

115. Pastur, L.A.: Spectral properties of random selfadjoint operators and matrices (a survey). Transl., Ser. 2, Am. Math. Soc., **188**, 153–195 (1999) [Russian original: Tr. St-Peterbg. Mat. Obshch., **4**, 222–286 (1996)]

116. Pastur, L., Figotin, A.: Spectra of random and almost-periodic operators. Springer, Berlin (1992)

117. Pulé, J.V., Scrowston, M.: Infinite degeneracy for a Landau Hamiltonian with Poisson impurities. J. Math. Phys., **38**, 6304–6314 (1997)

118. Pulé, J.V., Scrowston, M.: The spectrum of a magnetic Schrödinger operator with randomly located delta impurities. J. Math. Phys., **41**, 2805–2825 (2000)

119. del Rio, R., Jitomirskaya, S., Last, Y., Simon, B.: What is localization? Phys. Rev. Lett., **75**, 117–119 (1995)
120. del Rio, R., Jitomirskaya, S., Last, Y., Simon, B.: Operators with singular continuous spectrum IV: Hausdorff dimensions, rank one perturbations and localization. J. d'Analyse Math., **69**, 153–200 (1996)
121. Ruelle, D.: Statistical mechanics: rigorous results. Imperial College Press, London (1999)
122. Salomon, R.: Density of states for high Landau levels and random potential. Z. Physik B, **65**, 443–451 (1987)
123. Shklovskii, B.I., Efros, A.L.: Electronic properties of doped semiconductors. Springer, Berlin (1984) [Russian original: Nauka, Moscow (1979)]
124. Simon, B.: Schrödinger semigroups. Bull. Am. Math. Soc. (N.S.), **7**, 447–526 (1982); Erratum: *ibid.*, **11**, 426 (1984)
125. Simon, B.: Absence of ballistic motion. Commun. Math. Phys., **134**, 209–212 (1990)
126. Sommerfeld, A.: Zur Elektronentheorie der Metalle auf Grund der Fermischen Statistik. I. Teil: Allgemeines, Strömungs= und Austrittsvorgänge. Z. Physik, **47**, 1–42 (1928) [in German]
127. Spencer, T.C.: The Schrödinger equation with a random potential: a mathematical review. In: Osterwalder, K., Stora, R. (eds) Critical phenomena, random systems, gauge theories. Part II. Noth-Holland, Amsterdam (1986), pp. 895–943
128. Spencer, T.: Localization for random and quasiperiodic potentials. J. Stat. Phys., **51**, 1009–1019 (1988)
129. Störmer, H.L., Tsui, D.C., Gossard, A.C.: The fractional quantum Hall effect. Rev. Mod. Phys., **71**, S298–S305 (1999)
130. Stollmann, P.: Lifshitz asymptotics via linear coupling of disorder. Math. Phys. Anal. Geom., **2**, 279–289 (1999)
131. Stollmann, P.: Caught by disorder: bound states in random media. Birkhäuser, Boston (2001)
132. Stolz, G.: Localization for random Schrödinger operators with Poisson potential. Ann. Inst. Henri Poincaré, Phys. Théor., **63**, 297–314 (1995)
133. Stolz, G.: Strategies in localization proofs for one-dimensional random Schrödinger operators. Proc. Indian Acad. Sci., Math. Sci., **112**, 229–243 (2002)
134. Suslov, I. M.: Development of a $(4 - \epsilon)$-dimensional theory for the density of states of a disordered system near the Anderson transition. Phys. Usp., **41**, 441–467 (1998) [Russian original: Usp. Fiz. Nauk, **168**, 503–530 (1998)]
135. Sznitman, A.-S.: Brownian motion, obstacles and random media. Springer, Berlin (1998)
136. Tcheremchantsev, S.: How to prove dynamical localization. Commun. Math. Phys., **221**, 27-56 (2001)
137. Tip, A.: Absolute continuity of the integrated density of states of the quantum Lorentz gas for a class of repulsive potentials. J. Phys. A: Math. Gen., **27**, 1057–1069 (1994)
138. Ueki, N.: On spectra of random Schrödinger operators with magnetic fields. Osaka J. Math., **31**, 177–187 (1994)
139. Ueki, N.: Wegner estimates and localization for Gaussian random potentials. Publ. Res. Inst. Math. Sci., Kyoto Univ., **40**, 29–90 (2004)
140. Veselić, I.: Integrated density of states and Wegner estimates for random Schrödinger operators. Contemp. Math., **340**, 97–183 (2004)

141. Wang, Wei-Min: Microlocalization, percolation, and Anderson localization for the magnetic Schrödinger operator with a random potential. J. Funct. Anal., **146**, 1–26 (1997)
142. Warzel, S.: On Lifshits tails in magnetic fields. Logos, Berlin (2001) [PhD-Thesis, Universität Erlangen-Nürnberg]
143. Wegner, F.: Bounds on the density of states in disordered systems. Z. Physik B, **44**, 9–15 (1981)
144. Wegner, F.: Exact density of states for lowest Landau level in white noise potential. Superfield representation for interacting systems. Z. Physik B, **51**, 279–285 (1983)
145. Weidmann, J.: Lineare Operatoren in Hilberträumen (Teil II: Anwendungen). Teubner, Stuttgart (2003) [in German]
146. Weyl, H.: Das asymptotische Verteilungsgesetz der Eigenwerte linearer partieller Differentialgleichungen (mit einer Anwendung auf die Theorie der Hohlraumstrahlung). Math. Ann., **71**, 441–479 (1912) [in German]
147. Yoshioka, D.: The quantum Hall effect. Springer, Berlin (2002)
148. Zak, J.: Magnetic translation group. Phys. Rev., **134**, A1602–A1606 (1964)
149. Zallen, R.: The physics of amorphous solids. Wiley, New York (1983)
150. Ziman, J.: Models of disorder. Cambridge UP, Cambridge (1979)
151. Zittartz, J., Langer, J.S.: Theory of bound states in a random potential. Phys. Rev., **148**, 741–747 (1966)

The Parabolic Anderson Model[*]

Jürgen Gärtner[1] and Wolfgang König[1]

Technische Universität Berlin, Institut für Mathematik, MA7-5
Straße des 17. Juni 136, D-10623 Berlin, Germany
koenig@math.tu-berlin.de

Summary. This is a survey on the intermittent behavior of the parabolic Anderson model, which is the Cauchy problem for the heat equation with random potential on the lattice $\mathbb{Z}^d$. We first introduce the model and give heuristic explanations of the long-time behavior of the solution, both in the annealed and the quenched setting for time-independent potentials. We thereby consider examples of potentials studied in the literature. In the particularly important case of an i.i.d. potential with double-exponential tails we formulate the asymptotic results in detail. Furthermore, we explain that, under mild regularity assumptions, there are only four different universality classes of asymptotic behaviors. Finally, we study the moment Lyapunov exponents for space-time homogeneous catalytic potentials generated by a Poisson field of random walks.

Key words: Parabolic Anderson problem, heat equation with random potential, intermittency, Feynman-Kac formula, random environment.

1 Introduction and Heuristics

1.1 Evolution of Spatially Distributed Systems in Random Media

One of the often adequate and frequently used methods for studying the evolution of spatially distributed systems under the influence of a random medium is *homogenization*. After rescaling, the system, modeled by partial differential equations with random coefficients, is approached by a system with 'properly averaged' deterministic coefficients, see e.g. [ZKO94]. But there are simple and important situations when random systems exhibit effects which cannot be recovered by such deterministic approximations and related fluctuation corrections. This concerns, in particular, *localization* effects for non-reversible random walks in random environment [S82] and for the electron transport in disordered media [And58].

Another such effect is that of *intermittency*. Roughly speaking, intermittency means that the solution of the system develops pronounced spatial structures on islands located far from each other that, in one or another

---

[*]Partially supported by the German Science Foundation, Schwerpunkt project SPP 1033

sense, deliver the main output to the system. One of the sources of interest is magnetohydrodynamics and, in particular, the investigation of the induction equation with incompressible random velocity fields [Z84], [ZMRS87]. Another source are simple mathematical models such as the random Fisher-Eigen equation that have been used to derive caricatures of Darwinian evolution principles [EEEF84].

One of the simplest and most basic models exhibiting the effect of intermittency is the Cauchy problem for the spatially discrete heat equation with a random potential:

$$\partial_t u(t, x) = \kappa \Delta u(t, x) + \xi(t, x)u(t, x), \qquad (t, x) \in (0, \infty) \times \mathbb{Z}^d,$$
$$u(0, x) = u_0(x), \qquad\qquad\qquad\qquad x \in \mathbb{Z}^d. \qquad (1)$$

Here $\kappa > 0$ is a diffusion constant, Δ denotes the discrete Laplacian,

$$\Delta f(x) = \sum_{y:\, |y-x|=1} [f(y) - f(x)],$$

ξ is a space-time homogeneous ergodic random potential, and u_0 is a non-negative initial function. Problem (1) is often called *parabolic problem for the Anderson model* or *parabolic Anderson model* (abbreviated *PAM*). As simplest localized initial datum one may take $u_0 = \delta_0$, and as non-local initial datum $u_0 = \mathbb{1}$. In the latter case, the solution $u(t, \cdot)$ is spatially homogeneous and ergodic for each t. Let us remark that the solution $u(t, x)$ of (1) allows the interpretation as average number of particles at site x at time t for branching random walks in random media given a realization of the medium ξ, cf. [CM94] and the remarks in the next subsection.

1.2 The PAM with Time-Independent Potential

In the particular case when the potential $\xi(t, x) = \xi(x)$ is time-independent, the large-time behavior of the solution u to the PAM (1) is determined by the spectral properties of the Anderson Hamiltonian

$$\mathcal{H} = \kappa \Delta + \xi \qquad (2)$$

and therefore closely related to the mentioned localization of the electron transport. Namely, since (under natural assumptions on ξ) the upper part of the spectrum of $\mathcal{H}$ in $\ell^2(\mathbb{Z}^d)$ is a pure point spectrum [FMSS85], [AM93], the solution u admits the spectral representation

$$u(t, \cdot) = \sum_i e^{\lambda_i t} (v_i, u_0) v_i(\cdot) \qquad (3)$$

with respect to the random eigenvalues λ_i and the corresponding exponentially localized random eigenfunctions v_i. (For simplicity we ignore the possible occurrence of a continuous central part of the spectrum.) As t increases unboundedly, only summands with larger and larger eigenvalues will contribute to (3), and the corresponding eigenfunctions are expected to be localized more and more far from each other. Hence, for large t, the solution $u(t, \cdot)$ indeed looks like a weighted superposition of high peaks concentrated on distant islands.

A mathematically rigorous understanding of the nature of the spectrum of the Anderson Hamiltonian and the random Schrödinger operator is still far from being complete. For an overview about some recent developments we refer to the surveys [BKS04] and [LMW04] in this proceedings volume. The spectral results obtained so far do not yet seem directly applicable to answer the crucial questions about intermittency. A direct spectral approach clearly fails for space-time dependent potentials.

In this survey we present a part of the results about intermittency for the PAM which have been obtained by use of more intrinsic probabilistic methods. In the next subsections we stick to the PAM with time-independent potential and localized initial datum:

$$\begin{aligned}
\partial_t u(t,x) &= \kappa \Delta u(t,x) + \xi(x)u(t,x), & (t,x) &\in (0,\infty) \times \mathbb{Z}^d, \\
u(0,x) &= \delta_0(x), & x &\in \mathbb{Z}^d.
\end{aligned} \tag{4}$$

We assume throughout that $\xi = (\xi(x))_{x \in \mathbb{Z}^d}$ is a field of i.i.d. random variables with finite positive exponential moments. Under these basic assumptions, $u(t,x)$ has moments of all orders.

The solution u to (4) describes a random particle flow in $\mathbb{Z}^d$ in the presence of random sources (lattice sites x with $\xi(x) > 0$) and random sinks (sites x with $\xi(x) < 0$).[1] Two competing effects are present: the diffusion mechanism governed by the Laplacian, and the local growth governed by the potential. The diffusion tends to make the random field $u(t, \cdot)$ flat, whereas the random potential ξ has a tendency to make it irregular.

The solution u to (4) also admits a branching particle dynamics interpretation. Imagine that initially, at time $t = 0$, there is a single particle at the origin, and all other sites are vacant. This particle moves according to a continuous-time symmetric random walk with generator $\kappa\Delta$. When present at site x, the particle is split into two particles with rate $\xi_+(x)$ and is killed with rate $\xi_-(x)$, where $\xi_+ = (\xi_+(x))_{x \in \mathbb{Z}^d}$ and $\xi_- = (\xi_-(x))_{x \in \mathbb{Z}^d}$ are independent random i.i.d. fields ($\xi_-(x)$ may attain the value ∞). Every particle continues from its birth site in the same way as the parent particle, and their movements are independent. Put $\xi(x) = \xi_+(x) - \xi_-(x)$. Then, given ξ_- and ξ_+, the expected number of particles present at the site x at time t is equal to $u(t,x)$. Here the expectation is taken over the particle motion and over the splitting resp. killing mechanism, but not over the random medium (ξ_-, ξ_+).

A very useful standard tool for the probabilistic investigation of (4) is the well-known *Feynman-Kac formula* for the solution u, which (after time-

---

[1] Sites x with $\xi(x) = -\infty$ may be allowed and interpreted as ('hard') traps or obstacles, sites with $\xi(x) \in (-\infty, 0)$ are sometimes called 'soft' traps.

reversal) reads

$$u(t, x) = \mathbb{E}_0 \Big[\exp \Big\{ \int_0^t \xi(X(s)) \, \mathrm{d}s \Big\} \delta_x(X(t)) \Big], \qquad (t, x) \in [0, \infty) \times \mathbb{Z}^d, \quad (5)$$

where $(X(s))_{s \in [0,\infty)}$ is continuous-time random walk on $\mathbb{Z}^d$ with generator $\kappa \Delta$ starting at $x \in \mathbb{Z}^d$ under $\mathbb{E}_x$.

Our main interest concerns the large-time behavior of the random field $u(t, \cdot)$. In particular, we consider the total mass, i.e., the random variable

$$U(t) = \sum_{x \in \mathbb{Z}^d} u(t, x) = \mathbb{E}_0 \Big[\exp \Big\{ \int_0^t \xi(X(s)) \, \mathrm{d}s \Big\} \Big], \qquad t > 0. \quad (6)$$

Note that $U(t)$ coincides with the value $\hat{u}(t, 0)$ of the solution $\hat{u}$ to the parabolic equation (4) with initial datum $\hat{u}(0, \cdot) = \mathbb{1}$ instead of $u(0, \cdot) = \delta_0$. One should have in mind that, because of this, our considerations below also concern the large-time asymptotics of $\hat{u}$.

We ask the following questions:

1. What is the asymptotic behavior of $U(t)$ as $t \to \infty$?
2. Where does the main mass of $u(t, \cdot)$ stem from? What are the regions that contribute most to $U(t)$? What are these regions determined by? How many of them are there and how far away are they from each other?
3. What do the typical shapes of the potential $\xi(\cdot)$ and of the solution $u(t, \cdot)$ look like in these regions?

We call the regions that contribute the overwhelming part to the total mass $U(t)$ *relevant islands* or *relevant regions*. The notion of *intermittency* states that there does exist a small number of relevant islands which are far away from each other and carry asymptotically almost all the total mass $U(t)$ of $u(t, \cdot)$. See Section 1.3 for details.

This effect may also be studied from the point of view of *typical paths* $X(s)$, $s \in [0, t]$, giving the main contribution to the expectation in the Feynman-Kac formula (6). On the one hand, the random walker X should move quickly and as far as possible through the potential landscape to reach a region of exceptionally high potential and then stay there up to time t. This will make the integral in the exponent on the right of (6) large. On the other hand, the probability to reach such a distant potential peak up to t may be rather small. Hence, the first order contribution to $U(t)$ comes from paths that find a good compromise between the high potential values and the far distance. This contribution is given by the height of the peak. The second order contribution to $U(t)$ is determined by the precise manner in which the optimal walker moves within the potential peak, and this depends on the geometric properties of the potential in that peak.

It is part of our study to understand the effect of intermittency for the parabolic Anderson model in great detail. We distinguish between the so-called *quenched* setting, where we consider $u(t, \cdot)$ almost surely with respect to the medium ξ, and the *annealed* one, where we average with respect to ξ. It is clear that the quantitative details of the answers to the above questions strongly depend on the distribution of the field ξ (more precisely, on the upper tail of the distribution of the random variable $\xi(0)$), and that different phenomena occur in the quenched and the annealed settings.

It will turn out that there is a universal picture present in the asymptotics of the parabolic Anderson model. Inside the relevant islands, after appropriate vertical shifting and spatial rescaling, the potential ξ will turn out to asymptotically approximate a universal, non-random shape, V, which is determined by a characteristic variational problem. The absolute height of the potential peaks and the diameter of the relevant islands are asymptotically determined by the upper tails of the random variable $\xi(0)$, while the number of the islands and their locations are random. Furthermore, after multiplication with an appropriate factor and rescaling, also the solution $u(t, \cdot)$ approaches a universal shape on these islands, namely the principal eigenfunction of the Hamiltonian $\kappa\Delta + V$ with V the above universal potential shape. Remarkably, there are only four universal classes of potential shapes for the PAM in (4), see Section 4 for details.

For a general discussion we refer to the monograph by Carmona and Molchanov [CM94], the lectures by Molchanov [M94], and also to the results by Sznitman about the important (spatially continuous) case of bounded from above Poisson-like potentials summarized in his monograph [S98]. A discussion from a physicist's and a chemist's point of view in the particular case of trapping problems (see also Section 2.2 below), including a survey on related mathematical models and a collection of open problems, is provided in [HW94]. A general mathematical background for the PAM is provided in [GM90].

1.3 Intermittency

As before, let $\hat{u}$ denote the solution to the equation in (4) with initial datum $\hat{u}(0, \cdot) = \mathbb{1}$, but now with a homogeneous ergodic potential $\xi = (\xi(x))_{x \in \mathbb{Z}^d}$. Assume that all positive exponential moments of $\xi(0)$ are finite. Let $\mathrm{Prob}(\cdot)$ and $\langle \cdot \rangle$ denote probability and expectation w.r.t. ξ.

A first, rough, mathematical approach to intermittency consists in a comparision of the growth of subsequent moments of the ergodic field $\hat{u}(t, \cdot)$ as $t \to \infty$. Define

$$\Lambda_p(t) = \log \langle \hat{u}(t, 0)^p \rangle, \qquad p \in \mathbb{N},$$

and write $f \ll g$ if $\lim_{t \to \infty}[g(t) - f(t)] = \infty$.

Definition 1. For $p \in \mathbb{N} \setminus \{1\}$, the homogeneous ergodic field $\hat{u}(t, \cdot)$ is called *p-intermittent* as $t \to \infty$, if

$$\frac{\Lambda_{p-1}}{p-1} \ll \frac{\Lambda_p}{p}. \tag{7}$$

Note that, by Hölder's inequality, always $\Lambda_{p-1}/(p-1) \leq \Lambda_p/p$. If the finite moment Lyapunov exponents

$$\lambda_p = \lim_{t\to\infty} \frac{1}{t}\Lambda_p(t), \qquad p \in \mathbb{N},$$

exist, then the strict inequality $\lambda_{p-1}/(p-1) < \lambda_p/p$ implies p-intermittency. Such a comparison of the moment Lyapunov exponents has first been used in the physics literature to study intermittency, cf. [ZMRS87], [ZMRS88]. We will use this approach in Section 5.

To explain the meaning of Definition 1, assume (7) for some $p \in \mathbb{N} \setminus \{1\}$ and choose a level function ℓ_p such that $\Lambda_{p-1}/(p-1) \ll \ell_p \ll \Lambda_p/p$. Then, by Chebyshev's inequality,

$$\mathrm{Prob}\Big(\hat{u}(t,0) > e^{\ell_p(t)}\Big) \leq e^{-(p-1)\ell_p(t)} \left\langle \hat{u}(t,0)^{p-1} \right\rangle$$
$$= \exp\{\Lambda_{p-1}(t) - (p-1)\ell_p(t)\},$$

and the expression on the right converges to zero as $t \to \infty$. In other words, the density of the homogeneous point process

$$\Gamma(t) = \Big\{x \in \mathbb{Z}^d : \hat{u}(t,x) > e^{\ell_p(t)}\Big\}$$

vanishes asymptotically as $t \to \infty$. On the other hand,

$$\Big\langle \hat{u}(t,0)^p \, \mathbb{1}\big\{\hat{u}(t,0) \leq e^{\ell_p(t)}\big\}\Big\rangle \leq e^{p\ell_p(t)} = e^{p\ell_p(t) - \Lambda_p(t)} \left\langle \hat{u}(t,0)^p \right\rangle$$
$$= o\left(\langle \hat{u}(t,0)^p\rangle\right)$$

and, consequently,

$$\langle \hat{u}(t,0)^p\rangle \sim \Big\langle \hat{u}(t,0)^p \, \mathbb{1}\big\{\hat{u}(t,0) > e^{\ell_p(t)}\big\}\Big\rangle$$

as $t \to \infty$. Hence, by Birkhoff's ergodic theorem, for large t and large centered boxes B in $\mathbb{Z}^d$,

$$|B|^{-1} \sum_{x\in B} \hat{u}(t,x)^p \approx |B|^{-1} \sum_{x\in B\cap\Gamma(t)} \hat{u}(t,x)^p.$$

This means that the p-th moment $\langle \hat{u}(t,0)^p\rangle$ is 'generated' by the high peaks of $\hat{u}(t,\cdot)$ on the 'thin' set $\Gamma(t)$ and therefore indicates the presence of intermittency in the above verbal sense. Unfortunately, this approach does not reflect the geometric structure of the set $\Gamma(t)$. This set might consist of islands or, e.g., have a net-like structure.

Theorem 1. *If* $\xi = (\xi(x))_{x\in\mathbb{Z}^d}$ *is a non-deterministic field of i.i.d. random variables with* $\langle e^{t\xi(0)}\rangle < \infty$ *for all* $t > 0$*, then the solution* $\hat{u}(t,\cdot)$ *is* p-*intermittent for all* $p \in \mathbb{N} \setminus \{1\}$*.*

This is part of Theorem 3.2 in [GM90], where, for general homogeneous ergodic potentials ξ, necessary and sufficient conditions for p-intermittency of $\hat{u}(t,\cdot)$ have been given in spectral terms of the Hamiltonian (2).

1.4 Annealed Second Order Asymptotics

Let us discuss, on a heuristic level, what the asymptotics of the moments of $U(t)$ are determined by, and how they can be described. For simplicity we restrict ourselves to the first moment.

The basic observation is that, as a consequence of the spectral representation (3),

$$U(t) \approx e^{t\lambda_t(\xi)} \tag{8}$$

(in the sense of logarithmic equivalence), where $\lambda_t(\varphi)$ denotes the principal (i.e., largest) eigenvalue of the operator $\kappa\Delta + \varphi$ with zero boundary condition in the 'macrobox' $B_t = [-t, t]^d \cap \mathbb{Z}^d$. Hence, we have to understand the large-time behavior of the exponential moments of the principal eigenvalue of the Anderson Hamiltonian $\mathcal{H}$ in a large, time-dependent box.

It turns out that the main contribution to $\langle e^{t\lambda_t(\xi)} \rangle$ comes from realizations of the potential ξ having high peaks on distant islands of some radius of order $\alpha(t)$ that is much smaller than t. But this implies that $\lambda_t(\xi)$ is close to the principal eigenvalue of $\mathcal{H}$ on one of these islands. Therefore, since the number of subboxes of B_t of radius of order $\alpha(t)$ grows only polynomial in t and ξ is spatially homogeneous, we may expect that

$$\left\langle e^{t\lambda_t(\xi)} \right\rangle \approx \left\langle e^{t\lambda_{R\alpha(t)}(\xi)} \right\rangle$$

for R large as $t \to \infty$.

The choice of the scale function $\alpha(t)$ depends on asymptotic 'stiffness' properties of the potential, more precisely of its tails at its essential supremum, and is determined by a large deviation principle, see (15) below. In Section 2 we shall see examples of potentials such that $\alpha(t)$ tends to 0, to ∞, or stays bounded and bounded away from zero as $t \to \infty$. In the present heuristics, we shall assume that $\alpha(t) \to \infty$, which implies the necessity of a spatial rescaling. In particular, after rescaling, the main quantities and objects will be described in terms of the continuous counterparts of the discrete objects we started with, i.e., instead of the discrete Laplacian, the continuous Laplace operator appears etc. The following heuristics can also be read in the case where $\alpha(t) \equiv 1$ by keeping the discrete versions for the limiting objects.

The optimal behavior of the field ξ in the 'microbox' $B_{R\alpha(t)}$ is to approximate a certain (deterministic) shape φ after appropriate spatial scaling and vertical shifting. It easily follows from the Feynman-Kac formula (6) that

$$e^{H(t) - 2d\kappa t} \le \langle U(t) \rangle \le e^{H(t)},$$

where

$$H(t) = \log\langle e^{t\xi(0)} \rangle, \qquad t > 0, \tag{9}$$

denotes the *cumulant generating function* of $\xi(0)$ (often called *logarithmic moment generating function*). Hence, the peaks of $\xi(\cdot)$ mainly contributing to $\langle U(t) \rangle$ have height of order $H(t)/t$. Together with Brownian scaling this leads to the ansatz

$$\overline{\xi}_t(\cdot) = \alpha(t)^2 \left[\xi(\lfloor \cdot \alpha(t) \rfloor) - \frac{H(t)}{t} \right], \tag{10}$$

for the spatially rescaled and vertically shifted potential in the cube $Q_R = (-R, R)^d$. Now the idea is that the main contribution to $\langle U(t) \rangle$ comes from fields that are shaped in such a way that $\overline{\xi}_t \approx \varphi$ in Q_R, for some $\varphi \colon Q_R \to \mathbb{R}$, which has to be chosen optimally. Observe that

$$\overline{\xi}_t \approx \varphi \quad \text{in } Q_R \quad \Longleftrightarrow \quad \xi(\cdot) \approx \frac{H(t)}{t} + \frac{1}{\alpha(t)^2} \varphi\left(\frac{\cdot}{\alpha(t)}\right) \quad \text{in } B_{R\alpha(t)}. \tag{11}$$

Let us calculate the contribution to $\langle U(t) \rangle$ coming from such fields. Using (8), we obtain

$$\begin{aligned}
&\left\langle U(t) \, \mathbb{1}\{\overline{\xi}_t \approx \varphi \text{ in } Q_R\} \right\rangle \\
&\approx e^{H(t)} \exp\left\{ t\lambda_{R\alpha(t)}\left(\frac{1}{\alpha(t)^2} \varphi\left(\frac{\cdot}{\alpha(t)}\right) \right) \right\} \text{Prob}\left(\overline{\xi}_t \approx \varphi \text{ in } Q_R \right).
\end{aligned} \tag{12}$$

The asymptotic scaling properties of the discrete Laplacian, Δ, imply that

$$\lambda_{R\alpha(t)}\left(\frac{1}{\alpha(t)^2} \varphi\left(\frac{\cdot}{\alpha(t)}\right) \right) \approx \frac{1}{\alpha(t)^2} \lambda_R^c(\varphi), \tag{13}$$

where $\lambda_R^c(\varphi)$ denotes the principal eigenvalue of $\kappa\Delta^c + \varphi$ in the cube Q_R with zero boundary condition, and Δ^c is the usual 'continuous' Laplacian. This leads to

$$\left\langle U(t) \, \mathbb{1}\{\overline{\xi}_t \approx \varphi \text{ in } Q_R\} \right\rangle \approx e^{H(t)} \exp\left\{ \frac{t}{\alpha(t)^2} \lambda_R^c(\varphi) \right\} \text{Prob}\left(\overline{\xi}_t \approx \varphi \text{ in } Q_R \right). \tag{14}$$

In order to achieve a balance between the second and the third factor on the right, it is necessary that the logarithmic decay rate of the considered probability is $t/\alpha(t)^2$. One expects to have a large deviation principle for the shifted, rescaled field, which reads

$$\text{Prob}\left(\overline{\xi}_t \approx \varphi \text{ in } Q_R \right) \approx \exp\left\{ -\frac{t}{\alpha(t)^2} I_R(\varphi) \right\}, \tag{15}$$

where the scale $\alpha(t)$ has to be determined in such a way that the rate function I_R is non-degenerate. Now substitute (15) into (14). Then the Laplace method tells us that the exponential asymptotics of $\langle U(t) \rangle$ is equal to the one of $\langle U(t) \, \mathbb{1}\{\overline{\xi}_t \approx \varphi \text{ in } Q_R\}$ with optimal φ. Hence, optimizing on φ and remembering that R is large, we arrive at

$$\langle U(t) \rangle \approx e^{H(t)} \exp\left\{ -\frac{t}{\alpha(t)^2} \chi \right\}, \tag{16}$$

where the constant χ is given in terms of the characteristic variational problem

$$\chi = \lim_{R \to \infty} \inf_{\varphi : Q_R \to \mathbb{R}} \left[I_R(\varphi) - \lambda_R^c(\varphi) \right]. \tag{17}$$

The first term on the right of (16) is determined by the absolute height of the typical realizations of the potential and the second contains information about the shape of the potential close to its maximum in spectral terms of the Anderson Hamiltonian $\mathcal{H}$ in this region. More precisely, those realizations of ξ with $\overline{\xi}_t \approx \varphi_*$ in Q_R for large R and φ_* a minimizer in the variational formula in (17) contribute most to $\langle U(t) \rangle$. In particular, the geometry of the relevant potential peaks is hidden via χ in the second asymptotic term of $\langle U(t) \rangle$.

1.5 Quenched Second Order Asymptotics

Here we explain, again on a heuristic level, the almost sure asymptotics of $U(t)$ as $t \to \infty$. Because of (8), it suffices to study the asymptotics of the principal eigenvalue $\lambda_t(\xi)$.

Like for the annealed asymptotics, the main contribution to $\lambda_t(\xi)$ comes from islands whose radius is of a certain deterministic, time-depending order, which we denote $\widetilde{\alpha}(t)$. As $t \to \infty$, the scale function $\widetilde{\alpha}(t)$ tends to zero, one, or ∞, respectively, if the scale function $\alpha(t)$ for the moments tends to these respective values (see also (20) below). However, $\widetilde{\alpha}(t)$ is roughly of logarithmic order in $\alpha(t)$ if $\alpha(t) \to \infty$, hence it is *much* smaller than $\alpha(t)$.

The relevant islands ('microboxes') have radius $R\widetilde{\alpha}(t)$, where R is chosen large. Let $z \in B_t$ denote the (certainly random) center of one of these islands $\widetilde{B} = z + B_{R\widetilde{\alpha}(t)}$ meeting the two requirements (1) the potential ξ is very large in $\widetilde{B}$ and (2) ξ has an optimal shape within $\widetilde{B}$. This is further explained as follows. Let $h_t = \max_{B_t} \xi$ be the maximal potential value in the large box B_t. (Then h_t is a priori random, but well approximated by deterministic asymptotics, which can be deduced from asymptotics of $H(t)$.) Then $\xi - h_t$ is roughly of finite order within the relevant 'microbox' $\widetilde{B}$. Furthermore, $\xi - h_t$ should approximate a fixed deterministic shape in $\widetilde{B}$. Hence, we consider the shifted and rescaled field in the box $\widetilde{B}$,

$$\overline{\xi}_t(\cdot) = \widetilde{\alpha}(t)^2 \left[\xi\big(z + \cdot \widetilde{\alpha}(t)\big) - h_t \right], \qquad \text{in } Q_R = (-R, R)^d. \tag{18}$$

Note that

$$\overline{\xi}_t \approx \varphi \ \text{ in } Q_R \qquad \Longleftrightarrow \qquad \xi(z + \cdot) \approx h_t + \tfrac{1}{\widetilde{\alpha}(t)^2} \varphi\big(\tfrac{\cdot}{\widetilde{\alpha}(t)}\big) \quad \text{in } \widetilde{B} - z. \tag{19}$$

A crucial Borel-Cantelli argument shows that, for a given shape φ, with probability one, for any t sufficiently large, there does exist at least one box $\widetilde{B}$ having radius $R\widetilde{\alpha}(t)$ such that the event $\{\overline{\xi}_t \approx \varphi \text{ in } Q_R\}$ occurs if $I_R(\varphi) < 1$, where I_R is the rate function of the large deviation principle in (15). If $I_R(\varphi) > 1$, then this happens with probability 0. For the Borel-Cantelli argument to work, one needs the scale function $\widetilde{\alpha}(t)$ to be defined in terms of the annealed scale function $\alpha(t)$ in the following way:

$$\frac{\widetilde{\alpha}(t)}{\alpha(\widetilde{\alpha}(t))^2} = d \log t, \tag{20}$$

i.e., $\widetilde{\alpha}(t)$ is the inverse of the map $t \mapsto t/\alpha(t)^2$, evaluated at $d \log t$. Note that the growth of $\widetilde{\alpha}(t)$ is roughly of logarithmic order of the growth of $\alpha(t)$, i.e., if the annealed relevant islands grow unboundedly, then the quenched relevant islands also grow unboundedly, but with much smaller velocity.

Hence, with probability one, for all large t, there is at least one box $\widetilde{B}$ in which the potential looks like the function on the right of (19). The contribution to $\lambda_t(\xi)$ coming from one of the boxes $\widetilde{B}$ is equal to the associated principal eigenvalue

$$\lambda_{\widetilde{B}-z}\left(h_t + \tfrac{1}{\widetilde{\alpha}(t)^2}\varphi\left(\tfrac{\cdot}{\widetilde{\alpha}(t)}\right)\right) \approx h_t + \frac{1}{\widetilde{\alpha}(t)^2}\lambda_R^c(\varphi), \tag{21}$$

where we recall that $\lambda_R^c(\varphi)$ is the principal Dirichlet eigenvalue of the operator $\kappa\Delta^c + \varphi$ in the 'continuous' cube Q_R. Obviously, $\lambda_t(\xi)$ is asymptotically not smaller than the expression on the right of (21). In terms of the Feynman-Kac formula in (5), this lower estimate is obtained by inserting the indicator on the event that the random path moves quickly to the box $\widetilde{B}$ and stays all the time until t in that box.

It is an important technical issue to show that, asymptotically as $t \to \infty$, $\lambda_t(\xi)$ is also estimated from *above* by the right hand side of (21), if φ is optimally chosen, i.e., if $\lambda_R^c(\varphi)$ is optimized over all admissible φ and on R. This implies that the almost sure asymptotics of $U(t)$ are given as

$$\frac{1}{t}\log U(t) \approx \lambda_t(\xi) \approx h_t - \frac{1}{\widetilde{\alpha}(t)^2}\widetilde{\chi}, \qquad t \to \infty, \tag{22}$$

where $\widetilde{\chi}$ is given in terms of the characteristic variational problem

$$\widetilde{\chi} = \lim_{R \to \infty} \inf_{\varphi:\, Q_R \to \mathbb{R},\, I_R(\varphi) < 1} [-\lambda_R^c(\varphi)]. \tag{23}$$

This ends the heuristic derivation of the almost sure asymptotics of $U(t)$. Like in the annealed case, there are two terms, which describe the absolute height of the potential in the 'macrobox' B_t, and the shape of the potential in the relevant 'microbox' $\widetilde{B}$, more precisely the spectral properties of $\kappa\Delta + \xi$ in that microbox. The interpretation is that, for R large and φ_* an approximate minimizer in (23), the main contribution to $U(t)$ comes from a small box $\widetilde{B}$ in B_t, with radius $R\widetilde{\alpha}(t)$, in which the shifted and rescaled potential $\overline{\xi}_t$ looks like φ_*. The condition $I_R(\varphi_*) < 1$ guarantees the existence of such a box, and $\lambda_R(\varphi_*)$ quantifies the contribution from that box.

Let us remark that the variational formulas in (23) and (17) are in close connection to each other. In particular, it can be shown that the minimizers of (23) are rescaled versions of the minimizers of (17). This means that, up to rescaling, the optimal potential shapes in the annealed and in the quenched setting are identical.

1.6 Geometric Picture of Intermittency

In this section we explain the geometric picture of intermittency, still on a heuristic level.

The heuristics for the total mass of $u(t, \cdot)$ in Section 1.5 makes use of only *one* of the relevant islands $\widetilde{B}$ in which the potential is optimally valued and shaped. In order to describe the entire function $u(t, \cdot)$, one has to take into account a certain (random) number of such islands. Let $n(t)$ denote their number, and let $z_1, z_2, \ldots, z_{n(t)} \in B_t$ denote the centers of these relevant microboxes $B_1, B_2, \ldots, B_{n(t)}$, whose radii are equal to $R\widetilde{\alpha}(t)$. Then, almost surely,

$$U(t) = \sum_{x \in \mathbb{Z}^d} u(t, x) \approx \sum_{i=1}^{n(t)} \sum_{x \in B_i} u(t, x), \qquad \text{as } t \to \infty, \tag{24}$$

i.e., asymptotically the total mass of the random field $u(t, \cdot)$ stems only from the unions of the relevant islands, $B_1, \ldots, B_{n(t)}$. These islands are far away from each other. On each of them, the shifted and rescaled potential $\overline{\xi}_t$, see (18)), looks approximately like a minimizer φ_* of the variational problem in (23). In particular, it has an asymptotically *deterministic* shape. This is the universality in the potential landscape: the height and the (appropriately rescaled) shape of the potential on the relevant islands are deterministic, but their location and number are random.

The shape of the solution, $u(t, \cdot)$, on each of the relevant islands also approaches a universal deterministic shape, namely a time-dependent multiple of the principal eigenfunction of the operator $\kappa\Delta + \varphi_*$.

2 Examples of Potentials

2.1 Double-exponential Distributions

Consider a distribution which lies in the vicinity of the *double-exponential distribution*, i.e.,

$$\text{Prob}\,(\xi(0) > r) \approx \exp\{-e^{r/\varrho}\}, \qquad r \to \infty, \tag{25}$$

with $\varrho \in (0, \infty)$ a parameter. It turns out [GM98] that this class of potentials constitutes a critical class in the sense that the radius of the relevant islands stays finite as $t \to \infty$. This is related to the characteristic property of the double-exponential distribution that

$$\text{Prob}\left(\xi(x) > h\right) \approx \text{Prob}\left(\xi(y) > h - \varrho \log 2, \, \xi(z) > h - \varrho \log 2\right),$$

meaning that single-site potential peaks of height $h \gg 1$ occur with the same frequency as two-site potential peaks with height of the same order. Hence, no spatial rescaling is necessary, and we put $\alpha(t) = 1$. In Sections 3.1–3.3 below we shall describe our results for this type of potentials more closely.

For the boundary cases $\varrho = \infty$ and $\varrho = 0$ ('beyond' and 'on this side of' the double-exponential distribution, respectively), [GM98] argued that the boundary cases $\alpha(t) \downarrow 0$ and $\alpha(t) \to \infty$ occur. In other words, the fields beyond the double-exponential (which includes, e.g., Gaussian fields) are simple in the sense that the main contribution comes from islands consisting of single lattice sites. Unbounded fields that are in the vicinity of the case $\varrho = 0$ are called 'almost bounded' in [GM98].

2.2 Survival Probabilities

The case when the field ξ assumes the values $-\infty$ and 0 only has a nice interpretation in terms of survival probabilities and is therefore of particular importance. The fundamental papers [DV75] and [DV79] by Donsker and Varadhan on the Wiener sausage contain apparently the first substantial annealed results on the asymptotics for the parabolic Anderson model. In the nineties, the thorough and deep work by Sznitman (see his monograph [S98]), pushed the rigorous understanding of the quenched situation much further.

Brownian Motion in a Poisson Field of Traps We consider the continuous case, i.e., the version of (4) with $\mathbb{Z}^d$ replaced by $\mathbb{R}^d$ and the lattice Laplacian replaced by the usual Laplace operator. The field ξ is given as follows. Let $(x_i)_{i \in I}$ be the points of a homogeneous Poisson point process in $\mathbb{R}^d$, and consider the union $\mathcal{O}$ of the balls $B_a(x_i)$ of radius a around the Poisson points x_i. We define a random potential by putting

$$\xi(x) = \begin{cases} 0 & \text{if } x \notin \mathcal{O}, \\ -\infty & \text{if } x \in \mathcal{O}. \end{cases} \tag{26}$$

(There are more general versions of this type of potentials, but for simplicity we keep with that.) The set $\mathcal{O}$ receives the meaning of the set of 'hard traps' or 'obstacles'. Let $T_{\mathcal{O}} = \inf\{t > 0 \colon X(t) \in \mathcal{O}\}$ denote the entrance time into $\mathcal{O}$ for a Brownian motion $(X(t))_{t \in [0,\infty)}$. Then we have the Feynman-Kac representation

$$u(t, x) = \mathbb{P}_0\left(T_{\mathcal{O}} > t, \, X(t) \in dx\right) / dx,$$

i.e., $u(t, x)$ is equal to the sub-probability density of $X(t)$ on survival in the Poisson field of traps by time t for Brownian motion starting from the origin.

The total mass $U(t) = \mathbb{P}_0(T_{\mathcal{O}} > t)$ is the survival probability by time t. It is easily seen that the first moment of $U(t)$ coincides with a negative exponential moment of the volume of the Wiener sausage $\bigcup_{s \in [0,t]} B_a(X(s))$.

Donsker and Varadhan analyzed the leading asymptotics of $\langle U(t) \rangle$ by using their large deviation principle for Brownian occupation time measures. The relevant islands have radius of order $\alpha(t) = t^{1/(d+2)}$. To handle the quenched asymptotics of $U(t)$, Sznitman developed a coarse-graining scheme for Dirichlet eigenvalues on random subsets of $\mathbb{R}^d$, the so-called method of enlargement of obstacles (MEO). The MEO replaces the eigenvalues in certain complicated subsets of $\mathbb{R}^d$ by those in coarse-grained subsets belonging to a *discrete* class of much smaller combinatorial complexity such that control is kept on the relevant properties of the eigenvalue.

Qualitatively, the considered model falls into the class of bounded from above fields introduced in Section 2.3 with $\gamma = 0$.

Related potentials critically rescaled with time have been studied in particular by van den Berg et al. [BBH01] and by Merkl and Wüthrich [MW02].

Simple Random Walk among Bernoulli Traps This is the discrete version of Brownian motion among Poisson traps. Consider the i.i.d. field $\xi = (\xi(x))_{x \in \mathbb{Z}^d}$ where $\xi(x)$ takes the values 0 or $-\infty$ only. Again, $u(t, x)$ is the survival probability of continuous-time random walk paths from 0 to x among the set of traps $\mathcal{O} = \{y \in \mathbb{Z}^d \colon \xi(y) = -\infty\}$.

In their paper [DV79], Donsker and Varadhan also investigated the discrete case and described the logarithmic asymptotics of $\langle U(t) \rangle$ by proving and exploiting a large deviation principle for occupation times of random walks. Later Bolthausen [B94] carried out a deeper analysis of $\langle U(t) \rangle$ in the two-dimensional case using refined large deviation arguments. Antal [Ant94], [Ant95] developed a discrete variant of the MEO and demonstrated its value by proving limit theorems for the survival probability $U(t)$ and its moments.

2.3 General Fields Bounded from Above

In [BK01a] and [BK01b], a large class of potentials with $\mathrm{esssup}\, \xi(0) < \infty$ is considered. Assume for simplicity that $\mathrm{esssup}\, \xi(0) = 0$ and that the tail of $\xi(0)$ at 0 is given by

$$\mathrm{Prob}\left(\xi(0) > -x\right) \approx \exp\left\{-Dx^{-\frac{\gamma}{1-\gamma}}\right\}, \qquad x \downarrow 0, \qquad (27)$$

with $D > 0$ and $\gamma \in [0,1)$ two parameters. The case $\gamma = 0$ contains simple random walk among Bernoulli traps as a particular case. The cumulant generating function is roughly $H(t) \approx -\mathrm{const}\, t^\gamma$, and the annealed scale function is $\alpha(t) \approx t^\nu$ where $\nu = (1-\gamma)/(2+d-d\gamma)$. The power ν ranges from 0 to $1/(d+2)$ as the parameter γ ranges from 1 to 0.

It turns out in [BK01a] that the rate function I_R (see (15)) is given by

$$I_R(\varphi) = \text{const} \int_{Q_R} |\varphi(x)|^{-\frac{\gamma}{1-\gamma}} \, dx, \qquad (28)$$

where in the case $\gamma = 0$ we interpret the integral as the Lebesgue measure of the support of φ. The characteristic variational formula for the annealed field shapes in (17) has been analyzed in great detail in the case $\gamma = 0$. In particular it was shown that the minimizer is unique and has compact support, and it was characterized in terms of Bessel functions. However, in the general case $\gamma \in (0, 1)$, an analysis of (17) has not yet been carried out.

2.4 Gaussian Fields and Poisson Shot Noise

Two important particular cases in the *continuous* version of the parabolic Anderson model are considered in [GK00] and [GKM00] (see also [CM95] for first rough results). The continuous version of (4) replaces $\mathbb{Z}^d$ by $\mathbb{R}^d$ and the discrete lattice Laplacian by the usual Laplace operator. Unlike in the discrete case, where any distribution on $\mathbb{R}$ may be used for the definition of an i.i.d. potential, in the continuous case it is not easy to find examples of fields that can be expressed in easily manageable terms. Since a certain degree of regularity of the potential is required, the condition of spatial independence must be dropped.

In [GK00] and [GKM00], two types of fields are considered: a Gaussian field ξ whose covariance function B has a parabolic shape around zero with $B(0) = \sigma^2 > 0$, and a so-called Poisson shot-noise field, which is defined as the superposition of copies of parabolic-shaped *positive* clouds around the points of a homogeneous Poisson point process in $\mathbb{R}^d$ (in contrast to the trap case of Section 2.2). A certain (mild) assumption on the decay of the covariance function (respectively of the cloud) at infinity ensures sufficient independence between regions that are far apart.

Both fields easily develop very high peaks on small islands (the Poisson shot noise field is large where many Poisson points are close together). The annealed scale function is $\alpha(t) = t^{-1/4}$ for the Gaussian field and $\alpha(t) = t^{d/8} e^{-\sigma^2 t/4}$ for the Poisson field [GK00].

3 Results for the Double-exponential Case

In this section, we formulate our results on the large-time asymptotics of the parabolic Anderson model in the particularly important case of a double-exponentially distributed random potential, see Section 25. We handle the annealed asymptotics of the total mass $U(t)$ in Section 3.1, the quenched ones in Section 3.2, and the geometric picture of intermittency in Section 3.3. The material of the first two subsections is taken from [GM98], that of the last subsection from [GKM04].

3.1 Annealed Asymptotics

As before, we assume that $\xi = (\xi(x))_{x \in \mathbb{Z}^d}$ is a field of i.i.d. random variables. We impose the following assumption on the cumulant generating function of $\xi(0)$ defined by (9).

Assumption (H). *The function $H(t)$ is finite for all $t > 0$. There exists $\varrho \in [0, \infty]$ such that*

$$\lim_{t \to \infty} \frac{H(ct) - cH(t)}{t} = \varrho c \log c \qquad \text{for all } c \in (0, 1).$$

Note that the vicinity of the double-exponential distribution (25) corresponds to $\varrho \in (0, \infty)$. If $\varrho = \infty$, then the upper tail of the distribution of $\xi(0)$ is heavier than in the double exponential case, whereas for $\varrho = 0$ it is thinner.

Let $\mathcal{P}(\mathbb{Z}^d)$ denote the space of probability measures on $\mathbb{Z}^d$. We introduce the Donsker-Varadhan functional S_d and the entropy functional I_d on $\mathcal{P}(\mathbb{Z}^d)$ by

$$S_d(\mu) = \sum_{\substack{\{x,y\} \subset \mathbb{Z}^d \\ |x-y|=1}} \left(\sqrt{\mu(x)} - \sqrt{\mu(y)} \right)^2 \quad \text{and} \quad I_d(\mu) = - \sum_{x \in \mathbb{Z}^d} \mu(x) \log \mu(x),$$

respectively, and set

$$\chi_d = \inf_{\mu \in \mathcal{P}(\mathbb{Z}^d)} \left[\kappa S_d(\mu) + \varrho I_d(\mu) \right], \qquad \varrho \in [0, \infty]. \tag{29}$$

As before, let $U(t)$ denote the total mass of the solution $u(t, \cdot)$ to the PAM (4).

Theorem 2. *Let Assumption (H) be satisfied. Then, for any $p \in \mathbb{N}$,*

$$\langle U(t)^p \rangle = \exp \{ H(pt) - \chi_d pt + o(t) \} \qquad \text{as } t \to \infty.$$

It turns out that $\chi_d = 2d\kappa\chi_0(\varrho/\kappa)$, where $\chi_0 \colon [0, \infty) \to [0, 1)$ is strictly increasing and concave, $\chi_0(0) = 0$, and $\chi_0(\varrho) \to 1$ as $\varrho \to \infty$. Moreover, for $\varrho \in (0, \infty)$, each minimizer μ of the variational problem (29) has the form $\mu = \text{const } v^2$, where $v = v_1 \otimes \cdots \otimes v_d$ and each of the factor $v_1, \ldots, v_d$ is a positive solution of the equation

$$\kappa \Delta v + 2\varrho v \log v = 0 \qquad \text{on } \mathbb{Z}$$

with *minimal* ℓ^2-norm. Uniqueness of v modulo shifts holds for large ϱ/κ but is open for small values of this quantity.

Recall that $U(t) = \hat{u}(t,0)$, where $\hat{u}$ is the solution to (4), but with non-localized initial datum $\mathbb{1}$ instead of δ_0. A much deeper question is the computation of the asymptotics of the 'correlation'

$$c(t,x) = \frac{\langle \hat{u}(t,0)\hat{u}(t,x)\rangle}{\langle \hat{u}(t,0)^2\rangle}$$

of the spatially homogeneous solution $\hat{u}$ of the PAM. Assuming additional regularity of the cumulant generating function H and uniqueness modulo spatial shifts of the minimizer μ of the variational problem (29), it was shown in [GH99] that

$$\lim_{t\to\infty} c(t,x) = \frac{\sum_z v(z)v(z+x)}{\sum_z v(z)^2}.$$

This indicates that the second moment (considered as the limiting expression $\lim_{R\to\infty} |B_R|^{-1} \sum_{x\in B_R} \hat{u}^2(t,x)$) is generated by rare high peaks of the solution $\hat{u}(t,\cdot)$ *with shape $v(\cdot)$*.

3.2 Quenched Asymptotics

Here we again consider i.i.d. random potentials $(\xi(x))_{x\in\mathbb{Z}^d}$ in the vicinity of the double-exponential distribution (25) but formulate our assumptions in a different manner.

To be precise, let F denote the distribution function of $\xi(0)$. Provided that F is continuous and $F(r) < \infty$ for all $r \in \mathbb{R}$ (i.e., ξ is unbounded from above), we may introduce the non-decreasing function

$$\varphi(r) = \log \frac{1}{1 - F(r)}, \qquad r \in \mathbb{R}. \tag{30}$$

Its left-continuous inverse ψ is given by

$$\psi(s) = \min\{r \in \mathbb{R} : \varphi(r) \geq s\}, \qquad s > 0. \tag{31}$$

Note that ψ is strictly increasing with $\varphi(\psi(s)) = s$ for all $s > 0$. The relevance of ψ comes from the observation that ξ has the same distribution as $\psi \circ \eta$, where $\eta = (\eta(x))_{x\in\mathbb{Z}^d}$ is an i.i.d. field of exponentially distributed random variables with mean one.

We now formulate our main assumption.

Assumption (F). *The distribution function F is continuous, $F(r) < 1$ for all $r \in \mathbb{R}$, and, in dimension $d = 1$, $\int_{-\infty}^{-1} \log|r|\, F(\mathrm{d}r) < \infty$. There exists $\varrho \in (0,\infty]$ such that*

$$\lim_{s\to\infty} [\psi(cs) - \psi(s)] = \varrho \log c, \qquad c \in (0,1). \tag{32}$$

If $\varrho = \infty$, then ψ satisfies in addition

$$\lim_{s\to\infty} [\psi(s + \log s) - \psi(s)] = 0. \tag{33}$$

The crucial supposition (32) specifies that the upper tail of the distribution of $\xi(0)$ is close to the double-exponential distribution (25) for $\varrho \in (0, \infty)$ and is heavier for $\varrho = \infty$. Assumption (33) excludes too heavy tails. Note that (33) is fulfilled for Gaussian but not for exponential tails.

The reader easily checks that (32) implies that $\psi(t) \sim \varrho \log t$ as $t \to \infty$. Let

$$h_t = \max_{x \in B_t} \xi(x), \qquad t > 0, \tag{34}$$

be the height of the potential ξ in $B_t = [-t, t]^d \cap \mathbb{Z}^d$. It can be easily seen that, under Assumption (F), almost surely,

$$h_t = \psi(d \log t) + o(1) \qquad \text{as } t \to \infty. \tag{35}$$

Let us remark that it is condition (33) which ensures that the almost sure asymptotics of h_t in (35) is non-random up to order $o(1)$.

One of the main results in [GM98], Theorem 2.2, is the second order asymptotics of the total mass $U(t)$ defined in (6).

Theorem 3. *Under Assumption (F), with probability one,*

$$\log U(t) = t [h_t - \widetilde{\chi}_d + o(1)] \qquad \text{as } t \to \infty. \tag{36}$$

Here $0 \le \widetilde{\chi}_d \le 2d\kappa$. An analytic description of $\widetilde{\chi}_d$ is as follows. Define $I \colon [-\infty, 0]^{\mathbb{Z}^d} \to [0, \infty]$ by

$$I(V) = \begin{cases} \sum_{x \in \mathbb{Z}^d} e^{V(x)/\varrho}, & \text{if } \varrho \in (0, \infty), \\ |\{x \in \mathbb{Z}^d \colon V(x) > -\infty\}|, & \text{if } \varrho = \infty. \end{cases} \tag{37}$$

One should regard I as *large deviation rate function* for the fields $\xi - h_t$ (recall (15) and note that $\alpha(t) = 1$ here). Indeed, if the distribution of ξ is exactly given by (2.1), then we have

$$\text{Prob}\left(\xi(\cdot) - h > V(\cdot) \text{ in } \mathbb{Z}^d\right) = \exp\left\{-e^{h/\varrho} I(V)\right\}$$

for any $V \colon \mathbb{Z}^d \to [-\infty, 0]$ and any $h \in (0, \infty)$. For $V \in [-\infty, 0]^{\mathbb{Z}^d}$, let $\lambda(V) \in [-\infty, 0]$ be the top of the spectrum of the self-adjoint operator $\kappa\Delta + V$ in the domain $\{V > -\infty\}$ with zero boundary condition. In terms of the Rayleigh-Ritz formula,

$$\lambda(V) = \sup_{f \in \ell^2(\mathbb{Z}^d) \colon \|f\|_2 = 1} \langle (\kappa\Delta + V)f, f \rangle, \tag{38}$$

where $\langle \cdot, \cdot \rangle$ and $\| \cdot \|_2$ denote the inner product and the norm in $\ell^2(\mathbb{Z}^d)$, respectively. Then

$$-\widetilde{\chi}_d = \sup\left\{\lambda(V) \colon V \in [-\infty, 0]^{\mathbb{Z}^d}, I(V) \le 1\right\}. \tag{39}$$

This variational problem is 'dual' to the variational problem (29) and, in particular, $\widetilde{\chi}_d = \chi_d$.

3.3 Geometry of Intermittency

In this section we give a precise formulation of the geometric picture of intermittency, which was heuristically explained in Section 1.6.

We keep the assumptions of the last subsection. For our deeper investigations, in addition to Assumption (F), we introduce an assumption about the *optimal potential shape*.

Assumption (M). *Up to spatial shifts, the variational problem in* (39) *possesses a unique maximizer, which has a unique maximum.*

By V_* we denote the unique maximizer of (39) which attains its unique maximum at the origin. We will call V_* *optimal potential shape*. Assumption (M) is satisfied at least for large ϱ/κ. This fact as well as further important properties of the variational problem (39) are stated in the next proposition.

Proposition 1. *(a) For any* $\varrho \in (0, \infty]$, *the supremum in* (39) *is attained.*

(b) If ϱ/κ *is sufficiently large, then the maximizer in* (39) *is unique modulo shifts and has a unique maximum.*

(c) If Assumption (M) is satisfied, then the optimal potential shape has the following properties.

    (i) If $\varrho \in (0, \infty)$, *then* $V_* = f_* \otimes \cdots \otimes f_*$ *for some* $f_* \colon \mathbb{Z} \to (-\infty, 0)$. *If* $\varrho = \infty$, *then* V_* *is degenerate in the sense that* $V_*(0) = 0$ *and* $V_*(x) = -\infty$ *for* $x \neq 0$.

    (ii) The operator $\kappa\Delta + V_*$ *has a unique nonnegative eigenfunction* $w_* \in \ell^2(\mathbb{Z}^d)$ *with* $w_*(0) = 1$ *corresponding to the eigenvalue* $\lambda(V_*)$. *Moreover,* $w_* \in \ell^1(\mathbb{Z}^d)$. *If* $\varrho \in (0, \infty)$, *then* w_* *is positive on* $\mathbb{Z}^d$, *while* $w_* = \delta_0$ *for* $\varrho = \infty$.

We shall see that the main contribution to the total mass $U(t)$ comes from a neighborhood of the set of best local coincidences of $\xi - h_t$ with spatial shifts of V_*. These neighborhoods are widely separated from each other and hence not numerous. We may restrict ourselves further to those neighborhoods in which, in addition, $u(t, \cdot)$, properly normalized, is close to w_*.

Denote by $B_R(y) = y + B_R$ the closed box of radius R centered at $y \in \mathbb{Z}^d$ and write

$$B_R(A) = \bigcup_{y \in A} B_R(y) \tag{40}$$

for the 'R-box neighborhood' of a set $A \subset \mathbb{Z}^d$. In particular, $B_0(A) = A$.

For any $\varepsilon > 0$ and any sufficiently large $\varrho \in (0, \infty]$, let $r(\varepsilon, \varrho)$ denote the smallest $r \in \mathbb{N}_0$ such that

$$\|w_*\|_2^2 \sum_{x \in \mathbb{Z}^d \setminus B_r} w_*(x) < \varepsilon. \tag{41}$$

Note that $r(\varepsilon, \infty) = 0$, due to the degeneracy of w_∞. Given $f: \mathbb{Z}^d \to \mathbb{R}$ and $R > 0$, let $\|f\|_R = \sup_{x \in B_R} |f(x)|$.

The main result of [GKM04] is the following.

Theorem 4. *Let the Assumptions (F) and (M) be satisfied. Then there exists a random time-dependent subset $\Gamma^* = \Gamma^*_{t \log^2 t}$ of $B_{t \log^2 t}$ such that, almost surely,*

$$(i) \quad \liminf_{t \to \infty} \frac{1}{U(t)} \sum_{x \in B_{r(\varepsilon, \varrho)}(\Gamma^*)} u(t, x) \geq 1 - \varepsilon, \qquad \varepsilon \in (0, 1); \tag{42}$$

$$(ii) \quad |\Gamma^*| \leq t^{o(1)} \quad and \quad \min_{y, \widetilde{y} \in \Gamma^* : y \neq \widetilde{y}} |y - \widetilde{y}| \geq t^{1 - o(1)} \qquad as\ t \to \infty; \tag{43}$$

$$(iii) \quad \lim_{t \to \infty} \max_{y \in \Gamma^*} \left\| \xi(y + \cdot) - h_t - V_*(\cdot) \right\|_R = 0, \qquad R > 0; \tag{44}$$

$$(iv) \quad \lim_{t \to \infty} \max_{y \in \Gamma^*} \left\| \frac{u(t, y + \cdot)}{u(t, y)} - w_*(\cdot) \right\|_R = 0, \qquad R > 0. \tag{45}$$

Theorem 4 states that, up to an arbitrarily small relative error ε, the islands with centers in Γ^* and radius $r(\varepsilon, \varrho)$ carry the whole mass of the solution $u(t, \cdot)$. Locally, in an arbitrarily fixed R-neighborhood of each of these centers, the shapes of the potential and the normalized solution resemble $h_t + V_*$ and w_*, respectively. The number of these islands increases at most as an arbitrarily small power of t and their distance increases almost like t. Note that, for $\varrho = \infty$, the set $B_{r(\varepsilon, \varrho)}(\Gamma^*)$ in (42) is equal to Γ^* and, hence, the islands consist of single lattice sites.

It is an open problem under what assumptions on the potential ξ the number $|\Gamma^*|$ of relevant peaks stays bounded as $t \to \infty$; we have made no attempt to choose Γ^* as small as possible.

4 Universality

In this section we explain that, under some mild regularity assumptions on the tails of $\xi(0)$ at its essential supremum, there are only four universality classes of asymptotic behaviors of the parabolic Anderson model. Three of them have already been analyzed in the literature: the double-exponential distribution with $\varrho \in (0, \infty)$ respectively $\varrho = \infty$ [GM98], [GH99], [GK00], [GKM00], [GKM04]) (see Sections 2.1 and 3) and general fields bounded from above [BK01a], [BK01b], [S98], [Ant94], [Ant95] (see Section 2.3). A fourth and new universality class is currently under investigation, see [HKM04]. This class lies in the union of the boundary cases $\varrho = 0$ of the double-exponential distribution ('almost bounded' fields) and $\gamma = 1$ for the general bounded from above fields. Examples of distributions that fall into this class look somewhat odd, but it turns out that the optimal potential shape and the optimal shape of the solution are perfectly parabolic, respectively Gaussian, which makes this class rather appealing. In particular, the appearing variational formulas can be easily solved explicitly and uniquely.

We now summarize [HKM04]. Our basic assumption on the logarithmic moment generating function H in (9) is the following.

Assumption ($\hat{H}$): *There are a function $\hat{H}\colon (0,\infty) \to \mathbb{R}$ and a continuous auxiliary function $\eta\colon (0,\infty) \to (0,\infty)$ such that*

$$\lim_{t\uparrow\infty} \frac{H(ty) - yH(t)}{\eta(t)} = \hat{H}(y) \neq 0 \qquad \text{for } y \neq 1. \tag{46}$$

The function $\hat{H}$ extracts the asymptotic scaling properties of the cumulant generating function H. In the language of the theory of regular functions, the assumption is that the logarithmic moment generating function H is in the de Haan class, which does not leave many possibilities for $\hat{H}$:

Proposition 2. *Suppose that Assumption ($\hat{H}$) holds.*

(i) *There is a $\gamma \geq 0$ such that $\lim_{t\uparrow\infty} \eta(ty)/\eta(t) = y^\gamma$ for any $y > 0$, i.e., η is regularly varying of index γ. In particular, $\eta(t) = t^{\gamma+o(1)}$ as $t \to \infty$.*
(ii) *There exists a parameter $\rho > 0$ such that, for every $y > 0$,*
   (a) $\hat{H}(y) = \rho \dfrac{y - y^\gamma}{1 - \gamma}$ *if $\gamma \neq 1$,*
   (b) $\hat{H}(y) = \rho y \log y$ *if $\gamma = 1$.*

Our second regularity assumption is a mild supposition on the auxiliary function η. This assumption is necessary only in the case $\gamma = 1$ (which will turn out to be the critical case).

Assumption (K): *The limit $\eta_* = \lim_{t\to\infty} \eta(t)/t \in [0,\infty]$ exists.*

We now introduce a scale function $\alpha\colon (0,\infty) \to (0,\infty)$, by

$$\frac{\eta\big(t\alpha(t)^{-d}\big)}{t\alpha(t)^{-d}} = \frac{1}{\alpha(t)^2}. \tag{47}$$

The function $\alpha(t)$ turns out to be the annealed scale function for the radius of the relevant islands in the parabolic Anderson model. We can easily say something about the asymptotics of $\alpha(t)$:

Lemma 1. *Suppose that Assumptions ($\hat{H}$) and (K) hold. If $\gamma \leq 1$ and $\eta_* < \infty$, then there exists a unique solution $\alpha\colon (0,\infty) \to (0,\infty)$ to (47), and it satisfies $\lim_{t\to\infty} t\alpha(t)^{-d} = \infty$. Moreover,*

(i) *If $\gamma = 1$ and $0 < \eta_* < \infty$, then $\lim_{t\to\infty} \alpha(t) = 1/\sqrt{\eta_*} \in (0,\infty)$.*
(ii) *If $\gamma = 1$ and $\eta_* = 0$, then $\alpha(t) = t^{\nu+o(1)}$ as $t \to \infty$, where $\nu = (1 - \gamma)/(d + 2 - d\gamma)$.*

Now, under Assumptions ($\hat{H}$) and (K), we can formulate a complete distinction of the PAM into four cases:

(1) $\eta_* = \infty$ (in particular, $\gamma \geq 1$).
   This is the boundary case $\varrho = \infty$ of the double-exponential case. We have $\alpha(t) \to 0$ as $t \to \infty$, as is seen from (47), i.e., the relevant islands consist of single lattice sites.
(2) $\eta_* \in (0, \infty)$ (in particular, $\gamma = 1$).
   This is the case of the double-exponential distribution in Section 2.1. By rescaling, one can achieve that $\eta_* = 1$. The parameter ϱ of Proposition 2(ii)(b) is identical to the one in Assumption (H) of Section 3.1.
(3) $\eta_* = 0$ and $\gamma = 1$.
   This is the case of islands of slowly growing size, i.e., $\alpha(t) \to \infty$ as $t \to \infty$ slower than any power of t. This case comprises 'almost bounded' and bounded from above potentials. This class is the subject of [HKM04], see also below.
(4) $\gamma < 1$ (in particular, $\eta_* = 0$)
   This is the case of islands of rapidly growing size, i.e., $\alpha(t) \to \infty$ as $t \to \infty$ at least as fast as some power of t. Here the potential ξ is necessarily bounded from above. This case was treated in [BK01a]; see Section 2.3.

Let us comment on the class (3), which appears to be new in the literature and is under investigation in [HKM04]. One obtains examples of potentials (unbounded from above) that fall into this class by replacing ϱ in (25) by a sufficiently regular function $\varrho(r)$ that tends to 0 as $r \to \infty$, and other examples (bounded from above) by replacing γ in (27) by a sufficiently regular function $\gamma(x)$ tending to 1 as $x \downarrow 0$. According to Lemma 1, the scale function $\alpha(t)$ defined in (47) tends to infinity, but is slowly varying. The annealed rate function for the rescaled potential shape, I_R, introduced in (15) turns out to be

$$I_R(\varphi) = \text{const} \int_{Q_R} e^{\varphi(x)/\varrho} \, dx. \tag{48}$$

The characteristic variational problem for the annealed potential shape in (17), χ, turns out to be uniquely minimized by a parabolic function $\varphi_*(x) = \text{const} - \varrho\|x\|_2^2$, and the principal eigenfunction v_* of the operator $\kappa\Delta + \varphi_*$ is the Gaussian density $v_*(x) = \text{const}\, e^{-\varrho\|x\|_2^2}$.

5 Time-Dependent Random Potentials

In this section we study the intermittent behavior of the parabolic Anderson model (PAM) with a space-time homogeneous ergodic random potential ξ:

$$\partial_t u(t,x) = \kappa\Delta u(t,x) + [\xi(t,x) - \langle\xi(t,x)\rangle] u(t,x), \qquad (t,x) \in (0,\infty) \times \mathbb{Z}^d,$$
$$u(0,x) = 1, \qquad\qquad\qquad\qquad\qquad\qquad\qquad x \in \mathbb{Z}^d. \tag{49}$$

Note that for time-dependent potentials the direct connection to the spectral representation of the Anderson Hamiltonian (2) is lost. Our focus will be on the situation when the potential ξ is given by a field $\{Y_k(t); k \in \mathbb{N}\}$ of independent random walks on $\mathbb{Z}^d$ with diffusion constant ϱ in Poisson equilibrium with density ν:

$$\xi(t, x) = \gamma \sum_{k \in \mathbb{N}} \delta_{Y_k(t)}(x), \tag{50}$$

where γ denotes a positive coupling constant. Clearly

$$\langle \xi(t, x) \rangle = \nu\gamma.$$

This deterministic correction to the potential in (49) has been added for convenience to eliminate non-random terms.

The form (50) of the potential is motivated by the following particle model. Consider a system of two types of independent particles, A and B, performing independent continuous-time simple random walks with diffusion constants κ and ϱ and Poisson initial distribution with densities ν and 1, respectively. Assume that the B-particles split into two at a rate that is γ times the number of A-particles present at the same location and die at rate $\nu\gamma$. Hence, the A- and B-particles may be regarded as catalysts and reactants in a simple catalytic reaction model. Then the (spatially homogeneous and ergodic) solution $u(t, x)$ of the PAM (49) is nothing but the average number of reactants at site x at time t given a realization of the catalytic dynamics (50). Such a particle model (with arbitrary death rate) has been considered by Kesten and Sidoravicius [KS03]. We will come back to the results in [KS03] at the end of this section. We further refer to the overview papers by Dawson and Fleischmann [DF00] and by Klenke [Kl00] for continuum models with singular catalysts in a measure-valued context where questions different from ours have been addressed.

Our aim is to study the moment Lyapunov exponents

$$\lambda_p = \lim_{t \to \infty} \frac{1}{t} \log \langle u(t, 0)^p \rangle$$

as well as the quantities

$$\lambda_p^* = \lim_{t \to \infty} \frac{1}{t} \log \log \langle u(t, 0)^p \rangle$$

$(p = 1, 2, \dots)$ as functions of the model parameters. The phase diagram will turn out to be different in dimensions $d = 1, 2$, $d = 3$, and $d \geq 4$.

Definition 2. a) For $p \in \mathbb{N}$, we will say that the PAM (49)–(50) is *strongly p-catalytic* if $\lambda_p^* > 0$. Otherwise the PAM will be called *weakly p-catalytic*.

b) For $p \in \mathbb{N} \setminus \{1\}$ and $\lambda_p < \infty$, we will say that the PAM is *p-intermittent* if $\lambda_p/p > \lambda_{p-1}/(p-1)$.

We believe that strongly catalytic behavior is related to heavy tails of the Poisson distribution of the catalytic point process $\{Y_k(t); k \in \mathbb{N}\}$ and may occur if the main contribution to the p-th moment comes from realizations with a huge number of catalysts at the same lattice site. Recall that p-intermittency means that, for large t, the p-th moment, considered as $\lim_{R\to\infty} |B_R|^{-1} \sum_{x \in B_R} u(t,x)^p$, is 'generated' by high peaks of the solution $u(t, \cdot)$ located far from each other.

For potentials of the form

$$\xi(t,x) = \gamma \dot{W}_x(t)$$

with $(W_x(t))_{x \in \mathbb{Z}^d}$ being a field of i.i.d. (or correlated) Brownian motions and (49) understood as a system of Itô equations, the moment Lyapunov exponents have been shown by Carmona and Molchanov [CM94] to exhibit the following behavior. In dimensions $d = 1, 2$ there is p-intermittency for all $p \in \mathbb{N} \setminus \{1\}$ and all choices of the model parameters κ and γ. If $d \geq 3$, then there exist critical points $0 < c_2 < c_3 < \cdots$ such that p-intermittency holds if and only if $\kappa/\gamma < c_p$. For this model, the asymptotics of the almost sure ('quenched') Lyapunov exponent as $\kappa \to 0$ has been investigated in [CM94], [CMV96], [CKM01], and [CMS02].

In the following we present the results for catalytic potentials of the form (50) obtained in [GH04].

Our analysis of the moment Lyapunov exponents is based on the following probabilistic representation of the p-th moment which is easily derived from the Feynman-Kac formula for $u(t,0)$:

$$\langle u(t,0)^p \rangle = \mathbb{E}_0^{(p)} \exp \left\{ \nu\gamma \int_0^t \sum_{i=1}^p w(s, X_i(s)) \, ds \right\}, \tag{51}$$

where $\mathbb{E}_0^{(p)}$ denotes expectation with respect to p independent random walks $X_1, \ldots, X_p$ on $\mathbb{Z}^d$ with generator $\kappa\Delta$ starting at the origin, and w is the solution of the random initial value problem

$$\partial_t w(t,x) = \varrho\Delta w(t,x) + \gamma \sum_{i=1}^p \delta_{X_i(t)}(x)\,(1 + w(t,x)), \quad (t,x) \in (0,\infty) \times \mathbb{Z}^d,$$

$$w(0,x) = 0, \qquad\qquad\qquad\qquad\qquad\qquad x \in \mathbb{Z}^d.$$

One of the main difficulties in analyzing (51) is related to the observation that $w(t, X_i(t))$ depends in a nontrivial way *on the whole past* $\{X_j(s); 0 \leq s \leq t\}$, $j = 1, \ldots, p$, of our random walks (although our notation does not reflect this).

For $r \geq 0$, let $\mu(r)$ denote the upper boundary of the spectrum of the operator $\Delta + r\delta_0$ in $\ell^2(\mathbb{Z}^d)$. It is well-known that, in dimensions $d = 1, 2$, $\mu(r) > 0$ for all r and, in dimensions $d \geq 3$, $\mu(r) = 0$ for $0 \leq r \leq r_d$ and $\mu(r) > 0$ for $r > r_d$, where

$$r_d = 1/G_d(0)$$

and G_d denotes the Green function associated with the discrete Laplacian Δ.

Theorem 5. *For any choice of the parameters of the PAM (49)–(50), the limit λ_p^* exists, and*

$$\lambda_p^* = \varrho\,\mu(p\gamma/\varrho), \qquad p \in \mathbb{N}.$$

Hence, for $d = 1, 2$, the PAM (49)–(50) is always strongly p-catalytic, whereas for $d \geq 3$ this is true only if $p\gamma/\varrho$ exceeds the critical threshold r_d. Note that λ_p^* does not depend on κ nor on ν.

We next study the behavior of the moment Lyapunov exponents $\lambda_p = \lambda_p(\kappa)$ as a function of the diffusion constant κ in the weakly catalytic regime $0 < p\gamma/\varrho < r_d$ for $d \geq 3$. We will mainly describe their behavior for small and for large values of κ.

Theorem 6. *Let $d \geq 3$, $p \in \mathbb{N}$, and $0 < p\gamma/\varrho < r_d$. Then the limit λ_p exists and is finite for all κ and ν. Moreover, $\kappa \mapsto \lambda_p(\kappa)$ is strictly decreasing and convex on $[0, \infty)$ and satisfies*

$$\lim_{\kappa \downarrow 0} \frac{\lambda_p(\kappa)}{p} = \frac{\lambda_p(0)}{p} = \nu\gamma\,\frac{p\gamma/\varrho}{r_d - p\gamma/\varrho}.$$

Hence, in any dimension $d \geq 3$, the PAM (49)–(50) is p-intermittent in the weakly catalytic regime for small values of the diffusion constant κ.

To formulate the behavior of the moment Lyapunov exponents for $\kappa \to \infty$, we introduce the variational expression $\mathcal{P}$ for the polaron problem analyzed in [L77], [DV83], and [BDS93]:

$$\mathcal{P} = \sup_{\substack{f \in C_c^\infty(\mathbb{R}^3) \\ \|f\|_2 = 1}} \left[\left\| (-\Delta)^{-1/2} f^2 \right\|_2^2 - \|\nabla f\|_2^2 \right].$$

Theorem 7. *Let $d \geq 3$, $p \in \mathbb{N}$, and $0 < p\gamma/\varrho < r_d$. Then*

$$\lim_{\kappa \to \infty} \kappa \frac{\lambda_p(\kappa)}{p} = \begin{cases} \dfrac{\nu\gamma^2}{r_3} + \sqrt{p}\sqrt{\dfrac{\nu\gamma^2}{\varrho}}\mathcal{P}, & \text{if } d = 3, \\[2mm] \dfrac{\nu\gamma^2}{r_d}, & \text{if } d \geq 4. \end{cases}$$

In other words, in dimensions $d = 3$ for p in the weakly catalytic regime, the PAM (49)–(50) is p-intermittent also for large values of κ. We conjecture intermittent behavior for all κ. In dimensions $d \geq 4$, the leading term in the asymptotics of $\lambda_p(\kappa)/p$ as $\kappa \to \infty$ is the same for all p in the weakly catalytic regime. We in fact conjecture that there is even no intermittency in high dimensions for large κ.

Let us remark that in [KS03] Kesten and Sidoravicius obtained the following results for the above mentioned catalytic particle model with arbitrary death rate δ instead of $\nu\gamma$. If $d = 1, 2$, then for all choices of the parameters, the average number of B-particles per site tends to infinity faster than exponential. If $d \geq 3$, γ small enough and δ large enough, then the average number of B-particles per site tends to zero exponentially fast. The first of these two results is covered by Theorem 5, since the death rate δ does not affect λ_1^*. The second result is covered by Theorem 6, since $0 < \lambda_1 < \infty$ in the weakly catalytic regime and the death rate δ shifts λ_1 by $\gamma\nu - \delta$. Kesten and Sidoravicius further show that in all dimensions for γ large enough, conditioned on the evolution of the A-particles, there is a phase transition. Namely, for small δ the B-particles locally survive, while for large δ they become locally extinct. In [GH04] there are no results for the quenched situation. The analysis in [KS03] does not lead to an identification of Lyapunov exponents, but is more robust under an adaption of the model than the above analysis based on the Feynman-Kac representation.

References

[And58]   Anderson, P.M.: Absence of diffusion in certain random lattices. Phys. Rev. **109**, 1492–1505 (1958)

[Ant94]   Antal, P.: Trapping Problems for the Simple Random Walk. Dissertation ETH Zürich, No. 10759 (1994)

[Ant95]   Antal, P.: Enlargement of obstacles for the simple random walk. Ann. Probab. **23: 3**, 1061–1101 (1995)

[AM93]   Aizenman, M., Molchanov, S.: Localization at large disorder and extreme energies: an elementary derivation. Comm. Math. Phys. **157: 2**, 245–278 (1993)

[BBH01]   van den Berg, M., Bolthausen, E., den Hollander, F.: Moderate deviations for the volume of the Wiener sausage. Ann. of Math. (2) **153:2**, 355–406 (2001)

[BK01a]   Biskup, M., König, W.: Long-time tails in the parabolic Anderson model with bounded potential. Ann. Probab. **29;2**, 636–682 (2001)

[BK01b]   Biskup, M., König, W.: Screening effect due to heavy lower tails in one-dimensional parabolic Anderson model. J. Stat. Phys. **102:5/6**, 1253–1270 (2001)

[BKS04]   Böcker, S., Kirsch, W., Stollmann, P.: Spectral theory for nonstationary random potentials. In this proceedings volume.

[B94]   Bolthausen, E.: Localization of a two-dimensional random walk with an attractive path interaction. Ann. Probab. **22**, 875–918 (1994)

[BDS93]   Bolthausen, E., Deuschel, J.-D., Schmock, U.: Convergence of path measures arising from a mean field or polaron type interaction. Probab. Theory Relat. Fields **95**, 283–310 (1993)

[CKM01]   Carmona, R.A., Koralov, L., Molchanov, S.A.: Asymptotics for the almost-sure Lyapunov exponent for the solution of the parabolic Anderson problem. Random Oper. Stochastic Equations **9**, 77–86 (2001)

[CM94]   Carmona, R.A., Molchanov, S.A.: Parabolic Anderson problem and intermittency. Memoirs of the AMS **108**, nr. 518 (1994)

[CM95]   Carmona, R.A., Molchanov, S.A.: Stationary parabolic Anderson model and intermittency. Probab. Theory Relat. Fields **102**, 433–453 (1995)

[CMV96]   Carmona, R.A., Molchanov, S.A., Viens, F.: Sharp upper bound on the almost-sure exponential behavior of a stochastic partial differential equation. Random Oper. Stochastic Equations **4**, 43–49 (1996)

[CMS02]   Cranston, M., Mountford, T.S., Shiga, T.: Lyapunov exponents for the parabolic Anderson model. Acta Math. Univ. Comeniane **LXXI**, 163–188 (2002)

[DF00]   Dawson, D.A., Fleischmann, K.: Catalytic and mutually catalytic branching, in: Infinite dimensional stochastic analysis, 145–170, Ph. Clément, F. den Hollander, J. van Neerven and B. de Pagter, eds., R. Neth. Acad. Sci., Amsterdam (2000)

[DV75]   Donsker, M., Varadhan, S.R.S.: Asymptotics for the Wiener sausage. Comm. Pure Appl. Math. **28**, 525–565 (1975)

[DV79]   Donsker, M., Varadhan, S.R.S.: On the number of distinct sites visited by a random walk. Comm. Pure Appl. Math. **32**, 721–747 (1979)

[DV83]   Donsker, M., Varadhan, S.R.S.: Asymptotics for the polaron. Comm. Pure Appl. Math. **36**, 505–528 (1983)

[EEEF84]   Ebeling, W., Engel, A., Esser, B., Feistel, R.: Diffusion and reaction in random media and models of evolution processes. J. Stat. Phys. **37**, 369–385 (1984)

[FMSS85]   Fröhlich, J., Martinelli, F., Scoppola, E., Spencer, T.: Constructive proof of localization in the Anderson tight binding model. Comm. Math. Phys. **101**, 21–46 (1985)

[GH99]   Gärtner, J., den Hollander, F.: Correlation structure of intermittency in the parabolic Anderson model. Probab. Theory Relat. Fields **114**, 1–54 (1999)

[GH04]   Gärtner, J., den Hollander, F.: Intermittency in a catalytic random medium. In preparation.

[GK00]   Gärtner, J., König, W.: Moment asymptotics for the continuous parabolic Anderson model. Ann. Appl. Probab. **10**, 3, 192–217 (2000)

[GKM00]   Gärtner, J., König, W., Molchanov, S.A.: Almost sure asymptotics for the continuous parabolic Anderson model. Probab. Theory Relat. Fields **118:4**, 547–573 (2000)

[GKM04]   Gärtner, J., König, W., Molchanov, S.A.: Geometric characterization of intermittency in the parabolic Anderson model. In preparation.

[GM90]   Gärtner, J., Molchanov, S.A.: Parabolic problems for the Anderson model. I. Intermittency and related topics. Commun. Math. Phys. **132**, 613–655 (1990)

[GM98]   Gärtner, J., Molchanov, S.A.: Parabolic problems for the Anderson model. II. Second-order asymptotics and structure of high peaks. Probab. Theory Relat. Fields **111**, 17–55 (1998)

[GM00]   Gärtner, J., Molchanov, S.A.: Moment asymptotics and Lifshitz tails for the parabolic Anderson model. Canadian Math. Soc. Conference Proceedings 26, Gorostiza, L.G., Ivanoff, B.G. (eds.) AMS, Providence, 141–157 (2000)

[HKM04]   van der Hofstad, R., König, W., Mörters, P.: Universality for the parabolic Anderson model. In preparation.

[HW94]   den Hollander, F., Weiss, G.H.: Aspects of trapping in transport pro-
          cesses, in: Contemporary problems in Statistical Physics, G.H. Weiss,
          ed., SIAM, Philadelphia (1994)
[KS03]   Kesten, H., Sidoravicius, V.: Branching random walk with catalysts,
          Electr. J. Prob. **8**, Paper no. 5, 1–51 (2003)
[Kl00]   Klenke, A.: A review on spatial catalytic branching, in: Stochastic
          models, 245–263, CMS Conf. Proc. 26, Amer. Math. Soc. (2000)
[LMW04]   Leschke, H., Müller, P., Warzel, S.: A survey of rigorous results on
          random Schrödinger operators for amorphous solids. In this proceedings
          volume.
[L77]   Lieb, E.H.: Existence and uniqueness of the minimizing solution of
          Choquard's nonlinear equation. Studies in Applied Mathematics **57**,
          93–105 (1977)
[MW02]   Merkl, F., Wüthrich, M.V.: Infinite volume asymptotics for the ground
          state energy in a scaled Poissonian potential. Ann. Inst. H. Poincaré
          Probab. Statist. **38:3**, 253–284 (2002)
[M94]   Molchanov, S.A.: Lectures on random media. In: Bakry, D., Gill, R.D.,
          Molchanov, S.A., Lectures on Probability Theory, Ecole d'Eté de Prob-
          abilités de Saint-Flour XXII-1992, LNM 1581, pp. 242–411. Berlin,
          Springer (1994)
[S82]   Sinai, Ya.G.: Limit behavior of one-dimensional random walks in ran-
          dom environment. Theor. Probab. Appl. **27**, 247–258 (1982)
[S98]   Sznitman, A.-S.: Brownian motion, Obstacles and Random Media.
          Springer Berlin (1998)
[Z84]   Zel'dovich, Ya.B.: Selected Papers. Chemical Physics and Hydrody-
          namics (in Russian), Nauka, Moscow (1984)
[ZKO94]   Zhikov, V.V., Kozlov, S.M., Olejnik, O.A.: Homogenization of Differ-
          ential Operators and Integral Functionals. Springer, Berlin (1994)
[ZMRS87] Zel'dovich, Ya.B., Molchanov, S.A., Ruzmajkin, S.A., Sokolov, D.D.:
          Intermittency in random media. Sov. Phys. Uspekhi **30:5**, 353–369
          (1987)
[ZMRS88] Zel'dovich, Ya.B., Molchanov, S.A., Ruzmajkin, S.A., Sokolov, D.D.:
          Intermittency, diffusion and generation in a nonstationary random
          medium. Sov. Sci. Rev. Sect. C, Math. Phys. Rev. **7**, 1–110 (1988)

Random Spectral Distributions

Friedrich Götze[1] and Franz Merkl[2]

. Fakultät für Mathematik, Universität Bielefeld, Germany
goetze@mathematik.uni-bielefeld.de
. Department of Mathematics, University of Leiden, The Netherlands
merkl@math.leidenuniv.nl

Summary. We review recent results and new methods for the distributions of spectra of random matrices and prove a kind of Mock-Gaussian behavior for eigenvalues of unitary random matrices (CUE), using creation and annihilation operators from quantum field theory. The result is non-asymptotic and plays a key role in establishing the relation between the local distribution of the zeros of the zeta function and the universal asymptotic local distribution of eigenvalues of unitary matrix ensembles.

1 Asymptotic Approximation of Random Spectra

In this part, we shall review results on the limit distributions and the rate of convergence for the spectra of certain ensembles of random matrices. The spectral theory of random matrices originated from statistical models for the magnetic resonances of complex systems like heavy nuclei starting with the work of Landau and Smolochowski (1937) and Wigner [Wig55]. Since then numerous other applications and connections of these models to various branches of mathematics have been found. The following represents a rather incomplete list of fields

- Local empirical spectral distribution of the quantization of hyperbolic ('chaotic') dynamical systems, Bohigas-Gianonni-Schmit conjecture.
- Stability of master equations (networks, ecology)
- Combinatorics of representations of the symmetric group
- Nuclear growth models, monotone paths in random combinatorics
- Principle components, data reduction and discriminant analysis in statistics
- Distributions of eigenvalues of random unitary matrices and distribution of zeros of zeta– and L–functions in number theory.

Distributions of random unitary matrices in connection with the last topic will be discussed in more detail in section 2.

In the following, we shall discuss certain models of random matrices with independent entries introduced by Wigner in [Wig55].

1.1 Wigner and GUE Ensembles

Let W_N be a complex Hermitian matrix of order N whose elements $W_{lj} = X_{lj} + iY_{lj}$, $(l, j = 1, \ldots, N)$ are independent for $1 \leq l \leq j \leq N$.

Assume that the real and imaginary parts X_{lj} and Y_{lj} of the matrix elements are independent as well. Furthermore, we standardize the random variables such that $\mathbb{E}X_{kj} = \mathbb{E}Y_{kj} = 0$, $1 \leq k \leq j \leq N$, and let $Y_{ll} = 0$ for $l = 1, \ldots N$. Finally, we may assume that

$$\mathbb{E}X_{ll}^2 = 1, \quad l = 1, 2, \ldots, \quad \mathbb{E}X_{kj}^2 = \mathbb{E}Y_{kj}^2 = 1/2, \quad 1 \leq k < j.$$

This class is called *Wigner ensemble*.

If we assume furthermore that the random variables X_{ij} and Y_{ij} have a *Gaussian* distribution, this matrix model is called the *Gausssian unitary ensemble*, or GUE for short. Its densities are invariant under unitary transformations. The limit distributions of spectra of this matrix ensemble are universal in the sense that they are the limits of the spectra of a large class of non unitary invariant matrix ensembles as well.

The distribution of the N eigenvalues $\lambda_1, \lambda_2, \ldots, \lambda_N$ of the rescaled random matrix $\frac{1}{\sqrt{N}} W$ is given by

$$\phi_{\mathrm{GUE}}(\lambda) := \frac{1}{Z_N} \prod_{1 \leq i < j \leq N} |\lambda_i - \lambda_j|^2 \prod_{1 \leq j \leq N} \exp[-\frac{N}{2}\lambda_j^2], \tag{1}$$

where Z_N denotes a normalizing factor.

The derivation of this relation can be roughly described as follows: we start with a matrix diagonalization $\mathbf{W} = \mathbf{U}^{-1}\mathbf{\Lambda}\mathbf{U}$, where $\mathbf{U}$ denotes an $N \times N$ unitary matrix and $\mathbf{\Lambda}$ denotes a diagonal matrix. Then matrix calculus yields a transformation formula for the mapping $\mathbf{W} \mapsto \mathbf{\Lambda}$. Its Jacobian results in the interaction term $\Delta(\mathbf{\Lambda}) := \prod_{1 \leq i < j \leq N}(\lambda_i - \lambda_j)^2$ in the induced density. For details see the monographs by Mehta 1991, [Meh91] and Forrester, [For04]. The probability density ϕ_{GUE} vanishes on the set of Hermitean matrices where at least two eigenvalues coincide. This fact may be interpreted as "repulsion" of eigenvalues. For an interpretation of ϕ_{GUE} as a statistical mechanics partition function based on a logarithmic interaction $\log|\lambda_i - \lambda_j|$ of particles at positions λ_i and λ_j, see the aforementioned monographs.

If we replace the independent Gaussian random variables X_{ij}, Y_{ij} in the random matrix $\mathbf{W}$ by independent Brownian motions $X_{ij}(t), Y_{ij}(t), t \geq 0$, we get a Hermitean matrix valued Brownian motion process $\mathbf{H}_t := (W_{ij}(t))_{1 \leq i, j \leq N}$. Diagonalizing $\mathbf{H}_t$, that is

$$\mathbf{H}_t = \mathbf{U}_t^{-1}\mathbf{\Lambda}_t\mathbf{U}_t, \quad \mathbf{\Lambda}_t := \mathrm{diag}(\lambda_1(t), \ldots, \lambda_N(t)), \tag{2}$$

results in a diagonal matrix process of eigenvalues $\lambda_t := (\lambda_1(t), \ldots, \lambda_N(t))$, which is called Dyson's Brownian motion. It may be obtained alternatively as a solution to the following coupled system of stochastic differential equations

$$d\lambda_t = dW_t + \nabla_\lambda \log \Delta(\Lambda_t)\, dt,$$

where $W_t := (W_1(t), \ldots, W_N(t))$ denotes an i.i.d. vector of Brownian motions; see [Dys62]. Another interesting representation of Dyson's Brownian motion is given by a sequence of maximum path transformations of W_t with respect to the hyperplanes $\lambda_i = \lambda_j$, $i < j$, extending a one-dimensional construction by Pitman [Pit75]. This transformation yields in each step a process confined to another half-plane ($\lambda_i < \lambda_j$), and it results in a process confined to the Weyl chamber $\lambda_1 < \ldots < \lambda_N$, see [OY02] and [O'C03].

Since $\Delta(\lambda) = \det\left(\lambda_i^j, i = 1, \ldots, N; j = 0, \ldots, N-1\right)^2$ is the square of the so-called Vandermonde determinant, simple linear combinations of rows result in replacing the powers by Hermite polynomials, say $H_j(x)$. Finally, moving the Gaussian density factors into the determinant yields

$$\varphi_{\text{GUE}}(\lambda) = c_N \det\left(\varphi_i(\sqrt{N}\lambda_j), i = 0, \ldots, N-1; j = 1, \ldots, N\right)^2, \quad (3)$$

where $\varphi_j(x)$ denotes the j-th Hermite orthogonal function given by

$$\varphi_j(x) = \frac{1}{\sqrt{j! 2^j \sqrt{\pi}}} e^{-x^2/2} H_j(x).$$

Thus φ_{GUE} may be interpreted as the probability density of N fermionic particles in the quantization of a harmonic oscillator.

The average density of the eigenvalues, say $\phi_N(x)$, may be obtained as the one-dimensional marginal distribution of the symmetric density $\varphi_{\text{GUE}}(x)$. It is given (in the physical interpretation) by the average probability density of the wave functions $\varphi_j(x), 1 \le j \le N$

$$p_N(x) = \frac{1}{\sqrt{2N}} \sum_{j=0}^{N-1} \varphi_j^2(x\sqrt{N}/\sqrt{2}) \quad (4)$$

$$= \sqrt{N/2}\, \varphi_N^2(x\sqrt{N/2}) - \sqrt{(N+1)/2}\, \varphi_{N-1}(x\sqrt{N/2})\, \varphi_{N+1}(x\sqrt{N}/\sqrt{2}).$$

See, for example, Mehta [Meh91], Appendix A.8.

In the following, we shall consider the empirical spectral distribution function of the N random eigenvalues $\lambda_j, j = 1, \ldots N$, of a random matrix from the standardized GUE-ensemble in (1) given by

$$F_N(x) := \frac{1}{N} \sum_{j=1}^{N} I_{\{\lambda_j \le x\}}, \quad (5)$$

where I_B denotes the indicator of an event B.

Here, F_N converges in probability and a.s. to the so-called Wigner semicircular law given by the density $g(x)$ and distribution function $G(x)$ defined by

$$g(x) = \frac{1}{2\pi}\sqrt{4 - x^2}\, I_{\{|x| \le 2\}}, \quad G(x) = \int_{-\infty}^{x} g(u)\, du. \qquad (6)$$

This result is due to Wigner [Wig57].

We shall review recent results for the rate of convergence of the expected spectral distribution function $\mathbb{E}F_N(x)$ to the limit distribution function $G(x)$ in terms of the so-called Kolmogorov distance. In [GT04a], the following is shown.

Theorem 1. *There exists absolute positive constants C_1, C_2, and γ such that, for any $N \ge 1$, the Kolmogorov distance, say Δ_N, of the distribution functions is bounded by*

$$\Delta_N := \sup_x |\mathbb{E}F_N(x) - G(x)| \le C_1 N^{-1}. \qquad (7)$$

The density p_N satisfies

$$|p_N(x) - g(x)| \le \frac{C_2}{N(4 - x^2)}, \qquad (8)$$

for all x in the interval $-2 + \gamma N^{-2/3} \le x \le 2 - \gamma N^{-2/3}$.

The derivation of the bound (7) for the Kolmogorov distance needs the bound (8). The inequality (7) is an optimal bound, conjectured by Bai and Silverstein in 1993; see [Bai93]. It improves a less precise previous bound of order $O(N^{-2/3})$, see [GT02]). In the following, we shall outline some of the key ingredients for this result, using Stein's method.

Haageruup and Thorbjornsen [HT03] show that the characteristic function $f_N(t) = \int e^{itx} p_N(x)\, dx$ of $p_N(t)$ is given by

$$f_N(t) = \exp\{-\frac{t^2}{2N}\}\Phi(1 - N, 2; \frac{t^2}{N}), \qquad (9)$$

where $\Phi(\alpha, \beta; s)$ denotes the confluent hypergeometric function. It is well known that these special functions $\Phi(\alpha, \beta; s)$ satisfy a differential equation of second order, that is

$$s\frac{d^2\Phi}{ds^2} + (\beta - s)\frac{d\Phi}{ds} - \alpha\Phi = 0. \qquad (10)$$

Using (9) and (10), one derives the following differential equation for the density $p_N(x)$, passing from Fourier transforms to densities:

$$(4 - x^2)p_N'(x) + xp_N(x) + \frac{1}{N^2}p_N'''(x) = 0. \qquad (11)$$

Using this relation, we may write, for $|x| < 2$,

$$p_N(x) = \frac{\sqrt{4 - x^2}}{2}p_N(0) + \frac{\sqrt{4 - x^2}}{N^2}\int_0^x \frac{p_N'''(u)\, du}{(4 - u^2)^{\frac{3}{2}}}. \qquad (12)$$

Since $\mathbb{E}F_N(x)$ and $G(x)$ are symmetric distributions, we have $\mathbb{E}F_N(0) = G(0)$. This implies that for $|x| < 2$,

$$\mathbb{E}F_N(x) - G(x) =$$

$$\frac{1}{2}\left(p_N(0) - \frac{1}{\pi}\right)\int_0^x \sqrt{4-u^2}\, du + \frac{1}{N^2}\int_0^x \sqrt{4-u^2}\int_0^u \frac{p_N'''(s)\, ds}{(4-s^2)^{\frac{3}{2}}}\, du. \quad (13)$$

The first term in the right hand side may be estimated via $|p_N(0) - \frac{1}{\pi}| \le \frac{C}{N}$. The estimation of the second term is more involved. For details see [GT03a].

Note that in the limit $N \to \infty$, the differential equation (11) turns into the equation

$$(4 - x^2)p'(x) + xp(x) = 0, \quad (14)$$

which has again Wigner's semi-circle density $g(x)$ as solution.

Passing to the adjoint differential operator of $\mathcal{A} = (4 - x^2)D_x + x$ in the last equation, that is $(\mathcal{A}^* f)(x) = -(4 - x^2)f'(x) + 3xf(x)$, we may write a characterizing equation for the semi-circle distribution of Wigner in the spirit of C. Stein [Ste86] as follows. Let ξ denote a random variable with distribution function F. If the equation

$$\mathbb{E}\big((4 - \xi^2)f'(\xi) - 3\xi f(\xi)\big) = 0 \quad (15)$$

holds for all functions in a sufficiently large class of functions $\mathcal{C}$, then the distribution of ξ has to be the Wigner distribution, that is $F = G$. Here, we may take e.g.

$$\mathcal{C} := \{f : \mathbb{R} \to \mathbb{R} : f \in C^1(\mathbb{R} \setminus \{-2, 2\}), \quad \overline{\lim}_{|y| \to \infty} |yf(y)| < \infty;$$

$$\limsup_{y \to \pm 2} |4 - y^2||f'(y)| < \infty\},$$

where $C^1(\mathbb{R} \setminus \{-2, 2\})$ denotes the class of differentiable functions on $\mathbb{R} \setminus \{-2, 2\}$, with bounded derivatives on the all compact subsets of $(-2, 2)$.

Moreover, if (15) holds approximately only, F will only approximate G. A quantitative version of this fact is obtained using an explicit solution, say $f \in \mathcal{C}$, to the following inhomogeneous first order equation in the interval $(-2, 2)$

$$(4 - x^2)f'(x) - 3xf(x) = I_{\{x \le y\}} - F(y) =: \varphi(u) \quad (16)$$

with a function $\varphi(u)$ such that $\int_{-2}^2 \varphi(u)g(u)\, du = 0$. Note that a formal solution f of this equation is given by

$$f(x) := \frac{\text{sign}(4 - x^2)}{\sqrt{|4 - u^2|}^3}\int_{-2}^x \varphi(u)\sqrt{|4 - u^2|}\, du, \quad \text{for} \quad x \ne \pm 2, \quad (17)$$

such that f is continuous if and only if $\varphi(\pm 2) = 0$ holds.

1.2 Universality in the Wigner Ensemble of Matrices

We now apply this inverse representation to the eigenvalues of a hermitean $N \times N$ random matrix $\mathbf{W}_N$ from a Wigner ensemble introduced in (1) with eigenvalues $\lambda_1 \le \lambda_2 \le \cdots \le \lambda_N$. Using the notation (2), we have

$$\mathbf{W}_N = \mathbf{U}_N^{-1}\mathbf{\Lambda}_N \mathbf{U}_N, \quad f(\mathbf{W}_N) := \mathbf{U}_N^{-1} f(\mathbf{\Lambda}_N)\mathbf{U}_N, \quad \text{and}$$
$$f(\mathbf{\Lambda}_N) := \mathrm{diag}(f(\lambda_1), \ldots, f(\lambda_N)).$$

Then

$$\frac{1}{N}\mathbb{E}\,\mathrm{tr}(4\mathbf{I}_N - \mathbf{W}_N^2)f'(\mathbf{W}_N) - \frac{3}{N}\mathbb{E}\,\mathrm{tr}\,\mathbf{W}_N f(\mathbf{W}_N) = \mathbb{E}F_N(y) - F(y). \quad (18)$$

This type of identity is due to C. Stein and may be used to bound the error on the right hand side by "partial summation" arguments on the left hand side. To this end, we may use the fact that the trace of the product $\mathbf{W}_N f(\mathbf{W}_N)$ is linear in the *independent* entries of $\mathbf{W}_N$. Hence an expansion in X_{ij} and Taylor expansion of $f(\mathbf{W}_N)$ in the entries X_{ij} resp. X_{ij} yields expectations of products of independent random quantities which may be evaluated using the first two moments of the entries X_{ij}. This will finally result in the trace of the term $(4\mathbf{I}_N - \mathbf{W}_N^2)f'(\mathbf{W}_N)$. For details, see [GT03a]. The arguments work as well in cases where the entries of the matrix $\mathbf{W}_N$ are not independent, but satisfy martingale type conditions for conditional means and variances of the form

$$\mathbb{E}(X_{kl}|\mathcal{F}^{kl}) = 0, \quad \frac{1}{N^2}\sum_{1 \le j \le l \le N} \mathbb{E}\big|\mathbb{E}(X_{jl}^2|\mathcal{F}_j) - \sigma_{jl}^2\big| \to 0. \quad (19)$$

Here, $\mathcal{F}^{kl}$ resp. $\mathcal{F}^k$ denote the σ-fields generated by all random entries $X_{ij}, i \le j$ (except X_{kl}) resp. all random entries $X_{ij}, i \le j$ not in row or column k, l resp. not in row or column k. If in addition we assume Lindeberg's condition and some further technical moment condition

$$\lim_N \frac{1}{N^2} \sum_{j,l=1}^N \mathbb{E}X_{jl}^2 \mathbb{I}_{\{|X_{jl}|>\tau\sqrt{N}\}} = 0,$$

$$\lim_N \frac{1}{N^2} \sum_{1 \le j \le l \le N} |\mathbb{E}X_{ij}^2 - \sigma^2| = 0, \quad \text{for some} \quad \sigma^2 > 0,$$

then universality of the Wigner distribution in the Wigner ensemble holds, that is

$$\limsup_N \sup_x |\mathbb{E}F_N(x) - F(x\sigma^{-1})| = 0. \quad (20)$$

As an immediate application, we may consider the ensemble of real symmetric $n \times n$ matrices given by the uniform distribution in the *ball* or on the *sphere* of radius $\sqrt{n}$ in $\mathbb{R}^n$ with $n = N(N+1)/2$, that is the ensemble of matrices with

$$\sum_{1 \le l \le j \le N} (X_{lj}^{(N)})^2 \le n, \quad \text{or} \ = n \tag{21}$$

These classes of random matrices in a Hadamar ball or on a Hadamar sphere were introduced by Bronk in [Bro64] as a bounded trace ensembles. The eigenvalue density for the bounded trace ensemble is identical to the density of zeros of Hermite-like polynomials. Using this fact, Bronk proved the semi-circular law for such matrices. See also [Meh91], Ch. 19. The convergence result (20) above immediately applies and yields the convergence of the expected distribution of eigenvalues $\mathbb{E}F_N(x)$ to Wigner's law both for the ball and the spherical ensemble.

After the pioneering paper of Wigner [Wig57], who used a moment method to determine the limit distribution, a number of authors have studied the universality problem in the Wigner ensemble in greater generality, see e.g. Arnold [Arn71].

Universality of Local Distributions in the Wigner Ensemble The local distribution of eigenvalues in the bulk of the renormalized spectrum of a Wigner matrix, that is in the interval $(-2 + \epsilon, 2 - \epsilon)$, is expected to be universal in the following sense. Mehta [Meh91] conjectured that the empirical distribution of spacings of the increasing number of eigenvalues in a shrinking neighborhood $(a - N^{-\beta}, a + N^{-\beta})$ of $-2 < a < 2$ where $\beta < 1$ converges to the same universal limit (provided that the mean spacings are normalized to 1) as for matrices from a GUE ensemble. The latter has been computed in terms of the Fredholm determinant of the kernel $\sin(\pi(x - y))/(\pi(x - y))$ in $L^2(0, s)$. The shrinking neighborhood is necessary to guarantee a locally constant density for the eigenvalues. The same universal limit applies to the angle spacings of the N eigenvalues on the unit circle of unitary $N \times N$ matrices under the Haar measure on $U(N)$; this distribution of eigenvalues is called *circular unitary ensemble, or CUE*. Here the limit density of eigenvalues is constant on the unit circle. There are some indications for this universality in view of the results by Johannson [Joh01a] for mixture models between Wigner and GUE ensembles.

For the spacings on the boundary ± 2, the limit spacing distribution is different, see [TW98, TW02], and has been shown to be universal in a Wigner type class by Soshnikov [Sos02].

Rates of Convergence for Distribution Functions The best recent bounds on the rate of convergence in terms the Kolmogorov distance Δ_N in the Wigner ensemble are based on the so-called Stieltjes transforms of the random matrix $\mathbf{W}_N$. For random variables ξ with distribution G, the Stieltjes transform is given by

$$S_G(z) := \mathbb{E}\frac{1}{\xi - z}, \quad z \in \mathbb{C} \setminus \mathbb{R}.$$

In order to relate Stieltjes tranforms to our characterizing equation, choose $f(x) := (x - z)^{-1}$ in (15) and note that $S'_G(z) = \mathbb{E}(\xi - z)^{-2}$. Hence, we arrive at

$$0 = \mathbb{E}\frac{4 - \xi^2}{(\xi - z)^2} + 3\mathbb{E}\frac{\xi}{\xi - z},$$
$$0 = (z^2 - 4)S'_G(z) - zS_G(z) - 2,$$

with solution
$$S_G(z) = -\frac{1}{2}\left(z - \sqrt{z^2 - 4}\right). \tag{22}$$

Here, we choose the branch of $\sqrt{z^2 - 4}$ such that S_G is complex analytic in $\mathbb{C} \setminus [-2, 2]$, and $S_G(z) \to 0$ as $|z| \to \infty$.

This shows that the only distribution G satisfying (15) is one with a Stieltjes transform given above, which is the Stieltjes transform of Wigner's semicircle distribution F. Stieltjes transforms characterize distributions uniquely and allow, similar to using Fourier transforms, to estimate the approximation error in terms of integrals over the transform, These methods have been developed by Bai [Bai93] and Bai et al. [BMT02], Khorunzhy et al.,[KKP96], Pastur, [Pas96], Mehta, [Meh91] and Girko ([Gir89, Gir98, Gir02]). See also the survey by Bai [Bai99]. Using refined versions of Stieltjes transform inversion in the complex plane, Götze and Tikhomirov, [GT00, GT02, GT03b], proved that the Kolmogorov distance between the expected empirical distribution function, $\mathbb{E}F_N(x)$ and $G(x)$ as well as the expected Kolmogorov distance between the random distribution $F_N(x)$ and $G(x)$ are of order $O(N^{-1/2})$. More precisely, it has been shown that

$$\Delta_N := \sup_x |\mathbb{E}F_N(x) - F(x)| \leq c\max_{i,j}\left(\mathbb{E}|X_{ij}|^4\right)^{1/2} N^{-1/2} \tag{23}$$

and

$$\Delta_N^* := \mathbb{E}\sup_x |F_N(x) - F(x)| \leq c\max_{i,j}\left(\mathbb{E}|X_{ij}|^8\right)^{1/4} N^{-1/2}. \tag{24}$$

The bound for Δ_N is sharp for sparse matrices of the type $W_N = \left(X_{ij}\varepsilon_{ij}q_N^{-1/2}\right)$, $q_N := \mathbb{P}(\varepsilon_{ij}^{(N)} = 1)$, $Nq_N \to \infty$.

The rate of convergence to Wigner's law has been intensively studied in the last years. Bai (1993, 1999) proved assuming condition (1.1) that $\Delta_N = O(N^{-1/4})$ and that $\Delta_N = O(N^{-1/3})$. The above bound $O(N^{-1/2})$ assuming uniform bounded *fourth* moments has been shown in Götze and Tikhomirov [GT00, GT03b]. Bai et al. [BMT02] proved that $\Delta_N = O(N^{-1/2})$ assuming that $\sup_{i,j,N} \mathbb{E}|X_{ij}|^8 < \infty$. Concerning the rate of convergence in probability, he proved the rate $\Delta_N^* = O_P(N^{-\frac{2}{5}})$.

Girko [Gir89] proved a bound for the difference between the Stieltjes transforms of distribution functions $\mathbb{E}F_N(x)$ and $G(x)$ which yields bounds for Δ_N of order $O(N^{-1/6})$ (see inequality (3.12) in Girko (1989) and e.g. inequality

(4.26) in Bai [Bai93]). Girko [Gir98] stated as well that $\Delta_N = O(N^{-1/2})$. He used a different approach to that in Bai [Bai93] and Götze and Tikhomirov [GT00, GT02, GT03b]. Girko [Gir02] states again $\Delta_N = O(N^{-1/2})$ providing an update with extended arguments and some corrections to his previous paper.

1.3 Laguerre Ensembles and Universality

Many of the results on asymptotic spectral distributions carry over to the following class of random matrices, which we shall briefly mention here.

Let X_{jk} and Y_{jk}, $j \geq 1, k \geq 1$ denote two independent arrays of independent random variables with with $\mathbb{E}X_{jk} = \mathbb{E}Y_{jk} = 0$ and $\mathbb{E}X_{jk}^2 = \mathbb{E}Y_{jk}^2 = \frac{1}{2}$.

For an integer $1 \leq N \leq m$ consider a $N \times m$ matrix $\mathbf{B} = (B_{jk})$ with entries $B_{jk} = X_{jk} + iY_{jk}$, $1 \leq j \leq N$ and $1 \leq k \leq m$. Let

$$\mathbf{W} = \frac{1}{m}\mathbf{BB}^T. \tag{25}$$

Furthermore, let $\lambda_1, \ldots, \lambda_N$ denote the eigenvalues of the matrix $\mathbf{W}$. If the X_{ij}, Y_{ij} have a Gaussian distribution, we obtain the so-called Laguerre ensemble with joint distribution

$$\frac{1}{Z_N} \prod_{1 \leq i < j \leq N} (\lambda_i - \lambda_j)^2 \prod_{j=1}^{N} \lambda_j^{m-N} \exp\{-\lambda_j\}.$$

Define the spectral distribution function of the matrix $\mathbf{W}$ as

$$F_N^{(m)}(x) = \frac{1}{N} \sum_{j=1}^{N} \mathbb{I}_{\{\lambda_j \leq x\}}. \tag{26}$$

Assuming that $\lim_{N\to\infty} N/m(N) = y > 0$ exists, the expected distribution function $\mathbb{E}F_n(x)$ converges, see [MP67], to the Marchenko-Pastur distribution function $M_y(x)$ with a density given by

$$m_y(x) = \frac{1}{2\pi yx} \sqrt{-x^2 + 2(1+y) - (1-y)^2} \mathbb{I}_{\{[a,b]\}}(x), \tag{27}$$

where $a = (1 - \sqrt{y})^2$, $b = (1 + \sqrt{y})^2$. The matrix $\mathbf{W}$ is essentially an empirical covariance matrix. In case of Gaussian random variables, it is called Wishart matrix. Note that if $N > m(N)$, the $N \times N$ matrix $\mathbf{W}$ is of rank m only, so that $N-m$ eigenvalues have to be zero. In the limiting eigenvalue distribution, this generates a point mass at zero of size $y - 1$ for $y > 1$.

In Götze and Tikhomirov [GT04a], it has been shown analogously to (7) that

$$\Delta_N^{(m(N))} := \sup_x |\mathbb{E}F_N^{(m(N))}(x) - M_y(x)| \leq CN^{-1}, \tag{28}$$

when X_{ij} and Y_{ij} are *Gaussian* random variables and $|y - 1| > 0$. This result holds as well for the special case $y = 1$.

Universality. Analogously to (23) and (24), it has been shown in [GT01, GT04b], that with
$\gamma_q := \max_{l,j}(\mathbb{E}|X_{lj} + iY_{lj}|^q)^{2/q}$, the following bounds hold:

$$\sup_x |\mathbb{E}F_N(x) - G(x)| \leq c\gamma_8 N^{-1/2}, \quad \mathbb{E}\sup_x |F_N(x) - G(x)| \leq c\gamma_{12} N^{-1/2}.$$

We remark that the characterizing equation for $y < 1$ corresponding to (15) is given by

$$\mathbb{E}\left[(a - \xi)(b - \xi)f'(\xi) - 3\left(\xi + \frac{a+b}{2}\right)f(\xi)\right] = 0$$

for random variables $\xi \neq 0$.

Marchenko and Pastur [MP67], Pastur [Pas73], and Yin [Yin86] have shown that the convergence result holds in the i.i.d. case assuming $\mathbb{E}X_{ij}^2 = \sigma^2$ only. Wachter [Wac78] proved the result for independent X_{ij} with $\mathbb{E}X_{ij} = 0$, $\mathbb{E}X_{ij}^2 = 1$ and $\mathbb{E}|X_{ij}|^{2+\varepsilon} \leq C < \infty$, for any $\varepsilon > 0$. The convergence of $\Delta_n^{(m)}$ to zero under Lindeberg's condition is shown in [Bai99]. The rate $\Delta_N^{(m)} = O(N^{-\frac{1}{4}})$ was proved by Bai [Bai93], assuming uniform bounds on the fourth moments of X_{ij}, provided that y has a positive distance to 1. If y is close to 1, the limit density and the Stieltjes transform of the limit density have a singularity. In this case, the investigation of the rate of convergence is more difficult. Here, Bai [Bai93] has shown that $\Delta_N^{m(N)} = O(N^{-\frac{5}{48}})$.

Largest Eigenvalue of the Laguerre Ensemble and Universality The distribution of the largest eigenvalue in the Laguerre ensemble has been determined by Tracy and Widom and has important application in multivariate statistics and Data Analysis; see [TW02] and [Joh01b]. This distribution has been shown by [Sos02] to be universal in the class of Wigner type matrices **B**, introduced above.

2 Eigenvalues of CUE-Ensembles

2.1 Correlation Functions for CUE

Let $Z_1, \ldots, Z_m$ denote the eigenvalues (in random order) of a $m \times m$ random CUE matrix U, i.e. U is chosen with respect to the Haar measure on the unitary group $U(m)$. As mentioned before, the abbreviation CUE stands for "circular unitary ensemble".

The distribution of the spacings of these eigenvalues has been used successfully as a model to describe the empirical distribution of the high lying zeros of the Riemann zeta function and other L-functions. Roughly speaking, the empirical distribution of Riemann zeta zeros appears to be locally very similar to the eigenvalue distribution of a large random matrix, appropriately rescaled. A first indication in favor of this coincidence was given by Montgomery [Mon73]; see also [Gal85]. His result may be interpreted as the coincidence of Fourier-transformed two-point correlation functions, at least in some restricted domain. There is a strong numerical evidence in favor of the coincidence of Riemann zeta distributions and random matrix eigenvalue distributions. This evidence was provided by large-scale numerical computations of Riemann zeta zeros by Odlyzko; see e.g. [Odl01]. Montgomery's result was extended to three-point correlations by Hejhal [Hej94], and then to correlations of any order by Rudnick and Sarnak; see [RS96] and the summary in [RS94]. Rudnick and Sarnak's result also applies to a large class of L-functions. However, all these rigorous results only apply to a restricted class of test functions, corresponding to the above-mentioned restriction of the domain in Fourier space.

On a conjectural level, more precisely using Hardy-Littlewood conjectures, Bogomolny and Keating [BK95], [BK96] give some theoretical evidence that the connection between Riemann zeta zero statistics and random spectral distributions will be valid beyond the restricted domain in Fourier space.

There is a connection between the values of characteristic polynomials of a random matrices and values of the Riemann zeta function and its derivatives; for more details see [KS00], [HKO00], [HKO01], the review paper [KS03], and the classical paper [Sel46].

In the following, we examine the local eigenvalue distribution of CUE non-asymptotically, using the language of quantum field theory. The result obtained by these methods represents a key combinatorial step in establishing the connection between zeros of L-functions and the above mentioned zeros of characteristic polynomials.

According to a formula of Weyl (Theorems 7.4B–D in [Wey39]), the random variables $Z_1, \ldots, Z_m$ have the joint density

$$\rho(z_1, \ldots, z_m) = \frac{1}{m!} |\Omega_m(z_1, \ldots, z_m)|^2 \qquad (29)$$

with respect to the normalized Lebesgue measure on S_1^m, where Ω_m denotes the Vandermonde determinant

$$\Omega_m(z_1, \ldots, z_m) = \begin{vmatrix} 1 & 1 & \ldots & 1 \\ z_1 & z_2 & \ldots & z_m \\ \vdots & \vdots & \ddots & \vdots \\ z_1^{m-1} & z_2^{m-1} & \ldots & z_m^{m-1} \end{vmatrix}. \qquad (30)$$

The density (29) for CUE is analogous to the density for GUE described in formula (3), above. In the limit as the dimensions go to infinity, the local eigenvalue distributions of CUE and GUE coincide, up to scaling.

The random eigenvalues $Z_1, \ldots, Z_m$ of a CUE random matrix give rise to a random pure point measure

$$\mu_m = \delta_{Z_1} + \ldots + \delta_{Z_m}. \tag{31}$$

Here δ_x denotes the Dirac measure located at x. This section is concerned with n-th order correlation functions of the point measure μ_m consisting of m atoms. It is convenient to normalize μ_m by subtracting its expectation; therefore we set

$$\tilde{\mu}_m = \mu_m - \mathbb{E}[\mu_m]; \tag{32}$$

note that $\mathbb{E}[\mu_m]$ equals m times the normalized Lebesgue measure on S_1, by rotational symmetry. Thus the quantities of interest are

$$\mathbb{E}\left[\int_{S_1^n} f \, d\tilde{\mu}_m^n\right] \tag{33}$$

with appropriate test functions $f : S_1^n \to \mathbb{R}$. Now let

$$f(z_1, \ldots, z_n) = \sum_{k \in \mathbb{Z}^n} \hat{f}(k) z_1^{k_1} \ldots z_n^{k_n} \tag{34}$$

be a finite Laurent sum in n variables. We claim:

Theorem 2. *If f is symmetric in its n arguments, and if*

$$\hat{f}(k_1, \ldots, k_n) = 0 \ \text{whenever} \ |k_1| + \ldots + |k_n| > 2m, \tag{35}$$

then the following holds:
If n is even, then

$$\mathbb{E}\left[\int_{S_1^n} f \, d\tilde{\mu}_m^n\right] =$$

$$\frac{n!}{2^{n/2}(n/2)!} \sum_{k \in \mathbb{Z}^{n/2}} |k_1| \ldots |k_{n/2}| \hat{f}(k_1, \ldots, k_{n/2}, -k_1, \ldots, -k_{n/2}). \tag{36}$$

If n is odd, then

$$\mathbb{E}\left[\int_{S_1^n} f \, d\tilde{\mu}_m^n\right] = 0. \tag{37}$$

The theorem is similar in spirit to Theorem 6.5 in the article [HR02] by Rudnick and Hughes, Theorem 1 in Soshnikov's paper [Sos00], and to the combinatorial arguments given in section 4 of the article [RS96] by Rudnick and Sarnak; in the latter article the distribution of zeros of principal L-functions is compared with the distribution of eigenvalues of random matrices, using test functions with a similar constraint $|k_1| + \ldots + |k_m| \leq 2$ on their support of their Fourier transform as our condition (35) above.

Rudnick and Hughes use a lemma by Diaconis and Shashahani [DS94]; see also [DE01]. The combinatorial argument in [RS96] and [Sos00] relies on a lemma of Spitzer in [Spi56].

Here, we present an alternative approach, replacing Diaconis and Shahsha-hani's result by a argument from quantum field theory, in particular Wick's theorem. We do not use Spitzer's lemma here, rather we present an alternative combinatorial argument in Lemma 1, below.

2.2 Fock Space

Let $\mathcal{F}_1$ denote the space of finite Laurent sums

$$f = \sum_k \hat{f}(k)z^k, \tag{38}$$

endowed with the scalar product

$$\langle f, g \rangle = \frac{1}{2\pi} \int_{S_1} \overline{f(z)} g(z) \, |dz|. \tag{39}$$

Furthermore, let

$$\mathcal{F} = \bigoplus_{N=0}^{\infty} \mathcal{F}_N, \quad \text{where} \quad \mathcal{F}_N = \Lambda^N \mathcal{F}_1, \tag{40}$$

denote the fermionic Fock space (Grassmann algebra) over $\mathcal{F}_1$. In explicit terms, we view the elements of $\mathcal{F}_N$ as antisymmetric functions in N variables $z_1, \ldots, z_N$; thus one may represent $\mathcal{F}_N$ as the vector space spanned by the Slater determinants $\det((z_j^{k_i})_{i,j=1,\ldots,N})$ in the variables $z_1, \ldots, z_N$. The scalar product on $\mathcal{F}_N$ is given by

$$\langle \omega, \chi \rangle = \frac{1}{(2\pi)^N N!} \int_{S_1^N} \overline{\omega(z_1, \ldots, z_N)} \chi(z_1, \ldots, z_N) \, |dz_1| \ldots |dz_N|. \tag{41}$$

For $f \in \mathcal{F}_1$, we have the creation operator

$$a^*(f) : \mathcal{F}_N \to \mathcal{F}_{N+1}, \quad \omega \mapsto f \wedge \omega \tag{42}$$

and its adjoint, the annihilation operator $a(f) : \mathcal{F}_{N+1} \to \mathcal{F}_N$. On Slater determinants, they act as follows: If $\omega(z_1, \ldots, z_N) = \det((\phi_i(z_j))_{i,j=1,\ldots,N})$, then

$$[a^*(f)\omega](z_1, \ldots, z_{N+1}) = \begin{vmatrix} f(z_1) & \ldots & f(z_{N+1}) \\ \phi_1(z_1) & \ldots & \phi_1(z_{N+1}) \\ \vdots & \ddots & \vdots \\ \phi_N(z_1) & \ldots & \phi_N(z_{N+1}) \end{vmatrix} \tag{43}$$

and

$$[a(f)\omega](z_1, \ldots, z_{N-1}) = \frac{1}{2\pi} \int_{|z|=1} \overline{f(z)} \begin{vmatrix} \phi_1(z) & \phi_1(z_1) & \ldots & \phi_1(z_{N-1}) \\ \vdots & \vdots & \ddots & \vdots \\ \phi_N(z) & \phi_N(z_1) & \ldots & \phi_N(z_{N-1}) \end{vmatrix} |dz|. \tag{44}$$

We get the canonical anticommutation relations

$$a(g)a^*(f) + a^*(f)a(g) = \langle g, f \rangle \, \mathrm{id}, \qquad (45)$$

$$a(g)a(f) + a(f)a(g) = 0, \qquad (46)$$

$$a^*(g)a^*(f) + a^*(f)a^*(g) = 0. \qquad (47)$$

Abbreviating

$$a_k^* = a^*(z^k), \qquad (48)$$

$$a_k = a(z^k), \qquad (49)$$

we rewrite the Vandermonde determinant (30) as follows:

$$\Omega_m = a_0^* a_1^* \ldots a_{m-1}^* 1 \in \mathcal{F}_m, \qquad (50)$$

where the argument 1 is viewed as an element of $\mathcal{F}_0 = \mathbb{C}$. Consequently, using the abbreviation $I_m = \{0, 1, \ldots, m-1\}$, we have

$$a_k^* \Omega_m = 0 \quad \text{for } k \in I_m, \qquad (51)$$

$$a_k \Omega_m = 0 \quad \text{for } k \notin I_m. \qquad (52)$$

For a finite Laurent sum f as in (38), multiplication with f gives rise to the following operator:

$$Q(f) = \sum_{k,l \in \mathbb{Z}} \hat{f}(k-l) a_k^* a_l. \qquad (53)$$

We also use its normally ordered variant

$$:Q(f) := \sum_{k,l \in \mathbb{Z}} \hat{f}(k-l) :a_k^* a_l: \qquad (54)$$

with

$$:a_k^* a_l := a_k^* a_l - \mathrm{I}_{I_m}(k) \delta_{k,l} \, \mathrm{id} = \begin{cases} a_k^* a_l & \text{for } k \in I_m^c, \\ -a_l a_k^* & \text{for } k \in I_m. \end{cases} \qquad (55)$$

Note that the normal-ordering operation depends on the choice of m, which we view as fixed. The operator $Q(f)$ acts as follows:

$$[Q(f)\omega](z_1, \ldots, z_N) = \sum_{j=1}^N f(z_j)\omega(z_1, \ldots, z_N), \qquad (56)$$

and on the m-particle space $\mathcal{F}_m$, the operator $:Q(f):$ acts as

$$[:Q(f): \omega](z_1, \ldots, z_m) = \sum_{j=1}^m (f(z_j) - \langle f \rangle)\omega(z_1, \ldots, z_m), \qquad (57)$$

where $\langle f \rangle = (2\pi)^{-1} \int_{|z|=1} f(z)\,|dz|$ denotes the average of f. More generally, for any finite Laurent sum in n variables,

$$f(z_1, \ldots, z_n) = \sum_{k \in \mathbb{Z}^n} \hat{f}(k) z_1^{k_1} \ldots z_n^{k_n}, \tag{58}$$

we set

$$Q_n(f) = \sum_{k,l \in \mathbb{Z}^n} \hat{f}(k - l) a_{k_1}^* a_{l_1} a_{k_2}^* a_{l_2} \ldots a_{k_n}^* a_{l_n}, \tag{59}$$

$${:}Q_n(f){:} = \sum_{k,l \in \mathbb{Z}^n} \hat{f}(k - l) : a_{k_1}^* a_{l_1} :: a_{k_2}^* a_{l_2} : \ldots : a_{k_n}^* a_{l_n} :. \tag{60}$$

Note that the second operator is only partially normally ordered. We have for $\omega \in \mathcal{F}_N$:

$$[Q_n(f)\omega](z_1, \ldots, z_N) = \sum_\iota f((z_{\iota(j)})_{j=1,\ldots,n}) \omega(z_1, \ldots, z_N), \tag{61}$$

where the sum over ι runs over the set of all maps $\iota : \{1, \ldots, n\} \to \{1, \ldots, N\}$. Furthermore, in the special case $f(z_1, \ldots, z_n) = f_1(z_1) \ldots f_n(z_n)$, one has $Q_n(f) = Q(f_1) \ldots Q(f_n)$ and ${:}Q_n(f){:} =: Q(f_1) : \ldots : Q(f_n) :$. Since $(m!)^{-1}|\Omega_m(z_1, \ldots, z_m)|^2$ is the joint density of the CUE eigenvalues $Z_1, \ldots, Z_m$, we obtain

$$\mathbb{E}\left[\int_{S_1^n} f \, d\mu_m^n\right] = \langle \Omega_m, Q_n(f)\Omega_m \rangle, \tag{62}$$

$$\mathbb{E}\left[\int_{S_1^n} f \, d\tilde{\mu}_m^n\right] = \langle \Omega_m, :Q_n(f): \Omega_m \rangle. \tag{63}$$

Let γ_n denote the permutation group of $\{1, 2, \ldots, n\}$, $\gamma_n^{\text{no fp}} \subseteq \gamma_n$ denote the subset of all permutations without fixed points, and $\gamma_n^{\text{cycl}} \subseteq \gamma_n$ the subset of all cyclic permutations. We observe the following version of Wick's theorem:

$$\langle \Omega_N, a_{k_1}^* a_{l_1} \ldots a_{k_n}^* a_{l_n} \Omega_N \rangle = \sum_{\sigma \in \gamma_n} \text{sign}(\sigma) \prod_{j=1}^n e_m(\sigma, j, k_j) \delta_{k_j, l_{\sigma j}}, \tag{64}$$

$$\langle \Omega_N, :a_{k_1}^* a_{l_1} : \ldots : a_{k_n}^* a_{l_n} : \Omega_N \rangle = \sum_{\sigma \in \gamma_n^{\text{no fp}}} \text{sign}(\sigma) \prod_{j=1}^n e_m(\sigma, j, k_j) \delta_{k_j, l_{\sigma j}}, \tag{65}$$

where we have set

$$e_m(\sigma, j, k) := \begin{cases} \mathrm{I}_{I_m}(k) & \text{for } \sigma j \geq j, \\ \mathrm{I}_{I_m}(k) - 1 = -\mathrm{I}_{I_m^c}(k) & \text{for } \sigma j < j, \end{cases} \tag{66}$$

for $\sigma \in \gamma_n$, $1 \leq j \leq n$, and $k \in \mathbb{Z}$. To show (64,65), one permutes all factors a_k^* with $k \in I_m$ and all a_l with $l \in I_m^c$ to the right, but all a_k^* with $k \in I_m^c$ and all a_l with $l \in I_m$ to the left, using the canonical anticommutation relations and finally (51) and (52). Thus all pairs $(a_{k_i}^*, a_{l_j})$ with $i \leq j$ give rise to a factor $I_{I_m}(k_i)\delta_{k_i,l_j}$, and all pairs $(a_{l_j}, a_{k_i}^*)$ with $j < i$ give rise to a factor $-I_{I_m^c}(k_i)\delta_{k_i,l_j}$. Combining (59) with (64,65) yields

$$\langle \Omega_N, Q_n(f)\Omega_N \rangle = \sum_{\sigma \in \gamma_n} \text{sign}(\sigma) \sum_{k \in \mathbb{Z}^n} \hat{f}((k_j - k_{\sigma^{-1}j})_j) \prod_{j=1}^n e_m(\sigma, j, k_j). \quad (67)$$

$$\langle \Omega_N, :Q_n(f): \Omega_N \rangle = \sum_{\sigma \in \gamma_n^{\text{no fp}}} \text{sign}(\sigma) \sum_{k \in \mathbb{Z}^n} \hat{f}((k_j - k_{\sigma^{-1}j})_j) \prod_{j=1}^n e_m(\sigma, j, k_j). \quad (68)$$

We decompose any permutation σ into cycles $\sigma_1, \ldots, \sigma_d$, supported on a partition $A_1, \ldots, A_d$ of $\{1, \ldots, n\}$. Then we write the sum over σ in (67,68) as $\sum_d \sum_{A_1, \ldots, A_d} \sum_{\sigma_1} \cdots \sum_{\sigma_d}$, where $A_1, \ldots, A_d$ runs over all partitions of $\{1, \ldots, n\}$ in the case of (67), and over all partitions of $\{1, \ldots, n\}$ into sets of at least 2 elements in the case of (68). Furthermore, σ_i runs over the cyclic permutations of A_i. As a consequence of the following lemma, the sum over σ_i vanishes as soon as $|A_i| \geq 3$. Recall that $\gamma_n^{\text{cycl}} \subseteq \gamma_n$ denotes the set of all cyclic permutations of $\{1, 2, \ldots, n\}$.

Lemma 1. Let $n \geq 3$, and let $\hat{f} : \mathbb{Z}^n \to \mathbb{C}$ be supported in $\{x \in \mathbb{Z}^n : \sum_j |x_j| \leq 2m\}$. Then one has:

$$\sum_{\sigma \in \gamma_n^{\text{cycl}}} \sum_{k \in \mathbb{Z}^n} \hat{f}((k_j - k_{\sigma^{-1}j})_{j=1,\ldots,n}) \prod_{j=1}^n e_m(\sigma, j, k_j) = 0. \quad (69)$$

The lemma is proven in the next section. If we let A_i play the role of $\{1, \ldots, n\}$ in the lemma, we conclude that there remain only the contributions from partitions $A_1, \ldots, A_d$ with all $|A_i| \leq 2$ in the case of (67), and with all $|A_i| = 2$ in the case of (68). Let us now consider the case (68): For odd n, there are no such partitions, and for even n, there are $n!/(2^{n/2}(n/2)!)$ such partitions. Now take a permutation σ that transposes two indices j and j': $\sigma j = j'$, $\sigma j' = j$. Take $g : \mathbb{Z}^2 \to \mathbb{R}$, supported in $\{x \in \mathbb{Z}^2 : |x_1| + |x_2| \leq 2m\}$. Then we have

$$\sum_{k_j, k'_j \in \mathbb{Z}} g(k_j - k_{j'}, k_{j'} - k_j)e_m(\sigma, j, k_j)e_m(\sigma, j', k_{j'}) = -\sum_{k \in \mathbb{Z}} |k|g(k, -k). \quad (70)$$

To prove this, assume $j < j'$ without loss of generality. Then we have

$$e_m(\sigma, j, k_j)e_m(\sigma, j', k_{j'}) = -1 \text{ for } k_j \in I_m \text{ and } k_{j'} \in I_m^c,$$

and $e_m(\sigma, j, k_j)e_m(\sigma, j', k_{j'}) = 0$ otherwise. Hence, if we fix $k_j - k_{j'} = k$ with $|k| \leq m$, there are $|k|$ choices for $(k_j, k_{j'})$ with $e_m(\sigma, j, k_j)e_m(\sigma, j', k_{j'}) = -1$. These choices are $k_j = 0, 1, \ldots, |k| - 1$ for $k > 0$ and $k_j = m - 1, m -$

$2, \ldots, m - |k|$ for $k < 0$. The case $|k| > m$ is irrelevant for our purposes, since $g(k, -k) = 0$ holds in this case. This proves the claim (70).

We conclude for even n, symmetric functions $\hat{f}$ as in the claim of Theorem 2, and any permutation σ of n that decomposes into $n/2$ cycles of length 2:

$$\text{sign}(\sigma) \sum_{k \in \mathbb{Z}^n} \hat{f}((k_j - k_{\sigma^{-1}j})_{j=1,\ldots,n}) \prod_{j=1}^{n} e_m(\sigma, j, k_j)$$

$$= \sum_{k \in \mathbb{Z}^{n/2}} |k_1| \ldots |k_{n/2}| \hat{f}(k_1, \ldots, k_{n/2}, -k_1, \ldots, -k_{n/2}). \tag{71}$$

In view of (63) and (68), this finishes the proof of the Theorem.

2.3 Proof of the Combinatorial Lemma

In this section, we prove Lemma 1.

First, we observe for any $\sigma \in \gamma_n^{\text{cycl}}$: If $\hat{f}((k_j - k_{\sigma^{-1}j})_j) \neq 0$, then $\sum_j |k_j - k_{\sigma^{-1}j}| \leq 2m$, thus

$$\max_j k_j - \min_j k_j \leq m. \tag{72}$$

To see the last statement, take j_* with $k_{j_*} = \min_j k_j$, and j^* with $k_{j^*} = \max_j k_j$. Since σ is a cyclic permutation, we can take l with $0 \leq l < n$ such that $j^* = \sigma^l(j_*)$, $j_* = \sigma^{n-l}(j^*)$. Then, using telescope sums,

$$k_{j^*} - k_{j_*} = \sum_{j=1}^{l} [k_{\sigma^j j_*} - k_{\sigma^{j-1} j_*}], \tag{73}$$

$$k_{j_*} - k_{j^*} = \sum_{j=l+1}^{n} [k_{\sigma^j j_*} - k_{\sigma^{j-1} j_*}], \tag{74}$$

hence

$$2|k_{j^*} - k_{j_*}| \leq \sum_{j=1}^{n} |k_{\sigma^j j_*} - k_{\sigma^{j-1} j_*}| = \sum_{j=1}^{n} |k_j - k_{\sigma^{-1}j}| \leq 2m, \tag{75}$$

which proves (72).

Furthermore, we claim: If $\prod_j e_m(\sigma, j, k_j) \neq 0$, then

$$k_j \in I_m \text{ for at least one } k_j, \text{ and } k_{j'} \in I_m^c \text{ for at least one } k_{j'}. \tag{76}$$

Indeed, we have $\sigma j \geq j$ for at least one j, and $\sigma j' < j'$ for at least one j', using that σ is not the identity. In view of (66), this implies the claim (76). Summarizing (72) and (76), we get the following. If

$$\hat{f}((k_j - k_{\sigma^{-1}j})_j) \prod_{j=1}^{n} e_m(\sigma, j, k_j) \neq 0, \tag{77}$$

then one and only one of the following two cases holds:

"**case** $A_+(k)$": Either all k_j are contained in $[0, \infty)$, but not all in $[0, m-1]$ or all in $[m, \infty)$,

"**case** $A_-(k)$": or all k_j are contained in $(-\infty, m-1]$, but not all in $(-\infty, 0]$ or all in $[0, m-1]$.

To see this, note that we cannot have both, $\max_j k_j \geq m$ and $\min_j k_j \leq -1$, since $\max_j k_j - \min_j k_j \leq m$.

In order to prove the claim (69), we split the function $\hat{f}$ into several pieces, according to the signs of the arguments: With

$$s_1(x) = I_{\{x \geq 0\}}, \tag{78}$$

$$s_2(x) = I_{\{x < 0\}}, \tag{79}$$

we set for $\alpha \in \{1, 2\}^n$:

$$\hat{f}_\alpha(x_1, \ldots, x_n) = \hat{f}(x_1, \ldots, x_n) \prod_{j=1}^{n} s_{\alpha(j)}(x_j); \tag{80}$$

thus

$$\hat{f} = \sum_{\alpha \in \{1,2\}^n} \hat{f}_\alpha. \tag{81}$$

Hence it suffices to prove the claim (69) with f replaced by $\hat{f}_\alpha$:

$$\sum_{\sigma \in \gamma_n^{\mathrm{cycl}}} \sum_{k \in \mathbb{Z}^n} \hat{f}_\alpha((k_j - k_{\sigma^{-1}j})_{j=1,\ldots,n}) \prod_{j=1}^{n} e_m(\sigma, j, k_j) = 0. \tag{82}$$

Let $\alpha \in \{1, 2\}^n$ be fixed. We consider first the case $\alpha(1) = \alpha(n)$. For this case, we claim that the summands in (82) vanish:

$$\hat{f}_\alpha((k_j - k_{\sigma^{-1}j})_{j=1,\ldots,n}) \prod_{j=1}^{n} e_m(\sigma, j, k_j) = 0 \text{ for all } k_1, \ldots, k_n. \tag{83}$$

To prove this, suppose that (83) were false for $k_1, \ldots, k_n$. Then $s_{\alpha(1)}(k_1 - k_{\sigma^{-1}1}) = 1$ and $s_{\alpha(n)}(k_n - k_{\sigma^{-1}n}) = 1$, i.e.

either $k_1 \geq k_{\sigma^{-1}1}$ and $k_n \geq k_{\sigma^{-1}n}$ [this holds in the case $\alpha(1) = \alpha(n) = 1$],

or $k_1 < k_{\sigma^{-1}1}$ and $k_n < k_{\sigma^{-1}n}$ [this holds in the case $\alpha(1) = \alpha(n) = 2$].

$$\tag{84}$$

However, we have $\sigma(1) > 1$ and $\sigma(n) < n$; recall that σ is a cyclic permutation and thus has no fixed point. We conclude

$$e_m(\sigma, 1, k_1) = I_{I_m}(k_1), \tag{85}$$
$$e_m(\sigma, n, k_n) = -I_{I_m^c}(k_n). \tag{86}$$

Similarly, $\sigma^{-1}1 > 1$ and $\sigma^{-1}n < n$ is also valid; thus

$$e_m(\sigma, \sigma^{-1}1, k_{\sigma^{-1}1}) = -I_{I_m^c}(k_{\sigma^{-1}1}), \tag{87}$$
$$e_m(\sigma, \sigma^{-1}n, k_{\sigma^{-1}n}) = I_{I_m}(k_{\sigma^{-1}n}). \tag{88}$$

Since we assumed $\prod_{j=1}^n e_m(\sigma, j, k_j) \neq 0$, this implies $k_1, k_{\sigma^{-1}n} \in I_m$ and $k_n, k_{\sigma^{-1}1} \in I_m^c$. In the case $A_+(k)$, we conclude $k_1 \leq m - 1 < k_{\sigma^{-1}1}$ and $k_{\sigma^{-1}n} \leq m - 1 < k_n$, and in the case $A_-(k)$, we get $k_1 \geq 0 > k_{\sigma^{-1}1}$ and $k_{\sigma^{-1}n} \geq 0 > k_n$. In any case, this contradicts (84), and we conclude that (83) holds for $\alpha(1) = \alpha(n)$.

Next, we prove (82) for general $\alpha \in \{1, 2\}^n$. First, we claim that permuting the arguments of f in (69) does not change the whole expression: For $\tau \in \gamma_n$, we claim that

$$\sum_{\sigma \in \gamma_n^{\text{cycl}}} \sum_{k \in \mathbb{Z}^n} \tau \hat{f}((k_j - k_{\sigma^{-1}j})_{j=1,\dots,n}) \prod_{j=1}^n e_m(\sigma, j, k_j)$$
$$= \sum_{\sigma \in \gamma_n^{\text{cycl}}} \sum_{k \in \mathbb{Z}^n} \hat{f}((k_j - k_{\sigma^{-1}j})_{j=1,\dots,n}) \prod_{j=1}^n e_m(\sigma, j, k_j) \tag{89}$$

holds, where the permuted version τf of f is defined by

$$\tau \hat{f}(x_1, \dots, x_n) = \hat{f}(x_{\tau^{-1}1}, \dots, x_{\tau^{-1}n}). \tag{90}$$

This claim implies the claim (82) by the following consideration: Given $\alpha \in \{1, 2\}^n$, there are $i, j \in \{1, \dots, n\}$ with $i \neq j$ with $\alpha(i) = \alpha(j)$. Here we use the assumption $n \geq 3$ essentially. We permute the indices i and j to get them at the leftmost and rightmost position, respectively: We take $\tau \in \gamma_n$ with $\tau 1 = i$ and $\tau n = j$, hence $\alpha(\tau 1) = \alpha(\tau n)$. Note that $\tau \hat{f}_\alpha = (\tau f)_{\alpha \circ \tau}$, since

$$\tau \hat{f}_\alpha(x_1, \dots, x_n) = \hat{f}_\alpha(x_{\tau^{-1}1}, \dots, x_{\tau^{-1}n})$$
$$= \hat{f}(x_{\tau^{-1}1}, \dots, x_{\tau^{-1}n}) \prod_{j=1}^n s_{\alpha(j)}(x_{\tau^{-1}j})$$
$$= \tau \hat{f}(x_1, \dots, x_n) \prod_{j=1}^n s_{\alpha(\tau j)}(x_j)$$
$$= (\tau f)_{\alpha \circ \tau}(x_1, \dots, x_n). \tag{91}$$

We get the claim (82) also in the general case:

$$\sum_{\sigma\in\gamma_n^{\mathrm{cycl}}}\sum_{k\in\mathbb{Z}^n}\hat{f}_\alpha((k_j-k_{\sigma^{-1}j})_{j=1,\ldots,n})\prod_{j=1}^n e_m(\sigma,j,k_j)$$

$$\stackrel{(89)}{=}\sum_{\sigma\in\gamma_n^{\mathrm{cycl}}}\sum_{k\in\mathbb{Z}^n}\tau\hat{f}_\alpha((k_j-k_{\sigma^{-1}j})_{j=1,\ldots,n})\prod_{j=1}^n e_m(\sigma,j,k_j)$$

$$=\sum_{\sigma\in\gamma_n^{\mathrm{cycl}}}\sum_{k\in\mathbb{Z}^n}(\tau f)_{\alpha\circ\tau}((k_j-k_{\sigma^{-1}j})_{j=1,\ldots,n})\prod_{j=1}^n e_m(\sigma,j,k_j)\stackrel{(83)}{=}0.\qquad(92)$$

It remains to prove (89). It suffices to consider the case that the permutation τ is a transposition $\tau:i\leftrightarrow i+1$ of two neighboring indices, since these transpositions generate the permutation group γ_n.

We calculate for these transpositions $\tau=\tau^{-1}$, using the substitution $\tilde{k}_{\tau j}=k_j$ in the first step:

$$\sum_{k\in\mathbb{Z}^n}\tau\hat{f}((k_j-k_{\sigma^{-1}j})_j)\prod_j e_m(\sigma,j,k_j)$$

$$=\sum_{k\in\mathbb{Z}^n}\tau\hat{f}((k_{\tau j}-k_{\tau\sigma^{-1}j})_j)\prod_j e_m(\sigma,j,k_{\tau j})$$

$$=\sum_{k\in\mathbb{Z}^n}\hat{f}((k_j-k_{\tau\sigma^{-1}\tau j})_j)\prod_j e_m(\sigma,\tau j,k_j)\qquad(93)$$

Now

$$e_m(\sigma,\tau j,k_j)-e_m(\tau\sigma\tau,j,k_j)=\begin{cases}0\text{ for }\{\tau j,\sigma\tau j\}\neq\{i,i+1\}&\text{(case A)},\\1\text{ for }\tau j=i\text{ and }\sigma\tau j=i+1&\text{(case B)},\\-1\text{ for }\tau j=i+1\text{ and }\sigma\tau j=i&\text{(case C)}.\end{cases}$$
$$(94)$$

To prove this in case A, note that $\sigma\tau j\geq\tau j$ holds if and only if $\tau\sigma\tau j\geq j$, since the only elements which are reversed in order by the transposition τ are i and $i+1$. Thus

$$e_m(\sigma,\tau j,k_j)=\begin{cases}\mathrm{I}_{I_m}(k_j)&\text{for }\sigma\tau j\geq\tau j\\\mathrm{I}_{I_m}(k_j)-1&\text{for }\sigma\tau j<\tau j\end{cases}=e_m(\tau\sigma\tau,j,k_j)\qquad(95)$$

is valid. In the other two cases B and C, we observe

$$e_m(\sigma,\tau j,k_j)=\begin{cases}\mathrm{I}_{I_m}(k_j)&\text{for }\sigma\tau j\geq\tau j\\\mathrm{I}_{I_m}(k_j)-1&\text{for }\sigma\tau j<\tau j\end{cases}=\begin{cases}\mathrm{I}_{I_m}(k_j)&\text{in the case B},\\\mathrm{I}_{I_m}(k_j)-1&\text{in the case C},\end{cases}$$
$$(96)$$

and

$$e_m(\tau\sigma\tau,j,k_j)=\begin{cases}\mathrm{I}_{I_m}(k_j)&\text{for }\tau\sigma\tau j\geq j\\\mathrm{I}_{I_m}(k_j)-1&\text{for }\tau\sigma\tau j<j\end{cases}=\begin{cases}\mathrm{I}_{I_m}(k_j)&\text{in the case C},\\\mathrm{I}_{I_m}(k_j)-1&\text{in the case B}.\end{cases}$$
$$(97)$$

This implies the claim (94).

Combing (93) with (94), we get:

– If $\sigma i \neq i + 1$ and $\sigma(i + 1) \neq i$ (case A):

$$\sum_{k \in \mathbb{Z}^n} \tau \hat{f}((k_j - k_{\sigma^{-1}j})_j) \prod_j e_m(\sigma, j, k_j)$$

$$= \sum_{k \in \mathbb{Z}^n} \hat{f}((k_j - k_{\tau\sigma^{-1}\tau j})_j) \prod_j e_m(\tau\sigma\tau, j, k_j) \tag{98}$$

– If $\sigma i = i + 1$ (case B for the factor with index $j = i + 1$):

$$\sum_{k \in \mathbb{Z}^n} \tau \hat{f}((k_j - k_{\sigma^{-1}j})_j) \prod_j e_m(\sigma, j, k_j)$$

$$= \sum_{k \in \mathbb{Z}^n} \hat{f}((k_j - k_{\tau\sigma^{-1}\tau j})_j) \prod_j e_m(\tau\sigma\tau, j, k_j)$$

$$+ \sum_{k \in \mathbb{Z}^n} \hat{f}((k_j - k_{\tau\sigma^{-1}\tau j})_j) \prod_{j \neq i+1} e_m(\tau\sigma\tau, j, k_j). \tag{99}$$

– If $\sigma(i + 1) = i$ (case C for the factor with index $j = i$):

$$\sum_{k \in \mathbb{Z}^n} \tau \hat{f}((k_j - k_{\sigma^{-1}j})_j) \prod_j e_m(\sigma, j, k_j)$$

$$= \sum_{k \in \mathbb{Z}^n} \hat{f}((k_j - k_{\tau\sigma^{-1}\tau j})_j) \prod_j e_m(\tau\sigma\tau, j, k_j)$$

$$- \sum_{k \in \mathbb{Z}^n} \hat{f}((k_j - k_{\tau\sigma^{-1}\tau j})_j) \prod_{j \neq i} e_m(\tau\sigma\tau, j, k_j). \tag{100}$$

Let us sum the contributions in (89) coming from σ and from $\tau\sigma\tau$. We observe: If case B occurs for σ, i.e. if $\sigma i = i + 1$, then case C occurs for $\tau\sigma\tau$, i.e. $\tau\sigma\tau(i + 1) = i$. In this case, we get with the substitution $\tilde{k}_j = k_{\tau j}$:

$$\prod_{j \neq i+1} e_m(\tau\sigma\tau, j, \tilde{k}_j) = \prod_{j \neq i} e_m(\sigma, j, k_j). \tag{101}$$

Furthermore, we have in this case:

$$\sum_{\tilde{k}_{i+1} \in \mathbb{Z}} \hat{f}((\tilde{k}_j - \tilde{k}_{\tau\sigma^{-1}\tau j})_j) = \sum_{k_i \in \mathbb{Z}} \hat{f}((k_j - k_{\sigma^{-1}j})_j). \tag{102}$$

Indeed:

$$\sum_{\tilde{k}_{i+1} \in \mathbb{Z}} \hat{f}(\dots, \underbrace{\tilde{k}_i - \tilde{k}_{i+1}}_{i}, \underbrace{\tilde{k}_{i+1} - \tilde{k}_{\sigma^{-1}i}}_{i+1}, \dots, \underbrace{\tilde{k}_{\sigma(i+1)} - \tilde{k}_i}_{\sigma(i+1)}, \dots)$$

$$= \sum_{k \in \mathbb{Z}} \hat{f}(\dots, \underbrace{k_{i+1} - k}_{i}, \underbrace{k - k_{\sigma^{-1}i}}_{i+1}, \dots, \underbrace{k_{\sigma(i+1)} - k_{i+1}}_{\sigma(i+1)}, \dots)$$

$$= \sum_{k_i \in \mathbb{Z}} \hat{f}(\dots, \underbrace{k_i - k_{\sigma^{-1}i}}_{i}, \underbrace{k_{i+1} - k_i}_{i+1}, \dots, \underbrace{k_{\sigma(i+1)} - k_{i+1}}_{\sigma(i+1)}, \dots) \tag{103}$$

where we substituted $k_i = k_{i+1} - k + k_{\sigma^{-1}i}$.

Summing (99) and (100) with σ replaced by $\tau\sigma\tau$, the second terms on the right hand side in (99) and (100) cancel each other, and we get, again in the case $\sigma i = i + 1$,

$$\sum_{k \in \mathbb{Z}^n} \tau\hat{f}((k_j - k_{\sigma^{-1}j})_j) \prod_j e_m(\sigma, j, k_j)$$

$$+ \sum_{k \in \mathbb{Z}^n} \tau\hat{f}((k_j - k_{\tau\sigma^{-1}\tau j})_j) \prod_j e_m(\tau\sigma\tau, j, k_j)$$

$$= \sum_{k \in \mathbb{Z}^n} \hat{f}((k_j - k_{\sigma^{-1}j})_j) \prod_j e_m(\sigma, j, k_j)$$

$$+ \sum_{k \in \mathbb{Z}^n} \hat{f}((k_j - k_{\tau\sigma^{-1}\tau j})_j) \prod_j e_m(\tau\sigma\tau, j, k_j). \tag{104}$$

By the symmetry in $\tau\sigma\tau \leftrightarrow \sigma$, (104) also holds in the case $\sigma(i+1) = i$, since $\sigma i = i + 1$ if and only if $\tau\sigma\tau(i+1) = i$.

Using (94) in the case A again, we see that (104) holds also for *all* $\sigma \in \gamma_n^{\mathrm{cycl}}$. Summing over all $\sigma \in \gamma_n^{\mathrm{cycl}}$ proves (89). This finishes the proof of Lemma 1.

References

[Arn71]   L. Arnold. On Wigner's semicircle law for the eigenvalues of random matrices. *Z. Wahrscheinlichkeitstheorie und Verw. Gebiete*, 19:191–198, 1971.

[Bai93]   Z. D. Bai. Convergence rate of expected spectral distributions of large random matrices. I. Wigner matrices. *Ann. Probab.*, 21(2):625–648, 1993.

[Bai99]   Z. D. Bai. Methodologies in spectral analysis of large-dimensional random matrices, a review. *Statist. Sinica*, 9(3):611–677, 1999. With comments by G. J. Rodgers and Jack W. Silverstein; and a rejoinder by the author.

[BK95]   E. B. Bogomolny and J. P. Keating. Random matrix theory and the Riemann zeros. I. Three- and four-point correlations. *Nonlinearity*, 8(6):1115–1131, 1995.

[BK96]   E. B. Bogomolny and J. P. Keating. Random matrix theory and the Riemann zeros. II. *n*-point correlations. *Nonlinearity*, 9(4):911–935, 1996.

[BMT02]   Z. D. Bai, Baiqi Miao, and Jhishen Tsay. Convergence rates of the spectral distributions of large Wigner matrices. *Int. Math. J.*, 1(1):65–90, 2002.

[Bro64]   Burt V. Bronk. Accuracy of the semicircle approximation for the density of eigenvalues of random matrices. *J. Mathematical Phys.*, 5:215–220, 1964.

[DE01]   Persi Diaconis and Steven N. Evans. Linear functionals of eigenvalues of random matrices. *Trans. Amer. Math. Soc.*, 353(7):2615–2633 (electronic), 2001.

[DS94]   Persi Diaconis and Mehrdad Shahshahani. On the eigenvalues of random matrices. *J. Appl. Probab.*, 31A:49–62, 1994. Studies in applied probability.

[Dys62]   Freeman J. Dyson. A Brownian-motion model for the eigenvalues of a random matrix. *J. Mathematical Phys.*, 3:1191–1198, 1962.

[For04]   P. J. Forrester. Log-gases and random matrices. *Monograph*, http://www.ms.unimelb.edu.au/ matpjf.html, 2004.

[Gal85]   P. X. Gallagher. Pair correlation of zeros of the zeta function. *J. Reine Angew. Math.*, 362:72–86, 1985.

[Gir89]   V. L. Girko. Asymptotics of the distribution of the spectrum of random matrices. *Uspekhi Mat. Nauk*, 44(4(268)):7–34, 256, 1989.

[Gir98]   V. L. Girko. Convergence rate of the expected spectral functions of symmetric random matrices is equal to $O(n^{-\bullet/\bullet})$. *Random Oper. Stochastic Equations*, 6(4):359–408, 1998.

[Gir02]   V. L. Girko. Extended proof of the statement: convergence rate of the expected spectral functions of symmetric random matrices Ξ_n is equal to $O(n^{-\bullet/\bullet})$ and the method of critical steepest descent. *Random Oper. Stochastic Equations*, 10(3):253–300, 2002.

[GT00]   F. Götze and A. Tikhomirov. Rate of convergence to the semi-circular law. *Preprint 00-125, SFB 343, Universität Bielefeld*, 2000.

[GT01]   F. Götze and A. Tikhomirov. Rate of convergence to the Marchenko-Pastur law. *Preprint 01-020, Forschergruppe, Universität Bielefeld*, 2001.

[GT02]   F. Götze and A. Tikhomirov. Rate of convergence to the semi-circular law for the Gaussian unitary ensemble. *Theory Probab. Appl.*, 47(2):381–388, 2002.

[GT03a]   F. Götze and A. Tikhomirov. Limit theorems for the spectra of random matrices with martingale structure. *Preprint 03-018, Forschergruppe, Universität Bielefeld*, 2003.

[GT03b]   F. Götze and A. Tikhomirov. Rate of convergence to the semi-circular law. *Probab. Theory Related Fields*, 127(2):228–276, 2003.

[GT04a]   F. Götze and A. Tikhomirov. The rate of convergence for the spectra of GUE and LUE matrix ensembles. *Preprint 04-004, Forschergruppe, Universität Bielefeld*, 2004.

[GT04b]   F. Götze and A. Tikhomirov. Rate of convergence to the Marchenko-Pastur law in probability. *Bernoulli*, 10(3), 2004.

[Hej94]   Dennis A. Hejhal. On the triple correlation of zeros of the zeta function. *Internat. Math. Res. Notices*, (7):293ff., approx. 10 pp. (electronic), 1994.

[HKO00]   C. P. Hughes, J. P. Keating, and Neil O'Connell. Random matrix theory and the derivative of the Riemann zeta function. *R. Soc. Lond. Proc. Ser. A Math. Phys. Eng. Sci.*, 456(2003):2611–2627, 2000.

[HKO01]   C. P. Hughes, J. P. Keating, and Neil O'Connell. On the character-istic polynomial of a random unitary matrix. *Comm. Math. Phys.*, 220(2):429–451, 2001.

[HR02]   C.P. Hughes and Zeév Rudnick. Linear statistics of low-lying zeros of *L*-functions. Preprint, arXiv:math.NT/0208230, 2002.

[HT03]   Uffe Haagerup and Steen Thorbjørnsen. Random matrices with complex Gaussian entries. *Expo. Math.*, 21(4):293–337, 2003.

[Joh01a] Kurt Johansson. Universality of the local spacing distribution in certain ensembles of Hermitian Wigner matrices. *Comm. Math. Phys.*, 215(3):683–705, 2001.

[Joh01b] Iain M. Johnstone. On the distribution of the largest eigenvalue in principal components analysis. *Ann. Statist.*, 29(2):295–327, 2001.

[KKP96]   Alexei M. Khorunzhy, Boris A. Khoruzhenko, and Leonid A. Pastur. Asymptotic properties of large random matrices with independent entries. *J. Math. Phys.*, 37(10):5033–5060, 1996.

[KS00]   J. P. Keating and N. C. Snaith. Random matrix theory and $\zeta(1/2 + it)$. *Comm. Math. Phys.*, 214(1):57–89, 2000.

[KS03]   J. P. Keating and N. C. Snaith. Random matrices and *L*-functions. *J. Phys. A*, 36(12):2859–2881, 2003. Random matrix theory.

[Meh91]   M. L. Mehta. *Random matrices*. Academic Press Inc., Boston, MA, second edition, 1991.

[Mon73]   H. L. Montgomery. The pair correlation of zeros of the zeta function. In *Analytic number theory (Proc. Sympos. Pure Math., Vol. XXIV, St. Louis Univ., St. Louis, Mo., 1972)*, pages 181–193. Amer. Math. Soc., Providence, R.I., 1973.

[MP67]   V. A. Marčenko and L. A. Pastur. Distribution of eigenvalues in certain sets of random matrices. *Mat. Sb. (N.S.)*, 72 (114):507–536, 1967.

[O'C03]   Neil O'Connell. Random matrices, non-colliding processes and queues. In *Séminaire de Probabilités, XXXVI*, volume 1801 of *Lecture Notes in Math.*, pages 165–182. Springer, Berlin, 2003.

[Odl01]   A. M. Odlyzko. The 10··-nd zero of the Riemann zeta function. In *Dynamical, spectral, and arithmetic zeta functions (San Antonio, TX, 1999)*, volume 290 of *Contemp. Math.*, pages 139–144. Amer. Math. Soc., Providence, RI, 2001.

[OY02]   Neil O'Connell and Marc Yor. A representation for non-colliding random walks. *Electron. Comm. Probab.*, 7:1–12 (electronic), 2002.

[Pas73]   L. A. Pastur. Spectra of random selfadjoint operators. *Uspehi Mat. Nauk*, 28(1(169)):3–64, 1973.

[Pas96]   L. Pastur. Eigenvalue distribution of random matrices: some recent results. *Ann. Inst. H. Poincaré Phys. Théor.*, 64(3):325–337, 1996.

[Pit75]   J. W. Pitman. One-dimensional Brownian motion and the three-dimensional Bessel process. *Advances in Appl. Probability*, 7(3):511–526, 1975.

[RS94]   Zeév Rudnick and Peter Sarnak. The *n*-level correlations of zeros of the zeta function. *C. R. Acad. Sci. Paris Sér. I Math.*, 319(10):1027–1032, 1994.

[RS96]   Zeév Rudnick and Peter Sarnak. Zeros of principal *L*-functions and random matrix theory. *Duke Math. J.*, 81(2):269–322, 1996. A celebration of John F. Nash, Jr.

[Sel46]   Atle Selberg. Contributions to the theory of the Riemann zeta-function. *Arch. Math. Naturvid.*, 48(5):89–155, 1946.

[Sos00]   Alexander Soshnikov. The central limit theorem for local linear statistics in classical compact groups and related combinatorial identities. *Ann. Probab.*, 28(3):1353–1370, 2000.

[Sos02]   A. Soshnikov. A note on universality of the distribution of the largest eigenvalues in certain sample covariance matrices. *J. Statist. Phys.*, 108(5-6):1033–1056, 2002. Dedicated to David Ruelle and Yasha Sinai on the occasion of their 65th birthdays.

[Spi56]   Frank Spitzer. A combinatorial lemma and its application to probability theory. *Trans. Amer. Math. Soc.*, 82:323–339, 1956.

[Ste86]   Charles Stein. *Approximate computation of expectations.* Institute of Mathematical Statistics Lecture Notes—Monograph Series, 7. Institute of Mathematical Statistics, Hayward, CA, 1986.

[TW98]   Craig A. Tracy and Harold Widom. Correlation functions, cluster functions, and spacing distributions for random matrices. *J. Statist. Phys.*, 92(5-6):809–835, 1998.

[TW02]   Craig A. Tracy and Harold Widom. Distribution functions for largest eigenvalues and their applications. In *Proceedings of the International Congress of Mathematicians, Vol. I (Beijing, 2002)*, pages 587–596, Beijing, 2002. Higher Ed. Press.

[Wac78]   Kenneth W. Wachter. The strong limits of random matrix spectra for sample matrices of independent elements. *Ann. Probability*, 6(1):1–18, 1978.

[Wey39]   Hermann Weyl. *The Classical Groups. Their Invariants and Representations.* Princeton University Press, Princeton, N.J., 1939.

[Wig55]   E. P. Wigner. Characteristic vectors of bordered matrices with infinite dimensions. *Ann. of Math. (2)*, 62:548–564, 1955.

[Wig57]   E. P. Wigner. Characteristic vectors of bordered matrices with infinite dimensions. II. *Ann. of Math. (2)*, 65:203–207, 1957.

[Yin86]   Y. Q. Yin. Limiting spectral distribution for a class of random matrices. *J. Multivariate Anal.*, 20(1):50–68, 1986.

Part II

Stochastic in Population Models

Renormalization and Universality for Multitype Population Models

Andreas Greven

Mathematisches Institut, Bismarckstr. 1 $\frac{1}{2}$, D-91054 Erlangen
Tel. (0049 9131) 85-22454, Fax (0049 9131) 85-26214

Summary. We are concerned with spatial models of populations, where individuals have a type and a geographic location which both undergo a stochastic dynamic. Typical classes we consider are branching systems, state dependent branching systems, catalytic branching, mutually catalytic branching and we also touch on properties of Fleming-Viot models. Many features of such systems are universal in large classes of possible branching mechanism. Important features are the longtime behavior and the small-scale structure of spatial continuum limits of such systems and structural properties of the historical process associated with them.

To exhibit these universal features the systems are analysed by renormalizing them through rescaling space and time according to whole sequences of separating scales. The arising collection of limiting systems can be described in a simpler fashion namely as a Markov chain governed by parameters given as a function on the one-component state space. Two such objects arise, one for the longtime and one for the small-scale behavior of the continuum limit. Hence the universality classes of the stochastic system are associated with universality classes for certain nonlinear maps in function spaces, which can be analysed via analytical tools. We review the progress made from 1993 to 2003 and formulate the problems currently under investigation.

The techniques developed in the renormalization analysis also have many applications in the analysis of evolutionary models in population genetics.

Keywords: Branching, multitype population models, renormalization, universality, continuum limit, longtime behavior, clustering, equilibria, historical process

AMS-Subject Classification: 60K35

1 Introduction

1.1 Background and Motivation

Interacting spatial stochastic systems arise in many different contexts. Very prominent are dynamics of processes modelling a physical system in a potential with thermic noise or populations evolving stochastically. Both types of situations share common phenomena and mathematical methods to analyse them. One question which is common in both situations, is to find large classes of models where we observe the same large-scale behavior. This determination of universality classes for the behavior is important in order to be able to use such processes in applications where the detailed mechanism can often not be verfied but distinct qualitative behavior of a system is observed. We will focus here on population models, for the situation in systems in physics we refer the reader to section 1 of this volume.

In these notes we consider stochastic systems consisting of countably many spatial components, which interact linearly (modelling migration) and where the single components evolve randomly as one, multi- or infinite-dimensional diffusion with diffusion coefficient depending on the state of that component. These systems arise naturally in population genetics where one has spatial models of populations, i.e. populations of individuals of possibly different type located in a collection of colonies (geographic space), where individuals can migrate, leading to a linear interaction between the spatial components, and where the population in one spatial component fluctuates due to the change of generation. The population in a spatial component may be composed of individuals of one or two types, of finitely many types or of infinitely many types. The geographic space labelling the components is a discrete countable group like Z^d or the hierarchical group or their continuum versions, which is $\mathbb{R}^d$ in the first case.

There are two important classes of these models: *branching systems* on the one hand and *resampling systems* (most prominent representatives are Fleming-Viot processes) on the other hand (we shall focus in this survey on the first case mostly). In the multitype situation different from one-type models many open problems remain, among which is the unique characterization of the processes via the infinitesimal characteristics, the description of the longtime behavior in the complete class of models considered and clarifying the question which classes of processes allow for nontrivial spatial continuum limits.

The main interest is to understand the longtime behavior of the system, which is characterized by the competition between the migration and the fluctuation due to the change of generations. The first mechanism homogenizes configurations, the second one deregularizes by creating clumps or monotype cluster. This leads to dichotomies according to which of these forces wins and it is necessary to analyse the two resulting regimes in more detail. However this competing mechanism leads also to dichotomies for the small-scale behavior of spatial continuum limits of such systems and the resulting behavior is then characteristic for the continuum limit of the whole class of models.

It has been observed in the context of one-type models by Cox, Fleischmann and Greven [FG94], [FG96], [CFG96], Cox and Greven [CG94], Dawson, Greven, Vaillancourt [DGV95], Swart [Sw00] and for branching models by Klenke [K96], [K98], Winter [Wi02], Dawson and Greven [DG03], that many features of the large scale view in the longtime behavior are *universal* in the parameters regulating migration and fluctuations. A similar observation holds for the behavior of the continuum limits of these classes of models.

In a series of papers Dawson and Greven [DG93a]- [DG03]; Cox, Dawson and Greven [CDG03], Baillon, Clément, Greven and den Hollander [BCGH95], [BCGH97] and den Hollander and Swart [HS98] the attempt has been made to explain and analyse this phenomenon by the method of *renormalization analysis*. These considerations have recently been used by Dawson, Gorostiza and Wakolbinger [DGW03b] to analyse even multi-level branching systems, in which the role of the fluctuations is much more pronounced. The goal of these notes is to describe the main ideas of the approach of renormalization and state the main results.

The structure of this theory is roughly as follows. Using a collection of space-time scales we associate with the stochastic system a collection of *renormalized systems*. By then performing what we call the *hierarchical mean-field limit* we obtain a collection of limiting objects forming a Markov chain which we call the *interaction chain*:

- stochastic system in large scales ⟷ interaction chain.

The law of the (time-inhomogeneous) chain is described via the orbit of a certain nonlinear map acting on a function space, the points in this function space represent the possible local diffusion matrices in a single spatial component:

- interaction chain ⟷ orbit of nonlinear map.

The multiple large scale properties of the original system are reflected in the asymptotic behavior of the interaction chain. This behavior however can be reduced to properties of the orbit of the nonlinear map and the nature of this map is such that the orbit can be analyzed using

- fixed points, fixed shapes

of this nonlinear map. (A fixed shape g for a map $\mathcal{F}$ acting on a function space satisfies $\mathcal{F}(ag) = L(a)g$ for all $a \geq 0$ and some function L).

These distinguished diffusion functions correspond then on the level of the original stochastic system to certain classes of models which are "explicitly solvable" due to the duality relations they permit, which makes it easier to analyse them. In the case of models with one-dimensional components these special ones are interacting collections of Feller's diffusions, Fisher-Wright diffusions, critical Ornstein-Uhlenbeck diffusions and the parabolic Anderson model and in the case of two-dimensional $(\mathbb{R}^+)^2$-valued components, mutually catalytic branching, catalytic branching and independent multitype branching and their combinations are the typical examples.

The basic result which we derive is that in an asymptotic sense above mentioned distinguished models are characteristic for the behavior of models in their respective universality class. Behavior refers here to the properties in the *limit of large times* or the behavior in small scales in a rescaling leading to a *spatial continuum limit*. We can determine the universality class for a given diffusion matrix for *both* problems purely analytically by deciding whether the corresponding orbit of our nonlinear map approaches the fixed points or after rescaling fixed shapes corresponding to that class. That means:

- universality properties of stochastic systems and stochastic continuum limits $\longleftrightarrow$ universality properties of the orbit of the nonlinear map.

Finally we mention two other very important directions of thought. Namely first of all applications of the renormalization technique in *mathematical biology*, carried out by Dawson and Greven in [DG99] and [DG03b]. We can apply the renormalization scheme and the hierarchical mean-field limit approach presented in the sequel to models where we add to the neutral models considered here in this review as mechanisms mutation and selection. This allows to tackle the problem of modeling evolutions passing through chains of successive *quasi-equilibria*. This is a situation which arises in evolutionary models in mathematical biology, if one tries to understand the combined effects of migration, selection and mutation in the presence of random fluctuations due to the successive replacement of one generation by the next.

Another direction of thought is to include historical and genealogical aspects of the considered models into the problem of universal pattern of behavior, an aspect which has drawn recently much attention among probabilists since these are aspects which are of primary interest in application. See for example [DG96], [DG03] and [GLW03] for more details on the approach via renormalization.

1.2 The Models

The models we have considered are population growth models and arise as continuum limits of particle models (the latter we shall describe explicitly in a remark below). There are two principal types of models considered in population genetics namely on the one hand *Fisher-Wright* or *Fleming-Viot systems* and on the other hand *branching systems*. In the first case the size of the population at a spatial location is kept constant through time and only the type decomposition changes according to migration and change of generation. In the branching model the population size can fluctuate as well.

Both situations have been treated and we will focus here only on the second case in order to keep the presentation readable. For the first case we refer the reader to [DGV95]. We distinguish one type, multi-type and infinitely many type populations which we describe separately below.

I. The one-type component case
Consider the process $(X(t))_{t \geq 0}$ where the states at fixed times t are of the form

$$X = (x_i)_{i \in I} \in (\mathbb{R}^+)^I, \tag{1}$$

where I is a countable abelian group (like Z^d) and the dynamics is given by

$$dx_i(t) = \sum_{j \in I} a(i,j)(x_j(t) - x_i(t))dt + \sqrt{2g(x_i(t))}dw_i(t), \qquad i \in I, \tag{2}$$

where the ingredients $a(\cdot, \cdot), g(\cdot)$ and $w(\cdot)$ are as follows:

$$\begin{aligned} 0 \leq a(i,j) \leq c < \infty, &\qquad i,j \in I, \\ a(i,j) = a(0, j - i), &\qquad i,j \in I, \\ \sum_i a(0,i) < \infty, & \end{aligned} \tag{3}$$

$$g : [0, \infty) \longrightarrow [0, \infty) \tag{4}$$

g is locally Lipschitz, $g(x) > 0$ for $x > 0$, $g(0) = 0$

$$\left\{ (w_i(t))_{t \geq 0}, \; i \in I \right\} \quad \text{are independent standard Brownian motions.} \tag{5}$$

As initial states we chose typically random configurations, with a law which is translation invariant with $Ex_i(0) < \infty$.

The components of the system feel a *drift* towards the average of the surrounding components and a *random perturbation* given by increments of a Brownian motion magnified by a state-dependent factor $g(x_i)$.

Remark   If $Ex_i(0) = \theta$ for all $i \in I$, then $Ex_i(t) = \theta$ for all $t \geq 0$, i.e. the system is mean preserving.

Remark   Under the given assumptions there exists a unique strong solution of (2) , if we choose $X(0) \in \mathcal{E} \subseteq (\mathbb{R}^+)^I$, with $\mathcal{E} = \{\eta | \sum \eta(i)\gamma(i) < \infty\}$ where γ is a strictly positive summable function with $\gamma a \leq M\gamma$ for some $M < \infty$.

Remark   The dynamics is obtained as *diffusion limit* of a particle model. Define $h(x) = g(x)/x$. Consider a configuration $\eta = (\eta(i))_{i \in I} \in (\mathbb{N})^I$ of particles on I, where $\eta(i)$ denotes the number of particles at i. The particles at $i \in I$ have the following stochastic evolution:

- they migrate by carrying out independent continuous-time random walks with rate $\bar{a}(i,j)$ for a jump from i to j and $\bar{a}(\cdot, \cdot)$ is given by $\bar{a}(i,j) = a(j,i)$.
- die or split into two at exponential rate $h(\eta(i))$, choosing with probability $\frac{1}{2}$ each of the two possibilities.
- migrate and branch independently, different particles acting independently.

If we give every particle mass n^{-1}, speed up the branching by switching to the rate $nh(\eta(i)/n)$ and increase the initial number of particles to $n\eta_0$, then the corresponding process of total masses $(\eta_t^{(n)})_{t \geq 0}$ converges as $n \to \infty$ to the solution of the system of SDE's given above.

II.   The multi-type case

Here we start in the particle picture. Particles can have one of $\{1, 2, \ldots, k\}$-types but otherwise evolve as before independent of their type but with a branching rate $h(x)$ at a site x which depends on the population at site x. This leads in the diffusion limit (here the rate scaling might be more subtle than in I) to the following situation. Let

$$X = \left((x_i^{(\ell)})_{\ell=1,\ldots,k} \right)_{i \in I}, \tag{6}$$

with $x_i^{(\ell)}$ being the mass of the type ℓ in colony i. The dynamics is now given by the solution of

$$dx_i^{(\ell)}(t) = \sum_{j \in I} a(i,j) \left(x_j^{(\ell)}(t) - x_i^{(\ell)}(t) \right) dt \tag{7}$$

$$+ \sum_{\ell=1}^{k} \sqrt{h^\ell(x_i(t)) x_i^{(\ell)}(t)} dw_i^{(\ell)}(t), \text{ for } i \in I \text{ and } \ell \in \{1, \cdots, k\},$$

where the ingredients are as before and initial configurations are such that $(\bar{x}_i)_{i \in I} \in \mathcal{E}$ with $\bar{x}$ denoting the sum of the components.

Consider the case $k = 2$, the two-type case. The most important examples here are:

$$h^1(x) = b_1, \ h^2(x) = b_2 \quad \text{(independent two-type branching)} \tag{8}$$

$$h^1(x) = b_1 \, x^{(2)}, \ h^2(x) = b_2 \quad \text{(catalytic branching)} \tag{9}$$

$$h^1(x) = b_1 \, x^{(2)}, \ h^2(x) = b_2 x^{(1)} \quad \text{(mutually catalytic branching)} \tag{10}$$

and a model which can be defined for $k = 2$ as well as $k \in \mathbb{N}$:

$$h^i(x) = h(\bar{x}) \text{ with } i = 1, \cdots, k \text{ and }, \bar{x} = \sum_{\ell=1}^{k} x^{(\ell)} \tag{11}$$

$$\text{(total mass dependent branching)}.$$

Further classes arise by considering linear combinations of these mechanisms.

III.   The infinite-type case (measure-valued case)

If one wants to consider the case of a continuum of possible types, which one may label with $[0, 1]$, then the state of a single component becomes a Radon measure on $[0, 1]$ and their total masses satisfy the growth condition we had before and which is needed to get a proper state space, we call again $\mathcal{E}$:

$$X = (x_i)_{i \in I} \in \mathcal{E} \subseteq (\mathcal{M}([0, 1]))^I. \tag{12}$$

The stochastic evolution of the particle model is as before and again we can pass to a diffusion limit similarly as described for the one-type case. In order to describe the limiting dynamics one now works with a tool called *martingale problem.* In order to describe this approach return to the one-type situation.

Let f be a bounded function on $\mathcal{E}$. We look at the new function $S_t f$ given by

$$(S_t f)(\tilde{X}) = E[f(X(t))|X(0) = \tilde{X}). \tag{13}$$

One obtains this way a semigroup $(S_t)_{t\geq 0}$ acting on the bounded measurable functions on $\mathcal{E}$. This semigroup defines the finite dimensional distribution of a Markov process uniquely. The semigroup in turn can be determined by its generator. Call the generator of this semigroup G.

First note that by Ito's formula the system given in (2) has the generator G, which acts on the set of functions which depend in a twice continuously differentiable way on finitely many components, as follows:

$$(Gf)(x) \tag{14}$$

$$= \sum_i \left\{ \sum_j (a(i,j)x_j - x_i) \left(\tfrac{\partial}{\partial x_i} f \right)(x) + g(x) \left(\tfrac{\partial^2}{\partial x_i \partial x_i} f \right)(x) \right\}.$$

Denote the set of the test functions just described by $\mathcal{A}$ and note that they are measure determining. But rather than constructing $(S_t)_{t\geq 0}$ from G acting on $\mathcal{A}$ and then constructing via the finite dimensional distributions the law $\mathcal{L}((X(t))_{t\geq 0})$, we characterize this law directly on the space of path by using the operator G restricted to $\mathcal{A}$ and writing down functionals of the process, which are required to be martingales.

The distribution Q_X of the process starting in X, which is a law on $C([0,\infty), \mathcal{E})$, can be defined in a unique way (martingale problem) by requiring that the process

$$\left((f(X_t) - f(X_0)) - \int_0^t (Gf)(X_s)ds \right)_{t\geq 0}. \tag{15}$$

under Q_X is a martingale with continuous path for all $f \in \mathcal{A}$ and with $X_0 = X$ Q_X-a.s.. The (G, X)-martingale problem is called *wellposed* if there is exactly one solution.

In the measure-valued case consider functions $F(X) = \prod_{k=1}^n \varphi_k(\langle x_{i_k}, \Phi_k \rangle)$ with Φ_k functions on $[0,1]$ and $\varphi_k \in C_b^2(\mathbb{R})$. The rates for birth and death are given by $h(x_i)$ in colony i and then the generator G has the form (it suffices to consider variations $\delta_u, \delta_v, u, v \in [0,1]$):

$$(GF)(X) = \int_0^1 \sum_j a(i,j)(x_j(du) - x_i(du)) \frac{\partial F}{\partial x_i}(X)[u] \tag{16}$$

$$+ \int_0^1 \int_0^1 h(x_i)x_i(du)\delta_u(dv)\frac{\partial^2 F}{\partial x_i \partial x_i}(X)[u,v].$$

An important example, which was studied extensively in [DG03], [Pf03], is the case $h(x_i) = h(\bar{x}_i)$, the total mass dependent branching.

The wellposedness of this martingale problem is a serious task but in special cases, like h depends only on the total mass or for catalytic and mutually catalytic two-type branching it can be established.

2 Qualitative Properties of the Population Models

2.1 The Longtime Behavior

The longtime behavior of the system (2), (7) or (16) results from two *competing tendencies* caused by migration and fluctuations, namely:

- approach to a constant state (equal to the space average of the initial state), which are the invariant states if we put the diffusion matrix equal to zero.
- approach of the state $x_i \equiv 0$ or X monotype vector (depending on the diffusion matrix), which is the invariant state if $a(\cdot,\cdot) = \delta(\cdot,\cdot)$.

The outcome of this competition depends on $\hat{a}(i,j) = \frac{1}{2}(a(i,j) + a(j,i))$, namely migration winning or loosing depending on whether $\hat{a}$ is transient or recurrent. More precisely it was shown in ([CG94], [CFG96]):

Theorem 1. *(Ergodic theorem)*
Consider the one-type case in (2) with $g(\cdot)$ subquadratic.

(a)If $\hat{a}$ is transient and $\mathcal{L}(X(0))$ is translation invariant and ergodic, with $Ex_i(0) = \theta$, then

$$\mathcal{L}(X(t)) \underset{t\to\infty}{\Longrightarrow} \nu_\theta, \tag{17}$$

where ν_θ is an extremal invariant measure of the process with

$$E_{\nu_\theta}(x_i) = \theta \tag{18}$$

and which is translation invariant and ergodic.
(b)If $\hat{a}$ is recurrent and $\sup_{i\in I} Ex_i(0) < \infty$, then

$$\mathcal{L}(X(t)) \underset{t\to\infty}{\Longrightarrow} \delta_{\underline{0}}, \tag{19}$$

where $\underline{0}$ is the state which is identically 0.   □

The phenomena occurring in the two regimes described above, as well as the dichotomy expressed in Theorem 1, can be explained by means of a renormalization analysis and these ideas we want to present in these notes. But first we discuss the situation for multitype situations.

Here different regimes are possible. A regime where we obtain exactly the situation just described for the one-type case is the following result proved in [DG03]:

Theorem 2. *If h is a function of the total mass only, then the multitype version of (17) - (19) hold.* □

In other cases the situation is very different and we will give here only examples for these cases. Dawson and Perkins [DP98] proved that

Theorem 3. *([DP98]) Start the mutually catalytic branching super random walk process translation invariant and shift ergodic with intensity $\theta \in (\mathbb{R}^+)^2$. If $\widehat{a}$ is transient, then*

$$\mathcal{L}(X(t)) \underset{t \to \infty}{\Longrightarrow} \nu_\theta \ , \ \theta \in (\mathbb{R}^+)^2 \ , \tag{20}$$

where ν_θ is an extremal shift-invariant, shift-ergodic, invariant measure of the dynamic with coexistence of both types, i.e. for $i \in Z^d$:

$$\nu_\theta\{x_i^1 x_i^2 > 0\} > 0 \quad \text{if } \theta_1\theta_2 > 0. \tag{21}$$

If $\widehat{a}$ is recurrent then

$$\mathcal{L}(X(t)) \underset{t \to \infty}{\Longrightarrow} \mathcal{L}(B_{\underline{H}}), \tag{22}$$

where B_H is planar Brownian motion with variance (b_1, b_2) stopped when hitting the axes. □

We see that in this case we always have survival of the population but depending on the migration mechanism we have coexistence or monotype configurations locally for large times. Again very different behavior occurs in the case of catalytic models with two types, we refer the reader here to [GKW99] and [GKW02] due to lack of space.

The fundamental problem now is to determine all types of possible behavior of such models and to characterize all universality classes showing the same behavior. This question we treat in the sequel using the hierarchical mean-field limit.

After having obtained these universality classes the main task is to exhibit the finer structure of the longtime behavior. This means in the case of recurrent $\hat{a}$ to analyse the extension and height of the rare populated spots respectively to determine the extension of monotype regions. (Compare [FG94], [FG96], [K97] and [Wi02]). In the case of $\hat{a}$ transient this requires to give more detailed properties of the equilibrium states, one possibility being to study large blocks of the equilibrium configuration and to derive scaling limits of the fluctuations of these blocks. (Compare [CDG03], [Z01] and [Z02]).

2.2 Continuum Limit

A fundamental concept in the treatment of large space-time scale properties of interacting systems is given by the *spatial continuum limit*. Suppose that $I = Z^d$ and the number of types is k. We can embed finer and finer lattices εZ^d in the continuum $\mathbb{R}^d$ and we also know that we can scale the random walk on Z^d such that in the limit we obtain Brownian motion on $\mathbb{R}^d$. In this spirit we can define our interacting systems for a whole collection of site spaces namely $\{\varepsilon Z^d, \varepsilon > 0\}$ and view the system as a vector-valued measure on $\mathbb{R}^d$, by defining the value of the measure as the sum of the values corresponding to lattice points in this set. The random walks we choose such that

$$a_\varepsilon(0, \varepsilon x) = a(0, x). \tag{23}$$

Then if we scale the time, the mass and possibly the branching parameters suitably, we expect to obtain in the limit $\varepsilon \to 0$ for the rescaled system a limiting dynamic which has as states a k-tupel of Borel measures on $\mathbb{R}^d$.

A famous example is the case $k = 1, g(x) = bx$. Here we scale time by ε^{-2} (recall space is scaled by ε) in order to obtain the convergence of the migration transition semigroup to the semigroup of Brownian motion on $\mathbb{R}^d$. Furthermore we scale mass by ε^d. Then we obtain in the limit $\varepsilon \to 0$ the Dawson-Watanabe process (or often called super Brownian motion), a measure-valued process (i.e. values are in $\mathcal{M}(\mathbb{R}^d)$, the Borel measures on $\mathbb{R}^d$).

Similar rescaling procedures can be carried out for other systems of this type but large classes of dynamics result in the same continuum limit. The branching is distinguished here by the fact that no rescaling of the rates is required to get the spatial continuum limits. This is shared only by those dynamics with a diffusion function asymptotically like branching for large values. This distinguishes the branching regime over other processes like catalytic branching, mutually catalytic branching.

Much more subtle is the question whether a continuum limit exists for multitype models. Here the class of total mass dependent branching is again exhibiting no new effects. Everything else however is very difficult. Great efforts have been made for two type models, in particular for catalytic and for mutually catalytic branching. This already turn out to be quite challenging and no models in the same universality class can be treated, except these special cases. Again in this situation a rate rescaling is needed for processes not in the universality class of independent multitype branching. For mutually catalytic branching only the critical dimension $d = 2$ requires no rate rescaling.

A key question concerning the qualitative properties of the limiting process is now whether the states of this process have a *density* with respect to the Lebesgue measure or whether they are *singular*. Take again as example the Dawson-Watanabe process. It is well-known that for $d = 1$ we obtain states which are absolutely continuous, while for $d \geq 2$ we obtain singular states. Furthermore a lot is known about the character of the "singularity", here one studies the Hausdorff dimension and the precise scaling functions (logarithmic corrections) of the states of the process to describe this phenomenon.

Of course one would like to obtain these results also for other branching dynamics where similar dichotomies and phenomena of the continuum limits are expected. Here many difficulties arise and the transition in the behavior arises around $d = 2$. Therefore it is useful to study these questions for I being the hierarchical group Ω_N and its continuum version Ω_N^∞ (we introduce below) in the case $N \to \infty$ which approximates the case $I = Z^d$ around $d = 2$ in the sense that the behavior of the Greens functions for both cases become similar. For a detailed study of the potential theoretic properties of random walks on the hierarchical group we refer to Dawson, Gorostiza and Wakolbinger [DGW].

The techniques of renormalization analysis developed for the study of the longtime behavior can be extended and modified in order to introduce a *hierarchical mean-field continuum limit*. This object gives insight in the question of existence and qualitative properties of the continuum limit around $d = 2$ on the lattice and is the topic of section 6.

2.3 Historical Processes

Consider a branching particle system that is individuals branch and migrate. Then we can associate with every individual alive at time t its path of descent by tracing back its positions in space till its birth, then continuing with the positions of its father etc. The resulting path is continued for times smaller 0 and bigger t as the constant path. Now assign to every path the δ measure on that path and sum over all individuals alive at time t. In this way we obtain a measure on the space of path with positions in I, more precisely an element of $\mathcal{M}(D(\mathbb{R}, I))$. Varying t we obtain the historical process for the particle system.

Carrying out the diffusion limit explained in subsection 2.1 results on the level of paths of descent in the *historical process* of the interacting diffusions. This process can be characterized by a martingale problem. This works in the multitype setup for type-independent branching and for branching rates depending on the total mass, see [DG03]. Developing the conceptual analog of the historical process for catalytic or mutually catalytic models is still a task for the future. Here one has to deal with the fact that an individual has always *two* ancestors.

The above construction of the historical process allows to analyse the equilibrium states or the clusters of the system, by decomposing them into the clans of masses belonging to one founding ancestor. This way one can obtain a transparent description of the dependence structure of these systems and of the genealogical structure of monotype clusters. We do not have the space here to describe this important line of thought in more detail and refer the reader for more details and more references to [DG03], [GLW03] and for resampling systems [DGV95].

3 Renormalization Analysis and Hierarchical Mean-Field Limit

3.1 Multiple Space-Time Rescaling and the Hierarchical Mean-Field Limit

We proceed now in three steps. We begin by adapting the model for the purpose of a *multiple scale renormalization* by introducing a specific choice of the index set I and defining random walks on that set. Then we define in the second step the renormalized system and conclude in the third step by giving the basic limit result.

(i) Framework for a multiple space-time scale analysis: The process $(X_t^N)_{t\geq 0}$. In this section we introduce a rescaling of our interacting system with countably many components. For this purpose we begin by specifying the set of indices I to be the hierarchical group Ω_N, which is the index set suitable for population models and was introduced by Felsenstein and Sawyer [SF83] in population genetics. Ω_N is defined as:

$$\Omega_N = \{(\xi_0, \xi_1, \ldots) | \xi_i \in \{0, 1, \ldots, N-1\}, \ \xi_i \neq 0 \quad \text{finitely often}\} \qquad (24)$$

and the addition is defined as the componentwise addition modulo N. Interprete this as specifying one of N islands in one of N archipelagos etc.

On this group Ω_N we define the hierarchical distance to be

$$d(\xi, \xi') = \inf(n \mid \xi_i = \xi_i' \quad \text{for all } i \geq n). \qquad (25)$$

Then we can define the k-balls

$$B_k = \{\xi \in \Omega_N | d(\xi, 0) \leq k\}. \qquad (26)$$

Note that we have *as sets*, the following embedding:

$$\Omega_N \subseteq \Omega_M \subseteq \Omega_\infty \quad \text{for } N \leq M, \qquad (27)$$

allowing to focus on a fixed window as $N \to \infty$.

On the group Ω_N we define next an appropriate class of random walks for which the points on the surface of balls become exchangeable for the walk and whose rate to make a jump with which one can reach the surface of a larger ball scales in a suitable way. Precisely, we fix a sequence $(c_k)_{k\in\mathbb{N}}$ of positive numbers such that

$$\sum_{k=0}^{\infty} \frac{c_k}{N^k} < \infty \quad \text{for } N \geq 2. \qquad (28)$$

Our random walk will have rate c_{k-1}/N^{k-1} to make a jump to a point within a k-ball around its present position and then chooses the target point according to the uniform distribution on this ball. In other words the transition matrix $a_N(\cdot,\cdot)$ is given by

$$a_N(\xi,\eta) = \sum_{j=k}^{\infty} \frac{1}{N^j}\left(c_{j-1}/N^{j-1}\right) \quad \text{if } d(\xi,\eta) = k. \tag{29}$$

Remark   This random walk will need a time of order N^j to move a distance j away and hence the process giving the distance from the origin behaves like a process moving in a linear potential. Important is that the times N^j are for different j of different order of magnitude as $N \to \infty$, i.e. they separate.

Remark   The random walk just introduced can be viewed as a carricature of a random walk on Z^d with effective dimension 2 $(logN/(logN/c))$ which means for $N \to \infty$ around two dimensions. Using the scaling factor $N^{k/2}$ instead of N^k leads to a situation around $d = 4$, see [DGW03b].

Definition 1. (Ω_N-indexed system)
For every N define the process $X^N(t)$ based on Ω_N and according to the definition in (2), (7) or (16) replacing $a(\cdot,\cdot)$ by $a_N(\cdot,\cdot)$ given above in (29).   □

(ii) The renormalization scheme
The system $X^N(t)$ will now be analyzed using a collection of space-time scales which *separate* if $N \to \infty$.
 Begin with the spatial rescaling. Define

$$x_{\xi,k} = \frac{1}{N^k} \sum_{\xi' \in B_k(\xi)} x_{\xi'} \quad \text{with } X = (x_\xi)_{\xi \in \Omega_N}, \tag{30}$$

in other words $x_{\xi,k}$ is the *block average* of X in the k-ball around ξ.

 Next we consider the time scales

$$t \to N^j t, \quad j \in \mathbb{N}. \tag{31}$$

Often it is more convenient in order to get a simpler limiting object to use

$$t \longrightarrow s(N)N^j t \tag{32}$$

with $s(N)$ satisfying: $s(N) \to \infty$ as $N \to \infty$ and $s(N) = o(N)$. (We shall see later that with this choice we get equilibration on the j-the level.)
 Above ingredients allow us to define for every pair $(k,j) \in \mathbb{N}^2$ the following collection of rescaled functionals of the process $(X^N(t))_{t \geq 0}$:

$$x_{\xi,k}^N(s(N)N^j t). \tag{33}$$

Of particular interest will be the following collection defined for fixed j:

Definition 2. *(Finite-N interaction chain)*
We shall call

$$\{x_{\xi,k}^N(s(N)N^j t);\ k = 0, 1, \ldots, j+1\}, \tag{34}$$

the finite-N interaction chain of level j.   □

This way we look for N large at a system which evolved for a long time in a whole hierarchy of block averages and describe their joint distribution.

(iii) The basic limit result (hierarchical mean-field limit)
The idea is that for N sufficiently large the collection $\{x_{\xi,k}(tN^j),\ k = 0, \ldots, j+1\}$ becomes approximately Markov in k the latter running from $j+1$ down to 0 and that the transition kernel becomes simple if the block averages are based on many components. We refer to $N \to \infty$ in connection with $(X^N(t))_{t \geq 0}$ as the *hierarchical mean-field limit*. Here mean-field expresses the fact that if we consider only permutations of components on the surface of k-balls then the system is *exchangeable* under these permutations. On the other hand we always work with an *infinite* interacting system.

   The following result can be proven for models with one-dimensional components [DG93c], [DG96], for multi- or infinite-dimensional models in the case of resampling or branching models [DGV95], [DG96] or [CDG03] for mutually catalytic branching:

Theorem 4. *(Basic scaling result)*
Assume that $\mathcal{L}(X^N(0))$ is i.i.d. with $Ex_\xi^N = \theta$. Then:

(a) $\mathcal{L}(\{(x_{\xi,-k}(s(N)N^j t))_{k=-j-1,-j,\ldots,0}\}) \underset{N \to \infty}{\Longrightarrow} \mathcal{L}((M_k^{(j)})_{k=-j-1,-j,\ldots,0}).$

(b) $(M_k^j)_{k=-j-1,\ldots,0}$ *is a Markov chain, which is time inhomogeneous but at time $-k$ the transition kernel P_k is independent of j.*

(c) $\mathcal{L}((M_k^{(j)})_{k=-j-1,\ldots,0}) \underset{j \to \infty}{\Longrightarrow} \mathcal{L}((M_k^\infty)_{k \in Z-})$, *where $(M_\cdot^\infty)$ corresponds to an entrance law, that is a Markov chain with transition kernels $(P_k)_{k \in \mathbb{N}}$ at times $-k$.*   □

In the next subsection we shall see that we can actually identify fairly explicitly the law of
$(M_k^j)_{k=-j-1,\ldots,0}.$

Definition 3. *(Interaction chain)*
We call $(M_k^j)_{k=-j-1,\ldots,0}$ the interaction chain of level j and $(M_k^\infty)_{k \in Z-}$ the entrance law of the interaction chain.   □

   Calculation of the law of the interaction chain and its entrance law requires first the analysis of certain finite systems in slow and later in fast time scales, which amount to carrying out the *mean-field finite system scheme* (see Theorem 5 below). In Theorem 6 we identify the interaction chain by explicitly identifying the hierarchical mean-field limit dynamics on all finite levels j.

Remark   The multitype time scale analysis just described can be extended to the level of historical processes, see [DG96], [DGV95] and [DGW01].

3.2 Background on the Hierarchical Mean-Field Limit

In this subsection we explain how the limit $N \to \infty$ just taken above comes about and how the limiting objects can be identified. We proceed in three steps.

(i) A mean-field system and its McKean-Vlasov limit
In this section we will focus on a finite system with N-components, denoted $Y^N(t)$. In the case of a model with k-types they have the form

$$Y^N(t) = (y_i^N(t))_{i=1,...,N} \in ((\mathbb{R}^+)^k)^N \tag{35}$$

and solve the following system of stochastic differential equations:

$$dy_i^N(t) = c \left(\frac{1}{N} \sum_{j=1}^N (y_j^N(t) - y_i(t)) \right) dt + \sum_{\ell=1}^k \sqrt{h^\ell(y_i^N(t))} y_i^{N,\ell}(t) dw_i^\ell(t),$$

$$i \in \{1, \ldots, N\} \tag{36}$$

$$\mathcal{L}(Y^N(0)) \quad \text{is i.i.d. with } Ey_i(0) = \theta. \tag{37}$$

We shall use the same marginal distribution $\mathcal{L}(y_0^N)$ for all N. It will be no difficulty for the reader to write down the equations (using generators) for infinite-dimensional components.

Next consider the process $Y^\infty(t) = (y_i^\infty(t))_{i \in \mathbb{N}}$ with $y_i^\infty(t) \in [0,\infty)^k$ given as follows. All components $i = 1, 2, \cdots$ evolve independently according to the following SDE on $[0,\infty)^k$:

$$dz_t = c(\theta - z_t)dt + \sum_{\ell=1}^k z_t^{(\ell)} \sqrt{h^\ell(z_t)z_t^{(\ell)}} \, dw_t^{(\ell)}, \tag{38}$$

and $\mathcal{L}(Y^\infty(0))$ is i.i.d. with the same marginal as $\mathcal{L}(Y^N(0)), N \in \mathbb{N}$.

Then one can show that $(Y^\infty(t))_{t \geq 0}$ is the infinite system corresponding to $(Y^N(t))_{t \geq 0}$ *(McKean-Vlasov limit)* [DGV95], [DG96], [CDG03]:

$$\mathcal{L}((Y^N(t))_{t \geq 0}) \underset{N \to \infty}{\Longrightarrow} \mathcal{L}(Y^\infty(t))_{t \geq 0}). \tag{39}$$

(ii) The longtime behavior of the system $(Y^\infty(t))_{t \geq 0}$
For the process $(z_t)_{t \geq 0}$ we expect that there exists a unique invariant measure which we denote:

$$\Gamma_\theta^{c,h}. \tag{40}$$

Furthermore the process Y^∞ should be *ergodic*, i.e. for all initial states:

$$\mathcal{L}(Y^\infty(t)) \underset{t \to \infty}{\Longrightarrow} \left(\Gamma_\theta^{c,h} \right)^{\otimes \mathbb{N}}. \tag{41}$$

This holds always in the case of one type, but in the case of several or infinitely many types this has not been proved yet in general but holds for example for total mass-dependent, catalytic and mutually catalytic branching. The main difficulty in finite type situations rests with the fact that the well-posedness of the martingale problem has not been shown in the needed generality. Assuming the wellposedness a proof is given in Dawson, Greven, den Hollander and Swart [DGHS].

(iii) The mean-field finite system scheme
The next step is to consider $(Y^N(t))_{t\geq 0}$ in the faster time scale $t \to Nt$ and in the limit $N \to \infty$. The basic idea is that the laws of the empirical mean processes

$$(\bar{Y}^N(t))_{t\geq 0} = \left(\frac{1}{N} \sum_{i=1}^N y_i^N(Nt) \right)_{t\geq 0} \tag{42}$$

form a tight sequence in this rescaling. Namely consider first the case of finitely many types k and set $g^{(\ell)}(x) = xh^{(\ell)}(x)$ and then observe that for every fixed N the process $\bar{Y}^N$ is a martingale with values in $(\mathbb{R}^+)^k$ with stochastically bounded increasing process

$$\langle \bar{Y}^N \rangle_t = \left(\int_0^t \left(\frac{1}{N} \sum_{i=1}^N g^{(\ell)}(y_i^N(sN)) \right) ds \right)_{\ell=1,2,\cdots,k} . \tag{43}$$

Using the techniques from the finite system scheme (see [CG94], [CGS95]) it was shown in [DG93a], [DG96], [CGS95], [CDG03] (recall (38) for $(z_t)_{t\geq 0}$):

Theorem 5. *(Mean-field finite system scheme)*
For the one-type models we have:

$$\mathcal{L}((\bar{Y}^N(t))_{t\geq 0}) \underset{N\to\infty}{\Longrightarrow} \mathcal{L}((\tilde{z}_t)_{t\geq 0}) \tag{44}$$

where $\tilde{z} = \theta \in [0, \infty)$ satisfies

$$d\tilde{z}_t = \sqrt{g^*(\tilde{z}_t)}dw_t, \quad g^*(\theta) = \int_{[0,\infty)} \Gamma_\theta^{c,g}(d\theta')g(\theta), \tag{45}$$

and

$$\mathcal{L}(Y^N(tN)) \underset{N\to\infty}{\Longrightarrow} \int_{[0,\infty)} Q_t(\theta, d\theta')(\Gamma_{\theta'}^{c,g})^{\otimes \mathbb{N}}, \tag{46}$$

where

$$Q_t(\theta, \cdot) = \mathcal{L}(\tilde{z}_t | \tilde{z}_0 = \theta). \qquad \square \tag{47}$$

This should be viewed as follows. The macroscopic variable given by the empirical mean fluctuates only on time scales of order Nt and given a value θ' for this empirical mean the system relaxes into the equilibrium corresponding to the center of drift given by θ'. With this scenario, however, the macroscopic variable becomes Markov in the limit, and its volatility in the state θ' is given by the mean of $g(\cdot)$ under $\Gamma_{\theta'}^{c,g}$ which we denoted $g^*(\theta')$.

Remark   This result holds also for certain branching and resampling models with multi-dimensional or infinite-dimensional components namely: total mass-dependent branching, catalytic branching, mutually catalytic branching and the Fleming-Viot process ([DGV95], [CDG03], [DG03], [Pf03]). In this case we have to replace the diffusion function by a diffusion matrix valued function and Brownian motion by multi-dimensional Brownian motion but otherwise we get literally the same relations.

(iv) Identification of the interaction chain
With the analysis of the previous point we are able to analyse mean-field systems in the time scale Nt. The next step is to consider systems with j-levels. These are finite systems indexed with $\{0, 1, \cdots, N-1\}^j$ and the hierarchical interaction obtained by restricting the random walk kernel to jumps up to level $j+1$. These systems need to be analysed in time and space scales given by k-block averages and time scales tN^k for $k = 0, 1, \cdots, j+1$. From this point one then approximates our infinite hierarchical system by letting $j \to \infty$, compare for example [DGV95] for details. This allows then to determine the limit for $N \to \infty$ of the finite-N interaction chain. We need a new ingredient for this.

We define the transition kernels on $\mathbb{R}^+, (\mathbb{R}^+)^k$ respectively $\mathcal{M}([0,1))$ for one type, k-type and infinitely many type models:

$$K_{c,g}(\theta, d\theta') = \Gamma_\theta^{c,g}(d\theta'), \text{ (resp. } \Gamma_\theta^{c,G}). \tag{48}$$

We now want to identify the limiting objects in Theorem 4. The key fact is a multi-level generalization of Theorem 5 given in ([DG93c], [DG96], [CDG03]) which leads for one-type models to the result:

Theorem 6. *(Interaction chain)*
The interaction chain $(M_k^j)_{k=-j-1,\ldots,0}$ *is the time-inhomogeneous Markov chain with*

$$M_{-j-1}^j = \theta \tag{49}$$

and with transition kernels P_k *at time* $-k$ *given by*

$$P_k = K_{c_k,g_k}, \tag{50}$$

where the function g_k *on* $[0,\infty)$ *is defined recursively by*

$$g_{k+1}(\theta) = \int_{[0,\infty)} \Gamma_\theta^{c_k,g_k}(d\theta')g_k(\theta'). \qquad \square \tag{51}$$

This result can be generalized in situations with more than one type in the following two cases

– total mass-dependent branching (infinitely many types)
– mutually catalytic branching and catalytic branching.

We have to replace in this case g_k in (51) defined on $[0, \infty)$ by a diffusion matrix defined as function on $[0, \infty)^k$ respectively $\mathcal{M}([0,1])$.

4 Analysis of the Limiting Renormalized System

4.1 *The Dichotomy Stability versus Clustering*

(i) The dichotomy: stability versus clustering
The longtime behavior of our system (recall Theorem 1 and its multi-dimensional extensions) depends on whether we have long range or short range interaction, which corresponds to the transience respectively recurrence property of $\hat{a}$. In the case of long range interaction we have a decent equilibrium state of the system in the case of short range interaction we will see the approach to the traps. This means in one-type models and total mass dependent multitype branching systems convergence towards $\underline{0}$ and the forming of large peaks of mass on locations with increasingly rare density. In other multitype cases such as mutually catalytic branching, the process converges in the case of recurrent symmetrized migration to a configuration concentrated on monotype constant configurations.

This will be reflected in the structure of $(M_k^\infty)_{k \in Z^-}$ as follows. In the case of long-range interaction (transient $\hat{a}$) $(M_k^\infty)_{k \in Z^-}$ describes the joint distribution of the collection of block averages of systems in equilibrium for N large, while in case of short-range interaction (recurrent $\hat{a}$) the behavior of M_k^j as $j \to \infty$ reflects the formation of clusters.

We formulate a general result for the one-dimensional component case, analog statements hold in all other situations as well (see [DG96], [DGV95], [DG03]) but does *depend* on the universaltiy class we consider. We therefore make this statement explicit here only for mutually catalytic branching and total mass-dependent branching.

Theorem 7. *(Stability versus clustering: state-dependent case)*
Assume that we have a system with one-dimensional components as given in (2).

$$\text{If } \sum_{k=0}^{\infty} c_k^{-1} < \infty, \text{ then for } \quad \theta > 0: \tag{52}$$

$$M_k^\infty \in (0, \infty) \quad \text{and} \quad M_k^\infty \xrightarrow[k \to -\infty]{} \theta.$$

$$\text{If } \sum_{k=0}^{\infty} c_k^{-1} = +\infty, \text{ then:} \qquad (53)$$

$$M_k^{\infty} = 0 \quad \forall k \in Z^-. \qquad \Box$$

Remark   The analog statement holds for total mass-dependent branching which is in the same universality class as independent multitype branching.

Remark   The condition $\sum c_k^{-1} = +\infty$ or $< \infty$ is equivalent to

$$a_N(\cdot, \cdot) \quad \text{is recurrent respectively transient on } \Omega_N. \qquad (54)$$

Theorem 8. *(Stability versus clustering: mutually catalytic case)*
In the mutually catalytic case we get a similar dichotomy as in Theorem 7. In (52) we get for vectors θ with $\theta_1, \theta_2 > 0$, that

$$M_k^{\infty} \in (0, \infty)^2 \quad , M_k^{\infty} \underset{k \to -\infty}{\Longrightarrow} \theta \qquad (55)$$

and in (53):

$$M_k^{\infty} = B_H \quad , \quad k \in Z^-, \qquad (56)$$

where B_H is planar Brownian motion started in $\theta \in [0, \infty)^2$ stopped at the hitting time H of the axes. $\quad \Box$

We can now analyse the properties in the two regimes in more detail. The principal question in the recurrent and transient cases are the following.

– Cluster formation: What are height and extension in space of clusters, respectively what are the extensions of the monotype regions in space as time evolves?
– Structure of equilibrium states: Do the spatial averages of the configuration in equilibrium satisfy a central limit theorem with Gaussian limits, in what scaling?

Both these questions can be addressed with the techniques of the hierarchical mean-field limit. We will only describe the first point and refer the reader for the second point to [CDG03]. In catalytic models different behavior occurs, compare [GKW99] and a different type of analysis is needed.

Another point we cannot describe here is the construction of the *historical process for the interaction chain* and the analysis of its properties. See [DG96], [DGV95] and [DGW01] for details.

4.2 Cluster Formation

We shall focus here mainly on the case of a system with one-dimensional components taking values in $[0, \infty)$, (see [DG96]) which allows to exhibit essential features of the analysis. Analogous results can be worked out for the cases of $[0, 1]$ - (see [DG93c]) or $(-\infty, \infty)$-valued components. These cases cover the possible behavior in the one-dimensional case and the situation is wellunderstood. More subtle is the multi-dimensional and infinite dimensional case where only some universality classes have been examined so far, namely the two-type model given by mutually catalytic branching (see [CDG03]) and in the infinite dimensional case the Fleming-Viot process ([DGV95]).

We focus first here in (i) on the special case where for a constant $d > 0$

$$g(x) = dx \tag{57}$$

and later in paragraph (ii) we discuss the issue of universality.

(i) Finer analysis of the cluster formation
The basic phenomenon we want to describe is the following. The process has in the one-type case for large times most components very close to 0, but since it is mean preserving it follows that on rare patches of components very large values (peaks) develop. The questions are:

– How fast do the peaks grow?
– How fast do the patches expand in space, which exhibit that growth?

To analyse this it turns out that we have to distinguish two regimes. In the regime I, the peaks are of roughly the same height, so that the components in the cluster have asymptotically on large space scales an exchangeable structure. In the regime II very irregular shaped clusters with rough geometric shape occur and here most of the mass is concentrated in few very high peaks. In regime I, the spatial extension of clusters has a random order of magnitude (i.e. growth on different scales) in regime II, it has random size and grows on one scale.

Which of the two regimes occurs depends on the migration. The two situations correspond to two different types of recurrent random walks. If $\hat{a}(\cdot, \cdot)$ is very weakly recurrent, meaning a slowly growing (i.e. logarithmically) Greens function then we shall see regime I of the smoothly shaped clusters while in the case of strongly recurrent $\hat{a}(\cdot, \cdot)$ we see the sharp peaks.

We shall see that for c_k going to 0 exponentially fast we are in regime II while the cases where $c_k \to \infty$ (but still $\sum c_k^{-1} = +\infty$) and $c_k \to \text{const}$, for example lead to regime I. Even though the occurrence of at least two regimes of clustering is seen in other universality classes, the criteria in terms of the random walk often depend very much on the particular universality class, a good example for the possible structures is mutually catalytic branching described in [CDG03].

The cluster formation is reflected in the behavior or the sequence of increasing block averages as follows. If the clusters are of different orders of magnitude we will find block averages over blocks extending over a volume $f_j(\alpha)$ exhibiting different random values when $f_j(\alpha)$ varies from $-j-1$ to 0 as α moves from 0 to 1 and is chosen properly. In other words we can rescale the sequence of block averages over blocks of size 0 up to j such that in the limit we get a diffusion process. If the clusters are of one order of magnitude, which corresponds to the possible range of the random walk, but random size, then the block averages over blocks of size $j, j-1, \ldots, j-k$ should be random and become deterministic as $k \to \infty$. In other words the block averages over blocks of size $j, j-1, \ldots, j-k, 1-k, \ldots$ converge to a Markov chain.

Both cases require a different type of analysis. In the regime I we can look at the configuration conditioned on finding at least mass ε at 0 and this will put us in the center of a cluster. In the regime II we cannot proceed this way since we shall typically then see the boundary of a cluster. Here we should look at the cluster as seen from a randomly sampled member of the cluster, this will put us in the part of the cluster containing the overwhelming part of the population.

Regime I (diffusive clustering)
The idea is to take the process $M_k^{(j)}$ conditioned on the event $\{M_0^{(j)} \geq \varepsilon\}$, which means we put ourselves in the center of a cluster of values bounded away from 0. Recall that most components will be very small at large times. Since on the event $\{M_0^{(j)}) \geq \varepsilon\}$ the process becomes very large, we have to rescale the size of $M_k^{(j)}$ as a function of j and in order to obtain a nondegenerate limiting dynamics in k we have to rescale the time index k, corresponding to block sizes in the original system. Such a rescaling of the interaction chain corresponds to the original system as follows:

- *mass rescaling in j:*
   growth rate of the height of a component which is not close to 0, shortly the *cluster height*
- *time rescaling in k:*
   growth rate in space of the rare patches with large values, measured in the volume of the patch, shortly the *cluster extension*.

Both these quantities are intimately related with properties of the migration parameter $(c_k)_{k \in \mathbb{N}}$. We can have either high and small (in spatial extension) clusters or flat and extended clusters, depending on the behavior of c_k as $k \to \infty$.

Introduce the *cluster height scale function*

$$H(j) = \sum_{k=0}^{j} c_k^{-1}. \tag{58}$$

In order to describe the *cluster extension* define for each j a function $\alpha \to f_\alpha(j)$ satisfying:

$$
\begin{aligned}
&f_\alpha(j) && [0,1] \to [-j-1, 0] \\
&f_\alpha(j) && \text{non decreasing} \\
&f_0(j) = -j-1 \;\; f_1(j) = 0.
\end{aligned}
\tag{59}
$$

Now, depending on the size of the spatial extension of the cluster the function $f_\alpha(j)$ will take different shape. Namely for extended clusters small α will already correspond to large indices for k, for clusters of small extension α has to be large to reach large index values.

We distinguish three cases within the regime I of diffusive clustering which will all give the same limit process but with different rescaling. In particular this distinction is quantitative not qualitative, as is the distinction between regime I and regime II. Here are these three cases:

$$c_k \to \infty, \ c_k \to c \in (0, \infty), \ c_k \to 0 \text{ slower than exponential as } k \to \infty. \quad (60)$$

The scaling function $f_\alpha(j)$ can be explicitly given in many of these cases. If $c_k \to c$ then we define ($[\cdot]$ meaning the integer part)

$$f_\alpha(j) = [\alpha(j+1)] - j - 1. \quad (61)$$

In the case $c_k = \text{const } k^\beta$ with $\beta \in (-\infty, 1)$ we define

$$f_\alpha(j) = \left[\alpha^{\frac{1}{-\beta+1}}(j+1)\right] - j - 1. \quad (62)$$

We refrain here from giving a general device, more details can be found in [DG96].

Now we can define a stochastic process $(\widetilde{M_\alpha^j})_{\alpha \in [0,1]}$ by rescaling the conditioned interaction chain:($[x]$ denoting the integer part of $x \in \mathbb{R}$)

$$\mathcal{L}\left(\left(\widetilde{M_\alpha^j}\right)_{\alpha \in [0,1]}\right) = \mathcal{L}\left((H(j))^{-1} M_{[f_\alpha(j)]}^j \mid M_0^j \geq \varepsilon\right). \quad (63)$$

Theorem 9. *(diffusive clustering)*
Assume that the sequence $(c_k)_{k \in \mathbb{N}}$ satisfies $\sum c_k^{-1} = +\infty$, and in case c_k decays to zero then assume, in addition, it decays slower than exponential. Then for all $\varepsilon > 0$:

$$\mathcal{L}\left(\left(\widetilde{M_\alpha^j}\right)_{\alpha \in [0,1]}\right) \underset{j \to \infty}{\Longrightarrow} \mathcal{L}\left(\left(\widetilde{M_\alpha^\infty}\right)_{\alpha \in [0,1]}\right), \quad (64)$$

where $(\widetilde{M_\alpha^\infty})_{\alpha \in [0,1]}$ is a time-inhomogeneous diffusion process with initial point θ and with infinitesimal characteristics given by the operator:

$$G_\alpha = a_\alpha(x)\frac{\partial}{\partial x} + b_\alpha(x)\frac{\partial^2}{\partial x \partial x} \quad (65)$$

$$a_\alpha(x) = 2x\left(\frac{1-\alpha}{e^{x/1-\alpha} - 1}\right), \quad b_\alpha(x) = (2x/a_\alpha(x)). \quad \square$$

Remark   If, instead of conditioning, one uses different approaches, for example, blow up of the initial state (compare Klenke's work in [K97], [K98]) or if one looks at the Palm distribution (compare the work of Winter in [Wi02]) one obtains different limiting diffusions, which are, however, related in a canonical way among each other and to the above, namely one gets the Feller diffusion, Feller diffusion conditioned on survival and the Feller diffusion palmed.

Regime II (concentrated clusters)

In this regime the approach of conditioning is not appropriate (since then we view the cluster from a typical colony which is a boundary point with a smaller peak). Instead we view everything from the point of view of a randomly picked mass in the cluster, that is, viewed from a typical member of the population. This means we consider the Palm measure of the sequence (M_k^j) with respect to M_0^j. The Palm distribution of X taking values in $\mathbb{R}$ arises by *size-biasing* the law of the random variable X.

We focus on the case

$$c_k = c^k \quad \text{with } c < 1. \tag{66}$$

Define $(\widehat{M}_k^j)_{k=-j-1,\dots,0}$ by taking the Palm measure of $(M_k^j)_{k=-j-1,\dots,0}$ with respect to M_0^j, i.e.

$$P\left(\widehat{M}_{-j-1}^j \in A_1, \ \widehat{M}_{-j}^j \in A_2, \dots, \widehat{M}_0^j \in A_{j+1}\right) \tag{67}$$

$$= \theta^{-1} E\left(M_0^j \mathbb{1}\left(M_{-j-1}^j \in A_1, \dots, M_0^j \in A_{j+1}\right)\right).$$

Now we do again our mass rescaling to cope with the increasing height of the clusters:

$$\widetilde{M}_{-k}^j = c^j \widehat{M}_{-k}^j. \tag{68}$$

However since the cluster height is large but drops steeply if we focus on smaller windows of observation, i.e. windows of slightly smaller order of magnitude, we look at $\widetilde{M}_{-j-1+m}^j$ as process in m.

Theorem 10. *(concentrated clustering)*
Assume that $c_k = c^k$ with $c < 1$. Then

$$\mathcal{L}\left(\left(\widetilde{M}_{-j-1+m}^j\right)_{m\in\mathbb{N}}\right) \underset{j\to\infty}{\Longrightarrow} \mathcal{L}\left(\left(\widetilde{M}_m^\infty\right)_{m\in\mathbb{N}}\right), \tag{69}$$

where $(\widetilde{M}_m^\infty)_{m\in\mathbb{N}}$ is a time inhomogeneous Markov chain starting in θ and with transition kernels K_m at time m given by:

$$K_m(\theta, d\theta') = \Gamma_\theta^{c^{-m}, x}(d\theta') \tag{70}$$

and with the property

$$\widetilde{M}_m^\infty \underset{m\to\infty}{\longrightarrow} \widetilde{M}_\infty^\infty \quad \text{and } E\widetilde{M}_\infty^\infty = \theta. \qquad \square \tag{71}$$

(ii) Universality properties of the cluster formation

In the last paragraph we focused on the special case $g(x) = dx$. The natural question is to what extent these results can be generalized to a broader class of diffusion functions? The answer is that indeed we do have here a fairly large universality class for the pattern of cluster formation. The property we need here is the following:

$$g(x)/x \xrightarrow[x \to \infty]{} d^* \in (0, \infty). \tag{72}$$

In that case (see [DG96]):

Theorem 11. *(Universality of cluster formation)*
If (72) holds, then Theorems 9 and 10 still hold. □

The key point here is that the functions g_k defined in connection with the interaction chain converge to d^* for $k \to \infty$.

If we adapt the scaling in Theorem 9, 10 we can still get a version of Theorem 11 for functions g such that $a_k g_k(x) \to x$ for some suitable sequence a_k. Such a property can be verified for $g(x) \sim x^\alpha$ as $x \to \infty$ and $\alpha \in (0, 2)$ (see [BCGH97]). We shall discuss in subsection 5.2 the background of both these facts.

(iii) Multitype and infinite type cluster formation
The analysis of the cluster formation just given works also in the class of total mass-dependent branching processes the results being the obvious generalisations. For other classes of models like mutually catalytic branching and the Fleming-Viot model one obtains equivalent phenomena for the forming of monotype regions, namely diffusive clustering and concentrated clustering for the extensions of the regions where one type prevails, but these phenomena occur for different coefficients $(c_k)_{k \in \mathbb{N}}$. We do not have the space to explain this in detail and refer the reader to [DGV95] and [CDG03]. In catalytic models different type behavior occurs, compare [GKW99] and a different type of analysis is needed.

5 Universality

5.1 A Nonlinear Map and Its Orbit

In view of the transition kernel of the interaction chain (see (51)), which depends only on $(c_k)_{k \in \mathbb{N}}$ and on the sequence of diffusion functions, we can define the following nonlinear map $\mathcal{F}_c$ generating this sequence.

To define first the domain of this map consider in the one-type case the following class of functions

$$\mathcal{G} = \{g : \mathbb{R}^+ \to \mathbb{R}^+ | g \text{ satisfies (i) -(iii)}\} \tag{73}$$

$$(i) \quad g(0) = 0 \text{ and } g(x) > 0, \quad \text{for } x > 0$$
$$(ii) \quad g(x)/x^2 \xrightarrow[x \to \infty]{} 0$$
$$(iii) \quad g \text{ is locally Lipschitz.}$$

Then $\mathcal{F}_c$ is defined by an nonlinear integral transformation:

Definition 4. *(Nonlinear map $\mathcal{F}_c$)*

$$(\mathcal{F}_c g)(\theta) = \int_{\mathbb{R}^+} g(x) \Gamma_\theta^{c,g}(dx), \quad \theta \geq 0. \quad \square \tag{74}$$

It can be shown that $\mathcal{F}_c$ maps $\mathcal{G} \hookrightarrow \mathcal{G}$. This allows to iterate $\mathcal{F}_c$ and to consider the orbit

$$\{(\mathcal{F}_c)^n(g),\ n \in \mathbb{N}\}. \tag{75}$$

Then defining

$$\tilde{g}_d(x) = dx, \quad d \in (0, \infty) \tag{76}$$

we find functions in $\mathcal{G}$ which are *fixed points* and in fact we get all of them. It was shown in ([BCGH97]):

Theorem 12. *(Fixed points)*

$$\mathcal{F}_c \tilde{g}_d = \tilde{g}_d. \tag{77}$$

No other fixed points of $\mathcal{F}_c$ then the ones given in (76) exist in $\mathcal{G}$. □

These results can be proved using that $\Gamma_\theta^{c,d}(\cdot)$ has a density w.r.t. Lebesgue measure which can be given explicitly, namely its value in θ equals

$$\frac{1}{Z} \cdot \exp\left(-\int\limits_x^\theta \frac{\theta - y}{g(y)}\right). \tag{78}$$

We can define $\mathcal{F}_c$ also in the multi-dimensional and infinite-dimensional case by defining:

$$\mathcal{F}_c(H) = \int\limits_{[0,\infty)^k} \Gamma_\theta^{c,H}(dx) G(x) \tag{79}$$

where

$$G(x) = (x^{(\ell)} h^\ell(x) \delta_{\ell,m})_{\ell = 1, \cdots, k} \tag{80}$$

and $\Gamma_\theta^{c,H}$ is the invariant measure of the diffusion (38).

However there is in general no simple formula anymore giving $\Gamma_\theta^{c,G}$ like in (78) and much more important now a whole collection of new classes of fixed points arises. Namely consider the case $k = 2$, then besides the fixed point to be expected, where all types follow independent branching dynamics already in the case of two types other fixed points arise, which correspond for example to *catalytic branching, mutually catalytic branching*.

The multi-dimensional situation for branching models is in fact quite complex. Different from the case of resampling models, where one finds only one class of fixed points, namely the Fleming-Viot dynamics, we are faced now with a whole collection of classes of fixed points already in the case of two types and already in the case where we require that the diffusion function vanishes exactly either on both axes, one axe or the point 0. Thus the multi-dimensional components bring a qualitative change of the picture. One finds at least the following fixed points (see Cox, Dawson and Greven in [CDG03]):

- two-type independent branching
- catalytic branching
- mutually catalytic branching
- two-type joint branching in pairs.

These correspond to diffusion matrices in state $(u, v) \in [0, \infty)^2$:

$$\sigma^{(1)}(u, v) = \begin{pmatrix} u & 0 \\ 0 & v \end{pmatrix}, \quad \sigma^{(2)}(u, v) = \begin{pmatrix} u & 0 \\ 0 & uv \end{pmatrix} \text{ or } \begin{pmatrix} uv & 0 \\ 0 & v \end{pmatrix} \tag{81}$$

$$\sigma^{(3)} = \begin{pmatrix} uv & 0 \\ 0 & uv \end{pmatrix}, \quad \sigma^{(4)} = \begin{pmatrix} uv & uv \\ uv & uv \end{pmatrix}.$$

Further fixed points can now be constructed by forming linear combinations of these or by multiplying with diagonal matrices.

The situation is different if we study evolutions of relative frequencies of types of some population of fixed size, that is socalled multi-type resampling systems where independent of the number of types one finds a unique fixed shape, the Fisher-Wright respectively in infinite-type situations the Fleming-Viot process (see [DGV95]). For related work on other types of multi-dimensional diffusions see [HS98] and [Sw00].

5.2 Fixed Points, Fixed Shapes and Their Domain of Attraction

If one has identified fixed points (or fixed shapes) this is really useful only if they attract other orbits. For one type models one has a fairly complete picture. It turns out ([BCGH97]) that the orbit in (75) approaches for many initial points the fixed points in the following sense:

Theorem 13. *(Attracting orbit)*

(a) If $g(x)/x \to d$ as $x \to \infty$, then

$$(\mathcal{F}_c)^n(g) \longrightarrow \tilde{g}_d \quad \text{in sup-norm.} \tag{82}$$

(b) If $g(x) \sim x^\alpha$ as $x \to \infty$ for some $\alpha \in (0, 2)$ then there exists a sequence (a_n) such that pointwise

$$a_n(\mathcal{F}_c)^n(g) \longrightarrow \tilde{g}_1. \tag{83}$$

For $\alpha \in (0, 1)$ or $(1, 2)$:

$$a_n \sim n^{(1-\alpha)/2-\alpha} \qquad \square \tag{84}$$

This result shows that for the map $\mathcal{F}_c$ the orbits arising from different starting points g look asymptotically like the one arising from the fixed points $\tilde{g}_d$ or has this property after rescaling. Note that the diffusion function $\tilde{g}_d$ corresponds to interacting Feller's branching diffusions.

Remark   Analog results hold for resampling models. For two types, i.e. one-dimensional components (sum of frequencies is 1), one has the fixed shape

$g(x) = dx(1 - x)$ on $[0, 1]$ (Fisher-Wright) which is unique and attracts everything after rescaling. Only finer convergence properties depend on the behavior of g close to 0 or 1. See [BCGH95] for details.

The discussion on the set of fixed points for the two-type case in the previous section makes it very clear that the determination of the universality classes, i.e. the sets $\mathcal{A}(\sigma^*)$, σ^* fixed point with

$$\sigma \in \mathcal{A}(\sigma^*) \quad \text{implies } (\mathcal{F}_c)^n(\sigma) \xrightarrow[n \to \infty]{} \sigma^*, \tag{85}$$

is an extremely delicate matter in the multitype case since we have several centers of attraction to deal with. This is presently investigated by Dawson, Greven, den Hollander and Swart. Other types of models with multidimensional components and exhibiting also more complexity in their behavior have been studied by den Hollander and Swart [HS98].

6 Hierarchical Mean-Field Continuum Limit

This chapter goes through the stages described in sections 3-5 for the analysis of the longtime behavior, now for the question of the small-scale properties of the spatial continuum limit of interacting systems. Namely one interesting point of the previous analysis is that these ideas of renormalization can also be applied to construct a *hierarchical mean-field continuum limit* which allows to address the question of the existence of the spatial continuum limit and its qualitative small-scale properties (density versus singularity, etc.) by associating another time-inhomogeneous Markov chain, the so-called *small-scale characteristic* with our interacting system. These ideas were developed in [CDG03] for mutually catalytic branching even though the method is applicable to all the processes we have discussed here. We shall demonstrate this here for mutually catalytic branching.

6.1 Hierarchical Mean-Field Continuum Limit

We adhere here to the notation in the literature on the process of mutually catalytic branching and write for the system given via (18) and (10):

$$X(t) = (u_\xi(t), v_\xi(t))_{\xi \in \Omega_N}. \tag{86}$$

We start with the ingredients needed to formulate the problem of the continuum limit for the hierarchical group and then in two further steps carry out the hierarchical mean-field continuum limit.

(i)   Basic ingredients for the continuum limit
We first find the analogue for Ω_N of the continuum limit. For models indexed by Z^d, one passes to a model on $\mathbb{R}^d$ by passing through a sequence of approximations by rescaled systems indexed by εZ^d with $\varepsilon \to 0$. We introduce

the analogues of $Z^d, \varepsilon Z^d$ and $\mathbb{R}^d$ in our context as Ω_N, Ω_N^j ($j \in \mathbb{N}$ replaces ε), Ω_N^∞ as follows. Define:

$$\Omega_N^j = \left\{ (\xi_\ell)_{\ell \in Z, \ell \geq -j} \,\big|\, \xi_\ell \in \{0, 1, \ldots, N - 1\}, \exists \ell_0 : \xi_\ell = 0 \quad \forall \ell \geq \ell_0 \right\}, \quad (87)$$

$$\Omega_N^\infty = \left\{ (\xi_\ell)_{\ell \in Z} \,\big|\, \xi_\ell \in \{0, 1, \ldots, N - 1\}, \exists \ell_0 : \xi_\ell = 0 \quad \forall \ell \geq \ell_0 \right\}. \quad (88)$$

For notational convenience we identify from here on

$$\Omega_N \text{ with } \Omega_N^0. \quad (89)$$

The 0-element is always the 0-sequence and the hierarchical distance between two points in Ω_N^j or Ω_N^∞, which is an *ultrametric*, is defined as follows. Given two elements ξ, ξ' define:

$$k^* = k^*(\xi, \xi') = (\min(k \in Z : \xi_m = \xi'_m \,\forall m \geq k)). \quad (90)$$

Then we define the distance on Ω_N^j as follows and on Ω_N^∞ by putting $j = \infty$:

$$d(\xi, \xi') = \begin{cases} 2^{k^*} & \text{for } k^* < 0, \quad k^* > -j \\ k^* & \text{for } k^* \geq 0, \\ 0 & \text{for } k^* = -j. \end{cases} \quad (91)$$

This leaves us with the *continuum hierarchical group* Ω_N^∞ and a collection of approximating countable groups Ω_N^j. This set-up allows us to define for every fixed N a continuum limit. In addition however we have the parameter N at our disposal producing for $N \to \infty$ a situation close to the euklidian case with $d = 2$, precisely for fixed N we are analog to dimension 2 $(logN)/(logN - logc))$.

(ii)   The rescaling

As on the lattice we will start from a system indexed by Ω_N and associate it with systems on the "finer" group Ω_N^j which we then embed in Ω_N^∞ as follows. We first identify Ω_N^j with a subset of Ω_N^∞ and then extend the system from this subset to one defined on the whole set Ω_N^∞. For this purpose set the values in all points $(\xi_\ell)_{\ell \in Z}$ in Ω_N^∞, which differ only for $\ell < -j$, equal to the values of the corresponding element in Ω_N^j with the same letters for $l \geq -j$. Hence from a given system of interacting mutually catalytic diffusions on Ω_N^j we obtain the natural extension of this system to one on the finer lattice Ω_N^∞. This system (depending on the parameter j is denoted:

$$\left(U_t^j, V_t^j \right)_{t \geq 0} = \left(\left\{ \left(u_\xi^j(t), v_\xi^j(t) \right), \xi \in \Omega_N^\infty \right\} \right)_{t \geq 0}. \quad (92)$$

We can also view this random variable as a *pair of random measures* on Ω_N^∞. Again we denote all quantities after this embedding in Ω_N^∞ by the same symbols.

Next we construct in four steps the rescaled system. The main task is to introduce a *rescaling* of the system (92). This involves scaling the migration and the branching by introducing both *space-time rescalings* and *scalings of the parameters*. We now explain these four operations in Step 0 and then later use them to rescale the system in a next step to obtain a continuum limit. Then we introduce a sequence of time-space renormalizations of these objects in two further steps to get the hierarchical mean-field continuum limit.

Step 0   We begin with the key ingredients.

(α)   On lattice models one rescales the random walk in space (later in time) to obtain Brownian motion or α-stable processes. On the hierarchical group this works as follows. First, we define by *shrinking* the rescaled random walk on Ω_N^j associated with the sequence $(c_k)_{k \in \mathbb{N}}$ which defined the original walk on Ω_N. Here we introduce a convention to embed Ω_N in Ω_N^∞ by putting $\xi_k = 0$ for $k < 0$ and Ω_N^j in Ω_N^∞ by doing this for $k < -j$. We now base the shrunk jump mechanism on the above notion of distance and by assigning the same rates now to the jump by $\xi' = (\xi_{k-j})_{k \in Z}$ instead of the one given by $\xi = (\xi_k)_{k \in Z}$. Formally we proceed as follows.

Let τ_j denote the operation mapping Ω_N^∞ in Ω_N that shifts sequences in Ω_N^∞ exactly j places the right and then cutting off components with negative index. Then we define the "shrunken" random walk on Ω_N^j (and consequently on Ω_N^∞ by embedding) by replacing $a(\cdot, \cdot)$ by $a_j(\cdot, \cdot)$, with,

$$a_j(0, \xi) := a(0, \tau_j \xi), \quad \xi \in \Omega_N^\infty. \tag{93}$$

This shrinks size k jumps to size $(k - j)^+$ jumps.

(β)   In order to complete the space-time rescaling we must speedup time so that the distance travelled by the random walker with transitions governed by the $(c_k)_{k \in \mathbb{N}}$ has a nontrivial limit as $j \to \infty$ in the sense that it moves distance $O(1)$ in times $O(1)$. To do this we make the time change

$$t \to \frac{N^j}{c_j} t \ . \tag{94}$$

If we want to obtain scaling limits for the random walk based on $(\alpha) - (\beta)$ we need additional regularity assumptions on the random walk kernel:

$$(c_{j+k}/c_j) \underset{j \to \infty}{\longrightarrow} c_k^*, \quad \text{for all } k \in Z. \tag{95}$$

In fact $(c_k^*)_{k \in Z}$ has always the form

$$c_k^* = (c^*)^k , \ k \in Z \tag{96}$$

and for $c_k = c^k$ we get $c^* = c$.

Under assumption 95 the random walks on Ω_N^j can be embedded in Ω_N^∞ and converge to a *Lévy process* on Ω_N^∞ as $j \to \infty$ which is characterized by $(c_k^*)_{k \in Z}$. Such a Lévy process is a process with independent stationary increments and countably many small jumps, the latter are regulated by $(c_k^*)_{k \in Z}$- while the recurrence, transience properties are regulated by the large jumps and hence by $(c_k^*)_{k \in Z^+}$. In fact c_k^* has always the form and for $c_k = c^k$ we get $c^* = c$.

(γ) In addition (and different from the case of ordinary branching) we then have to scale the branching rate to obtain nontrivial rescaling limits of the interacting mutually catalytic diffusions. We scale the branching rate $\gamma = (\gamma_1, \gamma_2)$ by

$$\gamma \to c_j \cdot \gamma. \tag{97}$$

Note that only in the case $c_j \equiv const$ the rates remain unchanged.

(δ) For each $j \in \mathbb{N}$ we have the system (92) of mutually catalytic processes indexed by Ω_N^j in which the distance between sites is 2^{-j} (which as explained above can also be identified with balls of radius 2^{-j} in Ω_N^∞). In order to have a common framework to study the limit $j \to \infty$ we now introduce the collection of spatial averages over small balls of orders k from the unit-ball down to 2^{-j}-balls, denoted

$$(u_\xi^j, v_\xi^j) \longrightarrow (u_{\xi,k}^j, v_{\xi,k}^j), \quad k = -j, -j+1, ..., 0 \tag{98}$$

which are obtained by averaging the components over balls of radius 2^k around ξ.

We now define with the above ingredients in three steps a sequence of *rescaled* systems indexed with Ω_N^∞ which correspond to the systems indexed by Ω_N^j. First we make the appropriate choice of the parameters c_k, γ and the right time scaling to define the continuum limit $j \to \infty$. In order to analyse the resulting limit we introduce in two further steps a renormalization scheme for that object. We choose the appropriate time and space scale followed by a *rearrangement* of levels as follows.

Step 1 Recall the mechanism of mutually catalytic branching on the hierarchical group for a given sequence $(c_k)_{k \in \mathbb{N}}$. Starting from this we associate with the system indexed by Ω_N one indexed by Ω_N^j and then we consider a new system indexed by Ω_N^j and denoted by

$$(\tilde{u}_\xi^j(t), \tilde{v}_\xi^j(t))_{\xi \in \Omega_N^j}, \tag{99}$$

which is based on parameters obtained by the above rescalings of migration, i.e. the shrinking given in (α) (which induces a change of the coefficients c_k/N^k in the drift term of the system of SDE's) and of the branching as in (γ) (which induces a change of coefficients in the diffusive term). Next take the system in (99) and rescale time corresponding to the index j (as in (β)):

$$\left(\tilde{u}_{\xi,-j}^j\left(c_j^{-1}N^j t\right), \tilde{v}_{\xi,-j}^j\left(c_j^{-1}N^j t\right)\right)_{\xi \in \Omega_N^j}. \tag{100}$$

Note that this is the natural time scale for the system at spatial scale $O(1)$.

This completes now the construction of the rescaled system, indexed by the "continuum hierarchical group" Ω_N^∞ which, if we let $j \to \infty$ does the very same thing as is done in lattice models when we let the parameter $\varepsilon \to 0$, passing from the εZ^d to the $\mathbb{R}^d$-system, the spatial continuum limit.

Step 2   Next we begin introducing the renormalization scheme. First rescale space according to (δ), namely we consider the system at different spatial scales corresponding to the index k by taking *averages* over balls of radius 2^{-k}, (i.e. a set of elements $d(\xi', \xi) \leq 2^{-k}$ for some ξ) and in the natural time scale. Such a ball contains N^k elements and hence the natural time scale of such an average is obtained by speeding time by the factor of N^k (relative to the natural time scale of level j). Note that now we are counting from $-j$ upwards. In formulas we obtain recall (94) and (99) the collection:

$$\left(\tilde{u}_{\xi,-k}^j \left(c_j^{-1} N^{(j-k)} t \right), \, \tilde{v}_{\xi,-k}^j \left(c_j^{-1} N^{(j-k)} t \right) \right)_{\xi \in \Omega_N^j} := \tag{101}$$

$$\left(\tilde{u}_{\xi,-k}^{*,j}(t), \tilde{v}_{\xi,-k}^{*,j}(t) \right)_{\xi \in \Omega_N^j}, \qquad 0 \leq k \leq j.$$

Here we do not fix t but on each level we consider this object as a process in its natural time scale. In other words we consider the following *path-valued* collection:

$$\left\{ \left(\tilde{u}_{\xi,-j+k}^{*,j}(t), \tilde{v}_{\xi,-j+k}^{*,j}(t) \right)_{t \geq 0}, \, \xi \in \Omega_N^j, \, k = 0, 1, \ldots, j \right\}. \tag{102}$$

Step 3   Now we come to the last operation. We now have the rescaled object and the question arises in what sense we expect to obtain convergence as $j \to \infty$ and $N \to \infty$. The point is that we should view the object in (102) either as a pair of measure-valued processes for that purpose or as sequences of averaged densities. In any case this means, if we want to pass to the continuum limit and investigate its small scale properties we have to zoom from macroscopic windows of observation down to smaller ones. This implies that we have to *rearrange this sequence of block averages by transforming* the index as follows:

$$k \to j - k. \tag{103}$$

At the same time we embed the Ω_N^j system in Ω_N^∞ by giving it a constant value in 2^{-j}-blocks. (If we want to view everything as a process of pairs of measures, then we assign constant density with respect to Haar measure in 2^{-j}-balls, where the the *Haar measure* of a ball of radius 2^{-j} is N^{-j}). Then we finally arrive from the original Ω_N^∞-indexed system in (92) at the object:

$$\left\{ \left(u_{\xi,k}^{*,j,N}(t), v_{\xi,k}^{*,j,N}(t) \right)_{t \geq 0}, \, \xi \in \Omega_N^\infty \right\}_{k=0,1,\ldots,j} \tag{104}$$

which describes the average of the $(\tilde{u}_\xi^j(\cdot), \tilde{v}_\xi^j(\cdot))_{\xi \in \Omega_N^j}$ over a ball of radius 2^{-k} in the natural time scale (for such balls) and where we have added the superscript N to stress the dependence on N.

(iii)   The hierarchical mean-field continuum limit results
In order to pass to the continuum limit we would now have to let $j \to \infty$. The limiting object we could then try to simplify by considering the limit $N \to \infty$ and the behavior of the time-space renormalized object. Both these limits are fairly subtle. We shall consider here the *hierarchical-mean field continuum limit* by *first* letting $N \to \infty$ and then only *later* $j \to \infty$.

To define convergence of our various systems, we adopt the point of view that we deal with a sequence of process-valued fields indexed by Ω_N^∞, respectively Ω_∞^∞. (Alternatively we could take the point of view of pairs of measure-valued processes). In order to be able to state our convergence results of these objects corresponding to different index sets, one has to introduce a suitable convention, which we summarize at this point again.

Note that we can embed the sets $\Omega_N^j, \Omega_\infty^j, \Omega_\infty^\infty$ into each other. The set Ω_N^j is subset of Ω_∞^j for each N. Taking an element $\xi \in \Omega_N^j$ and continuing the sequence to the left by 0, we get an element in Ω_N^∞. By continuing a field indexed by Ω_N^j and embedded first in Ω_N^∞ by putting it constant in all ξ in Ω_N^j which differ only in digits to the left of $-j$ and then putting it equal to 0 in all elements containing a digit bigger than N, we get a canonical field indexed by Ω_∞^∞.

Now we can use the analysis described in subsection 4 to obtain results for j-level systems in time scales of the form $t_0 + N^{-1}t, 1 + \ldots + t_{j-1}N^{-(k-1)} + N^{-k}t$ with $k = 1, \cdots, j$ which are now the building blocks of the rescaling on the level j if we observe block averages of size k. The point is that the analysis carried out in subsection 4 depends only on the relation between one scale to the next and hence the main work has been done to prove [CDG03]:

Theorem 14. *(Basic limit result for hierarchical mean-field spatial continuum limit)*
(a)   *We have existence of the two successive limits $N \to \infty$ and then $j \to \infty$:*

$$\mathcal{L}\left[\left\{\left\{\left(u_{\xi,k}^{*,j,N}(t), v_{\xi,k}^{*,j}(t)\right)_{t\geq 0}\right\}_{\xi\in\Omega_N^\infty}\right\}_{k=0,1,\ldots,j}\right] \underset{N\to\infty}{\Longrightarrow} \tag{105}$$

$$\mathcal{L}\left[\left\{\left(\left(M_{\xi,k}^{*,j}(t)\right)_{t\geq 0}\right)_{\xi\in\Omega_\infty^\infty}\right\}_{k=0,\ldots,j}\right] \tag{106}$$

$$\mathcal{L}\left[\left\{\left(\left(M_{\xi,k}^{*,j}(t)\right)_{t\geq 0}\right)_{\xi\in\Omega_\infty^{*,\infty}}\right\}_{k=0,1,\ldots,j}\right] \underset{j\to\infty}{\Longrightarrow} \tag{107}$$

$$\mathcal{L}\left[\left\{\left(\left(M_{\xi,k}^{*,\infty}(t)\right)_{t\geq 0}\right)_{\xi\in\Omega_\infty^{*,\infty}}\right\}_{k\in Z+}\right]. \qquad (108)$$

(b)   We can identify the limit on the r.h.s. of (107) above explicitly. For fixed ξ and fixed sequence $(t_k)_{k=0,1,...}$ of strictly positive time points the processes evaluated at these times, define a sequence $M_{\xi,k}^{*,\infty}$, $k = 0,1,\ldots.$ We have:

$$(M_{\xi,k}^{*,\infty})_{k=0,1,...} \text{ is a Markov chain on } (\mathbb{R}^+ \times \mathbb{R}^+), \qquad (109)$$

the transition kernel is given by (recall (95) for c_k^*)

$$K_k(\theta,\cdot) = \Gamma_\theta^{c_k^*,\gamma} \qquad (110)$$

and the initial state is $\mathcal{L}\left[(u_{\xi,0}(t_0), v_{\xi,0}(t_0)\right].$ □

We finally coin a terminology for the analogue of the interaction chain for the small-scale behavior of the system in the hierarchical mean-field continuum limit (abbreviated HMFCL).

Definition 5. *(HMFCL-small scale characteristics)*
The term small-scale characteristics *of the interacting mutually catalytic diffusions will be used to denote the Ω_∞^∞-indexed random field of random processes:*

$$\left(\left(M_{\xi,k}^{*,\infty}(t)\right)_{t\geq 0}, \quad \xi\in\Omega_\infty^{*,\infty}\right). \quad □ \qquad (111)$$

The study of properties of this object will be the main point of subsections 6.2 - 6.3.
Remark   We have simplified the HMFCL here by blending out the term describing the fluctuations of the rescaled process and its limit. This finer picture can be found in [CDG03].

6.2 Dichotomy for the Continuum Limit

Turn now to the spatial continuum limit of the interacting system described in subsection 2.2. and investigate its *small scale properties*. As already mentioned, there is also a dichotomy present in the local behavior of the continuum limit. For the Dawson-Watanabe process on $\mathbb{R}^d$ this is absolute continuity versus singularity of the states (corresponding to $d = 1$ versus $d \geq 2$). For mutually catalytic branching this has a somewhat different form (as it has for the longtime behavior).

We describe now how in this model the dichotomy of locally separated populations versus locally overlapping is reflected in the behavior of the small-scale characteristic of mutually catalytic branching. Return to the ordinary continuum limit and suppose it exists and that both populations have densities. Then the most important dichotomy is whether or not the continuum limit has densities which are positive for both populations at a given location as opposed to being positive on disjoint sets and hence the populations being spatially segregated. We will have to see here how such a property translates to one for the hierarchical mean-field continuum limit. This runs as follows:

Theorem 15. *(Overlap versus segregation of populations in the continuum limit)*

(a) *For every ξ:*

$$\left(M_{\xi,\infty}^{*,\infty}(t) \right)_{t \geq 0} \quad \text{is constant in time.} \tag{112}$$

(b) *Case $\sum\limits_{k=0}^{-\infty} (c_k^*)^{-1} < \infty$ (overlap).*
Fix ξ and assume that $M_{\xi,0}^{*,\infty}(0) \in int(\mathbb{R}^+)^2$, then

$$M_{\xi,\infty}^{*,\infty}(t) \in int(\mathbb{R}^+)^2 \tag{113}$$

for almost all $t \geq 0$, with positive probability.

(c) *Case $\sum\limits_{k=0}^{-\infty} (c_k^*)^{-1} = +\infty$ (segregation).*
Fix $\xi \in \Omega_\infty^{*,\infty}$. Then with $(B_s)_{s \geq 0}$ planar Brownian motion with $B_0 = M_{\xi,0}^{*,\infty}(t)$, H is the hitting time of the axes, we get:

$$M_{\xi,\infty}^{*,\infty}(t) \in \partial(\mathbb{R}^+)^2, \quad \mathcal{L}\left(M_{\xi,\infty}^{*,\infty}(t) \right) = \mathcal{L}(B_H). \qquad \square \tag{114}$$

Once we know that we have the dichotomy described above we can ask for finer properties. The most important question in the regime of overlap is to describe the distribution of the product of the two densities, which drives the local fluctuations. In the segregated regime we have no densities and we somehow have to describe the behavior of the overlap accumulated in larger block and how it shrinks as we zoom in on a point.

6.3 Hot Spot Formation

We focus here for a finer analysis of the segregated regime. In order to carry out this finer analysis analog to the analysis of the cluster formation we described for the longtime behavior, we would like to investigate the behavior of the small-scale characteristic on events where the population still coexists by "time" k which corresponds to a k-block in the original system. Here as in subsection 4.2 again the concept of size-biasing is useful. Only now we size-bias at time k by the function uv, i.e. the product of the components of $M_{\xi,k}^{*,\infty}(t)$ and to let $k \to \infty$. This means we study the probability measure on sequences $((u_1, v_1), \ldots, (u_k, v_k))$ given by

$$P_k^{(uv)}(A) = \frac{1}{\theta_1 \theta_2} \int_L u_k v_k 1_A d\mu, \quad A \in ((\mathbb{R}^+)^2)^k, \; L = ((\mathbb{R}^+)^2)^k \tag{115}$$

with

$$(\theta_1, \theta_2) = M_{\xi,0}^{*,\infty}(t), \quad (\theta_1, \theta_2) \in \text{int}(\mathbb{R}^+)^2. \tag{116}$$

Note however that uv is a harmonic function for *all* kernels of the form $K(\theta, \bullet) = \Gamma_\theta^{c,\gamma}(\bullet)$. Recall that the *h-transform* of a Markov chain with transition kernel P is generated by the kernel

$$(h(x))^{-1}P(x, dy)h(y), \quad h \text{ positive harmonic.} \tag{117}$$

In the time inhomogeneous case the same construction works, if we have a function which is harmonic for all kernels P_k. This means that

$$\left\{ P_k^{(uv)} \right\}_{k \in \mathbb{N}} \quad \text{is a consistent family generating a law } P^{(uv)} \tag{118}$$

$$\text{on } ((\mathbb{R}^+)^2)^{\mathbb{N}},$$

$$P^{(uv)} = h\text{-transform of } \mathcal{L}\left[\left(M_{\xi,i}^{*,\infty} \right)_{i \in \mathbb{N}} \right] \tag{119}$$

and its projections on $\{0, 1, \ldots, k\}$ give the size-biased law w.r.t. uv and with respect to the final time k.

The task is then to introduce as a function of k a scaling of the expression uv and to study this random variable at a fixed time i under the law of this h-transform $P_k^{(u,v)}$ as $k \to \infty$. This would describe the spatial extension, frequency and heat content of the hot spot.

For classical branching, i.e. the small-scale characteristic corresponding to the *super process* we obtain in the limit the Feller diffusion with constant immigration. Precisely we get in the forthcoming work of Dawson, Greven and Zähle [DGZ] in that simpler case

$$(P_{\lfloor \alpha k \rfloor}^u)_{\alpha \in [0,1]} \underset{K \to \infty}{\Longrightarrow} \mathcal{L}\left((z_\alpha)_{\alpha \in [0,1]} \right) \tag{120}$$

with

$$dz_t = 1dt + \sqrt{z_t}dw_t. \tag{121}$$

In the context of mutually catalytic branching some new phenomena are exhibited which are also treated in [DGZ].

6.4 Universality

A very natural question if one carries out the continuum limit is to find all possible processes which can occur in the limit and to identify all processes leading to the same continuum limit which are in the class of models we have described in subsection 1. It is then of course expected that one gets a great reduction in complexity. How does this problem appear if we pass to the HMFCL? Can we identify all limiting objects which can occur and the class of processes giving the same limit?

If one looks a bit more careful how the small-scale characteristic arises, one discovers that the rescaling procedure will give the same object for all diffusion matrices such that indeed the orbit under $\mathcal{F}_{c_k^*}$ converges as $k \to \infty$ to a limit which is one of our fixed points that is combinations of branching, catalytic branching and mutually catalytic branching. Therefore we are again back to an analytical problem, which is of the very same nature as the one we have to study for the longtime behavior. Namely identify orbits of the nonlinear map $\mathcal{F}_{c_k}^{(k)}$ as a function of the sequence $(c_k)_{k \in \mathbb{N}}$ only that here the sequence arises as $(c_k^*)_{k \in \mathbb{N}}$. However for $c_k = c^k$ we have $c_k^* = c^{-k}$ and hence the problem is not very different. This leads to the conjecture that the continuum limits (with diffusion operator vanishing exactly if the population vanishes, one population vanishes or both populations vanish) which can occur are combinations of branching, catalytic and mutually catalytic branching parallel to the situation for the longtime behavior. This is followed up in [DGHS].

A fundamental question arising next is whether the HMFCL is the limit of the continuum limits for $N \to \infty$, in other words are the limits $j \to \infty$ and $N \to \infty$ exchangeable. We expect yes, this is addressed in ongoing work [DGZ]. There it is at least proved that this is true for the branching case.

References

[BCGH95]   J.B. Baillon, Ph. Clément, A. Greven, F. den Hollander. *On the attracting orbit of a nonlinear transformation arising from normalisation of hierarchically interacting diffusion, Part I: The compact case*, Canadian Journal of Mathematics, Vol. **47**, 3-27 (1995).

[BCGH97]   J.B. Baillon, Ph. Clément, A. Greven, F. den Hollander. *On the attracting orbit of a nonlinear transformation arising from renormalisation of hierarchically interacting diffusion. Part 2: The noncompact case*, Journal of Functional Analysis, vol. **146**, No. 1, 236-298 (1997).

[CDG03]   J.T. Cox, D.A. Dawson, A. Greven: *Mutually catalytic super branching random walks: Large finite systems and renormalization analysis.* Memoirs of AMS, Vol. **171**, No. 809, (2004).

[CFG96]   J.T. Cox, K. Fleischmann, A. Greven: *Comparison of interacting diffusions and an application to their ergodic theory.* Probability Theory rel. Fields, **105**, 513-528, (1996).

[CG94]   J.T. Cox, A. Greven: *The finite systems scheme: An abstract theorem and a new example.* Proceedings of CRM Conference on Measure valued processes, Stochastic partial differential equation and interacting Systems. CRM and Lecture Notes Series of the American Mathematical Society. Vol. **5**, 55-68 (1994).

[CGS95]   J.T. Cox, A. Greven, T. Shiga: *Finite and infinite systems of interacting diffusions.* Prob. Theory and Rel. Fields, **103**, 165-197 (1995).

[DG93a]   D.A. Dawson, A. Greven: *Multiple Time Scale Analysis of Hierarchically Interacting Systems*, Stochastic Processes, A Festschrift in Honour of Gopinath Kallianpur. S. Cambanis, J.K. Gosh, R.L. Karandikar, P.K. Sen eds. Springer, 51-50 (1993).

[DG93b]   D.A. Dawson, A. Greven: *Multiple Scale Analysis of Interacting Diffusions*, Prob. Theory and Rel. Fields **95**, 467-508 (1993).

[DG93c]   D.A. Dawson, A. Greven: *Hierarchical models of interacting diffusions: Multiple time scales, Phase transitions and cluster formation*, Prob. Theor. and Rel. Fields, **96**, 435-473 (1993).

[DG96]   D.A. Dawson, A. Greven: *Multiple space-time scale analysis for interacting branching models*, EJP **Vol 1**, paper 6 (1996).

[DG99]   D.A. Dawson, A. Greven: *Hierarchically interacting Fleming-Viot processes with selection and mutation* EJP, vol. **4**, paper 14, 1-84 (1999).

[DG03]   D.A. Dawson, A. Greven: *State dependent multi-type spatial branching processes and their longtime behavior*, EJP, Vol. **8**, no. 4, 1-93 (2003).

[DG03b]   D.A. Dawson, A. Greven: *On the effect of migration in spatial Fleming-Viot models with selection and mutation.* (In preparation 2003).

[DGHS]   D. A. Dawson, A. Greven, F. den Hollander, J. Swart: *The renormalization transformation for two-type branching models*, in preparation (2003).

[DGV95]   Dawson D.A., A. Greven, J. Vaillancourt: *Equilibria and Quasi-equilibria for infinite systems of Fleming-Viot processes.* Transactions of the American Math. Society, Vol. **347**, Number 7, 2277-2361, (1995).

[DGW01]   D.A. Dawson, L.G. Gorostiza and A. Wakolbinger: *Occupation time fluctuations in branching systems.* J. Theoretical Probab., Vol.**14**, 729-796, (2001).

[DGW]   D.A. Dawson, L.G. Gorostiza and A. Wakolbinger: *Hierarchical Random Walks*, to be published in Fields Institute Communications and Monograph Series, AMS (2004).

[DGW03b]   D.A. Dawson, L.G. Gorostiza and A. Wakolbinger: *Hierarchical Equilibria of Branching Populations*, [ESI preprint 1393], [ArXiv math.PR/0310229] (2003).

[DGZ]   D. A. Dawson, A. Greven, I. Zähle: *Continuum limits of multitype population models and renormalization.* In preparation (2003).

[DP98]   D.A. Dawson and E.A. Perkins: *Long-time behavior and coexistence in a mutually catalytic branching model*, Ann. of Probab., Vol. **26**, 1088-1138, (1998).

[EF]   Etheridge, A.M., Fleischmann, K.: *Compact interface property for symbiotic branching.* Preprint (2003)

[FG94]   Fleischmann, K., A. Greven: *Diffusive clustering in an infinite system of hierarchically interacting Fisher-Wright diffusions.* Prob. Theory Rel. Fields. **98**, 517-566 (1994).

[FG96]   Fleischmann, K., A. Greven: *Time-space analysis of the cluster-formation in interacting diffusions* EJP, Vol. **1**, Paper 6 (1996).

[HS98]   Hollander, F. den, J. M. Swart: *Renormalization of hierarchically interacting isotropic diffusions*, J. Stat. Phys. **93**, 243-291 (1998).

[GKW99]   A. Greven, A. Klenke, A. Wakolbinger, *The longtime behavior of branching random walk in a catalytic medium*, EJP, Vol **4**, paper 12, 80, pages (electronic), (1999).

[GKW01]   A. Greven, A. Klenke, A. Wakolbinger, *Interacting Fisher-Wright Diffusions in a Catalytic Medium*, Prob. Theory and Rel. Fields, Vol. **120**, No. 1, 85-117, (2001).

[GKW02]   A. Greven, A. Klenke, A. Wakolbinger, *Interacting diffusions in a random medium: comparison and longtime behavior* Stochastic Process. Appl., Vol. **98**, 23–41, (2002).

[GLW03]   A. Greven, V. Limic and A. Winter, *Representation Theorems for Interacting Moran models, Interacting Fisher–Wright Diffusions and Applications*, manuscript 2003.

[K96]   Klenke, A.: *Different clustering regime in systems of hierarchically interacting diffusions*, Ann. Probab. **24(2)**, 660-697 (1996).

[K97]   Klenke, A.: *Multiple scale analysis of clusters in spatial branching models*, Ann. Probab. **25(4)**, 1670-1711 (1997).

[K98]   Klenke, A.: *Clustering and invariant measures for spatial branching models with infinite variance*, Ann. Probab. **26(3)**, 1057-1087 (1998).

[Pf03]   P. Pfaffelhuber: *State-dependent interacting multitype branching systems*, PHD-thesis, Erlangen, Germany (2003).

[SF83]   Sawyer, S., J. Felsenstein: *Isolation by distance in a hierarchically clustered population* J. Appl. Probab. **20**, 1-10 (1983).

[Sw00]   Swart, J.M.: *Clustering of linearly interacting diffusions and universality of their long-time distribution*, Prob. Th. Relat. Fields, Vol. **118**, 574-594 (2000).

[Wi02]   Winter, A.: *Multiple scale analysis of branching under the Palm distribution*, Electron. J. of Probab., Vol. **7**, no. 13, 74 pages, (2002).

[Z01]   I. Zähle: *Renormalization of the Voter Model in Equilibrium*, Annals of Probability, Vol. **29**, No. 3, 1262–1302 (2001).

[Z02]   I. Zähle: *Renormalizations of branching random walks in equilibrium*, Electronic Journal of Probability, Vol. **7**, paper 7, 1-57 (2002).

Stochastic Insertion-Deletion Processes and Statistical Sequence Alignment

Dirk Metzler[1], Roland Fleißner[2],
Anton Wakolbinger[3], and Arndt von Haeseler[4,5]

- FB Biologie und Informatik, Goethe-Universität
  Robert-Mayer-Straße 11-15, D-60054 Frankfurt, Germany
  metzler@informatik.uni-frankfurt.de
- Department of Mathematics, University of Idaho
  P.O. Box 441103. Moscow, ID 83844-1103, USA
  fleissne@uidaho.edu
- FB Mathematik, Goethe-Universität
  Robert-Mayer-Straße 10, D-60325 Frankfurt am Main, Germany
  wakolbin@math.uni-frankfurt.de
- Bioinformatik, Heinrich-Heine-Universität
  Universitätsstr. 1, Geb. 25.02.02, D-40225 Düsseldorf, Germany
  haeseler@cs.uni-duesseldorf.de
- Forschungszentrum Jülich
  John von Neumann-Institut für Computing (NIC), Forschungsgruppe
  Bioinformatik
  D-52425 Jülich, Germany

Summary. The reconstruction of the history of a set of sequences is a central problem in molecular evolutionary biology. Typically this history is summarized in a phylogenetic tree. In current practice the estimation of a phylogenetic tree is a two-step procedure: first a multiple alignment is computed and subsequently a phylogenetic tree is reconstructed, based on the alignment. However, it is well known that the alignment and the tree reconstruction problem are intertwined. Thus, it is of great interest to estimate alignment and tree simultaneously. We present here a stochastic framework for this joint estimation. We discuss a variant of the Thorne-Kishino-Felsenstein model, having equal rates of insertions and of deletions of sequence fragments, for $\ell \geq 2$ sequences related by a phylogenetic tree. Finally, we review novel approaches to tree reconstruction based on insertion-deletion models.

1 Introduction

Sequence Evolution and Alignments Biological (DNA or amino acid) sequences change over time due to the random process of evolution. Two ingredients play a major role:

(i) a process of *substitutions*, which in biological parlance replace a nucleotide or an amino acid by another one, or, more formally, change the *labels* of the *positions* in the sequence,

(ii) a process of *insertions* and *deletions* (briefly called *indel process*) of single positions or (more generally) sequence fragments.

The evolutionary process leading to observed sequences is modelled by a stochastic process indexed by a binary tree $\mathcal{T}$. The states of the process are the *labelled sequences* $(a_1, ..., a_n)$, $n \in \mathbb{N}_0, a_i \in \mathcal{A}$, where $\mathcal{A}$ is a finite *alphabet*. (In case of DNA, $\mathcal{A} = \{\text{A}, \text{C}, \text{G}, \text{T}\}$.) We will refer to a_i as the label of position no. i in the sequence. The sequences are thought to evolve from an ancestral sequence placed in an inner node which figures as the *root* of $\mathcal{T}$. What is observed are the sequences appearing at the *leaves* of $\mathcal{T}$. The edge lengths of $\mathcal{T}$ are measured in units of "evolutionary distance" (or *time* for simplicity). $\mathcal{T}$ is called *phylogenetic tree*, and the graph structure (or "tree topology") of $\mathcal{T}$ is called the *phylogeny* of the observed sequences. Positions in the observed sequences which are connected by a path of the evolution process without insertion or deletion are called *homologous*. An *alignment* of the observed sequences is an array each of whose lines consist of one of the sequences, possibly augmented by "gaps", such that positions assumed to be homologous appear in one and the same column.

It is a widespread practice in biology to construct in a first step an "optimal" alignment of the observed sequences through some scoring rule which penalizes mismatches (i.e. aligned positions carrying different labels) and gaps ([5]), and in a second step to infer, on the basis of the aligned positions without gaps, the underlying phylogenetic tree ([17]). With the advent of powerful computer technology it became more popular to incorporate some model for the substitution process, and to estimate a maximum likelihood tree that relates the observed sequences. To model the substitution process one usually considers a Markovian jump dynamics on $\mathcal{A}$ which acts independently from position to position ([19]).

However, the reliability of an alignment which optimizes some score is hard to judge. Also, basing the analysis on a single alignment may result in underestimating the variability of the parameter estimates of the substitution process and of the reconstructed phylogeny. Finally, ignoring the gaps and their clues about insertion and deletion events may lead to a considerable loss of valuable information ([15]). All these problems call for an explicit stochastic modelling of the insertion-deletion process.

Outline of the Paper In section 2 we will review various models of insertion-deletion processes and discuss a number of their basic properties. We will start from the nowadays classical models of Thorne, Kishino and Felsenstein for the insertion and deletion of single positions ("TKF", [21]) and of indivisible fragments ("TKF2", [22]). As will become clear, both models can be extended to the case of equal rates of insertion and deletion. We will show that although these dynamics have no equilibrium distribution on the sequences of finite length, reversibility is guaranteed, which appeases a caveat in the recent monograph [2] of J. Felsenstein. These models are building blocks for describing the evolution of sequences along a phylogenetic tree, and for obtaining a "multiple statistical alignment" of ℓ observed sequences.

In section 3 we turn to tree-indexed insertion-deletion processes, which form the mathematical basis for multiple statistical alignment. Indeed, a remarkable progress in multiple statistical alignment (based on the TKF model) was recently made by the group around J. Hein in Oxford. The basic idea is to think of the random multiple alignment as being generated by $\mathcal{T}$-indexed "events" which are brought into a well-defined order. Stimulated by recent communication with G. Lunter and I. Miklós in Oxford, we describe how this can be understood through a reading of the (now $\mathcal{T}$-indexed) indel genealogy "from left to right" and we show how to extend this approach from the TKF model to the fragment insertion-deletion model defined in [11].

Finally, in section 4 we will review some novel approaches to tree reconstruction based on insertion-deletion models.

2 Sequence Evolution Models with Insertion and Deletion

2.1 Stochastic Indel Dynamics

In the prototype insertion-deletion ("indel") model suggested by Thorne, Kishino and Felsenstein [21] in 1991 (see subsection 2.2), single positons are inserted with rate λ and deleted with rate μ, independently of each other. The assumption $\lambda < \mu$ is a tribute to the existence of a reversible equilibrium distribution. We will refer to this as the TKF model for short. It is assumed that the rate of insertions and deletions is not influenced by the labels of the positions, which makes it possible to construct the substitution process "on top" of the indel process.

Consider first the case of $\ell = 2$ observed sequences. It turns out that one can read a nice "genealogical" structure into the TKF dynamics, which enriches the insertion-deletion path of the evolutionary process and renders a forest of Galton-Watson trees with immigration, see subsection 2.3. Reading the forest from left to right allows to generate the alignment by a Markov chain with three states $\binom{p}{p}, \binom{-}{p}, \binom{p}{-}$, where $\binom{p}{p}$ stands for a homologous pair of positions, and $\binom{-}{p}$ and $\binom{p}{-}$ denote positions appearing in only one of the two observed sequences. This chain is "hidden" in the sense that its path cannot be observed from the data. What *is* observed is the result of all the states' *emissions*. E.g. for the sequence of states $\binom{p}{p}, \binom{p}{-}, \binom{p}{p}, \binom{-}{p}, \binom{-}{p}$ the result of emissions could be $\binom{AGT}{ACGG}$. This hidden Markov structure allows to apply dynamic programming, which is a powerful tool not only for computing likelihoods but also for the sampling of alignments from the conditional distribution, given the observed sequences. Alternating the sampling of alignments with a Metropolis-Hastings algorithm for sampling the evolution parameters gives a Markov chain Monte Carlo method which makes it possible to assess the variability of the joint estimation of parameters and alignments [12].

In [11], a variant of the TKF model was introduced which assumes *equal* insertion and deletion rates $\lambda = \mu$. In the present paper, this will be referred to as the cTKF model (c for critical). The cTKF model is a special case of the fragment insertion deletion ("FID") model ([11]), which in turn is a "critical" variant of an indel model introduced by Thorne et al. in [22]. This "TKF2" model assumes insertion (at rate λ) and deletion (at rate $\mu > \lambda$) of (indivisible) *fragments* whose lengths are independent, geometrically distributed with expectation $(1 - \rho)^{-1}$. In his recent monograph [2], J. Felsenstein quotes the FID model (with reference to [11]) with the caveat: " It is not clear that equality [of λ and μ] is tenable, as the resulting model of sequence-length variation then has no equilibrium distribution, and reversibility of the process is not guaranteed." However, as we will see in Proposition 1, these doubts can be remedied. In fact, the FID process which takes a sequence of length n_1 into a sequence of length n_2 can be thought of as "cut out" from an indel process on *infinite sequences*, with the condition that the development between the two sequences is flanked by an (invisible) homologous pair to the left, and another one to the right of the observed sequences.

The assumption made in FID (and TKF2) that inserted pieces are unbreakable entities is unrealistic. Abandoning this assumption, one arrives at a more general insertion-deletion model (introduced in [11] as "GID"), which however is computationally much less tractable. In [11], the FID and GID models were compared with regard to robustness of estimates. Computer simulations showed that estimation procedures for the parameters which are based on the FID assumptions also work well when applied to data generated without the fragmentation restrictions.

The GID model, in turn, is related to the class of *long indel models* recently studied by Miklós et al [13] (see also [14]).

2.2 TKF Bridges

In the pioneering paper [21], Thorne, Kishino and Felsenstein introduced the following evolution model for finite sequences (where a *sequence* consists of *positions*, arranged in linear order).

Definition 1. *(TKF(λ, μ) dynamics for finite sequences)*
Each position is deleted at rate μ and between any two neighbouring positions (as well as to the left and to the right of the current sequence), a new position is inserted at rate λ.

The *insertion-deletion path* records the ordering of all the positions alive at any time s. In particular, the indel path keeps track of the times of insertions and deletions. For fixed parameters $\lambda, \mu > 0$ we will write $\mathrm{TKF}_{n,[0,t]}$ for the distribution of an indel path starting from a sequence of length n and evolving over time t (or "evolutionary distance") according to the $\mathrm{TKF}(\lambda, \mu)$-dynamics.

Remark 1. For $\mu > \lambda$, the $\mathrm{TKF}(\lambda, \mu)$ dynamics has a reversible equilibrium, the random sequence length L in equilibrium being geometrically distributed with weights $\gamma^n(1 - \gamma)$, $n \in \mathbb{N}_0$, where $\gamma = \lambda/\mu$. Consequently, L has expectation $\lambda/(\mu - \lambda)$.

Thorne et al. [21] consider only the case $\mu > \lambda$. In this case let us write

$$\mathrm{TKF}_{[0,t]} = \sum_{n \geq 0} \gamma^n (1 - \gamma) \mathrm{TKF}_{n,[0,t]}$$

for the equilibrium distribution of an indel path evolving in the time interval $[0, t]$ under the $\mathrm{TKF}(\lambda, \mu)$-dynamics. The following definition makes sense also for general $\lambda, \mu > 0$.

Definition 2. *Let $\lambda, \mu, t > 0$. For given natural numbers n_1, n_2, we define the TKF bridge $\mathrm{TKF}_{n_1,n_2;[0,t]}$ as the distribution of the $TKF(\lambda, \mu)$ process, conditioned to take a sequence of length n_1 into a sequence of length n_2 over time t.*

The TKF bridge for $\lambda = \mu$ is a special case of the *fragment insertion deletion model* introduced in [11], see subsection 2.5 below. Although for $\lambda \geq \mu$ the TKF process has no equilibrium on the finite sequences, the distribution of the TKF bridge does not depend on the chosen direction of time, as the corollary to the following proposition tells.

Proposition 1. *For $\lambda, \mu > 0$ and $\gamma = \lambda/\mu$, the measure*

$$M := \sum_{n \geq 0} \gamma^n \, TKF_{n,[0,t]}$$

is invariant under time reversal (where the insertions and deletions in an indel path interchange roles).

Proof. Since, given the birth and death times, the birth and death positions within the current sequence are exchangeable, it suffices to consider the process $L = (L_s)_{0 \leq s \leq t}$ of sequence lengths. Under the $\mathrm{TKF}(\lambda, \mu)$ dynamics, this is a birth and death process on $\mathbb{N}_0$ with birth rates $b(n) := (n + 1)\lambda$, $n \geq 0$, and death rates $d(n) := n\mu$, $n \geq 1$, hence

$$b(n) = \gamma d(n + 1), \quad n \geq 0. \tag{1}$$

Write

$$P_s(m, n) := \mathbb{P}[L_s = n | L_0 = m]$$

for the transition probability of L, and define the measure g on $\mathbb{N}_0$ by

$$g(n) = \gamma^n, n \geq 0 \tag{2}$$

The detailed balance condition (1) guarantees reversiblity of the measure g:

$$g(m)P_s(m,n) = g(n)P_s(n,m), \ m,n \in \mathbb{N}_0, s \geq 0. \tag{3}$$

This extends to reversibility of the measure

$$M(L \in (.)) = \sum_{m \geq 0} g(m)\mathbb{P}[L \in (.)|L_0 = m] :$$

$$M(L_0 = m, L \in B, L_t = n) = M(L_0 = n, L \in R(B), L_t = m). \tag{4}$$

Here, B is a measurable set of $\mathbb{N}_0$-valued (right continuous) paths with jump size 1, and R is the time reversal operator mapping a path $(x_s)_{0 \leq s \leq t}$ into the right continuous modification of its time reversal $(x_{t-s})_{0 \leq s \leq t}$.

Corollary 1. *For all $\lambda, \mu > 0$ the bridge $TKF_{n_1,n_2;[0,t]}$ with parameters λ, μ equals the time reversal of the bridge $TKF_{n_2,n_1;[0,t]}$ with parameters λ, μ.*

Proof. For any (measurable) set A of indel paths,

$$\gamma^{n_1}\mathbb{P}[L_t = n_2|L_0 = n_1]TKF_{n_1,n_2;[0,t]}(A) = M(A; L_0 = n_1, L_t = n_2),$$

where M is the measure defined in Proposition 1. Hence the claim follows from the reversibility of M.

Let us in particular single out the "critical" case $\lambda = \mu$. In this case we will speak of the cTKF(λ) (instead of the TKF(λ, λ) dynamics, and denote the measure M in Proposition 1 by

$$\mathrm{cTKF}_{[0,t]} := \sum_{n \geq 0} \mathrm{TKF}_{n;[0,t]}.$$

Proposition 1 then specializes to

Corollary 2. *For each $\lambda > 0$, the (infinite) measure $cTKF_{[0,t]}$ with parameter λ is invariant under time reversal.*

2.3 A Genealogy of Positions

Following Thorne et al [21], we upgrade an indel path with a genealogy of positions by decreeing that each inserted position *is born by its left neigbour*. A position which inserted to the very left of the sequence is declared to be born by an invisible, immortal position to its left. Equivalently, one can interprete an insertion of this type as an *immigration* at the left end of the current sequence. Thus, the TKF(λ, μ) process turns into a (binary, continuous time)

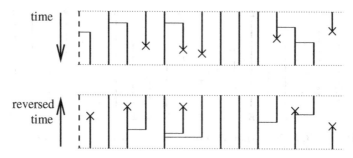

Fig. 1. Galton-Watson forest for a TKF-process. Applying the TKF-convention to the time-reversed realization changes the alignment (as seen at the right end) but not the homology structure. The immigrant (leftmost line in the forward-time figure) is considered to be born by an immortal position to its left (dashed line)

Galton-Watson process with birth rate and immigration rate equal to λ, and death rate equal to μ.

This can be illustrated with a graphical construction (see Figure 1, upper part): Let time run downwards. If a newly inserted position has a left neigbour in the sequence (that is, does not appear at the left end of the sequence), then draw a horizontal line at the time of insertion between the newly inserted position and its left neighbour. If a new insertion happens at the left end of the sequence, then draw a horizonal line to the invisible, immortal position thought to sit left of the sequence.

In this way the $\mathrm{TKF}_{n_1,n_2;[0,t]}$ distribution gives rise to a Galton-Watson forest with immigration, starting with n_1 mother positions at time 0, and conditioned to a total number of n_2 positions living at t. The picture can be reversed: at each deletion, draw a horizontal line to the current left neighbour, or to the invisible, immortal position if the deletion happens at the left end. Corollary 1 then turns into a statement about conditional Galton-Watson forests: a forest generated by $\mathrm{TKF}_{n_1,n_2;[0,t]}$ (with parameters $\lambda, \mu > 0$), when reversed in the described manner, equals in distribution a forest generated by $\mathrm{TKF}_{n_2,n_1;[0,t]}$, with the same parameters λ, μ (see Figure 1).

2.4 Reading an Indel Forest from Left to Right

Let us read the branches of an indel forest over the time interval $[0,t]$ from left to right, and write $\binom{p}{p}$ for a position which is conserved over the time interval $[0,t]$, $\binom{p}{-}$ for a position present at time 0 and deleted before time t, and $\binom{-}{p}$ for a position inserted after time 0 and surviving till time t. The indel forest generated by $\mathrm{cTKF}_{[0,t]}$ (or by $\mathrm{TKF}_{[0,t]}$ if $\lambda < \mu$) leads to a Markov chain with state space $\binom{p}{p}, \binom{p}{-}, \binom{-}{p}$. The transition probabilities of this chain can be phrased (and easily computed, see [11]) in terms of a binary Galton-Watson process.

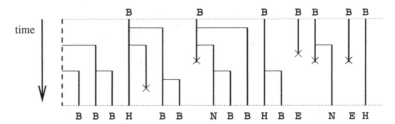

Fig. 2. $\{B, H, N, E\}$-notation of an alignment for a pair of sequences

For further use, let us think of time 0 as corresponding to the *root*, and time t to the *daughter node* of a "tree" consisting of a single edge.

Following a clever convention introduced in [9] we write a B for every position in the root, and the following symbols in the daughter node: H for $\binom{p}{p}$, N for $\binom{p-}{-p}$, E for $\binom{p}{-}$ not followed by $\binom{-}{p}$, and B for $\binom{-}{p}$ not preceded by $\binom{p}{-}$. Whereas H means that the position survives, N means that the position dies but leaves a (leftmost) child. In both cases H and N we will address the successor position as the *heir* of the ancestor position. The symbol B in the daughter node stands for either an immigrant position or a descendant position which is not an heir (these will be called *β-children* for short). For example, the *indel history* $\binom{--p-p\,p--}{pp\,pp\,--pp}$ translates into $X = \binom{B\ \ BB}{BBHBENB}$, or $(BBHBENB)$ for short (see Figure 2). Every X starts with an initial block $X^{(0)}$ (of immigrants) of the form $\binom{}{B...B}$, followed by n blocks $X^{(1)}, ..., X^{(n)}$, where n is the length of the sequence in the root. Each of these blocks is of type H, N or E, that is, of the form $\binom{B}{HB...B}$, $\binom{B}{NB...B}$ or $\binom{B}{E}$. A block of the latter type means that the ancestral position is extinct. In the type H and type N blocks, the number k of B's in the second line is geometrically distributed with probability weight $(1 - \pi_t)\pi_t^k$, where the parameter π_t depends also on λ and μ. In other words, right to any H, N or B in the second line, the current block is continued (with a B in the second line) with probability π_t; otherwise (as long as the current number of blocks is $\leq n$, a new block starts. Independently of what is to its left, a new block is of type H, N or E with probability π_t^H, π_t^N and π_t^E, where these three probabilites depend on λ and μ, and sum to 1. Under the infinite measure $cTKF_{[0,t]}$ every n is assigned mass 1, whereas under the probability measure $TKF_{[0,t]}$, n gets mass $\gamma^n(1 - \gamma)$ (where $\gamma = \lambda/\mu$). The latter can of course also be realized by an independent stopping, which, after each n, continues with a new block with probability γ and jumps to an **End** state with probability $1 - \gamma$.

Note that an alignment under the cTKF dynamics can be produced by an $\{H, N, E, B\}$ valued Markov chain with transition probabilities

$$\pi_t^{HB} = \pi_t^{NB} = \pi_t^{BB} = \pi_t; \quad \pi_t^{HH} = \pi_t^{NH} = \pi_t^{BH} = (1 - \pi_t)\pi_t^H$$

$$\pi_t^{HN} = \pi_t^{NN} = \pi_t^{BN} = (1 - \pi_t)\pi_t^N; \quad \pi_t^{HE} = \pi_t^{NE} = \pi_t^{BE} = (1 - \pi_t)\pi_t^E;$$

$$\pi_t^{EH} = \pi_t^H, \quad \pi_t^{EN} = \pi_t^N, \quad \pi_t^{EE} = \pi_t^E, \quad \pi_t^{EB} = 0.$$

This chain starts in the initial distribution giving weight π_t to B and weights $(1 - \pi_t)\pi_t^H, (1 - \pi_t)\pi_t^N, (1 - \pi_t)\pi_t^E$ to H, N and E, respectively.

2.5 A Fragment Insertion-Deletion Model

Let us consider the analogue of the cTKF dynamics (with parameter λ) acting on indivisible *fragments* whose lengths are independent and geometrically distributed with expectation $(1 - \rho)^{-1}$, $\rho \in [0, 1)$. The resulting indel dynamics on the finite sequences will be called *fragment insertion deletion dynamics* with parameters λ and ρ , abbreviated as FID(λ, ρ). The indel histories will be coded in a similar way as described in the previous subsection. For example, imagine a fragment of length 4 which survives and leaves one daughter fragment with length 3, say. Then the "second line" of the indel history reads as $(HHHHBBB)$. In case the mother fragment dies and leaves one daughter fragment with length 3, we would have $(EEENBB)$. Thanks to the properties of the geometric distribution, under the FID measure the alignment builds up as a Markov chain with transition probabilites π_t^{IJ}, $I, J \in \{H, N, E, B\}$, which are only slightly more complicated than those in the cTKF case, see [11] for details. Notably, for $\rho > 0$ one has $\pi_t^{NN} = \pi_t^{HN} = \pi_t^{BN}$ but $\pi_t^{HH} > \pi_t^{BH} = \pi_t^{NH}$, and $\pi_t^{BB} = \pi_t^{NB} > \pi_t^{HB}$, because compared to the cTKF model there is some additional probability for a position to be in the same fragment as its left neighbour.

3 Tree Indexed Indel Processes

3.1 Multiple TKF Bridges

Let $\mathbb{T}$ be a finite binary tree with ℓ leaves. Its sets of edges, nodes and leaves will be denoted by $\mathcal{E}, \mathcal{N},$ and $\mathcal{L}$. The edges ϵ are labelled by positive numbers (*lengths*) t_ϵ. The *branch* b_ϵ is represented as $\{\epsilon\} \times (0, t_\epsilon)$, and the *(labelled) tree* $\mathcal{T}$ is represented as

$$\mathcal{T} = \mathcal{N} \cup \bigcup_{\epsilon \in \mathcal{E}} b_\epsilon,$$

equipped with the obvious tree distance along the branches.

For the moment, one of the nodes of $\mathbb{T}$ is distinguished as the *root* of $\mathbb{T}$ and denoted by r. This choice assigns to each edge ϵ a direction (from the root to the leaves), and to each node $\nu \neq r$ its *mother node* and its *mother edge*. We will write $t(\nu)$ for the length of the mother edge of ν, and $\mathcal{T}_r$ for the pair $(\mathcal{T}, r)$.

Fix $0 < \lambda \leq \mu$, and consider the TKF(λ, μ) indel dynamics on the finite sequences.

Definition 3. *The* indel *process indexed by* $\mathcal{T}_r$ *starts with an ancestral sequence at the root and lets a copy of this sequence evolve independently according to the TKF(λ, μ) dynamics along each branch leading away from the root. In every inner node (different from the root), the process continues with two identical copies which evolve independently along the two branches descending from this node.*

The $\mathcal{T}_r$-indexed indel process induces a distribution $\text{TKF}_{n;\mathcal{T},r}$ on the $\mathcal{T}$-indexed indel paths starting with length n in the root r. As in section 2 we put

$$\text{cTKF}_{\mathcal{T}} = \sum_{n \geq 0} \text{TKF}_{n;\mathcal{T},r} \quad \text{if} \quad \lambda = \mu,$$

$$\text{TKF}_{\mathcal{T}} = \sum_{n \geq 0} \gamma^n (1 - \gamma) \text{TKF}_{n;\mathcal{T},r} \quad \text{if} \quad \lambda < \mu.$$

Note that due to Proposition 1 these measures indeed do not depend on the choice of the root. The same is true for the *multiple TKF bridge* $\text{TKF}_{n_1,\dots,n_\ell;\mathcal{T}}$ which arises by conditioning the tree-indexed indel process to produce given sequence lengths $n_1, \dots, n_\ell$ in the leaves.

3.2 Decomposing a Tree Indexed Indel Path into Heirs Lines

Let us consider a $\mathcal{T}_r$-indexed indel path w, and the $\mathbb{T}_r$-indexed indel history $\tilde{w}$ induced by w along the nodes of $\mathbb{T}$. As described in subsection 2.4, each position p at a node $\nu \in \mathcal{N}, \nu \neq r$, is either an immigrant or an heir or a β-child, depending on its genealogy along the mother edge of ν. Immigrant positions and positions at the root are called *founder positions*. Founder positions are marked with B, whereas heirs are marked by H or N, depending whether they are survivors or not (see subsection 2.4).

For a fixed node ν, $\mathbb{T}(\nu)$ denotes the subtree of $\mathbb{T}$ which is rooted in ν. Let p be a position at node ν which is marked by B. Following the heirs (and heirs of heirs...) descending from p along the nodes of $\mathbb{T}(\nu)$ we obtain a mapping from the nodes of $\mathbb{T}(\nu)$ which assigns a B to ν and an H, N, E or "-" to the other nodes of $\mathbb{T}(\nu)$ (where the convention is such that any node descending from a node which carries an E is assigned a "-"). We call this mapping the *tree indexed heirs line* ("tihl") e initiated by p, and say that the node ν is the *origin* r_e of the tihl e.

Let $\mathbb{T}_e$ be the subtree of $\mathbb{T}(\nu)$ consisting of all the nodes to which e assigns a B, H or N. Denote by supp(e) the *support* of e, i.e. the set of all those nodes which are connected with p by an heirs line, or in other words, the set of all those nodes of $\mathbb{T}(\nu)$ to which e assigns H or N. The set supp(e) $\cup \{r_e\}$ will be called the *rooted support* of e, and by $\overline{\text{supp}}(e)$ we will denote the *extended support* of e, i.e. the set of all those nodes of $\mathbb{T}(\nu)$ to which e assigns H, N or E.

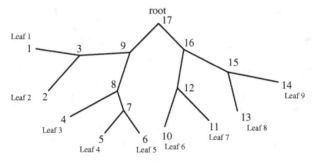

Fig. 3. Example for a total order "$\leq$" on the nodes of a tree: For each node ν the nodes in the left subtree stemming from ν have lower ranks than those in the right subtree

A tihl initiated by a founder position will be called a *founder* tihl. If p is a β-child, then it has a left neighbour p', and we call the tihl e' to which p' belongs the *mother* of the tihl e initiated by p, and e a *daughter* of e'. Every tihl in the tree-indexed indel history $\tilde{w}$ has an ancestral line tracing back to a founder tihl, and in this sense we obtain a "family decomposition" of all the tihls in $\tilde{w}$.

The idea is now to build up the tree-indexed indel history $\tilde{w}$ successively through the tihls it consists of. To this end we introduce an order on the nodes of $\mathbb{T}$ as follows. First we put $\nu_2 \preceq \nu_1$ if the node ν_1 is on the path from ν_2 to the root. Following [9] we fix a total order $\leq$ on the nodes of the tree which extends the partial order $\preceq$. We say that ν_1 has a *smaller rank* than ν_2 if $\nu_1 < \nu_2$ in the total order, see Figure 3. The first tihl to be filled in is the one initiated by the leftmost founder at the node with the smallest possible rank. Now proceed inductively: as long as there remain tihl's which are daughters of the ones already filled in, the next tihl to be filled in is the one among all these which is initiated by the leftmost β-child at the node with the smallest possible rank. After completion of a family of tihl's, proceed with the tihl initiated by the leftmost founder at the node with the next smallest possible rank.

3.3 Building an Indel History by a Markov Chain of Tree Indexed Heirs Lines and Sets of Active Nodes

Consider the measure $\mathrm{cTKF}_{\mathcal{T}}$ (with parameter λ) or the measure $\mathrm{TKF}_{\mathcal{T}}$ (with parameters $\lambda < \mu$) on the $\mathcal{T}_r$-indexed indel paths w. By the mapping described in subsection 3.2, this measure is transported into a measure on the finite sequences of tihls. Any such sequence of tihls starts with a block of tihl families founded by immigrants, and continues with n tihl families founded by positions at the root. Again (cf. subsection 2.4), n gets mass $(\lambda/\mu)^n(1 - \lambda/\mu)$ under the probability measure $\mathrm{TKF}_{\mathcal{T}}$, whereas under the infinite measure $\mathrm{cTKF}_{\mathcal{T}}$ every n is assigned mass 1. This can be expressed

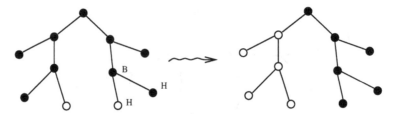

Fig. 4. Transition from one `soan` to the next via a `tihl`. Active nodes are drawn black. We assume the same type of order as shown in Figure 3

in a unified way as follows: Independently after the completion of each `tihl` family, a new `tihl` family starts with probability $\gamma = \lambda/\mu$.

The process of `tihls` is not Markov. However, one may keep track in parallel of the *set of active nodes* ("soans"), and in fact the process of pairs of `soans` and `tihls` *is* Markov. We will call a node ν *active* if it is a candidate for the origin of the next `tihl` to be inserted. Initially, all nodes of $\mathbb{T}$ are active. Each newly inserted `tihl` re-activates all nodes in its support, and de-activates all those nodes which have a smaller rank than its origin and do not belong to its support. Formally, for a set $\mathcal{S} \subseteq \mathcal{N}$, and a `tihl` e with $r_e \in \mathcal{S}$, we put

$$[\mathcal{S}, e] := (\mathcal{S} \setminus \{\sigma \in \mathcal{N} | \sigma < r_e\}) \cup \operatorname{supp} e. \tag{5}$$

For a `tihl` e, we write

$$p(e) = \pi_{t(r_e)} \prod_{\sigma \in \overline{\operatorname{supp}}(e)} \pi_{t(\sigma)}^{e(\sigma)},$$

where we put

$$\pi_{t(r)} := \gamma.$$

Given the current set of active nodes is $\mathcal{R}$, the probability that the next step leads to the `tihl` e is

$$P(\mathcal{R}, e) = p(e) \prod_{\sigma \in \mathcal{R} : \sigma < r_e} (1 - \pi_{t(\sigma)}), \tag{6}$$

whereas the probability that the process is stopped (i.e. jumps to an extra state **End**) is $P(\mathcal{R}, \textbf{End}) = \prod_{\sigma \in \mathcal{R}} (1 - \pi_{t(\sigma)})$. (Since the root r in any case belongs to $\mathcal{R}$, this latter transition probability is zero in the cTKF model.)

Let us consider the **soan-tihl-** *process* $Y = (\mathcal{S}_0, e_1, \mathcal{S}_1, e_2, \mathcal{S}_2, ...)$ which starts from $\mathcal{S}_0 = \mathcal{N}$ and whose dynamics is given by the transition probability (6) and the rule $\mathcal{S}_{i+1} = [\mathcal{S}_i, e_{i+1}]$, see (5). Note that the transition probability on the **soans** is

$$P(\mathcal{R}, \mathcal{S}) = \sum_{e : \mathcal{S} = [R, e]} P(\mathcal{R}, e).$$

For a `tihl` e, and $j = 1, ..., \ell$, let us write $v_e(j) = 1$ if the leaf l_j belongs to supp e, and $v_e(j) = 0$ otherwise. We define

$$\mathbf{v}_e = (v_e(1), ..., v_e(\ell)).$$

For a realization $e_1, e_2, ...$ of the `tihl` process, let $\bar{m}$ be such that $e_{\bar{m}+1} = \text{End}$ if the process is stopped, and $\bar{m} = \infty$ otherwise. In order to count how many positions are emitted into the leaves by the first m `tihls` in the process Y, we put

$$K_j(m) = \sum_{i=1}^{m \wedge \bar{m}} v_{e_i}(j), \; j = 1, ..., \ell; \quad \mathbf{K}(m) = (K_1(m), ... K_\ell(m)).$$

The multiple counting process $\mathbf{K}$ will be important in the next subsections; note that it is adapted to the process Y in the sense that $\mathbf{K}(m)$ can be read off from $(Y_1, ..., Y_m)$.

3.4 Generating Labelled Sequences in the Leaves

As a model for the *substitution process*, consider a time homogeneous Markov process M on a finite set $\mathcal{A}$ of *letters* in a reversible equilibrium α. For any `tihl` e, this induces a $\mathbb{T}_e$-indexed $\mathcal{A}$-valued process M^e in the following way: Start in r_e in the equilibrium α. Along an edge ϵ of $\mathbb{T}_e$ leading to a node which carries an H, the process develops (for the time t_ϵ) according to the substitution dynamics, whereas in a node carrying an N the process starts independently in distribution α.

By the process M^e, a `tihl` e assigns a random letter to each element of $\mathcal{L}_e := \mathcal{L} \cap \text{supp}\, e$. The joint distribution of these letters will be denoted by α^e. Given a sequence $(e_1, e_2, ...)$ we assume that the processes $M^{e_1}, M^{e_2}, ...$ are independent.

Let $\mathbf{A}^j(m) = (A_1^j, ..., A_{K_j(m)}^j)$ be the sequence of letters assigned to leaf l_j by the first m `tihls` $e_1, ..., e_m$ in the process Y, and put $\mathbf{A}(m) = (\mathbf{A}^1(m), ..., \mathbf{A}^\ell(m))$. The process $\mathbf{A}(m), m = 1, 2, ..$ is a variant of a multiple hidden Markov model in the sense of [7]: the role of the hidden Markov chain is played by the soan-tihl process, which in each step emits letters into some subset of $\mathcal{L}$.

In the sequel, we denote by h the `tihl` which originates in the root r and assigns H to all $\nu \in \mathcal{N} \setminus \{r\}$.

Definition 4. *Fix* $\mathbf{k} = (k_1, ..., k_\ell) \in \mathbb{N}_0^\ell$ *and* $\mathbf{a_k} = (\mathbf{a}_{k_1}^1, ..., \mathbf{a}_{k_\ell}^\ell)$, *where* $\mathbf{a}_{k_j}^j = (a_1^j, ..., a_{k_j}^j)$ *is an* $\mathcal{A}$-*valued sequence of length* k_j *for* $j = 1, ..., \ell$.

a) (case TKF, i.e. $\gamma < 1$*) Let*

$$P(\mathbf{k}) = \mathbb{P}[\mathbf{A}(m) = \mathbf{a_k}; e_{m+1} = \text{End} \quad \text{for some } m \in \mathbb{N}_0].$$

For $\mathcal{S} \subseteq \mathcal{N}$, let

$$P_{\mathcal{S}}(\mathbf{k}) = \mathbb{P}[\mathbf{A}(m) = \mathbf{a_k}; \, \mathcal{S}_m = \mathcal{S}; \, e_{m+1} = \mathbf{End} \quad \text{for some } m \in \mathbb{N}_0].$$

b) (case cTKF, i.e. $\gamma = 1$) Let

$$P(\mathbf{k}) = \mathbb{P}[\mathbf{A}(m) = \mathbf{a_k}; e_{m+1} = h \quad \text{for some } m \in \mathbb{N}_0].$$

For $\mathcal{S} \subseteq \mathcal{N}$, let

$$P_{\mathcal{S}}(\mathbf{k}) = \mathbb{P}[\mathbf{A}(m) = \mathbf{a_k}; \, \mathcal{S}_m = \mathcal{S}; \, e_{m+1} = h \quad \text{for some } m \in \mathbb{N}_0].$$

For a `tihl` e, and $\mathbf{k}, \mathbf{a_k}$ as in Definition 4, put

$$q(e) := \prod_{\sigma \in \text{supp} \, e} (1 - \pi_{t(\sigma)}); \quad \vartheta(e, \mathbf{k}) = \alpha^e((a_{k_j}^j)_{j \in \mathcal{L}_e}).$$

In words: $\vartheta(e, \mathbf{k})$ is the probability that the `tihl` e emits the letters $a_{k_j}^j$ into all leaves l_j which it reaches. The next lemma gives a recursion as well as the initial condition ($\mathbf{k} = \mathbf{0} = (0, ..., 0)$) for the $p_{\mathcal{S}}(\mathbf{k})$. In the TKF case this is in [8, 9]; we include a proof for convenience.

Lemma 1. *i) For $\mathbf{k}, \mathbf{a_k}$ as in Definition 4, and all $\mathcal{S} \subseteq \mathcal{N}$,*

$$P_{\mathcal{S}}(\mathbf{k}) = \sum_{(\mathcal{R},e): \, \mathcal{S} = [\mathcal{R},e]} p(e)q(e)P_{\mathcal{R}}(\mathbf{k} - \mathbf{v}_e)\vartheta(e, \mathbf{k}). \tag{7}$$

ii) In the TKF case (see Def. 4 a))

$$P_{\mathcal{N}}(\mathbf{0}) = P(\mathcal{N}, \mathbf{End}) = \prod_{\sigma \in \mathcal{N}} (1 - \pi_{t(\sigma)}),$$

and in the cTKF case (see Def. 4 b))

$$P_{\mathcal{N}}(\mathbf{0}) = P(\mathcal{N}, h) = \prod_{\sigma \in \mathcal{N} \setminus \{r\}} (1 - \pi_{t(\sigma)}).$$

Proof. i) Because of (6) we have

$$\mathbb{P}[\mathbf{A}(m) = \mathbf{a_k}, \, \mathcal{S}_m = \mathcal{S}] = \sum_{(\mathcal{R},e): \, \mathcal{S} = [\mathcal{R},e]} \mathbb{P}[\mathbf{A}(m-1)$$
$$= \mathbf{a_{k-v_e}}; \, \mathcal{S}_{m-1} = \mathcal{R}]$$
$$\prod_{\sigma \in \mathcal{R}: \sigma < r_e} (1 - \pi_{t(\sigma)})p(e)\vartheta(e, \mathbf{k}). \tag{8}$$

For all $\mathcal{R}$ and e such that $[\mathcal{R}, e] = \mathcal{S}$, one checks easily that

$$P(\mathcal{S}, \mathbf{End}) \prod_{\sigma \in \mathcal{R}: \sigma < r_e} (1 - \pi_{t(\sigma)}) = P(\mathcal{R}, \mathbf{End}) \, q(e) \tag{9}$$

in the TKF case, and

$$P(\mathcal{S}, h) \prod_{\sigma \in \mathcal{R}: \sigma < r_e} (1 - \pi_{t(\sigma)}) = P(\mathcal{R}, h)\, q(e) \tag{10}$$

in the cTKF case. Multiplying both sides of (8) with $P(\mathcal{S}, \mathtt{End})$ (or $P(\mathcal{S}, h)$), summing over m and using (9) (or (10)) we obtain the assertion.

ii) Because of the rule (5) it is impossible to have $\mathcal{S}_n = \mathcal{N}$ for some $n \geq 1$ and tihls $e_1, ..., e_n$ with $\mathbf{v}_{e_1} = ... \mathbf{v}_{e_n} = 0$ (since any such tihl deactivates some leaf which can be re-activated only with some leaf e with $\mathbf{v}_e \neq 0$). On the other hand, any tihl e with $\mathbf{v}_e \neq 0$ leads to an increase of some coordinates of $\mathbf{k}$. Thus, the only contribution to $p_{\mathcal{N}}(\mathbf{0})$ comes from jumping from $\mathcal{N}$ to $\mathtt{End}$ (in the TKF case) or h (in the cTKF case) without any other tihl in between.

3.5 Computing Multiple Alignment Likelihoods

We proceed by explaining the method of Lunter et al. [9], which, as we saw in the previous subsection, generalizes easily to the cTKF case.

Noting that

$$P_\mathcal{S}(\mathbf{0}) = \sum_{(\mathcal{R}, e): \mathcal{S} = [\mathcal{R}, e]} p(e)q(e)P_\mathcal{R}(\mathbf{0}); \quad \mathcal{S} \neq \mathcal{N}, \tag{11}$$

we obtain from (7) and (11) by summing over all $\mathcal{S}$:

$$P(\mathbf{k}) = \sum_e p(e)q(e)P(\mathbf{k} - \mathbf{v}_e)\vartheta(e, \mathbf{k}). \tag{12}$$

$$P(\mathbf{0}) = P_\mathcal{N}(\mathbf{0}) + \sum_{e: \mathbf{v}_e = 0} p(e)q(e)P(\mathbf{0}) \tag{13}$$

Note that $P(\mathbf{k})$ and $P(\mathbf{0})$ appear on both sides of the equations. It is, however, straightforward to turn them into a recursion:

$$P(\mathbf{k}) = \frac{\sum_{e: \mathbf{v}_e \neq \mathbf{0}} p(e)q(e)\vartheta(e, \mathbf{k})P(\mathbf{k} - \mathbf{v}_e)}{1 - \sum_{e: \mathbf{v}_e = \mathbf{0}} p(e)q(e)} \tag{14}$$

$$P(\mathbf{0}) = \frac{P_\mathcal{N}(\mathbf{0})}{1 - \sum_{e: \mathbf{v}_e = \mathbf{0}} p(e)q(e)} \tag{15}$$

Lunter et al. [9] make the computation still more efficient by using the accelerated chain $Y_{\tau(1)}, Y_{\tau(2)}, Y_{\tau(3)}, \ldots$ where $\tau(1) = 1$ and $\tau(m)$, $m > 1$ is the first time when either $\mathtt{End}$ occurs, or a tihl e which *overlaps* with one of the tihls $e_{\tau(m-1)}, ..., e_{\tau(m)-1}$ in the sense that their rooted supports intersect, see Figure 5. Thus the set of tihls occurring between $\tau(m-1)$ to

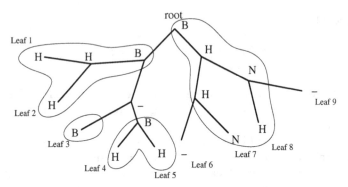

Fig. 5. A set of non-overlapping tihls

$\tau(m)-1$ are non-overlapping. Each *set of non-overlapping* tihls (called *a set of nested events* in [9]) can be ordered, say $e_1 < e_2 < \ldots < e_j$, corresponding to the ordering of the $r(e_i)$ in the total order on $\mathcal{N}$. (This is precisely the order in which the non-overlapping tihls occur in the process Y.) Call a set $\mathcal{E}$ of non-overlapping tihls *silent* if $\mathbf{v}_e = \mathbf{0}$ for all $e \in \mathcal{E}$.

Let $\mathfrak{E}_0$ ($\mathfrak{E}_1$) denote the family of all silent (non-silent) sets of non-overlapping tihls, and put $\mathfrak{E} := \mathfrak{E}_0 \cup \mathfrak{E}_1$. Similar as before but with an additional inclusion-exclusion argument, one can sum over the probabilities of all sets of non-overlapping tihls that occur between $\tau(m-1)$ and $\tau(m)-1$:

$$P(\mathbf{k}) = \sum_j (-1)^{j-1} \sum_{\{e_1,\ldots,e_j\} \in \mathfrak{E}} P(\mathbf{k} - \sum_i \mathbf{v}_{e_i}) \cdot \prod_i p(e_i) \cdot q(e_i) \cdot \vartheta(e_i, \mathbf{k}) \tag{16}$$

This leads to the recursion

$$P(\mathbf{k}) = \frac{\sum_j (-1)^{j-1} \sum_{\{e_1,\ldots,e_j\} \in \mathfrak{E}_1} P(\mathbf{k} - \sum_i \mathbf{v}_{e_i}) \cdot \prod_i p(e_i) \cdot q(e_i) \cdot \vartheta(e_i, \mathbf{k})}{1 - \sum_j (-1)^{j-1} \sum_{\{e_1,\ldots,e_j\} \in \mathfrak{E}_0} \prod_i p(e_i) \cdot q(e_i)} \tag{17}$$

Each $\mathcal{E} = \{e_1, \ldots, e_j\} \in \mathfrak{E}$ induces a mapping f from $\mathcal{N}$ to $\mathcal{Z} = \{-, B, H, N, E\}$ as follows: Put $f(\sigma) = e_i(\sigma)$ if $\sigma \preceq r(e_i)$ for some $i = 1, \ldots, j$, and $f(\sigma) = -$ otherwise. Lunter et al. [9] realized that the r.h.s. of (17) can be computed efficiently by grouping the elements $\mathcal{E} = \{e_1, ..., e_j\} \in \mathfrak{E}$ with respect to $\mathbf{V} = \mathbf{V}(\mathcal{E}) = \sum \mathbf{v}_{e_i}$. For $\mathbf{V} \in \{0,1\}^{\mathcal{L}}$ put

$$\mathcal{L}_{\mathbf{V}} =: \{l_\iota \in \mathcal{L} | \mathbf{V}_\iota = 1\}; \quad \mathfrak{E}_{\mathbf{V}} := \{\mathcal{E} = \{e_1, ..., e_j\} | \sum \mathbf{v}_{e_i} = \mathbf{V}\}.$$

Following [9], one can then compute for each node $\sigma \in \mathcal{N}$ and each $Z \in \mathcal{Z}$ the contribution to $P(\mathbf{k})$ from sets $\mathcal{E}$ of non-overlapping tihls with $\mathbf{V}(\mathcal{E}) = \mathbf{V}$ that assign a Z to σ:

$$F(\mathbf{V}, \sigma, Z) := \sum_j (-1)^{j-1} \sum_{\{e_1,\ldots,e_j\}} P(\mathbf{k} - \mathbf{V}) \cdot \prod_i p(e_i) \cdot q(e_i) \cdot \vartheta_\sigma(e_i, \mathbf{k}),$$

where the sum is taken over all $\mathcal{E} = \{e_1, ..., e_j\} \in \mathfrak{E}_\mathbf{V}$ that include an e_i with $e_i(\sigma) = Z$, and $\vartheta_\sigma(e_i, \mathbf{k})$ is the probability that, given the tihl e_i, the substitution process along e_i produces the labels $a_{k_{l_\iota}}^\iota$ in all the leaves l_ι which stem from σ and belong to $\mathcal{L}_\mathbf{V}$. The value $F(\mathbf{V}, \sigma, Z)$ can be computed directly from the corresponding values of the daughters of σ. Thus, using the dynamic programming idea of Felsenstein's pruning algorithm [2], one can efficiently compute the values of all nodes, starting in the leaves and ending in the root. The numerator in (17) then results as $\sum_{\mathbf{V}\neq\mathbf{0}} F(\mathbf{V}, r, B)P(\mathbf{k}-\mathbf{V})$, whereas the sum in the denominator of (17) equals $F(\mathbf{0}, r, B)$. However, one still has to do this for all $\mathbf{k} \leq (n_1, ..., n_\ell)$, so the time complexity is essentially the product of all sequence lengths n_ι and thus is exponential in the number ℓ of input sequences.

3.6 Extension to Fragment Insertions and Deletions

When extending the TKF2 model (see subsection 2.1) or the FID model along a tree, one can either impose indivisibility of the fragments along the whole tree (as suggested in [3] for the TKF2 model), or one can allow that the fragmentation changes from edge to edge, which we assume in the sequel.

In the TKF and cTKF model, as soon as a tihl is born in some node of the tree, it grows independently of the indel history "to its left". This is different in the FID model: if the left neighbour position at some node σ carries an H, this enhances the probability of an H in the tihl at σ. Consequently, the soan-tihl process defined in subsection 3.3 is not Markov any more. To obtain a Markovian sequence, we have to keep track not only if a node is active, but also to remember if the previous (when looking back to the left) position carried a B, H or N. Our new state space now consists of all mappings φ from $\mathcal{N}$ to $\{B, H, N, E, B^-, H^-, N^-\}$. Let us put $S_\varphi := \{\sigma \in \mathcal{N} \mid \varphi(\sigma) \in \{B, H, N\}\}$. We say that a node σ is φ-active if $\sigma \in S_\varphi$, and φ-inactive otherwise. For a pair (ψ, φ) of such mappings we say that φ activates a ψ-inactive node σ if $\psi(\sigma) = \varphi(\sigma)^-$, and we say that φ de-activates a $\sigma \in S_\psi$ if $\varphi(\sigma) = \psi(\sigma)^-$.

The initial state is $\varphi_0 = h$, which maps r to B and all other nodes to H. The update rule replacing (5) is now

$$[\psi, e] = \varphi$$

where φ de-activates all nodes in $(S_\psi \cap \{\sigma \mid \sigma < r_e\}) \setminus \operatorname{supp}(e)$, re-activates all ψ-inactive nodes in $\operatorname{supp}(e)$, and sets $\varphi(r_e) = B$ and $\varphi(\sigma) = \psi(\sigma)$ for all $\sigma > r_e$. For $U \in \{B, H, N\}$ and $W \in \{H, N\}$ we put $\pi_t^{U^- B} = 0$ and $\pi_t^{U^- W} = \pi_t^{UW}$. Given the current state is ψ, the probability that the next inserted tihl is e equals

$$P(\psi, e) = p_\psi(e) \prod_{\min S_\psi \leq \sigma < r_e} (1 - \pi_{t(\sigma)}^{\psi(\sigma)B}),$$

where

$$p_\psi(e) = \pi_{t(r_e)}^{\psi(r_e)B} \prod_{\sigma \in \overline{\mathrm{supp}}(e)} \pi_{t(\sigma)}^{\psi(\sigma)e(\sigma)}.$$

Note that $\pi_t^{BU} = \pi_t^{NU}$ and $\pi_t^{B^-U} = \pi_t^{N^-U}$ for all $U \in \{B, H, N, E\}$. As a consequence, one can in fact restrict the domain of the functions φ to $\{B, H, E, B^-, H^-\}$. This decreases the number of functions φ, which is favourable for computational purposes.

Defining $P_\varphi(\mathbf{k})$ in analogy to Definition 4, we obtain in a similar way as in Lemma 1 for $(\mathbf{k}, \varphi) \neq (\mathbf{0}, \varphi_0)$ the "forward equation"

$$P_\varphi(\mathbf{k}) = \sum_{(\psi, e):\, \varphi = [\psi, e]} P_\psi(\mathbf{k} - \mathbf{v}_e) p_\psi(e) \vartheta(e, \mathbf{k}) \cdot \tag{18}$$

$$\cdot \frac{p_\varphi(h)}{p_\psi(h)} \cdot \frac{1 - \pi_{t(r_e)}^{\varphi(r_e)B}}{1 - \pi_{t(r_e)}^{\psi(r_e)B}} \prod_{\sigma \in \mathrm{supp}(e)} (1 - \pi_{t(\sigma)}^{\varphi(\sigma)B})$$

and the start condition

$$P_{\varphi_0}(\mathbf{0}) = p_{\varphi_0}(h) \prod_{\sigma \neq r} (1 - \pi_{t(\sigma)}^{HB}).$$

In order to turn (18) into a recursion one has to solve it for the vector $(P_\varphi(\mathbf{k}))_\varphi$, since $\mathbf{v}_e$ can be equal to $\mathbf{0}$. This seems only tractable as long as the number ℓ of input sequences is small. An alternative is to neglect all tihls e with $\mathbf{v}_e = 0$, which seems legitimate if the indel rates are low. Another approach, which is briefly discussed in the next section, is to apply algorithms like simulated annealing or Markov chain Monte Carlo methods that assign sequences to the internal nodes (cf. [4], [7]).

4 Indel Models and Tree Reconstruction

Based on a stochastic insertion-deletion model, one can ask for the joint estimation of a multiple alignment and a phylogenetic tree by maximizing the probability of the alignment and the likelihood of the tree. *Exact* computations seem hopeless for large data sets. There are however promising heuristic approaches to this problem.

Thorne and Kishino [20] used the TKF model to get maximum-likelihood estimates of the pairwise evolutionary distances between the sequences in a data set from which they then reconstructed a neighbor-joining tree [16]. Although this approach might not be the best way to estimate the sequences' phylogeny, such a tree can be used as a guide tree for progressive statistical alignment and thus is a convenient starting point for further analysis. *Progressive statistical alignment* was introduced in [7] for the case of the TKF model: Given a tree $\mathcal{T}$, distinguish one of its nodes as the root and proceed from the leaves of $\mathcal{T}$ towards the root in the following way. Infer the most probable alignment for every pair of neighbouring nodes. Then, estimate a sequence at their parent node together with the indel history of this parent and its two children, which is compatible to the alignment of the offspring sequences. The sequence at the parent node is then aligned to its sibling, and so on. At the end of this procedure one has sequences for every node of the tree together with their indel history. This indel history and these inferred sequences depend, however, on the order in which the tree is traversed. In order to decrease this effect it is convenient to visit every node and every branch in random order, thereby emitting a new sequence at the respective node and optimizing the indel history along the respective branch. The sampling procedures for sequences at internal nodes use mutiple HMMs, where the observable emissions are the offspring sequences and the sequences at the neighbouring nodes, respectively.

In the heuristic outlined in [4] we use simulated annealing to approach a phylogenetic tree and an indel history whose posterior probability given a set of sequences $\mathbf{A}$ is maximal. We start with the progressive alignment procedure described above and then modify the trees and indel histories in the following way: Inspired by the tree sampling procedure proposed in [10], we change the edge lengths in the current tree $\mathcal{T}^i$ independently by a small random amount, which results in a proposed tree $\mathcal{T}^*$. For short internal edges, the proposed length may come out negative, which we interpret as proposed change in the tree topology. Whenever this situation occurs one has to sample new sequences and a new indel history for the newly introduced internal edge, conditioned on the neighbouring sequences and their current alignment. Again, this is done with a multiple HMM, where the observable emissions are the sequences at the neighbouring positions. The indel history is further refined by producing a new sequence for each of the tree's internal nodes and by optimizing the pairwise alignments along the tree's branches. Thus, altering the tree also results in proposing a new indel history $\tilde{w}^*$. The proposal is accepted with probability $\min\left\{1, \left(\frac{p_{\mathcal{T}^*}(\mathbf{A}, \tilde{w}^*)}{p_{\mathcal{T}^i}(\mathbf{A}, \tilde{w}^i)}\right)^{c(i)}\right\}$ where $p_{\mathcal{T}}(\mathbf{A}, \tilde{w})$ denotes the likelihood of $\mathcal{T}$, $\tilde{w}^i$ is the indel history in step i, and $c(i)$ increases with i.

In [4] also some alternative proposal chains are suggested which move faster through tree space, one of them based on nearest-neighbour interchanges (cf. [17]).

The estimation procedure becomes rather time consuming when applied to a data set consisting of a larger number of sequences. A way out might be to resort to a heuristics for tree reconstruction based on trees with four leaves only, known as quartet puzzling ([18]). This can be done in the following way: Find for every quartet of sequences the maximum-likelihood tree topology with the help of the recursion in section 3.5. Then, use the quartet puzzling algorithm to repeatedly combine all these quartet trees into a tree for the entire data set. Compute a consensus tree of all these intermediate trees and use this consensus tree as a guide tree for progressive statistical alignment Finally, optimize the branch lengths and the indel history by simulated annealing.

It is a challenging task and object of ongoing research to further improve the optimization heuristics for multiple alignment and phylogenetic trees. We have no doubt that this is most adequately done in the framework of stochastic insertion-deletion models and statistical alignment.

Acknowledgement Our project was embedded into the DFG program "Interacting stochastic systems of high complexity". Financial support by the DFG and stimulating scientific exchange with other groups in the program, in particular those in Berlin, Erlangen and Frankfurt, is gratefully acknowledged. We would especially like to thank Matthias Birkner, Gerton Lunter and István Miklós for fruitful discussions.

References

1. J. Felsenstein, Evolutionary trees from DNA sequences: a maximum likelihood approach. *J. Mol. Evol.* 17:368-376, 1981.
2. J. Felsenstein, *Inferring Phylogenies*, Sinauer Associates, 2004.
3. R. Fleißner, Sequence alignment and phylogenetic inference, PhD Thesis, Universität Düsseldorf, 2003.
4. R. Fleißner, D. Metzler, A. von Haeseler, Simultaneous statistical multiple alignment and phylogeny reconstruction, Preprint.
5. O. Gotoh, Multiple sequence alignments: algorithms and applications. *Adv. Biophys.* 36:159-206, 1999.
6. J. Hein, C. Wiuf, B. Knudsen, M.B. Møller, G. Wibling, Statistical alignment: Computational properties, homology testing and goodness-of-fit. *J. Mol. Biol.*, 302:265-179, 2000.
7. I. Holmes, W. J. Bruno, Evolutionary HMMs: a Bayesian approach to multiple alignment, *Bioinformatics* 17:803-820, 2001.
8. G.A. Lunter, I. Miklós, J.L. Jensen, A. Drummond, J. Hein, Bayesian phylogenetic inference under a statistical insertion-deletion model, Proceedings of the Third International Workshop on Algorithms in Bioinformatics, Budapest, G. Benson, R.D.M. Page (eds), *Lecture Notes in Comp. Sc.* 2812, pp. 228-244, Springer 2003.
9. G.A. Lunter, I. Miklós, Y.S. Song, J. Hein, An efficient algorithm for statistical multiple alignment on arbitrary phylogenetic trees. *J. Comp. Biol.* 10(6):869-889, 2003.
10. B. Mau, M. A. Newton, Phylogenetic inference for binary data on dendrograms using Markov chain Monte Carlo, *J. Computational and Graphical Statistics* 6:122-131. 1997.
11. D. Metzler, Statistical alignment based on fragment insertion and deletion models, *Bioinformatics* 19:490-499, 2003.

12. D. Metzler, R. Fleißner, A. Wakolbinger, A. von Haeseler, Assessing variability by joint sampling of alignments and mutation rates, *J. Mol Evol.* 53:660-669, 2001.

13. I. Miklós, G.A. Lunter, I. Holmes, A "long indel" model for evolutionary sequence alignment. *Mol. Biol. Evol.* (accepted)

14. I. Miklós, Z. Toroczkai, Z. An improved model for statistical alignment. In: *Algorithms in bioinformatics* (O. Gascuel and B. M. E.Moret, eds) pp. 1-10, Springer.

15. G. J. Mitchison, A probabilistic treatment of phylogeny and sequence alignment, *J. Mol. Evol.* 49:11-22, 1999.

16. N. Saitou, M. Nei, The neighbour-joining method: A new method for reconstructing phylogenetic trees. *Mol. Biol. Evol.* 4:406-425, 1987.

17. D. L. Swofford, G. L. Olsen, P. J. Waddell, D. M. Hillis, Phylogenetic Inference, in: *Molecular Systematics* (D. M. Hillis, C. Moritz, B. K. Mable, eds) pp. 407-514, Sinauer Accociates, Sunderland.

18. K. Strimmer, A. von Haeseler, Quartet puzzling: A quartet maximum likelihood method for reconstructing tree topologies. *Mol. Biol. Evol.* 13:964-969, 1996.

19. S. Tavaré, Some probabilistic and statistical problems on the analysis of DNA sequences, *Lec. Math. Life Sci.* 17, 57-86, 1986.

20. J. L. Thorne, H. Kishino, Freeing phylogenies from artifacts of alignment. *Mol. Biol. Evol.* 9:1148-1162, 1992.

21. J. L. Thorne, H. Kishino, J. Felsenstein, An evolutionary model for maximum likelihood alignment of DNA sequences, *J. Mol. Evol.* 33:114-124, 1991.

22. J. L. Thorne, H. Kishino, J. Felsenstein, Inching towards reality: an improved likelihood model for sequence evolution. *J. Mol. Evol.* 34:3-16, 1992.

Branching Processes in Random Environment – A View on Critical and Subcritical Cases

Matthias Birkner[1], Jochen Geiger[2], and Götz Kersting[3]

. Weierstraß-Institut für Angewandte Analysis und Stochastik
Mohrenstr. 39, D-10117 Berlin, Germany
birkner@wias-berlin.de
. Fachbereich Mathematik, Technische Universität Kaiserslautern
Postfach 3049, D-67653 Kaiserslautern, Germany
jgeiger@mathematik.uni-kl.de
. Fachbereich Mathematik, Universität Frankfurt
Fach 187, D-60054 Frankfurt am Main, Germany
kersting@math.uni-frankfurt.de

Summary. Branching processes exhibit a particularly rich longtime behaviour when evolving in a random environment. Then the transition from subcriticality to supercriticality proceeds in several steps, and there occurs a second 'transition' in the subcritical phase (besides the phase-transition from (sub)criticality to supercriticality). Here we present and discuss limit laws for branching processes in critical and subcritical i.i.d. environment. The results rely on a stimulating interplay between branching process theory and random walk theory. We also consider a spatial version of branching processes in random environment for which we derive extinction and ultimate survival criteria.

MSC 2000 subject classifications. Primary 60J80, Secondary 60G50, 60F17
Key words and phrases. Branching process, random environment, random walk, conditioned random walk, Spitzer's condition, functional limit theorem

1 Introduction

Branching processes in random environment is one of the topics, which we have studied in the project "Verzweigende Populationen: Genealogische Bäume und räumliches Langzeitverhalten" (Grant Ke 376/6) within the DFG-Schwerpunkt "Interagierende stochastische Systeme von hoher Komplexität". It is representative in that a central aspect of the whole project were probabilistic constructions of genealogical trees and the interplay between branching processes and random walks. These concepts turn out to be significant for branching processes in random environment in a rather specific way. This is due to the fact that from a methodical point of view the subject of branching processes in random environment is substantially influenced by the theory of random walks. Many of the results rely on a stimulating interplay between branching process theory on the one hand and fluctuation theory of random walks on the other hand. Here we present limit theorems for critical and subcritical branching processes in i.i.d. random environment and give a detailed explanation of this relationship.

For classical Galton-Watson branching processes it is assumed that individuals reproduce independently of each other according to some given offspring distribution. In the setting of this paper the offspring distribution varies in a random fashion, independently from one generation to the other. A mathematical formulation of the model is as follows. Let Δ be the space of probability measures on $\mathbb{N}_0$ which equipped with the metric of total variation becomes a Polish space. Let Q be a random variable with values in Δ. Then an infinite sequence $\Pi = (Q_1, Q_2, \ldots)$ of i.i.d. copies of Q is said to form a *random environment*. A sequence of $\mathbb{N}_0$-valued random variables $Z_0, Z_1, \ldots$ is called a *branching process in the random environment* Π, if Z_0 is independent of Π and if given Π the process $Z = (Z_0, Z_1, \ldots)$ is a Markov chain with

$$\mathcal{L}\big(Z_n \mid Z_{n-1} = z, \Pi = (q_1, q_2, \ldots)\big) \;=\; \mathcal{L}\big(\xi_1 + \cdots + \xi_z\big) \qquad (1)$$

for every $n \geq 1, z \in \mathbb{N}_0$ and $q_1, q_2, \ldots \in \Delta$, where $\xi_1, \xi_2, \ldots$ are i.i.d. random variables with distribution q_n. In the language of branching processes Z_n is the nth generation size of the population and Q_n is the distribution of the number of children of an individual at generation $n - 1$. For convenience, we will assume throughout that $Z_0 = 1$ a.s.

As it turns out the asymptotic properties of Z are first of all determined by its *associated random walk* $S = (S_0, S_1, \ldots)$. This random walk has initial state $S_0 = 0$ and increments $X_n = S_n - S_{n-1}, n \geq 1$ defined as

$$X_n \;:=\; \log \sum_{y=0}^{\infty} y \, Q_n(\{y\}) \,,$$

which are i.i.d. copies of the logarithmic mean offspring number

$$X \;:=\; \log \sum_{y=0}^{\infty} y \, Q(\{y\}) \,.$$

We will assume that X is a.s. finite. Due to (1) and our assumption $Z_0 = 1$ a.s. the conditional expectation of Z_n given the environment Π,

$$\mu_n \;:=\; \mathbf{E}[Z_n \mid \Pi] \,,$$

can expressed by means of S as

$$\mu_n \;=\; e^{S_n} \quad \mathbf{P}\text{-}a.s.$$

According to fluctuation theory of random walks (compare Chapter XII in [Fe71]) one may distinguish three types of branching processes in random environment. First, S can be a random walk with positive drift, which means that $\lim_n S_n = \infty$ a.s. In this case $\mu_n \to \infty$ a.s. and Z is called a *supercritical* branching process. Second, S can have negative drift, i.e., $\lim_n S_n = -\infty$ a.s. Then $\mu_n \to 0$ a.s. and Z is called *subcritical*. Finally, S may be an oscillating random walk meaning that $\limsup_n S_n = \infty$ a.s. and at the same time $\liminf_n S_n = -\infty$ a.s., which implies $\limsup_n \mu_n = \infty$ a.s. and $\liminf_n \mu_n = 0$ a.s. Then we call Z a *critical* branching process. Our classification extends the classical distinction of branching processes in random environment introduced in [AK71, SW69]. There it is assumed that the random walk has finite mean. Then Z is supercritical, subcritical or critical according as $\mathbf{E}X > 0$, $\mathbf{E}X < 0$ or $\mathbf{E}X = 0$. Only recently the requirement that the expectation of X exists could be dropped (see [AGKV04, DGV03, VD03]).

The distinction plays a similar role as for ordinary branching processes: If Z is a (non-degenerate) critical or subcritical branching process in random environment, then the population eventually becomes extinct with probability 1. This fact is an immediate consequence of the first moment estimate

$$\mathbf{P}\{Z_n > 0 \mid \Pi\} \leq \mathbf{P}\{Z_m > 0 \mid \Pi\} \leq \mu_m \quad \text{for all } m \leq n,$$

which implies

$$\mathbf{P}\{Z_n > 0 \mid \Pi\} \leq \min_{m \leq n} \mu_m = \exp\left(\min_{m \leq n} S_m\right). \tag{2}$$

Thus in critical and subcritical cases $\mathbf{P}\{Z_n > 0 \mid \Pi\} \to 0$ a.s. and consequently, $\mathbf{P}\{Z_n \to 0\} = 1$. Other than for classical Galton-Watson branching processes the converse is not always true. Also for supercritical branching processes the random fluctuations of the environment can have the effect that the entire population dies out within only a few generations. A criterion that excludes such catastrophes is the following integrability condition on the conditional probability of having no children (see Theorem 3.1 in [SW69])

$$\mathbf{E} \log \left(1 - Q(\{0\})\right) > -\infty.$$

Finding criteria for ultimate survival becomes a challenging problem for branching processes in random environment with a spatial component. In a branching random walk in space-time i.i.d. random environment individuals have a spatial position (in a countable Abelian group) and move as independent random walkers. Each individual at generation $n - 1$ located at x uses offspring law $Q_{n,x}$, where the $Q_{n,x}$ form an i.i.d. random field. The fact that individuals living at different locations use independent offspring distributions leads to a smoothing in comparison to the non-spatial case, and thus one might expect that ultimate survival is easier.

The topic of branching processes in random environment has gone through quite a development. For fairly long time research was restricted to the special case of offspring distributions with a linear fractional generating function (which means that given the event $\{Q \geq 1\}$ the offspring distribution Q is geometric with random mean) and to the case where the associated walk has zero mean, finite variance increments. Under these restrictions fairly explicit (albeit tedious) calculations of certain Laplace transforms are feasible. Later the advantages of methods from the theory of random walks have been recognized. General offspring distributions, however, have become accessible only recently.

In this paper we focus on the longtime behavior of critical and subcritical processes (except for the part on spatial branching processes), where differences to classical branching processes are especially striking. Our discussion is based on a formula for the probability of non-extinction, which is derived in the next section. It is this formula that allows to determine the exact asymptotic magnitude of the probability of non-extinction for general offspring distributions (see [GK00, GKV03, GL01]). The linear fractional case could be treated long before (see [Af80, Ko76]).

In Section 3 we discuss the limiting behaviour of branching processes in critical random environment under a general assumption known as Spitzer's condition in the theory of random walks. Section 4 is devoted to the transition from criticality to subcriticality, which takes place in several steps. If one considers the conditioned branching process given the event $\{Z_n > 0\}$ rather than the unconditioned process, then the transition from (super)criticality to subcriticality occurs within the subcritical phase. This so-called intermediate subcritical case is especially intriguing since it exhibits subcritical as well as supercritical behaviour alternating in time. In Section 5 we derive criteria for the a.s. extinction of a branching random walk in space-time i.i.d. random environment. It is shown that a transient (symmetrized) individual motion can be strong enough to completely counteract the correlations between different individuals introduced by the environment, while a recurrent motion cannot.

Within the DFG-Schwerpunkt "Interagierende stochastische Systeme von hoher Komplexität" related projects are the study of the longtime behaviour of population models in a stationary situation by Greven and by Höpfner and Löcherbach (see their articles in this volume). Also our results display phenomena known from statistical physics (see the models studied in the section "Disordered media" of this volume).

Acknowledgement The presented results were obtained in the course of the project "Verzweigende Populationen: Genealogische Bäume und räumliches Langzeitverhalten" and its successor, the Russian-German research project "Branching processes in random environment", sponsored by the DFG and the Russian Foundation of Basic Research (Grant 436 Rus 113/683). The results on *spatial* branching processes in random environment were obtained by the first author who had studied the corresponding infinite system in his PhD thesis [Bi03]. We thank the DFG and the Russian Foundation of Basic Research for their lasting support.

2 A Formula for the Survival Probability

In this section we investigate the relationship between the conditional survival probabilities $\mathbf{P}\{Z_n > 0 \mid \Pi\}$, $n \geq 0$ and the associated random walk $(S_n)_{n \geq 0}$. An essential observation will be that the estimate (2) not only gives an upper bound but also the right impression of the magnitude of the survival probability,

$$\mathbf{P}\{Z_n > 0 \mid \Pi\} \approx \min_{m \leq n} \mu_m = \exp(\min_{m \leq n} S_m). \tag{3}$$

This relation is plausible, however, trying to elaborate it directly to a precise mathematical statement is not a great promise. Instead we reformulate (3). The upper bound (which is estimate (2)) says that $\mathbf{P}\{Z_n > 0 \mid \Pi\}$ becomes particularly small, whenever this is true for some μ_m, $m \leq n$. To put it another way $1/\mathbf{P}\{Z_n > 0 \mid \Pi\}$ gets large, whenever one of the quantities $1/\mu_0, \ldots, 1/\mu_n$ is large. It would be particularly useful, if this dependence could be expressed in a linear fashion with coefficients which can be controlled sufficiently well. Thus we are looking for a formula of the form

$$\frac{1}{\mathbf{P}\{Z_n > 0 \mid \Pi\}} = \sum_{k=0}^{n} \frac{A_{k,n}}{\mu_k}$$

with 'tractable' quantities $A_{k,n} \geq 0$, $0 \leq k \leq n$. In the case of offspring distributions with a linear fractional generating function such a formula had been known for a long time. For general offspring distributions such a representation has been obtained only recently in [GK00].

Here we provide two different approaches to the desired formula. Following [GK00] we first proceed in a purely analytical manner and obtain analytical expressions and estimates for the $A_{k,n}$. Then we will present an alternative derivation of the formula which is obtained by means of a probabilistic construction of the conditional family tree produced by the branching process. This second approach, while leading to the same coefficients $A_{k,n}$, allows a probabilistic interpretation of these terms.

The analytical approach is a straightforward calculation. Consider the (random) generating functions

$$f_k(s) := \sum_{y=0}^{\infty} s^y Q_k(\{y\}), \quad 0 \leq s \leq 1,$$

$k = 1, 2, \ldots$ and their compositions

$$f_{k,n}(s) := f_{k+1}(f_{k+2}(\cdots f_n(s) \cdots)), \quad 0 \leq k \leq n,$$

with the convention $f_{n,n}(s) := s$. It follows from (1) and our assumption $Z_0 = 1$ a.s. that $f_{0,n}$ is the conditional generating function of Z_n given the environment Π, i.e.,

$$f_{0,n}(s) = \mathbf{E}[s^{Z_n} \mid \Pi] \quad a.s., \qquad 0 \le s \le 1. \tag{4}$$

Using a telescope type of argument we deduce

$$\frac{1}{1 - f_{0,n}(s)} = \frac{1}{\mu_n(1-s)} + \sum_{k=0}^{n-1} g_{k+1}(f_{k+1,n}(s))\frac{1}{\mu_k}, \qquad 0 \le s < 1,$$

with

$$g_k(s) := \frac{1}{1 - f_k(s)} - \frac{1}{f_k'(1)(1-s)}, \qquad 0 \le s < 1.$$

By (4), we have $\mathbf{P}\{Z_n > 0 \mid \Pi\} = 1 - f_{0,n}(0)$, so that we end up with the formula

$$\frac{1}{\mathbf{P}\{Z_n > 0 \mid \Pi\}} = \sum_{k=0}^{n} \frac{A_{k,n}}{\mu_k} \tag{5}$$

with random coefficients

$$A_{k,n} := g_{k+1}(f_{k+1,n}(0)), \ 0 \le k < n \quad \text{and} \quad A_{n,n} := 1.$$

Apparently, this identity has first been utilized by Jirina [Ji76]. By convexity of the f_k, the coefficients $A_{k,n}$ are non-negative. Geiger and Kersting proved the upper bound (see Lemma 2.1 in [GK00])

$$g_k(s) \le \eta_k, \qquad 0 \le s < 1, \tag{6}$$

where η_k is the standardized second factorial moment of Q_k,

$$\eta_k := \sum_{y=0}^{\infty} y(y-1)Q_k(\{y\}) \Big/ \Big(\sum_{y=0}^{\infty} yQ_k(\{y\}) \Big)^2. \tag{7}$$

Consequently, the $A_{k,n}$, $k < n$ satisfy the estimate

$$0 \le A_{k,n} \le \eta_{k+1}, \tag{8}$$

which allows to control their magnitude and to exploit the representation of the survival probability in (5).

Thus we succeeded in deriving a formula of the desired form, its probabilistic meaning still has to be revealed. This is achieved by the probabilistic approach which we discuss next. Our starting point is the identity

$$\frac{1}{\mathbf{P}\{Z_n > 0 \mid \Pi\}} = \frac{\mathbf{E}[Z_n \mid Z_n > 0, \Pi]}{\mathbf{E}[Z_n \mid \Pi]} = \frac{\mathbf{E}[Z_n \mid Z_n > 0, \Pi]}{\mu_n}, \tag{9}$$

which reduces the calculation of the non-extinction probability at n to that of the conditional mean of Z_n. For the latter it is essential to view the branching process as a mechanism to generate a random family tree rather than a mere sequence of generation sizes. We think of the family tree as a rooted ordered

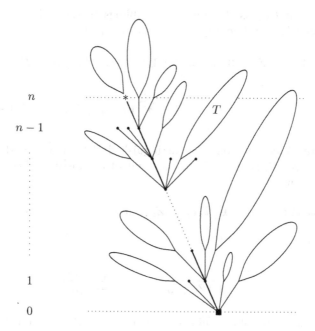

tree with the distinguishable offspring of each individual ordered from left to right. This allows to decompose the family tree along the ancestral line of the left-most individual $*$ at generation n. (In the illustration above the bold line marks the distinguished line of descent starting with the founding ancestor ■.)

This picture corresponds to a lucid probabilistic construction of the conditional family tree. The construction was originally devised and investigated for classical Galton-Watson branching processes in [Ge99]. However, the construction works as well for branching processes in varying (deterministic) environment, and, hence, for branching processes in random environment when conditioning on Π and the event $\{Z_n > 0\}$. Note that because of the distinguished role of $*$ as the left-most individual of generation n the subtrees to the left of $*$'s line of descent stay below level n. On the other hand it is natural to expect – and it does follow from the results in [Ge99] – that the subtrees to the right of the distinguished ancestral line remain unaffected by the conditioning event $\{Z_n > 0\}$, i.e., they evolve like ordinary branching processes given Π, independent of other parts of the tree. Thus, if T is one of the subtrees to the right with root in generation $k \in \{1, \ldots, n\}$ and $Z_{n,T}$ its number of individuals at generation n, then

$$\mathbf{E}[Z_{n,T} \mid Z_n > 0, \Pi] \;=\; e^{S_n - S_k} \;=\; \frac{\mu_n}{\mu_k}\;.$$

Moreover, if $R_{k,n}$ is the number of siblings to the right of $*$'s ancestor at generation k, then linearity of expectation yields

$$\mathbf{E}[Z_n \mid Z_n > 0, \Pi] = 1 + \mathbf{E}\left[\sum_T Z_{n,T} \mid Z_n > 0, \Pi\right]$$

$$= 1 + \sum_{k=0}^{n-1} \mathbf{E}[R_{k+1,n} \mid Z_n > 0, \Pi]\frac{\mu_n}{\mu_{k+1}} \ ,$$

where the 1 comes from the distinguished individual $*$ and the first sum extends over all subtrees T to the right of the distinguished ancestral line. Thus, putting

$$\tilde{A}_{k,n} := \exp(-X_{k+1})\,\mathbf{E}[R_{k+1,n} \mid Z_n > 0, \Pi] \ , \quad k < n \ , \quad \text{and} \quad \tilde{A}_{n,n} := 1 \ ,$$

we end up with (recall (9))

$$\frac{1}{\mathbf{P}\{Z_n > 0 \mid \Pi\}} = \sum_{k=0}^{n} \frac{\tilde{A}_{k,n}}{\mu_k} \ . \tag{10}$$

It is just a matter of careful calculation to show that representations (5) and (10) agree.

Clearly, $\tilde{A}_{k,n} \geq 0$. An upper estimate for $\tilde{A}_{k,n}$ may be derived as follows. The construction of the conditional family tree in [Ge99] shows that the ancestor of $*$ at generation k belongs to a family of a size, which is stochastically bounded by the so-called size-biased distribution

$$\tilde{Q}_k(\{y\}) := \frac{y\,Q_k(\{y\})}{\exp(X_k)} \ , \quad y \geq 0.$$

It follows

$$\mathbf{E}[R_{k,n} \mid Z_n > 0, \Pi] \leq \sum_y y\,\tilde{Q}_k(\{y\}) - 1$$

$$= \exp(-X_k) \sum_y y(y-1)Q_k(\{y\}) \ .$$

Thus

$$0 \leq \tilde{A}_{k,n} \leq \exp(-2X_{k+1})\sum_y y(y-1)Q_{k+1}(\{y\}),$$

which is relation (8). We note that even though the estimate (8) can be verified in a purely analytical manner it took the probabilistic interpretation to find it.

3 Criticality

As explained in the introduction a branching population in a (sub)critical random environment eventually becomes extinct with probability 1. A fundamental question for these processes is the following: If the population survives until some late generation n, in which way does this event occur. One can imagine several ways: The population might have been lucky to find an extraordinarily favourable environment, in which chances for survival are high. Or the population evolved in a typical environment, still by good luck it managed to avoid extinction. In order to weigh these alternatives the formula

$$\mathbf{P}\{Z_n > 0 \mid \Pi\} = \left(\sum_{k=0}^{n} A_{k,n} e^{-S_k} \right)^{-1}$$

from the last section proves useful. For the moment let us neglect effects coming from the $A_{k,n}$, which are of secondary order. Then, as explained above, the probability of survival is high for those unlikely environments, for which $\min(S_0, \ldots, S_n)$ is close to 0, otherwise the probability is extremely small.

In this section we look at critical branching processes, i.e., at the case of an oscillating associated random walk. Then the probability of the event $\{\min(S_0, \ldots, S_n) \geq 0\}$ is typically of order $n^{-\gamma}$ for some $\gamma > 0$. Thus exponentially small probabilities are negligible and one has to take only those ways of survival into account, where the associated random walk stays away from low negative values (for precise results compare Theorem 1 below and its corollary). These heuristic considerations suggest a program of research, which has been initiated by Kozlov [Ko76] and followed up by several authors, see [Af93, Af97, Af01b, DGV03, GK00, Ko95, Va02].

The results of this section and parts of the discussion are adapted from the recent paper [AGKV04]. As an overall assumption let us adopt **Spitzer's condition** from fluctuation theory of random walks: Assume that there exists a number $0 < \rho < 1$ such that

$$\frac{1}{n} \sum_{m=1}^{n} \mathbf{P}\{S_m > 0\} \to \rho .$$

This general condition guarantees that S is a non-degenerate oscillating random walk. Not striving for greatest generality we restrict ourselves here to two alternative sets of transparent further assumptions. Our first set of assumptions strengthens the condition on the associated random walk and adds some integrability condition on the standardized second factorial moment η (recall the definition in (7)).

Assumption (A). *Let the distribution of X belong without centering to the domain of attraction of a stable law λ with index $\alpha \in (0, 2]$. The limit law*

λ *is not a one-sided stable law, i.e.,* $0 < \lambda(\mathbb{R}^+) < 1$. *Further let for some* $\epsilon > 0$

$$\mathbf{E}\,(\log^+ \eta)^{\alpha+\epsilon} < \infty\,.$$

Note that $\alpha = 2$ is the case of a non-degenerate zero mean, finite variance random walk.

If we assume specific types of distribution for Q, then we can relax Assumption (A). In particular, we can deal with the three most important special cases of offspring distributions (playing a prominent role for classical branching processes, too).

Assumption (B). *The random offspring distribution Q is a.s. a binary, a Poisson or a geometric distribution on $\mathbb{N}_0$ (with random mean).*

The first of our results on the longtime behaviour of branching processes in critical random environment concerns the asymptotic behaviour of the survival probability at n (for the proofs of the results of this section we refer to the paper [AGKV04]).

Theorem 1. *Assume (A) or (B). Then there exists a number $0 < \theta < \infty$ such that, as $n \to \infty$,*

$$\mathbf{P}\{Z_n > 0\} \sim \theta\,\mathbf{P}\{\min(S_0,\dots,S_n) \geq 0\}\,.$$

This theorem gives evidence for our claim that the behaviour of Z is primarily determined by the random walk S. Only the constant θ depends on the fine structure of the random environment.

Since under Spitzer's condition the asymptotic behaviour of the probability on the right-hand side above is well-known, we obtain the following corollary.

Corollary 1. *Assume (A) or (B). Then, as $n \to \infty$,*

$$\mathbf{P}\{Z_n > 0\} \sim \theta n^{\rho-1} l(n)\,,$$

where $l(1), l(2), \dots$ is a sequence varying slowly at infinity.

The next theorem shows that conditioned on the event $\{Z_n > 0\}$ the process $Z_0, Z_1, \dots, Z_n$ exhibits 'supercritical behaviour'. Supercritical branching processes (whether classical or in random environment) obey the growth law $Z_n/\mu_n \to W$ a.s., where W is some typically non-degenerate random variable. In our situation this kind of behaviour can no longer be formulated as a statement on a.s. convergence, since the conditional probability measures depend on n.

Instead, we define for integers $0 \leq r \leq n$ the process $X^{r,n} = (X_t^{r,n})_{0 \leq t \leq 1}$ given by

$$X_t^{r,n} := \frac{Z_{r+[(n-r)t]}}{\mu_{r+[(n-r)t]}}\,, \quad 0 \leq t \leq 1\,. \tag{11}$$

Theorem 2. *Assume (A) or (B). Let $r_1, r_2, \ldots$ be a sequence of natural numbers such that $r_n \leq n$ and $r_n \to \infty$. Then, as $n \to \infty$,*

$$\mathcal{L}(X^{r_n, n} \mid Z_n > 0) \implies \mathcal{L}((W_t)_{0 \leq t \leq 1}),$$

where the limiting process is a stochastic process with a.s. constant paths, i.e., $\mathbf{P}\{W_t = W \text{ for all } t \in [0, 1]\} = 1$ for some random variable W. Furthermore,

$$\mathbf{P}\{0 < W < \infty\} = 1.$$

The symbol $\implies$ denotes weak convergence w.r.t. the Skorokhod topology in the space $D[0, 1]$ of càdlàg functions on the unit interval. Again the growth of Z is in the first place determined by the random walk (namely, the sequence $(\mu_k)_{0 \leq k \leq n}$). The fine structure of the random environment is reflected only in the distribution of W.

Thus the properties of S determine the behaviour of Z in the main. On the other hand one has to take into account that the properties of the random walk change drastically, when conditioned on the event $\{Z_n > 0\}$. As explained above one expects that S conditioned on $\{Z_n > 0\}$ behaves just as S given the event $\{\min(S_0, \ldots, S_n) \geq 0\}$, i.e., like a random walk conditioned to stay positive for a certain period of time (a so-called random walk meander). The next theorem confirms this expectation. Here we need Assumption (A).

Theorem 3. *Assume (A). Then there exists a slowly varying sequence $\ell(1), \ell(2), \ldots$ such that, as $n \to \infty$,*

$$\mathcal{L}\left(\left(n^{-\frac{1}{\alpha}} \ell(n) S_{[nt]}\right)_{0 \leq t \leq 1} \,\middle|\, Z_n > 0\right) \implies \mathcal{L}(L^+),$$

where L^+ denotes the meander of a stable Lévy process with index α.

Shortly speaking the meander $L^+ = (L_t^+)_{0 \leq t \leq 1}$ is a stable Lévy process L conditioned to stay positive for $0 < t \leq 1$ (for details we refer to [AGKV04]). In view of Theorem 2 the assertion of Theorem 3 is equivalent to the following result.

Corollary 2. *Assume (A). Then, as $n \to \infty$,*

$$\mathcal{L}\left(\left(n^{-\frac{1}{\alpha}} \ell(n) \log Z_{[nt]}\right)_{0 \leq t \leq 1} \,\middle|\, Z_n > 0\right) \implies \mathcal{L}(L^+)$$

for some slowly varying sequence $\ell(1), \ell(2), \ldots$

4 A Transition within the Subcritical Phase

We have seen in Theorem 2 that critical branching processes in random environment exhibit supercritical behaviour, when conditioned on non-extinction.

This phenomenon does not vanish instantly in the subcritical phase, but persists for processes in the 'vicinity' of criticality. The transition from criticality to subcriticality proceeds in several steps in a fashion, which is not known for ordinary branching processes. From now on we assume that the conditional mean offspring number has finite moments of all orders t,

$$\mathbf{E}\exp(tX) < \infty , \quad t \geq 0 .$$

In particular, X^+ has finite mean and subcriticality corresponds to

$$\mathbf{E}X < 0 .$$

In this case the probability of non-extinction at generation n no longer decays at a polynomial rate (as in the critical case) but at an exponential rate. Dealing with exponentially small probabilities it is natural to consider the 'tilted' measures $\widehat{\mathbf{P}}_\beta$, $\beta \geq 0$ given by

$$\widehat{\mathbf{E}}_\beta[\phi(Q_1,\ldots,Q_n,Z_1,\ldots,Z_n)] := \gamma^{-n}\mathbf{E}[\phi(Q_1,\ldots,Q_n,Z_1,\ldots,Z_n)e^{\beta S_n}]$$

for non-negative test functions ϕ, where

$$\gamma = \gamma_\beta := \mathbf{E}[e^{\beta X}] .$$

For the probability of survival at n we obtain

$$\mathbf{P}\{Z_n > 0\} = \gamma^n\widehat{\mathbf{E}}_\beta[\mathbf{P}\{Z_n > 0 \mid \Pi\}e^{-\beta S_n}] , \quad n \geq 0 .$$

For a suitable choice of β let us proceed in a heuristic fashion and replace $\mathbf{P}\{Z_n > 0 \mid \Pi\}$ by the upper bound $\exp(\min(S_0,\ldots,S_n))$, which, as we have argued in Section 2, is typically of the same order. Thus we consider

$$\mathbf{E}[e^{\min(S_0,\ldots,S_n)}] = \gamma^n\widehat{\mathbf{E}}_\beta[e^{\min(S_0,\ldots,S_n)-\beta S_n}] . \tag{12}$$

Following common strategies from large deviation theory we like to choose β in such a way that the expectation on the right-hand side of (12) no longer decays (nor grows) at an exponential rate. To keep the quantity $\min(S_0,\ldots,S_n)-\beta S_n$ bounded from above, we only consider the range $\beta \leq 1$. There are three different scenarios where $\min(S_0,\ldots,S_n) - \beta S_n$ is close to 0 with sufficiently large probability:

(S1) $\widehat{\mathbf{E}}_\beta X = 0$ with $\beta < 1$. Then $\min(S_0,\ldots,S_n)-\beta S_n$ takes a value close to 0 if and only if both $\min(S_0,\ldots,S_n)$ and S_n are close to 0. It is known that for a zero mean, finite variance random walk the probability of the event $\{S_0,\ldots,S_{n-1} \geq 0, S_n \leq 0\}$ (the probability for a random walk excursion of length n) is of order $n^{-3/2}$. Thus in this case one expects

$$\widehat{\mathbf{E}}_\beta[e^{\min(S_0,\ldots,S_n)-\beta S_n}] \approx n^{-3/2} .$$

(S2) $\widehat{\mathbf{E}}_\beta X = 0$ with $\beta = 1$. Then $\min(S_0, \ldots, S_n) - S_n$ is close to 0, if $\min(S_0, \ldots, S_n)$ and S_n are close to each other, which essentially means that the path $S_0, \ldots, S_n$ attains its minimum close to the end. For a zero mean, finite variance random walk the probability of the event $\{S_0, \ldots, S_{n-1} \geq S_n\}$ is of order $n^{-1/2}$. Thus in this case one expects

$$\widehat{\mathbf{E}}_1[e^{\min(S_0, \ldots, S_n) - S_n}] \approx n^{-1/2} .$$

(S3) $\widehat{\mathbf{E}}_1 X < 0$. Then it is to be expected that

$$\widehat{\mathbf{E}}_1[e^{\min(S_0, \ldots, S_n) - S_n}] \approx \text{const} ,$$

since for a random walk with negative drift the quantity $\min(S_0, \ldots, S_n) - S_n$ is of constant order.

Each of these scenarios might occur for branching processes in random environment. Our discussion above shows that the we have to distinguish three cases depending on where the parameter $\widehat{\mathbf{E}}_1 X = \gamma^{-1}\mathbf{E}[Xe^X]$ is located with respect to the value 0 (note that $\widehat{\mathbf{E}}_\beta X$ increases with β). In the sequel we will discuss each case in detail. Some of the results to follow are contained in manuscripts submitted for publication, some are proved in special situations and other are part of research in progress. We seize the opportunity and give a general prospect of the results rather than to go into technical details or to state precise integrability conditions (for single results see [Af80, Af98, Af01a, Af01b, De88, d'SH97, FV99, GKV03, GL01, Liu96]).

4.1 The weakly subcritical case. First we assume

$$\mathbf{E}X < 0 , \quad \mathbf{E}[Xe^X] > 0 . \tag{13}$$

Then there exists a number $0 < \beta < 1$ such that

$$\mathbf{E}[Xe^{\beta X}] = 0 ,$$

which is our choice for the parameter of the tilted measure. Thus

$$\widehat{\mathbf{E}}_\beta X = 0 ,$$

and we are in the scenario described under (S1). In accordance with our discussion there the survival probability obeys the following asymptotics.

Result 4.1. *Assume (13). Then there exists a number $0 < \theta < \infty$ such that, as $n \to \infty$,*

$$\mathbf{P}\{Z_n > 0\} \sim \theta n^{-3/2}\gamma^n .$$

Moreover, our heuristic arguments from (S1) suggest that only those environments give an essential contribution to the probability of non-extinction whose associated random walk is (close to) an excursion. The following result confirms this expectation.

Result 4.2. *Assume (13) and let $\sigma^2 := \widehat{\mathbf{E}}_\beta X^2$. Then, as $n \to \infty$,*

$$\mathcal{L}\Big((\sigma n^{-\frac{1}{2}} S_{[nt]})_{0 \le t \le 1} \Big| Z_n > 0\Big) \implies \mathcal{L}(B^e),$$

where B^e denotes a standard Brownian excursion of length 1, i.e., a Brownian motion $B = (B_t)_{0 \le t \le 1}$ given the event $\{B_t \ge 0$ for all $0 \le t < 1, B_1 = 0\}$.

This result is in some sense similar to Theorem 3 so that again one might expect that the branching process exhibits supercritical behaviour, as long as the random walk excursion is still far away from 0. The following result, which is the analogue of Theorem 2, says that this is indeed true. In fact, this behaviour persists until just before the random walk's return to 0. Recall the definition of the process $X^{r,n}$ from (11).

Result 4.3. *Assume (13). Let $r_1, r_2, \ldots$ be a sequence of natural numbers such that $r_n \le n/2$ and $r_n \to \infty$. Then, as $n \to \infty$,*

$$\mathcal{L}\big(X^{r_n, n-r_n} \big| Z_n > 0\big) \implies \mathcal{L}((W_t)_{0 \le t \le 1}),$$

where the limiting process is a stochastic process with a.s. constant paths, i.e., $\mathbf{P}\{W_t = W$ for all $t \in [0,1]\} = 1$ for some random variable W. Furthermore,

$$\mathbf{P}\{0 < W < \infty\} = 1.$$

4.2 The intermediate subcritical case. This case is where supercritical behaviour resolves into subcritical behaviour on the event $\{Z_n > 0\}$. Here we assume

$$\mathbf{E}X < 0, \quad \mathbf{E}[Xe^X] = 0. \tag{14}$$

This time we choose $\beta = 1$, so that

$$\widehat{\mathbf{E}}_\beta X = 0.$$

Now (S2) is the relevant scenario and the behaviour of the process changes in a remarkable fashion.

Our first result again describes the exact asymptotics of the non-extinction probability at n.

Result 4.4. *Assume (14). Then there exists a number $0 < \theta < \infty$ such that, as $n \to \infty$,*

$$\mathbf{P}\{Z_n > 0\} \sim \theta n^{-1/2} \gamma^n.$$

As suggested in the discussion of (S2) the event of survival is essentially carried by those environments whose associated random walk path has a late minimum. This is expressed by the following limit law.

Result 4.5. *Assume (14). Then, as $n \to \infty$,*

$$\mathcal{L}\left(\left(\sigma n^{-\frac{1}{2}} S_{[nt]}\right)_{0 \le t \le 1} \Big| Z_n > 0\right) \implies \mathcal{L}(B^l),$$

where B^l denotes a standard Brownian motion $B = (B_t)_{0 \le t \le 1}$ conditioned on the event $\{B_t \ge B_1$ for all $0 \le t \le 1\}$.

The change from Brownian excursion B^e to conditional Brownian motion B^l implies a change in the behaviour of Z which is unique for branching processes. Note that B^l is not built of a single excursion but contains countably many local excursions. At the same time B^l has (uncountably) many local minima. For this reason we distinguish two different types of epochs. The moment $k \in \{0, 1, \ldots, n\}$ is of the first type if the random walk S reaches a new minimum around k. Such moments are particularly difficult to survive and an unconditioned population would die out in the long run when repeatedly facing such bottlenecks. Given the event $\{Z_n > 0\}$ it is natural to expect that at such epochs the population consists of only very few indviduals (as in the strongly subcritical case discussed below). The other type of epoch is if the last minimum of S before time k is quite some time in the past and the next minimum is still quite far in the future. Then the random walk is in the midth of a local excursion. On such time stretches the population again exhibits supercritical behaviour and grows to a large size (as in the preceding weakly subcritical case). Thus it is reasonable to expect that sub- and supercritical behaviour alternate in the course of time.

It requires some effort to convert these heuristics into a precise mathematical statement. Still we formulate a result corresponding to Theorem 2 and Result 4.3. Keeping track of the successive minima of S the right normalization of Z_k is seen to be

$$\tilde{\mu}_k := \frac{\mu_k}{\min_{j \le k} \mu_j} = \exp\left(S_k - \min_{j \le k} S_j\right)$$

(rather than μ_k). (3) shows that one might alternatively use

$$\bar{\mu}_k := \frac{\mu_k}{\mathbf{P}\{Z_k > 0 \mid \Pi\}} = \mathbf{E}[Z_k \mid Z_k > 0, \Pi].$$

Now recall that an excursion interval of the conditioned Brownian process B^l is an interval $e = (a, b) \subset [0, 1]$ of maximal length such that $B^l_a = B^l_b$ and $B^l_t > B^l_a$ for all $t \in e$. There are countably many excursion intervals $e_1, e_2, \ldots$ which we assume to be enumerated in some order. Write $j(t) := j$, if $t \in e_j$.

Result 4.6. *Assume (14) and let $0 < t_1 < \cdots < t_k < 1$. Then, as $n \to \infty$,*

$$\mathcal{L}\left(\left(\frac{Z_{[nt_1]}}{\tilde{\mu}_{[nt_1]}}, \dots, \frac{Z_{[nt_k]}}{\tilde{\mu}_{[nt_k]}}\right) \,\Big|\, Z_n > 0\right) \xrightarrow{w} \mathcal{L}\left((W_{j(t_1)}, \dots, W_{j(t_k)})\right) ,$$

where $W_1, W_2, \dots$ are independent of B^l and i.i.d. copies of some random variable W satisfying

$$\mathbf{P}\{0 < W < \infty\} = 1 .$$

Thus the ith and jth component of the limiting random vector are identical, if t_i and t_j belong to the same excursion interval of B^l, otherwise they are independent. This result expresses the alternation between sub- and super-critical behaviour described above.

4.3 The strongly subcritical case. Finally, we come to the case, where super-critical behaviour vanishes completely. Now let

$$\mathbf{E}X < 0 , \quad \mathbf{E}[Xe^X] < 0 . \tag{15}$$

We choose $\beta = 1$ again. Then (other than in the intermediate subcritical case)

$$\widehat{\mathbf{E}}_\beta X < 0$$

and we are in the situation captured in scenario (S3).

Result 4.7. *Assume (15). Then there exists a number $0 < \theta \leq 1$ such that, as $n \to \infty$,*

$$\mathbf{P}\{Z_n > 0\} \sim \theta\gamma^n .$$

In scenario (S3) the behaviour of S given non-extinction at n is governed by the law of large numbers. This is the content of the following result.

Result 4.8. *Assume (15). Then, as $n \to \infty$,*

$$\mathcal{L}\left(\left(n^{-1}S_{[nt]}\right)_{0\leq t\leq 1} \,\Big|\, Z_n > 0\right) \implies \mathcal{L}\left((t\,\widehat{\mathbf{E}}_1 X)_{0\leq t\leq 1}\right) .$$

In particular, local random walk excursions vanish in the scaling limit and Z no longer exhibits any supercritical behaviour on the event $\{Z_n > 0\}$. Instead our last result shows that the population stays small throughout the time interval from 0 to n.

Result 4.9. *Assume (15) and let $0 < t_1 < \cdots < t_k < 1$. Then, as $n \to \infty$,*

$$\mathcal{L}\left((Z_{[nt_1]}, \dots, Z_{[nt_k]}) \,\big|\, Z_n > 0\right) \xrightarrow{d} (W_1, \dots, W_k) ,$$

where $W_1, W_2, \dots$ are i.i.d. copies of some random variable W satisfying

$$\mathbf{P}\{1 \leq W < \infty\} = 1 .$$

5 Spatial Branching Processes in Space-Time i.i.d. Random Environment

In this section we consider a model of a spatial branching process in random environment. Now individuals live on a countable Abelian group G. Again we start with a single founding ancestor who at time 0 is located at position $0 \in G$. Individuals at generation $n-1$ located at x have independent offspring according to the random distribution $Q_{n,x}$. Each child independently moves to y with probability $p(x, y) = p(y - x)$, where p is a given (irreducible) random walk kernel. Let $Z_n(x)$ be the number of individuals at x in generation n, and $Z_n := \sum_x Z_n(x)$ the population size at generation n.

The random offspring distributions $Q_{n,x}$, $n \in \mathbb{N}$, $x \in G$ are assumed i.i.d. Given the environment $\Pi = (Q_{n,x})$ individuals branch and move independently. For a probability measure $q = (q_y)_{y \in \mathbb{N}_0}$ we denote the first and second moments as

$$m_1(q) := \sum_y y q_y, \quad m_2(q) := \sum_y y^2 q_y.$$

We will assume $\mathbf{E} m_2(Q) < \infty$. A quantity of particular interest is

$$m := \mathbf{E} m_1(Q), \tag{16}$$

the mean number of offspring per individual. For *deterministic* Q, where $(Z_n)_{n \geq 0}$ is a classical Galton-Watson process, the case $m = 1$ is critical in the sense that $Z_n \to 0$ a.s., if $m \leq 1$ (and $q_1 < 1$), whereas $Z_n \to \infty$ with positive probability if $m > 1$. On the other hand, we have seen in previous sections that in a *non-spatial* scenario with $G = \{0\}$, the criterion for criticality is $\mathbf{E} \log m_1(Q) = 0$. We shall see that the model considered here is in a certain sense intermediate between these two cases. In principle, it is a special case of branching processes in random environment with infinitely many types.

Let $\mathcal{F}_n := \sigma(Z_k(x), Q_{k+1,x}, x \in G, k \leq n)$. One easily checks that

$$M_n := \frac{Z_n}{m^n}, \quad n = 0, 1, \dots \tag{17}$$

is an $(\mathcal{F}_n)$-martingale.

Let X and Y be two independent p-random walks on G. Note that then $X - Y$ is again a random walk, we denote its transition matrix by $\tilde{p}(x) := \sum_y p(y) p(x - y)$. Let

$$\alpha := \frac{\mathbf{E} m_1(Q)^2}{\left(\mathbf{E} m_1(Q)\right)^2} \quad (\in (1, \infty)), \tag{18}$$

$$\alpha_* := \sup \left\{ a > 0 : \mathbf{E}_{(0,0)} \left[a^{\#\{i \geq 1: Y_i = X_i\}} \, | X \right] < \infty \text{ a.s.} \right\}. \tag{19}$$

Remark. Obviously, $\alpha_* \geq \left(\mathbf{P}_{(0,0)}(\exists k \geq 1 : X_k = Y_k) \right)^{-1}$. This inequality is in fact strict in many cases, as

$$\alpha_* = 1 + \left(\sum_{n \geq 1} \exp(-H(p^n)) \right)^{-1}$$

whenever $\sup_{n \geq 1, x \in G} p^n(x)/\tilde{p}^n(0) < \infty$, where $H(p^n)$ is the entropy of $p^n(0, \cdot)$, see Theorem 5 in [Bi03].

Theorem 4. a) If $m \leq 1$ we have $Z_n \to 0$ a.s., in particular, $M_\infty = 0$.
b) Assume that $\mathrm{Var}(m_1(Q)) > 0$ and that there exists a sequence (C_n) of finite subsets of G satisfying

$$\sum_n |C_n|^{-1} = \infty \quad and \quad \lim_{n \to \infty} \sum_{y \in C_n} p^n(y, 0) = 1. \tag{20}$$

Then $M_\infty = 0$, irrespective of the value of m.
c) If $m > 1$, $\tilde{p}$ is transient, $\alpha < \alpha_*$, and

$$\liminf_{k \to \infty} m^k p^k(0, X_k) > 0 \quad a.s., \tag{21}$$

then the family (M_n) is uniformly integrable, hence, $\mathbf{E}\, M_\infty = 1$. In particular, $Z_n \to \infty$ with positive probability in this case.

Remark. Condition (20) in b) implies that $\tilde{p}$ is recurrent. It is satisfied for $G = \mathbb{Z}$ if $\sum_x |x| p(x) < \infty$, and for $G = \mathbb{Z}^2$ if $\sum_x ||x||^2 p(x) < \infty$ (see [Li85], p. 450).

   Condition (21) is satisfied (in fact, the left-hand side is ∞) whenever p satisfies a local CLT and a LIL, so e.g. for simple random walk on $\mathbb{Z}^d$.

Sketch of proof. a) Follows easily by comparison with a classical Galton-Watson process with offspring generating function $\bar{\varphi}(s) := \mathbf{E}[\sum_y s^y Q(\{y\})]$: A Jensen-type argument gives

$$\mathbf{E}[s^{Z_n}] \geq \underbrace{\bar{\varphi} \circ \cdots \circ \bar{\varphi}(s)}_{n \text{ times}}, \quad s \in [0, 1],$$

and it is well-known that the right-hand side tends to 1 as $n \to \infty$ because $\bar{\varphi}'(1-) \leq 1$. Thus $Z_n \to 0$ in probability, which together with $\{Z_n = 0\} \subset \{Z_m = 0, \forall m \geq n\}$ proves the claim.
b) Note that $\zeta_n(x) := m^{-n} \mathbf{E}[Z_n(x) \mid \Pi]$ satisfies

$$\zeta_{n+1}(x) = \sum_y \zeta_n(y) \frac{m_1(Q_{n+1,y})}{m} p(y, x), \quad n \in \mathbb{N}_0, x \in G, \tag{22}$$

so (ζ_n) is a discrete-time version of a Potlatch process (cf. e.g. Chapter IX in [Li85]), starting from $\zeta_0(\cdot) = \delta(0, \cdot)$. One can easily adapt the proof of Theorem IX.4.5 in [Li85] to this discrete-time setting to obtain that $\mathbf{E}[M_n \mid \Pi] = \sum_x \zeta_n(x) \to 0$ a.s. as $n \to \infty$ if there is a sequence (C_n)

satisfying (20). This implies that $M_n \to 0$ in probability by the (conditional) Markov inequality. As $M_n \to M_\infty$ a.s. we obtain $\mathbf{P}(M_\infty = 0) = 1$.

c) It is well-known that uniform integrability of $(M_n)_{n \in \mathbb{N}_0}$ is equivalent to tightness of the family of the corresponding size-biased laws, see e.g. Lemma 9 in [Bi03]. We use a stochastic representation of the size-biasing of M_n to obtain the criterion. We can view $\mathbf{P}$ as a measure on {ordered, spatially embedded trees} $\times$ {space-time fields of offspring distributions}. We think of size-biasing of Z_n as picking "uniformly" an individual at time n from a(n infinite) forest of independent trees (grown in independent environments), and then looking at the tree the chosen individual belongs to. Let us denote the chosen (spatially embedded, ordered) tree by τ with selected ancestral line λ, denote the space-time field of offspring laws by $Q_{n,x}$, $x \in G, n \in \mathbb{N}$. Technically, we construct a measure $\tilde{\mathbf{P}}$ on {infinite, ordered, spatially embedded trees} $\times$ {ancestral lines} $\times$ {space-time fields of offspring distributions} with the property

$$
\begin{aligned}
&\tilde{\mathbf{P}}\big(\tau|_n = t, \lambda|_n = a, Q_{k,x} \in B_{x,k}, x \in A, 1 \le k \le n+1\big) \\
&= \frac{1}{m^n} \mathbf{P}\big(\tau|_n = t, Q_{k,x} \in B_{x,k}, x \in A, 1 \le k \le n+1\big)
\end{aligned}
\tag{23}
$$

for any $n \in \mathbb{N}$ and spatially embedded, ordered, rooted tree t of height at most n, ancestral line $a \subset t$ of length n, finite $A \subset G$ and measurable $B_{x,k} \subset \mathcal{M}_1(\mathbb{N}_0)$. (τ, λ, Π) under $\tilde{\mathbf{P}}$ arises as follows:

Let $Y = (Y_j)$ be a p-random walk starting from $Y_0 = 0$. This will be the spatial embedding of the selected line. Given Y, the field Π has independent coordinates, and the law of $Q_{j+1,x}$ is $\mathcal{L}(Q)$ if $x \ne Y_j$, whereas it is $\mathcal{L}(\hat{Q})$ if $x = Y_j$, where $\tilde{\mathbf{P}}(\hat{Q} \in d\nu) = (m_1(\nu)/m)\mathbf{P}(Q \in d\nu)$. Given this, let $\hat{K}_0, \hat{K}_1, \dots$ be independent with $\tilde{\mathbf{P}}(\hat{K}_i = k | Y, \Pi) = (k/m_1(Q_{i+1,Y_i}))Q_{i+1,Y_i}(\{k\})$. The individual at generation i along the selected line will have $\hat{K}_i$ children (note that $\hat{K}_i \ge 1$ always), and we choose uniformly one of them to continue the distinguished line. The spatial embedding of the chosen child will then be Y_{i+1}, her siblings take each an independent p-step from Y_i. Finally, all the siblings branching off from the selected line form independent populations in the given space-time medium Π they see. Then straightforward calculation gives (23). We omit the details, see [Bi03], p. 69f where an analogous proof is carried out for a related construction. By summing over all possible ancestral lines of length n we obtain from (23) that

$$
\tilde{\mathbf{P}}(Z_n = k) = \frac{k}{m^n} \mathbf{P}(Z_n = k).
$$

We have to show that the distributions of M_n under $\tilde{\mathbf{P}}$ form a tight family. In order to do so it suffices to show that

$$
\sup_{n \in \mathbb{N}} \tilde{\mathbf{E}}\big[M_n \,|\, Y\big] < \infty \quad \text{a.s.}
\tag{24}
$$

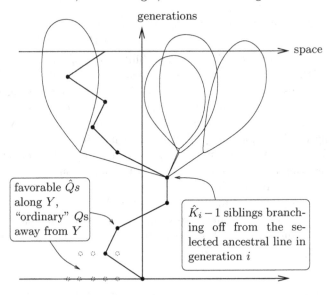

To prove (24) we compute

$$\tilde{\mathbf{E}}\left[\frac{Z_n}{m^n}\bigg|Y\right] = \frac{1}{m^n}\left\{1 + \mathbf{E}[\hat{K}-1]\sum_{k=0}^{n-1}\mathbf{E}_{(Y_k,k)}\left[\left(\mathbf{E}[m_1(\hat{Q})]\right)^{\#\{k<i<n:Y_i=X_i\}}\times\right.\right.$$

$$\left.\left.m^{\#\{k<i<n:Y_i\neq X_i\}}\bigg|Y\right]\right\}$$

$$= \frac{1}{m^n} + \sum_{k=0}^{n-1}\frac{\mathbf{E}[\hat{K}-1]}{m^{k+1}}\mathbf{E}_{(Y_k,k)}\left[\alpha^{\#\{k<i<n:Y_i=X_i\}}\bigg|Y\right].$$

Note that assumption (21) implies that

$$A := \inf_{k\in\mathbb{N}}m^{k+1}p_k(0,Y_k)\in(0,\infty]\quad\text{a.s.,}\tag{25}$$

which allows to estimate

$$\frac{1}{m^k}\mathbf{E}_{(Y_k,k)}\left[\alpha^{\#\{k<i<n:Y_i=X_i\}}\bigg|Y\right]\leq\frac{1}{A}\mathbf{E}_{(0,0)}\left[\mathbf{1}(Y_k=X_k)\alpha^{\#\{k<i<n:Y_i=X_i\}}\bigg|Y\right],$$

yielding

$$\tilde{\mathbf{E}}\left[\frac{Z_n}{m^n}\bigg|Y\right]\leq\frac{1}{m^n}+\frac{\mathbf{E}[\hat{K}-1]}{A}\mathbf{E}_{(0,0)}\left[\sum_{k=0}^{n-1}\mathbf{1}(Y_k=X_k)\alpha^{\#\{k<i<n:Y_i=X_i\}}\bigg|Y\right]$$

$$\leq 1 + \frac{\mathbf{E}[\hat{K}-1]}{A}\mathbf{E}_{(0,0)}\left[\#\{i\geq 0:Y_i=X_i\}\alpha^{\#\{i\geq 0:Y_i=X_i\}}\bigg|Y\right]$$

uniformly in n. The right-hand side is a.s. finite because $\alpha<\alpha_*$. $\square$

Let us note in concluding that in the spatial scenario, the picture is by far less complete than for the non-spatial branching processes discussed in Sections 2 to 4. Even the question of tractable criteria for criticality seems open. Theorem 4 shows that if $\tilde{p}$ is transient and the variance of the mean offspring number is small enough in comparison with a threshold that depends only on the motion, then the classical dichotomy for Galton-Watson processes (without random environments) holds: $m \leq 1$ implies almost sure extinction of a single family, while $m > 1$ implies survival with positive probability, and in this case the population grows exponentially.

This naturally leaves us with some questions: First, is the threshold value α_* given in Theorem 4 (c) sharp in the sense that $\alpha > \alpha_*$ would imply $M_\infty = 0$? Arguments from [CSY03] can be adapted to our scenario to show that even for transient $\tilde{p}$ we will have $M_\infty = 0$, if $\mathbf{E}[m_1(Q) \log m_1(Q)]/\mathbf{E}[m_1(Q)]$ is sufficiently large, see Theorem 2.3 (a) there, but there is this a wide gap between the two criteria. In principle, a route to check this would be to directly analyse the distributions of M_n under $\tilde{P}$ (recall that we only checked boundedness of some conditional expectation in the proof of Theorem 4 (c)). This looks like a very hard problem.

Second, even if the growth rate of the population is not captured by m and hence $M_\infty = 0$, this need not necessarily (and in general will not) imply a.s. extinction of a single family: Comparison with the non-spatial case shows at least that $\mathbf{E}[\log(m_1(Q))] > 0$ entails a positive probability of non-extinction irrespective of p.

References

[Af80]   Afanasyev, V.I.: Limit theorems for a conditional random walk and some applications. Diss. Cand. Sci., MSU, Moscow (1980)

[Af93]   Afanasyev, V.I.: A limit theorem for a critical branching process in random environment. Discrete Math. Appl., **5**, 45–58 (in Russian) (1993)

[Af97]   Afanasyev, V.I.: A new theorem for a critical branching process in random environment. Discrete Math. Appl., **7**, 497–513 (1997)

[Af98]   Afanasyev, V. I.: Limit theorems for a moderately subcritical branching process in a random environment. Discrete Math. Appl., **8**, 35–52 (1998)

[Af01a]   Afanasyev, V.I.: Limit theorems for intermediately subcritical and strongly subcritical branching processes in a random environment. Discrete Math. Appl., **11**, 105–132 (2001)

[Af01b]   Afanasyev, V.I.: A functional limit theorem for a critical branching process in a random environment. Discrete Math. Appl., **11**, 587–606 (2001)

[AGKV04]   Afanasyev, V.I., Geiger, J., Kersting, G., Vatutin, V.A.: Criticality for branching processes in random environment. To appear in Ann. Prob. (2004)

[AK71]   Athreya, K.B., Karlin, S.: On branching processes with random environments: I, II. Ann. Math. Stat., **42**, 1499–1520, 1843–1858 (1971)

290   Matthias Birkner, Jochen Geiger, and Götz Kersting

[Bi03]   Birkner, M.: Particle systems with locally dependent branching: long-
            time behaviour, genealogy and critical parameters. PhD thesis, Johann
            Wolfgang Goethe-Universität, Frankfurt. (2003)
            http://publikationen.stub.uni-frankfurt.de/volltexte/2003/314/

[CSY03]   Comets, F., Shiga, T., Yoshida, N.: Directed polymers in random envi-
            ronment: path localization and strong disorder. Bernoulli, **9**, 705–723
            (2003)

[De88]   Dekking, F.M.: On the survival probability of a branching process in a
            finite state i.i.d. environment. Stochastic Processes Appl., **27**, 151–157
            (1988)

[d'SH97]   D'Souza, J.S., Hambly, B.M.: On the survival probability of a branch-
            ing process in a random environment. Adv. Appl. Probab., **29**, 38–55
            (1997)

[DGV03]   Dyakonova, E.E., Geiger, J., Vatutin, V.A.: On the survival probabil-
            ity and a functional limit theorem for branching processes in random
            environment. To appear in Markov Process. Relat. Fields. (2003)

[Fe71]   Feller, W.: An Introduction to Probability Theory and Its Applications,
            Vol. II. Wiley, New York (1971)

[FV99]   Fleischmann, K., Vatutin, V.A.: Reduced subcritical Galton-Watson
            processes in a random environment. Adv. Appl. Probab., **31**, 88–111
            (1999)

[Ge99]   Geiger, J.: Elementary new proofs of classical limit theorems for
            Galton–Watson processes. J. Appl. Probab., **36**, 301–309 (1999)

[GK00]   Geiger, J., Kersting, G.: The survival probability of a critical branching
            process in a random environment. Theory Probab. Appl., **45**, 517–525
            (2000)

[GKV03]   Geiger, J., Kersting, G., Vatutin, V.A.: Limit theorems for subcritical
            branching processes in random environment. Ann. Inst. Henry Poincaré
            Probab. Stat., **39**, 593–620 (2003)

[GL01]   Guivarc'h, Y., Liu, Q.: Propriétés asymptotiques des processus de
            branchement en environnement aléatoire. C. R. Acad. Sci. Paris Sér. I
            Math., **332**, no. 4, 339–344 (2001)

[Ji76]   Jirina, M.: Extinction of non-homogeneous Galton-Watson processes.
            J. Appl. Probab., **13**, 132–137 (1976)

[Ko76]   Kozlov, M.V.: On the asymptotic behavior of the probability of non-
            extinction for critical branching processes in a random environment.
            Theory Probab. Appl., **21**, 791–804 (1976)

[Ko95]   Kozlov, M.V.: A conditional function limit theorem for a critical
            branching process in a random medium. Dokl.-Akad.-Nauk., **344**, 12–
            15 (in Russian) (1995)

[Li85]   Liggett, T.M.: Interacting particle systems. Springer, New York (1985)

[Liu96]   Liu, Q.: On the survival probability of a branching process in a random
            environment. Ann. Inst. Henry Poincaré Probab. Stat., **32**, 1–10 (1996)

[SW69]   Smith, W.L., Wilkinson, W.E.: On branching processes in random en-
            vironments. Ann. Math. Stat., **40**, 814–827 (1969)

[Va02]   Vatutin, V.A.: Reduced branching processes in a random environment:
            The critical case. Teor. Veroyatn. Primen., **47**, 21–38 (in Russian)
            (2002)

[VD03]   Vatutin, V.A., Dyakonova, E.E.: Galton-Watson branching processes in random environment, I: Limit theorems. Teor. Veroyatn. Primen., **48**, 274–300 (in Russian) (2003)

Part III

Stochastic Analysis

Thin Points of Brownian Motion Intersection Local Times[*]

Achim Klenke

Fachbereich Mathematik und Informatik, Johannes-Gutenberg Universität Mainz
Staudingerweg 9, D-55099 Mainz, Germany
math@aklenke.de

Summary. Let ℓ be the projected intersection local time of two independent Brownian paths in $\mathbb{R}^d$ for $d = 2, 3$. We determine the lower tail of the random variable $\ell(B(0,1))$, where $B(0,1)$ is the unit ball. The answer is given in terms of intersection exponents, which are explicitly known in the case of planar Brownian motion. We use this result to obtain the multifractal spectrum, or spectrum of thin points, for the intersection local times.

Keywords: Brownian motion, intersection of Brownian paths, intersection local time, Wiener sausage, lower tail asymptotics, intersection exponent, Hausdorff measure, thin points, Hausdorff dimension spectrum, multifractal spectrum.

1 Introduction

Intersections of Brownian motion paths have been studied for a long time in probability and in the physics literature as they are (i) interesting in their own right, and (ii) share fundamental properties of more complex models in statistical physics (see Aizenman [Aiz85]). Here we compute the multifractal spectrum of the intersection local time of two Brownian motions in $\mathbb{R}^d$, $d = 2, 3$. As this spectrum is non-trivial this result contrasts with the predictions of the (non-rigorous) multifractal formalism from statistical physics. Another example for failure of this formalism has been found before by Perkins and Taylor [PT98].

Consider two Brownian motions $(W_t^1)_{t \in [0,1]}$ and $(W_t^2)_{t \in [0,1]}$ in $\mathbb{R}^d$, $d = 2, 3$, both started in the origin. Let

$$S := W_{[0,1]}^1 \cap W_{[0,1]}^2 = \{x \in \mathbb{R}^d : x = W_s^1 = W_t^2 \quad \text{for some } s, t \in [0,1]\}$$

denote the set of intersection points of the paths of these motions. This is the object we study in this work.

Let us start with a crude consideration. Brownian motion paths are generically two-dimensional objects (if $d \geq 2$). A naive co-dimension argument

---

[*] Joint Work with Peter Mörters (Bath)

suggests that S has Hausdorff dimension 2 if $d = 2$ and 1 if $d = 3$. Furthermore $S = \{0\}$ (almost surely) iff $d \geq 4$ (see [Kho03, Per96a]). In fact, Le Gall ([LG87, LG89]) showed that the exact Hausdorff gauge function is

$$\psi(r) := \begin{cases} r^2 \left[\log(1/r)(\log \log \log(1/r) \right]^2, & \text{if } d = 2, \\ r \left[\log \log(1/r) \right]^2, & \text{if } d = 3. \end{cases} \tag{1}$$

This is a statement about the typical points in the set S. Hence one may ask the questions: Are there points that are particularly "thick", that is, points $x \in S$ where S is denser than usual in the sense that the local dimension is smaller than $4 - d$? Are there points that are particularly "thin" with the local Hausdorff dimension exceeding the value $4 - d$? Questions of this type have been studied in fractal geometry for certain deterministic and random objects by means of the multifractal spectrum. Roughly speaking, this spectrum is a function $f : [0, \infty) \to [0, d]$, where $f(a)$ is the dimension of the set of such points $x \in S$ where S has local dimension a. We make this precise in a moment.

The answer that we give to this question is: yes there are thin points in the sense that $f(a) > 0$ for an interval of values of a exceeding $4-d$. We can in fact compute the function f exactly in terms of the non-intersection exponents for Brownian motions. (As this f is a convex function in this interval, it could not be found by the multifractal spectrum formalism that, by using Legendre transforms, always predicts concave spectra.) On the other side, it was found by Dembo et al [DPRZ02] for $d = 2$ and by König and Mörters [KM02] for $d = 3$ that $f(a) = 0$ for all values $a < 4 - d$, hence there are no thick points in S.

In the next sections we will explain in more detail the intersection local time, the multifractal spectrum, and non-intersection exponents and will present our results as well as sketches of the proofs.

2 Intersection Local Time

First of all we have to make precise how we measure the concentration of the set S. We do so by considering the so-called intersection local time ℓ which is defined as the ψ-Hausdorff measure (see (1)) of the set S. We can construct ℓ also via an approximation scheme using Wiener sausages

$$S_\varepsilon^i := \bigcup_{t \in [0,1]} B(W_t^i, \varepsilon), \qquad \varepsilon > 0, \ i = 1, 2,$$

where $B(x, r)$ is the open ball around $x \in \mathbb{R}^d$ with radius r. If we let $S_\varepsilon := S_\varepsilon^1 \cap S_\varepsilon^2$ then $S = \bigcap_{\varepsilon > 0} S_\varepsilon$ and one can show (see Le Gall [LG87, LG89]) that for any measurable set $A \subset \mathbb{R}^d$

$$\lim_{\varepsilon \to 0} h(\varepsilon)^{-1} \lambda(A \cap S_\varepsilon) = \ell(A), \qquad \text{almost surely,}$$

where λ is Lebesgue measure on $\mathbb{R}^d$ and h is the normalizing function defined by

$$h(\varepsilon) = \begin{cases} \pi^2 \log(1/\varepsilon)^{-2}, & \text{if } d = 2, \\ 4\pi^2 \varepsilon^2, & \text{if } d = 3. \end{cases}$$

It is this random measure ℓ that we study here.

3 The Multifractal Spectrum

Let us discuss the notion of the multifractal spectrum first for a general deterministic or random locally finite, fractal measure μ. The value $f(a)$ of the multifractal spectrum is the Hausdorff dimension of the set of points x with local dimension

$$\lim_{r \downarrow 0} \frac{\log \mu(B(x,r))}{\log r} = a. \tag{2}$$

Here we may need to replace the limit by liminf or limsup in order to obtain a nontrivial spectrum. Examples of rigorously verified multifractal spectra for measures arising in probability theory are the occupation measures of stable subordinators, see [HT97], the states of super-Brownian motion, see [PT98], and the harmonic measure on a Brownian path [Law98a].

The multifractal formalism is based on a counting scheme for balls of radius r with $\mu(B(x,r)) \approx r^a$. For random μ and atypical a this number of balls can be computed with a large deviations type method using Legendre transforms. Hence this formalism always predicts concave functions f. (See, e.g. Mörters [Mör03] for a short description of this method.)

Here we are interested in the spectrum of ℓ. From [KM02] and [DPRZ02] we know that $f(a) = 0$ for all $a < 4 - d$. In fact, for $d = 3$, König and Mörters [KM02] show existence of thick points in a more subtle notion with logarithmic corrections rather than algebraic ones. Here we consider the sets of thin points, defined via limsups in (2).

Definition 1. *Denote by*

$$T(a) = \left\{ x \in S : \limsup_{r \downarrow 0} \frac{\log \ell(B(x,r))}{\log r} \geq a \right\},$$

$$T^s(a) = \left\{ x \in S : \limsup_{r \downarrow 0} \frac{\log \ell(B(x,r))}{\log r} = a \right\}$$

the sets of a-thin points, respectively strictly a-thin points in S.

The aim is to compute $f(a) = \dim_H(T(a))$ and $\dim_H(T^s(a))$ for $a \geq 4-d$. In order to give an explicit formula we have to introduce the non-intersection exponents for paths of Brownian motions.

4 Non-intersection Exponents

Consider $M + N$ Brownian motions $W^1, \ldots, W^{M+N}$ in $\mathbb{R}^d$, the first M of them coloured blue, the other N motions coloured red. Start these Brownian motions (independently) uniformly distributed on the unit sphere $\partial B(0,1)$ and stop them upon leaving $B(0, R)$ for some large R that we let go to infinity. What is the probability that none of the red paths intersects with any of the blue paths? More formally, we define $\tau_R^i = \inf\{t \geq 0 : W_t^i \notin B(0, R)\}$ and

$$\mathcal{B}^1(R) = \bigcup_{i=1}^{M} W^i([0, \tau_R^i]) \quad \text{and} \quad \mathcal{B}^2(R) = \bigcup_{i=M+1}^{M+N} W^i([0, \tau_R^i]).$$

It is easy, using subadditivity, to show that there exists a constant $\xi_d(M, N)$ such that

$$\mathbf{P}\left[\mathcal{B}^1(R) \cap \mathcal{B}^2(R) = \emptyset\right] = R^{-\xi_d(M,N)+o(1)}, \quad \text{as } R \uparrow \infty. \tag{3}$$

The numbers $\xi_d(M, N)$ are called the *intersection exponents*, although *non-intersection exponents* might seem more appropriate.

Recently, Lawler, Schramm and Werner ([LSW01b, LSW01c, LSW02], see [LSW01a] for a survey) could verify a physicists' conjecture that in the plane some of these exponents should be rational. They have shown that

$$\xi_2(M, N) = \frac{\left(\sqrt{24M+1} + \sqrt{24N+1} - 2\right)^2 - 4}{48}. \tag{4}$$

This gives $\xi_2(2,2) = 35/12$. As the proof of (4) is based on conformal invariance, there is no analogue in $d = 3$. The only value known in dimension $d = 3$ is $\xi_3(1,2) = \xi_3(2,1) = 1$. Indeed, there is no reason why $\xi_3(2,2)$ should be a rational number. The known bounds show that

$$2 = 2\xi_3(2,1) > \xi_3(2,2) > \xi_3(2,1) = 1,$$

where the strict inequalities follow from the strict concavity of $\lambda \mapsto \xi_3(2, \lambda)$ established in [Law98b].

Finally, we define for these packages of Brownian motions the *projected intersection local time*

$$\ell_{M,N} := \sum_{i=1}^{M} \sum_{j=M+1}^{M+N} \ell^{i,j},$$

where $\ell^{i,j}$ is the intersection local time of $W_{[0,1]}^i$ with $W_{[0,1]}^j$.

5 Results

We have now collected all the ingredients needed to formulate our theorems. The main statement is the following.

Theorem 1. *Suppose ℓ is the intersection local time of two Brownian motions in $\mathbb{R}^d$, for $d = 2, 3$, starting in the origin and running for one unit of time.*

(i) In $d = 2$ we have

$$\mathbf{P}\big[T(a) \neq \emptyset\big] > 0 \iff \mathbf{P}\big[T^s(a) \neq \emptyset\big] = 1 \iff 2 \leq a \leq \frac{2\xi_2(2,2)}{\xi_2(2,2) - 2}.$$

Moreover, for these values of a, almost surely,

$$\dim T(a) = \dim T^s(a) = 2\,\frac{\xi_2(2,2)}{a} + 2 - \xi_2(2,2). \tag{5}$$

(ii) In $d = 3$ we have

$$\mathbf{P}\big[T(a) \neq \emptyset\big] > 0 \iff \mathbf{P}\big[T^s(a) \neq \emptyset\big] = 1 \iff 1 \leq a \leq \frac{\xi_3(2,2)}{\xi_3(2,2) - 1}.$$

Moreover, for these values of a, almost surely,

$$\dim T(a) = \dim T^s(a) = \frac{\xi_3(2,2)}{a} + 1 - \xi_3(2,2). \tag{6}$$

For the proof of this statement we need to count (for very small $r > 0$) the points $x \in S$ with $\ell(B(x,r)) \approx r^a$. An essential step is to compute the probability that given the point x is in S, the paths of W^1 and W^2 do not intersect too much, at least in a small neighbourhood $B(x,r)$ of x. We split the paths W^1 and W^2 into the parts before first hitting x and after. One can show that for the purpose of estimating the desired probability it is good enough to replace the (time-reversed) incoming paths by Brownian motion paths started in x. (The outgoing paths are so by the strong Markov property.) Hence we have to compute for two pairs of Brownian motion paths the probability that the paths of the first pair do not intersect too much with the paths of the second pair. However, roughly speaking, this is approximately the probability $\mathbf{P}[\ell_{2,2}(B(0,r)) \leq r^a]$. Computing this quantity is a large deviation problem. It turns out that the best strategy for the pairs of Brownian motions is to "behave normally" until first leaving $B(0, r^{a/(4-d)})$ and then to not intersect the paths of the other colour. As the paths live for one time unit the latter probability is of the same order of magnitude as not intersecting before leaving $B(0,1)$ and is hence $\approx r^{a\xi_d(2,2)/(4-d)}$.

All these heuristics to compute $\mathbf{P}[\ell_{2,2}(B(0,r)) \leq r^a]$ can be made precise. We give the more general result for packages of M and N Brownian motions. Note that by Brownian scaling it is no loss of generality to consider $B(0,1)$ instead of $B(0,r)$.

Theorem 2.
$$\lim_{\delta \downarrow 0} \frac{\log \mathbf{P}\big[\ell_{M,N}(B(0,1)) < \delta\big]}{-\log \delta} = -\frac{\xi_d(M,N)}{4-d}.$$

Note that the algebraic lower tail of $\ell_{M,N}(B(0,1))$ contrasts significantly with the behaviour of the upper tail $\mathbf{P}[\ell_{M,N}(B(0,1)) > \delta] \approx \exp(-\theta\delta^{-1/2})$ as $\delta \to \infty$ for some $\theta > 0$, as shown by König and Mörters [KM02]. Here lies the reason that there do exist *thin* points but there do not exist *thick* points.

6 Sketches of Proofs

Theorem 2, best strategy for the Brownian motions Let us first consider the proof of Theorem 2. As indicated above, the proof of the lower bound is not too difficult. We start M red Brownian motions and N blue Brownian motions in the origin and let each of them run normally until time $\tau^i(r^{a/(4-d)})$ when the ith motion first leaves $B(0, r^{a/(4-d)})$. As the set S has dimension $4 - d$ the amount of intersection local that is produced is of order r^a. By the strong Markov property we can start afresh these motions that are now independently uniformly distributed on $\partial B(0, r^{a/(4-d)})$ and condition them not to hit paths of the other colour before leaving $B(0, 1)$. The probability for this is of order $r^{a\xi_d(M,N)/(4-d)}$. We have neglected here that paths should also not be allowed to intersect with the pieces of the paths of the other colour that were produced before first leaving $B(0, r^{a/(4-d)})$. However, this can be fixed by a simple decoupling argument.

Theorem 2, lower bound The more intriguing problem is that of the lower bound. This is not quite unexpected but let us highlight one problem that is typical for the intersection local times. For simplicity assume $M = N = 1$. In order to show that the strategy described above is optimal, we need to show that if the two Brownian motion paths intersect, this intersection contributes with probability one to the intersection local time. In other words, we need to show that $\mathbf{P}[\ell(x, r) > 0 | x \in S] = 1$ for any $x \in \mathbb{R}^d$ and any $r > 0$. This would be easy if the hitting time of the Brownian paths was a stopping time. However it is not, at least not simultaneously for both motions. However, given W^1, the first hitting time of $W^1_{[0,1]}$ is a stopping time for W^2. As points are polar while the set $W^1_{[0,1]}$ is regular for W^2 (almost surely), given $x \in S$, as $\varepsilon \downarrow 0$ there are arbitrarily many boxes E of side length ε^3 that have mutual distance at least ε^2 and distance to x at most ε, such that both W^1 and W^2 hit E. Denote this event by $A(E)$. Consider the event $B(E)$ that $\ell(E) > 0$ and that both motions do not enter E after first hitting E and then first leaving a larger box of side length $2\varepsilon^3$. By scaling arguments the conditional probability of $B(E)$ given $A(E)$ is at least δ for some $\delta > 0$ that is independent of ε and independent of the paths before hitting E and after leaving the larger box. Hence if there are, for some $\varepsilon > 0$, at least N such boxes, the probability for $\ell(B(x, \varepsilon)) = 0$ is at most $(1 - \delta)^N$.

Heuristics for Theorem 1 We fix the value $a \geq 4 - d$. We first give a heuristic argument: Very roughly, as $\dim S = 4 - d$, about r^{-a} balls of radius $r^{a/(4-d)}$

are needed to cover S. For any one of these the probability that it contains a point $x \in S$ with $\log \ell(B(x,r)) \leq a \log r$ is $r^{-\xi(1-a/(4-d))}$, with $\xi = \xi_d(2,2)$, by an argument similar to that leading to Theorem 2. Picking just these balls for all radii $r^a = 2^{-na}$, $n \geq m$ gives a covering

$$B(x_1, r_1^{a/(4-d)}), B(x_2, r_2^{a/(4-d)}), B(x_3, r_3^{a/(4-d)}), \ldots$$

of the thin points such that

$$\mathbf{E}\left[\sum_{i=1}^{\infty} r_i^{a\alpha/(4-d)}\right] = \sum_{n=m}^{\infty} 2^{-na\alpha/(4-d)} 2^{na} 2^{n\xi(1-a/(4-d))},$$

which is finite if $a\alpha - a(4-d) - \xi(4-d-a) > 0 \iff \alpha > (4-d)\xi/a + 4 - d - \xi$. In fact, this argument can be made rigorous. This leads to the upper bounds (in (5) and (6)) for the Hausdorff dimension of the set $\mathcal{T}(a)$ of a-thin points.

The percolation limit set For the lower bounds in (5) and (6), we need more subtle methods. We start with explaining a standard procedure to measure the Hausdorff dimension of certain sets. Consider two sets $A, \Gamma \subset \mathbb{R}^d$ with Hausdorff dimensions α and $d - \gamma$ respectively. If A and Γ are sufficiently nice, then $\dim(A \cap \Gamma) = \alpha - \gamma$ (where negative values indicate that the set is empty). Thus $A \cap \Gamma = \emptyset$ if $\gamma > \alpha$ and $A \cap \Gamma \neq \emptyset$ if $\gamma < \alpha$. If we can vary γ then we can use Γ as a test set for the dimension, just by observing whether or not $A \cap \Gamma = \emptyset$. A good choice for Γ is a percolation limit set, constructed as follows: Fix $\gamma \in [0, d]$. For $n \in \mathbb{N}_0$ let

$$\mathcal{D}_n := \left\{x + [0, 2^{-n}]^d, x \in \{0, 2^{-n}, \ldots, 1 - 2^{-n}\}^d\right\}$$

be the set of dyadic cubes of side length 2^{-n} in $[0,1]^d$. We consider $B \in \mathcal{D}_{n+1}$ as a child of $A \in \mathcal{D}_n$ if $B \subset A$. Now construct inductively a branching process by considering $[0,1]^d \in \mathcal{D}_0$ as alive. For any $A \in \mathcal{D}_n$ that is alive let any child be alive with probability $2^{-\gamma}$, independently of everything else. Denote $\mathcal{P}_n := \{A \in \mathcal{D}_n : A \text{ is alive }\}$ and $\Gamma_n = \Gamma_n(\gamma) = \bigcup_{A \in \mathcal{P}_n} A$. By construction $\dim \Gamma(\gamma) = d - \gamma$ (given $\Gamma(\gamma) \neq \emptyset$). Now one can show (see e.g. [Per96b] for a proof): For any Borel set $A \subset [0,1]^d$ the following properties hold

(i) if $\dim A < \gamma$, then $\mathbf{P}\left[A \cap \Gamma(\gamma) \neq \emptyset\right] = 0$,
(ii) if $\dim A > \gamma$, then $\mathbf{P}\left[A \cap \Gamma(\gamma) \neq \emptyset\right] > 0$.

Decoupling leads to the lower bound in Theorem 1 In order to show the lower bound in Theorem 1 we proceed similarly as described above for the proof that an intersection of paths implies positive intersection local time. Though the argument here is drastically more involved. The first step is to show that there are sufficiently many "admissible" cubes of order k, that is cubes $A \in \mathcal{P}_k$ which are hit by both W^1 and W^2 but which are not hit by either motion after first leaving a slightly larger box of side length $2^{-k(1-\varepsilon)}$. In fact, we show that the number of these cubes is at least of order $2^{k(4-d-\gamma)}$. (A subtle problem is here that planar Brownian motion is recurrent and is thus very likely to return to a box once visited.) By construction the various k–admissible boxes are decoupled. Thus it remains to show that for some of these boxes E, say, there exists an $x \in E \cap S$ with $\ell(B(x, 2^{-k}) \backslash B(x, 2^{-ak})) = 0$ (a so-called k-thin point). We show that the probability for E to contain such a point, given that E is admissible, is at least of order $2^{-k(4-d-\gamma-\xi)(1-a/(4-d))}$. As there are at least $2^{k(4-d-\gamma)}$ admissible boxes, we do have k–thin points for all sufficiently large k if $\gamma < 4 - d - \xi - (4 - d)\xi/a$. This shows that $\mathbf{P}[\mathcal{T}(a) \cap \Gamma(\gamma) \neq \emptyset] > 0$ if $\gamma < 4 - d - \xi - (4 - d)\xi/a$ which in turn implies (using an additional 0–1 argument) the lower bounds for $\mathcal{T}(a)$ in (5) and (6).

Acknowledgement We thank the *Mathematisches Forschungsinstitut Oberwolfach* where at the Miniworkshop "Stochastische Prozesse in zufälligen Medien" in May 2002 our joint work was initiated. We further thank the DFG Schwerpunkt programme "Stochastically interacting systems of high complexity" for support. For financial support we also thank the ESF scientific programme "Random dynamics in spatially extended systems" (RDSES), and the Nuffield Foundation.

References

[Aiz85]   Michael Aizenman, *The intersection of Brownian paths as a case study of a renormalization group method for quantum field theory*, Comm. Math. Phys. **97** (1985), no. 1-2, 91–110. MR 86h:60156

[DPRZ02]   Amir Dembo, Yuval Peres, Jay Rosen, and Ofer Zeitouni, *Thick points for intersections of planar sample paths*, Trans. Amer. Math. Soc. **354** (2002), no. 12, 4969–5003 (electronic). MR 2003m:60216

[HT97]   Xiaoyu Hu and S. James Taylor, *The multifractal structure of stable occupation measure*, Stochastic Process. Appl. **66** (1997), no. 2, 283–299. MR 98m:60117

[Kho03]   Davar Khoshnevisan, *Intersections of Brownian motions*, Expo. Math. **21** (2003), no. 2, 97–114. MR 1 978 059

[KM02]   Wolfgang König and Peter Mörters, *Brownian intersection local times: upper tail asymptotics and thick points*, Ann. Probab. **30** (2002), no. 4, 1605–1656. MR 2003m:60230

[Law98a]   Gregory F. Lawler, *The frontier of a Brownian path is multifractal*, (unpublishes) (1998).

[Law98b]   Gregory F. Lawler, *Strict concavity of the intersection exponent for Brownian motion in two and three dimensions*, Math. Phys. Electron. J. **4** (1998), Paper 5, 67 pp. (electronic). MR 2000e:60134

[LG87]   J.-F. Le Gall, *The exact Hausdorff measure of Brownian multiple points*, Seminar on stochastic processes, 1986 (Charlottesville, Va., 1986), Progr. Probab. Statist., vol. 13, Birkhäuser Boston, Boston, MA, 1987, pp. 107–137. MR 89a:60188

[LG89]   Jean-François Le Gall, *The exact Hausdorff measure of Brownian multiple points. II*, Seminar on Stochastic Processes, 1988 (Gainesville, FL, 1988), Birkhäuser Boston, Boston, MA, 1989, pp. 193–197.

[LSW01a]   Gregory F. Lawler, Oded Schramm, and Wendelin Werner, *The dimension of the planar Brownian frontier is 4/3*, Math. Res. Lett. **8** (2001), no. 4, 401–411. MR 2003a:60127b

[LSW01b]   Gregory F. Lawler, *Values of Brownian intersection exponents. I. Half-plane exponents*, Acta Math. **187** (2001), no. 2, 237–273. MR 2002m:60159a

[LSW01c]   Gregory F. Lawler, *Values of Brownian intersection exponents. II. Plane exponents*, Acta Math. **187** (2001), no. 2, 275–308. MR 2002m:60159b

[LSW02]   Gregory F. Lawler, *Values of Brownian intersection exponents. III. Two-sided exponents*, Ann. Inst. H. Poincaré Probab. Statist. **38** (2002), no. 1, 109–123. MR 2003d:60163

[Mör03]   P. Mörters, *Intersection exponents and the multifractal spectrum for measures of Brownian paths*, Fractal Geometry and Stochastics III (Umberto Mosco Christoph Bandt and Martina Zähle, eds.), vol. to appear, Birhäuser, 2003.

[Per96a]   Yuval Peres, *Intersection-equivalence of Brownian paths and certain branching processes*, Comm. Math. Phys. **177** (1996), no. 2, 417–434.

[Per96b]   Yuval Peres, *Remarks on intersection-equivalence and capacity-equivalence*, Ann. Inst. H. Poincaré Phys. Théor. **64** (1996), no. 3, 339–347.

[PT98]   Edwin A. Perkins and S. James Taylor, *The multifractal structure of super-Brownian motion*, Ann. Inst. H. Poincaré Probab. Statist. **34** (1998), no. 1, 97–138.

Coupling, Regularity and Curvature

Karl-Theodor Sturm

Institut für Angewandte Mathematik, Universität Bonn
Wegelerstrasse 6, 53115 Bonn, Germany
sturm@uni-bonn.de

1 Introduction

The purpose of this paper is to outline various connections between stochastic evolution processes on Euclidean or Riemannian spaces M and certain dynamics on the space $\mathcal{P}(M)$ of probability measures on M. The space $\mathcal{P}(M)$ will always be equipped with the L^θ-Wasserstein distance. Hence, contraction properties in $\mathcal{P}(M)$ mean coupling properties.

The processes to be considered are linear as well as nonlinear diffusions and martingales. Diffusions describe how the randomness of a stochastic system increases in time under the influence of some underlying noise. Martingales on nonlinear spaces more or less describe the converse procedure, namely, how one can filter out successively the noise in order to find the deterministic expectation of a given random variable. Both classes of processes play fundamental roles in Euclidean as well as in Riemannian settings. The crucial point in this paper will be to focus not on the pathwise construction and investigation of these evolution processes but on the evolution of robust quantities on the level of distributions. A major advantage of analysing the dynamics of distributions instead of studying the (stochastic) dynamics of particles is that the former exhibit a much greater deal of stability than pathwise constructions, such as stochastic development and stochastic parallel transport.

Our goal is to study the relation between dynamics on $\mathcal{P}(M)$ and stochastic processes on M. And we will try to show how curvature conditions on M determine contraction properties for the dynamics on $\mathcal{P}(M)$ and how the latter in turn determine contraction, convergence and regularity properties for the stochastic processes on M.

Linear diffusion equations on $\mathbb{R}^n$ or on a manifold M give rise to stochastic diffusion processes which describe the random motion of single particles or, in the same manner, the evolution of an initial distribution of particles. Nonlinear diffusion equations also lead to flows of probability measures. They may be interpreted as distributions of underlying systems of interacting stochastic particles. However, nonlinear diffusion equations can not be modeled by random motions of single particles.

Already half a century ago, K. Ito regarded flows of probability measures as a basic model for stochastic evolution processes. See e.g. the recent monograph [Str03] of D. Stroock on "Markov processes from K. Ito's perspective". However, at that time no appropriate tools had been available to construct and to investigate such flows. For the linear case (on Euclidean as well as on Riemannian spaces), this problem was resolved by the theory of stochastic differential equations, initiated by K. Ito at that time. This not only yields the flow of probability measures but the much more sophisticated random motion of the underlying particles.

For the nonlinear case, we will present a rather recent approach which allows to define a large class of nonlinear diffuions on M as gradient flows of probability measures w.r.t. appropriate free energy functionals on $\mathcal{P}(M)$.

An entirely different point of view is needed to understand the role of martingales on M. Whereas diffusions describe how mass spreads out in time, martingales describe the reverse procedure, namely, how one can find a center of mass for a given distribution. This leads to the concept of barycenter maps.

Let us briefly outline the role of Brownian motions and martingales in the theory of harmonic maps. For more details we refer to [Ken98]. Harmonic maps $f : M \to N$ between smooth Riemannian manifolds are critical values of the nonlinear energy functional $E(f) = \int_M ||df||^2 dm$. They may also be characterized as solutions to the corresponding Euler-Lagrange equation.

Ishihara's characterization states that a map $f : M \to N$ is harmonic if and only if for each convex function $\varphi : N_0 \to \mathbb{R}$ (defined on some subset N_0 of $\varphi(M)$) the function $\varphi \circ f : M_0 \to \mathbb{R}$ (defined on $M_0 = f^{-1}(N_0)$) is subharmonic. In this sense, the minimal requirements to build up a theory of harmonic maps between singular spaces is that the domain space M is a harmonic space and that the target space is a geodesic space: on M we need to know the subharmonic functions, on N the convex functions. However, this only allows to define harmonic maps. In order to construct them one needs more structure on the domain space. One possibility is to require the domain space to be a metric measure space. This allows to construct a Dirichlet form on it. For instance, one could derive the associated process from a rescaled random walk, cf. [Stu98]. Another possibility is to assume the domain space M to be the state space of a Markov process $(X_t)_t$ and the target space N to be a metric space of nonpositive curvature (in the sense of Alexandrov). The latter allows to develop a theory of martingales on N, [Stu02], [Stu03].

However, for the following discussion let us restrict to smooth Riemannian manifolds M and N. The probabilistic characterization due to Bismut states that f is harmonic if and only if for each (stopped) Brownian motion $(X_t)_t$ on M the process $(Y_t)_t = (f(X_t))_t$ on N is a martingale. The nonlinear evolution $(t, x) \mapsto f_t(x)$ of a given map $f : M \to N$ towards a harmonic map can be described as follows: for each t and x, $f_t(x)$ is the starting point Y_0 of a martingale $(Y_s)_s$ in N with terminal value $Y_t = f(X_t)$ where $(X_s)_s$ is a Brownian motion in M with starting point $X_0 = x$. In other, more robust terms this evolution can be described as follows

$$f_t(x) = \lim_{n \to \infty} \left(P_{t/n} \right)^n f(x) \quad \text{with} \quad P_t f(x) = b(f_* p_t(x, .)).$$

Here $p_t : M \to \mathcal{P}(M)$ denotes the transition semigroup of Brownian motion on M and $b : \mathcal{P}(N) \to N$ the so-called barycenter map on N.

Regularity results for harmonic maps typically depend on lower bounds for the Ricci curvature of M and on upper bounds for the sectional curvature of N. We will see that the Ricci curvature of M is bounded from below by K if and only if

$$d_\theta^W (p_t(x, .), p_t(y, .)) \le e^{-Kt} d(x, y)$$

($\forall x, y \in M$) and that the sectional curvature of N is nonpositive if and only if

$$d(b(\mu), b(\nu)) \le d_\theta^W (\mu, \nu)$$

($\forall \mu, \nu \in \mathcal{P}_\theta(N)$), both spaces of probability measures being equipped with the L^θ-Wasserstein distance for some $\theta \in [1, \infty[$. Both properties together imply

$$\text{Lip}(P_t f) \le e^{-Kt} \text{Lip}(f)$$

for all maps $f : M \to N$. This is a basic example of a gradient estimate for the nonlinear evolution of harmonic maps. Usually, its derivation is based on Bochner's formula (if analysts derive it) or on Bismut's formula (if probabilists do it). Here it is based on robust coupling properties for heat kernels and barycenters which allow to extend it to much more general situations (nonsymmetric, nonlocal, nonsmooth, infinite dimensional).

Let us briefly summarize the following sections. In section 2 we introduce the space $\mathcal{P}^\theta(N)$ of probability measures over a metric space (N, d), equipped with the L^θ-Wasserstein distance d_θ^W between probability measures as a natural metric.

Sections 3 and 4 are devoted to contraction properties for maps $b : \mathcal{P}^\theta(N) \to N$, called barycenter maps. The crucial ingredients will be upper curvature bounds on N in the sense of Alexandrov which generalize upper bounds for the sectional curvature.

In section 5 we discuss contraction properties for the heat semigroup on a Riemannian manifold M, regarded as a family of maps $p_t : M \to \mathcal{P}^\theta(M)$. Here lower bounds on the Ricci curvature of M will play the essential role.

Finally, in section 6 we study nonlinear diffusions on $\mathbb{R}^n$ or on a Riemannian manifold M as gradient flows of appropriate free energy functionals on $\mathcal{P}^\theta(M)$. Contraction properties for these nonlinear diffusions will be derived from convexity properties of the free energy functional. In particular, we present extensions of the Bakry-Emery criterion to nonlinear equations.

2 The Space of Probability Measures

Let (N, d) be a metric space and let $\mathcal{P}(N)$ denote the set of all probability measures p on N (equipped with its Borel σ-algebra $\mathcal{B}(N)$) with separable support $\operatorname{supp}(p) \subset N$. For $1 \le \theta < \infty$, $\mathcal{P}^\theta(N)$ will denote the set of $p \in \mathcal{P}(N)$ with $\int d^\theta(x, y)p(dy) < \infty$ for some (hence all) $x \in N$, and $\mathcal{P}^\infty(N)$ will denote the set of all $p \in \mathcal{P}(N)$ with bounded support. Obviously, $\mathcal{P}^\infty(N) \subset \mathcal{P}^\theta(N) \subset \mathcal{P}^1(N)$.

Given $p, q \in \mathcal{P}(N)$ we say that $\mu \in \mathcal{P}(N^2)$ is a *coupling* of p and q iff its *marginals* are p and q, that is, iff $\forall A \in \mathcal{B}(N)$

$$\mu(A \times N) = p(A) \quad \text{and} \quad \mu(N \times A) = q(A). \tag{1}$$

One such coupling μ is the product measure $p \otimes q$. Couplings μ of p and q are also called *transportation plans* from p to q. If p is the distribution of points at which a good is produced and q is the distribution of points where it is consumed, then each coupling μ of p and q gives a plan how to transport the production to the consumer and $d^W(p, q)$ describes the smallest cost of such a transportation process.

Definition 1. *For $\theta \in [1, \infty[$, we define the L^θ-Wasserstein distance distance d_θ^W on $\mathcal{P}^\theta(N)$ by*

$$d_\theta^W(p, q) = \inf \left\{ \int_N \int_N d^\theta(x, y)\mu(dxdy) : \mu \in \mathcal{P}(N^2) \text{ is coupling of } p \text{ and } q \right\}^{1/\theta}.$$

In probabilistic language,

$$d_\theta^W(p, q) = \inf \left[\mathbb{E}\, d^\theta(X, Y) \right]^{1/\theta},$$

where the infimum is over all probability spaces $(\Omega, \mathcal{A}, \mathbb{P})$ and all measurable maps $X : \Omega \to N$ and $Y : \Omega \to N$ with separable ranges and distributions $\mathbb{P}_X = p$ and $\mathbb{P}_Y = q$.

See e.g. [Dud89], [RR98], [Stu01], [Vil03].

Remark 1. If (N, d) is complete then so is $(\mathcal{P}^\theta(N), d_\theta^W)$. If (N, d) is a geodesic space, then so is $(\mathcal{P}^\theta(N), d_\theta^W)$. If (N, d) is separable then so is $(\mathcal{P}^\theta(N), d_\theta^W)$.

The case $\theta = 2$ is of particular importance. In this case, even curvature bounds carry over from N to $\mathcal{P}^\theta(N)$.

Definition 2. *We say that a metric space (N, d) has nonnegative curvature iff*

$$\frac{1}{2k} \sum_{i,j=1}^{k} d^2(y_i, y_j) \le \sum_{i=1}^{k} d^2(z, y_i)$$

for all $k \in \mathbb{N}$ and all $z, y_1, \ldots, y_k \in N$.

Obviously, the latter is equivalent to

$$\frac{1}{2} \int_N \int_N d^2(x,y)\, q(dx)\, q(dy) \leq \int_N d^2(z,x)\, q(dx) \qquad (2)$$

for all $z \in N$ and all probability measures $q \in \mathcal{P}^2(N)$.

The probabilistic interpretation of this property is as follows: N has non-negative curvature if and only if each pair of "randomly chosen points" X, Y in N (i.e. each pair of N-valued iid random variables) is seen from each point $z \in N$ "in average" under an angle $\angle(z; X, Y) \leq 90°$. More precisely, $\mathbb{E} \cos \angle(z; X, Y) \geq \cos 90°$.

One easily verifies that (N, d) has nonnegative curvature if and only if its metric completion $(\overline{N}, \overline{d})$ has nonnegative curvature.

If (N, d) is a geodesic metric space, then our definition of nonnegative curvature can be shown to be equivalent to nonnegative curvature in the sense of A. D. Alexandrov [Stu99]. Hence, in particular, for Riemannian manifolds it is equivalent to nonnegative sectional curvature. The property "nonnegative curvature" carries over from a metric space (N, d) to the space of probability measures over this spaces [Ott01], [Stu01].

Proposition 1. *Let (N, d) be a metric space. The space $(\mathcal{P}^2(N), d_2^W)$ of probability measures equipped with the L^2-Wasserstein distance has nonnegative curvature if and only if the underlying space (N, d) has nonnegative curvature.*

3 Probability Measures on Metric Spaces of Nonpositive Curvature

The next two sections are devoted to two generalizations of the class of Cartan-Hadamard manifolds. The first generalization is the class of metric spaces with nonpositive curvature with nonpositive curvature in the sense of Alexandrov. The second one (to be presented in the next section) will be the class of metric spaces which admit a contracting barycenter map. We summarize some of the basic results and refer to [Stu02] and [Stu03] for more details and further references.

Definition 3. *A metric space (N, d) is called* global NPC space *if it is complete and if for all $k \in \mathbb{N}$ and all $y_1, \ldots, y_k \in N$*

$$\frac{1}{2k} \sum_{i,j=1}^{k} d^2(y_i, y_j) \geq \inf_{z \in N} \sum_{i=1}^{k} d^2(z, y_i).$$

or, equivalently, if for all probability measures $q \in \mathcal{P}^2(N)$

$$\frac{1}{2} \int_N \int_N d^2(x,y)\, q(dy)\, q(dx) \geq \inf_{z \in N} \int_N d^2(z,x)\, q(dx).$$

Here "NPC" stands for "nonpositive curvature" in the sense of A. D. Alexandrov. Global NPC spaces are also called *Hadamard spaces*. If (N, d) is a global NPC space then it is a geodesic space. Even more, for any pair of points $x_0, x_1 \in N$ there exists a unique geodesic $x : [0, 1] \to N$ connecting them. For $t \in [0, 1]$ the intermediate points x_t depend continuously on the endpoints x_0, x_1. Finally, for any $z \in N$

$$d^2(z, x_t) \leq (1 - t)d^2(z, x_0) + td^2(z, x_1) - t(1 - t)d^2(x_0, x_1). \tag{3}$$

Our definition of nonpositive curvature is completely analogous to our definition of nonnegative curvature in Definition 2. Note, however, that there are two substantial differences between upper and lower curvature bounds. Firstly, if a complete metric space has globally curvature ≤ 0 then it is necessarily a geodesic space. This is not the case for complete metric spaces with (global) curvature ≥ 0. Secondly, if a complete geodesic space has "locally" curvature ≥ 0 then it has already "globally" curvature ≥ 0. The analogous statement is not true for complete geodesic spaces with local/global curvature ≤ 0.

Example 1. Examples of global NPC spaces are

– complete, simply connected Riemannian manifolds with nonpositive sectional curvature;
– trees and, more generally, Euclidean Bruhat-Tits buildings;
– Hilbert spaces;
– L^2-spaces of maps into such spaces;
– Finite or infinite (weighted) products of such spaces;
– Gromov-Hausdorff limits of such spaces.

See e.g. [Ale51], [BGS85], [Bal95], [BH99], [BBI01], [EF01] [Gro81/99], [Jos94], [Jos97], [KS93].

Proposition 2. *Let (N, d) be a global NPC space and fix $y \in N$. For each $q \in \mathcal{P}^1(N)$ there exists a unique point $z \in N$ which minimizes the uniformly convex, continuous function $z \mapsto \int_N [d^2(z, x) - d^2(y, x)]q(dx)$. This point is independent of y; it is called* barycenter *(or, more precisely, canonical barycenter or d^2-barycenter) of q and denoted by*

$$b(q) = \operatorname*{argmin}_{z \in N} \int_N [d^2(z, x) - d^2(y, x)] \, q(dx).$$

Moreover,

$$\int_N [d^2(z, x) - d^2(b(q), x)]q(dx) \geq d^2(z, b(q)). \tag{4}$$

If $q \in \mathcal{P}^2(N)$ then $b(q) = \operatorname{argmin}_{z \in N} \int_N d^2(z, x)q(dx)$.

Inequality (3.2) is called *variance inequality*. If in addition to nonpositive curvature we also assume that the curvature is bounded from below, say by $-K^2$, and that the space is geodesically complete then the following *reverse variance inequality* holds [Stu03]: For each $q \in \mathcal{P}^2(N)$ and for each $z \in N$

$$\int \left[d^2(z,x) - d^2(z,b(q)) - d^2(b(q),x) \right] q(dx)$$
$$\leq \frac{2K^2}{3} \int \left[d^4(z,b(q)) + d^4(b(q),x) \right] q(dx).$$

For a L^1-random variable $X : \Omega \to N$ we define its *expectation* by

$$\mathbb{E}\,X := b(\mathbb{P}_X) = \operatorname*{argmin}_{z \in N} \mathbb{E}\left[d^2(z,X) - d^2(y,X)\right].$$

That is, $\mathbb{E}\,X$ is the unique minimizer of the function $z \mapsto \mathbb{E}\left[d^2(z,X) - d^2(y,X)\right]$ on N (for each fixed $y \in N$). The variance inequality then reads as follows:

$$\mathbb{E}\left[d^2(z,X) - d^2(\mathbb{E}\,X, X)\right] \geq d^2(z,\mathbb{E}\,X)$$

for all $z \in N$. In the classical case $N = \mathbb{R}$, the corresponding *equality* should be well known after the first lessons in probability theory.

Our approach to barycenters and expectations is based on the classical point of view of [Gau1809]. He defined the expectation of a random variable (in Euclidean space) to be the uniquely determined point which minimizes the L^2-distance ("Methode der kleinsten Quadrate"). In the context of metric spaces, this point of view was successfully used by [Car28], [Fre48], [Kar77], and many others, under the name of barycenter, center of mass or center of gravity. Iterations of barycenters on Riemannian manifolds were used by [Ken90], [EM91] and [Pic94]. [Jos94] applied these concepts on global NPC spaces.

For other probabilistic approaches, see [Dos49], [Her91], [ESH99].

Another natural way to define the "expectations" $\mathbb{E}\,Y$ of a random variable Y is to use (generalizations of) the law of large numbers. This requires to give a meaning to $\frac{1}{n} \sum_{i=1}^{n} Y_i$. Our definition below only uses the fact that any two points in N are joined by unique geodesics. Our law of large numbers for global NPC spaces gives convergence towards the expectation defined as minimizer of the L^2-distance [Stu02].

Theorem 1 (Law of Large Numbers). *Let $(Y_i)_{i\in\mathbb{N}}$ be a sequence of independent, identically distributed bounded random variables $Y_i : \Omega \to N$ on a probability space $(\Omega, \mathcal{A}, \mathbb{P})$ with values in a global NPC space (N, d). Define their mean values $S_n : \Omega \to N$ by induction on $n \in \mathbb{N}$ as follows:*

$$S_1(\omega) := Y_1(\omega) \quad and \quad S_n(\omega) := \left(1 - \frac{1}{n}\right) S_{n-1}(\omega) + \frac{1}{n} Y_n(\omega),$$

where the RHS should denote the point $\gamma_{1/n}$ on the geodesic $\gamma : [0,1] \to N$ connecting $\gamma_0 = S_{n-1}(\omega)$ and $\gamma_1 = Y_n(\omega)$.

Then for $\mathbb{P}$-almost every $\omega \in \Omega$

$$S_n(\omega) \to \mathbb{E}\, Y_1 \qquad for\ n \to \infty$$

("strong law of large numbers").

Remark 2. (i) In strong contrast to the linear case, the mean value S_n will in general strongly depend on permutations of the iid variables $Y_i, i = 1 \ldots, n$. The distribution $\mathbb{P}_{S_n}$ is of course invariant under such permutations. But even $\mathbb{E}\, S_n$ in general depends on $n \in \mathbb{N}$. The law of large numbers only yields that $\mathbb{E}\, S_1 = \lim_{n\to\infty} \mathbb{E}\, S_n$.

(ii) It might seem more natural to define the mean value of the random variables $Y_1, \ldots, Y_n$ as the barycenter of these points, more precisely, as the barycenter of the uniform distribution on these points, i.e.

$$\overline{S}_n(\omega) := b\left(\frac{1}{n}\sum_{i=1}^{n} \delta_{Y_i(\omega)}\right).$$

In this case we also obtain a law of large numbers. Indeed, it is much easier to derive (and it holds for more or less arbitrary choices of $b(.)$), see Proposition 3. However, it is also of much less interest: we will obtain convergence of $\overline{S}_n(\omega)$ towards $b(\mathbb{P}_{Y_1})$, the barycenter of the distribution of Y_1, but to define $\overline{S}_n$ we already have to use $b(.)$.

(iii) Of course there are many other ways to define a mean value $\tilde{S}_n$ of the random variables $Y_1, \ldots, Y_n$ which do not depend on the a priori knowledge of $b(.)$. And indeed for many of these choices one can prove that $\tilde{S}_n$ converges almost surely to a point $\tilde{b}$ (which only depends on the distribution of Y_1). For instance, define $S_{n,1} := Y_n$ and recursively $S_{n,k+1}$ to be the midpoint of $S_{2n-1,k}$ and $S_{2n,k}$. Then $\tilde{S}_n(\omega) := S_{1,n}(\omega)$ converges for a.e. ω as $n \to \infty$ towards a point $\tilde{b} = \tilde{b}(\mathbb{P}_{Y_1})$. (Note that in the flat case, $S_{1,n} = 2^{-n}\sum_{i=1}^{2^n} Y_i$.) Another example is given by the mean value in the sense of [ESH99] which will be described as Example 6.

However, no choice of $\tilde{S}_n$ other than S_n is known to the author where one obtains convergence towards a point which can be characterized "extrinsically", like in our case as the minimizer of the function $z \mapsto \mathbb{E}d^2(z, Y_1)$.

The Law of Large Numbers yields a simple proof of Jensen's inequality. The key ingredient is the inequality

$$\varphi(S_n) \leq \frac{1}{n}\sum_{i=1}^{n} \varphi(Y_i)$$

which holds for any convex function $\varphi : N \to \mathbb{R}$.

Theorem 2 (Jensen's inequality). *For any global NPC space (N, d), any lower semicontinuous convex function $\varphi : N \to \mathbb{R}$ and any $q \in \mathcal{P}^1(N)$*

$$\varphi(b(q)) \leq \int_N \varphi(x) q(dx),$$

provided the RHS is well-defined.

The above RHS is well-defined if either $\int \varphi^+ \, dq < \infty$ or $\int \varphi^- \, dq < \infty$. In particular, it is well-defined if φ is Lipschitz continuous. Applying Jensen's inequality to the convex function $(x, y) \mapsto d(x, y)$ on the global NPC space $N \times N$ yields

Theorem 3 (Fundamental Contraction Property). *For all $\theta \in [1, \infty[$ and all $p, q \in \mathcal{P}^\theta(N)$:*

$$d(b(p), b(q)) \leq d_1^W(p, q) \leq d_\theta^W(p, q). \tag{5}$$

4 Barycenters

In the previous section, we saw that nonpositive curvature in the sense of Alexandrov implies the existence of a canonical barycenter map. It turns out that many results require nonpositive curvature only because they rely on contraction properties of this map. The existence of such a contracting map itself may be regarded as a far reaching generalization of nonpositive curvature.

Let us fix a complete metric space (N, d) and a number $\theta \in [1, \infty[$.

Definition 4. *A L^θ-barycenter contraction is a map $b : \mathcal{P}^\theta(N) \to N$ such that*

- $b(\delta_x) = x$ *for all $x \in N$;*
- $d(b(p), b(q)) \leq d_\theta^W(p, q)$ *for all $p, q \in \mathcal{P}^\theta(N)$.*

Obviously, a L^θ-barycenter contraction is a $L^{\theta'}$-barycenter contraction for each $\theta' \geq \theta$,

Example 2. For each global NPC space the canonical barycenter yields a L^θ-barycenter contraction ($\forall \theta$).

Actually, also partly the converse holds. If there exists a L^θ-barycenter contraction on (N, d) then (N, d) is a geodesic space: For each pair of points $x_0, x_1 \in N$ we can define a geodesic $t \mapsto x_t$ connecting x_0 and x_1 by $x_t := b((1-t)\delta_{x_0} + t\delta_{x_1})$. Given any four points $x_0, x_1, y_0, y_1 \in N$, the function $t \mapsto d(x_t, y_t)$ is convex. In particular, the geodesic $t \mapsto x_t$ depends continuously on x_0 and x_1. However, it is not necessarily the only geodesic connecting x_0 and x_1.

If geodesics in N are unique then the existence of a L^θ-barycenter contraction implies that $d : N \times N \to \mathbb{R}$ is convex. Thus N has globally "nonpositive curvature" in the sense of Busemann.

Corollary 1. *Let N be a complete, simply connected Riemannian manifold and let d be a Riemannian distance. Then (N, d) admits a L^θ-barycenter contraction b if and only if N has nonpositive sectional curvature.*

Indeed, if (N, d) admits a L^θ-barycenter contraction then so does (N_0, d) for each closed convex $N_0 \subset N$. Hence, geodesics in N_0 are unique and thus $t \mapsto d(\gamma_t, \zeta_t)$ is convex for any pair of geodesics γ and ζ in N_0. This implies that N has nonpositive curvature.

Example 3. Let $(N, \|.\|)$ be a (real or complex) Banach space and put $d(x, y) := \|x - y\|$. Then $\mathcal{P}^\theta(N)$ is the set of Radon measures p on N satisfying $\int_N \|x\|^\theta p(dx) < \infty$. For each $p \in \mathcal{P}^\theta(N)$, the identity $x \mapsto x$ on N is *Bochner integrable* and

$$b(p) := \int_N x\, p(dx)$$

defines a barycenter contraction on (N, d).

Example 4. Let I be a countable set and for each $i \in I$, let (N_i, d_i) be a complete metric space with L^1-barycenter contraction b_i and "base" point $o_i \in N_i$. Given $\eta \in [1, \infty]$, define a complete metric space (N, d) with base point $o = (o_i)_{i \in I}$ by

$$N := \left\{ x = (x_i)_{i \in I} \in \bigotimes_{i \in I} N_i : d(x, o) < \infty \right\}, \quad d(x, y) := \left[\sum_{i \in I} d_i^\eta(x_i, y_i) \right]^{\frac{1}{\eta}}$$

provided $\eta < \infty$ or by $d(x, y) = \sup_{i \in I} d_i(x_i, y_i)$ if $\eta = \infty$. One can define a barycenter contraction b on $\mathcal{P}^1(N)$ by

$$b(p) := (b_i(p_i))_{i \in I}$$

where $p_i \in \mathcal{P}^1(N_i)$ with $p_i : A \mapsto p(\{x = (x_j)_{j \in I} \in N : x_i \in A\})$ denotes the projection of $p \in \mathcal{P}^1(N)$ onto the i-th factor of N.

For instance, this applies to $N = \mathbb{R}^n, n \geq 2$ with the usual notion of barycenter but with "unusual" metric $d(x, y) = \sup\{|x_i - y_i| : i = 1, .., n\}$. In this case, geodesics are not unique, e.g. each curve $t \mapsto (t, \varphi_2(t), ..., \varphi_n(t))$ with $\varphi \in \mathcal{C}^1(\mathbb{R}), \varphi_i(0) = \varphi_i(1) = 0$ and $|\varphi_i'| \leq 1$ is a geodesic connecting $(0, 0, ..., 0)$ and $(1, 0, ..., 0)$.

Remark 3. Each barycenter map b on a complete metric space (N, d) gives rise to a whole family of barycenter maps b_n, $n \in \mathbb{N}$ (which in general do not coincide with b, see Example below). Let b be a L^θ-barycenter contraction and $\Phi : N \times N \to N$ be the "midpoint map" induced by b, i.e. $\Phi(x, y) = b(\frac{1}{2}\delta_x + \frac{1}{2}\delta_y)$. Define a map $\Xi : \mathcal{P}^1(N) \to \mathcal{P}^1(N)$ by

$$\Xi(q) := \Phi_*(q \otimes q).$$

Then Ξ is a contraction with respect to d_θ^W. Thus for each $n \in \mathbb{N}$

$$b_n(q) := b(\Xi^n(q))$$

defines a contracting barycenter map $b_n : \mathcal{P}^\theta(N) \to N$.

Example 5. Define the *tripod* by gluing together 3 copies of $\mathbb{R}_+$ at their origins, i.e.

$$N = \{(i, r) : i \in \{1, 2, 3\}, r \in \mathbb{R}_+\}/\sim \quad \text{where } (i, r) \sim (j, s) :\Leftrightarrow r = s = 0.$$

It can be realized as the subset $\{r \cdot \exp(\frac{l}{3}2\pi i) \in \mathbb{C} : r \in \mathbb{R}_+, l \in \{1, 2, 3\}\}$ of the complex plane, however, equipped with the (non-Euclidean!) intrinsic metric

$$d((i, r), (j, s)) = \begin{cases} |r - s|, & \text{if } i = j \\ r + s, & \text{else.} \end{cases}$$

Then (N, d) is a complete metric space of globally nonpositive curvature and according to Example 1 there exists a canonical barycenter map b. Derive from that the barycenter map $b_1 = b(\Xi(.))$ as above. Then the maps b and b_1 do not coincide. Indeed, choose $q = \frac{1}{2}\delta_{(1,1)} + \frac{1}{4}\delta_{(2,1)} + \frac{1}{4}\delta_{(3,1)}$. Then $\Xi(q) = \frac{1}{4}\delta_{(1,1)} + \frac{1}{16}\delta_{(2,1)} + \frac{1}{16}\delta_{(3,1)} + \frac{5}{8}\delta_o$. Hence, $b(q) = (1, 0)$ and $b_1(q) = b(\Xi(q)) = (1, \frac{1}{8})$.

Example 6 (Barycenter Map of Es-Sahib & Heinich [ESH99]). Let (N, d) be a locally compact complete separable metric space with *negative curvature in the sense of Busemann* or a global NPC space in the sense of Alexandrov (see next section). Then one can define recursively for each $n \in \mathbb{N}$ a unique map $\beta_n : N^n \to N$ satisfying

- $\beta_n(x_1, \ldots, x_1) = x_1$
- $d(\beta_n(x_1, \ldots, x_n), \beta_n(y_1, \ldots, y_n)) \leq \frac{1}{n}\sum_{i=1}^n d(x_i, y_i)$
- $\beta_n(x_1, \ldots, x_n) = \beta_n(\check{x}_1, \ldots, \check{x}_n)$
  where $\check{x}_i := \beta_{n-1}(x_1, \ldots, x_{i-1}, x_{i+1}, \ldots, x_n)$.

This map is symmetric and satisfies $d(z, \beta_n(x_1, \ldots, x_n)) \leq \frac{1}{n}\sum_{i=1}^n d(z, x_i)$ for all $z \in N$. Given any $p \in \mathcal{P}^1(N)$ let $(Y_i)_i$ be an independent sequence of maps $Y_i : \Omega \to N$ on some probability space $(\Omega, \mathcal{A}, \mathbb{P})$ with distribution $\mathbb{P}_{Y_i} = p$ and define $\tilde{S}_n(\omega) := b_n((Y_1(\omega), \ldots, Y_n(\omega))$. Then there exists a point $\beta(p) \in N$ such that $\tilde{S}_n(\omega) \to \beta(p)$ for $\mathbb{P}$-a.e. ω and $n \to \infty$. The map

$\beta : \mathcal{P}^1 \to N$ is easily seen to be a L^1-contracting barycenter map. Note, however, that in general, $\beta\left(\frac{1}{n}\sum_{i=1}^n \delta_{x_i}\right) \neq \beta_n(x_1, \ldots, x_n)$.

Moreover, we emphasize that on non-flat Riemannian manifolds as well as on trees this barycenter map β is different from the canonical one, defined via minimizing the L^2-distance (see previous section). For instance, let (N, d) be the tripod and let $p = \frac{1}{2}\delta_{(1,1)} + \frac{1}{4}\delta_{(2,1)} + \frac{1}{4}\delta_{(3,1)}$. Then with the latter choice $b(p) = (1,0)$, whereas an easy calculation shows that the previous choice yields $\beta(p) = (1, 1/6)$.

Proposition 3 (Empirical Law of Large Numbers). *Let (N, d) be a complete metric space with a contracting barycenter map $b : \mathcal{P}^\theta(N) \to N$ and fix $p \in \mathcal{P}^\infty(N)$. Moreover, let $(\Omega, \mathcal{A}, \mathbb{P})$ be a probability space and $(X_i)_{i \in \mathbb{N}}$ be an independent sequence of measurable maps $X_i : \Omega \to N$ with identical distribution $\mathbb{P}_{X_i} = p$. Define the "barycentric mean value" $s_n : \Omega \to N$ by $s_n(\omega) := b\left(\frac{1}{n}\sum_{i=1}^n \delta_{X_i(\omega)}\right)$. Then for $\mathbb{P}$-almost every $\omega \in \Omega$*

$$s_n(\omega) \longrightarrow \beta(p) \qquad as \ n \to \infty.$$

5 Transport Inequalities and Gradient Estimates

In previous sections, we studied contraction properties of the canonical barycenter maps $b : \mathcal{P}(N) \to N$ on spaces N with some kind of upper curvature bounds (e.g. nonpositive curvature in the sense of Alexandrov). Now we will be concerned with the reverse situation. We will study contraction properties of "canonical" maps $p_t : M \to \mathcal{P}(M)$ on spaces M with appropriate lower curvature bounds. To be more specific, we will restrict ourselves (mostly) to Riemannian manifolds and we will choose $(p_t)_t$ to be the heat semigroup. The appropriate curvature bounds will turn out to be lower bounds for the Ricci curvature.

We will present various equivalences between transportation inequalities (for volume measures, heat kernels, Brownian motions), gradient estimates for the heat semigroup and lower bounds for the Ricci curvature, These are joint results with Max-K. von Renesse. Details can be found in [RS03].

In the sequel, (M, g) always is assumed to be a complete smooth Riemannian manifold with dimension n, Riemannian distance $d(x, y)$ and Riemannian volume $m(dx)$. Here and henceforth, $p_t(x, y)$ always denotes the heat kernel on M, i.e. the minimal positive fundamental solution to the heat equation $(\Delta - \frac{\partial}{\partial t})p_t(x, y) = 0$. It is smooth in (t, x, y), symmetric in (x, y) and satisfies $\int_M p_t(x, y)m(dy) \leq 1$. Hence, it defines a subprobability measure $p_t(x, dy) := p_t(x, y)m(dy)$ as well as operators $p_t : \mathcal{C}_c^\infty(M) \to \mathcal{C}^\infty(M)$ and $p_t : L^2(M) \to L^2(M)$ which are all denoted by the same symbol. Given $\mu \in \mathcal{P}^\theta(M)$ and $t > 0$ we define a new measure $\mu p_t \in \mathcal{P}^\theta(M)$ by $\mu p_t(A) = \int_A \int_M p_t(x, y)\mu(dx)m(dy)$.

Brownian motion on M is by definition the Markov process with generator $\frac{1}{2}\Delta$. Thus its transition (sub-)probabilities are given by $p_{t/2}$.

If the Ricci curvature of the underlying manifold M is bounded from below then all the $p_t(x,.)$ are probability measures. If the latter holds true we say that the heat kernel and the associated Brownian motion are conservative. It means that the Brownian motion has infinite lifetime.

Our first main result in this section deals with robust versions of gradient estimates.

Theorem 4. *For any complete smooth Riemannian manifold M and any $K \in \mathbb{R}$ the following properties are equivalent:*

(i) $\mathrm{Ric}(M) \geq K$,

  which always should be read as: $\mathrm{Ric}_x(v,v) \geq K|v|^2$ *for all $x \in M, v \in T_xM$.*

(ii) For all $f \in C_c^\infty(M)$, all $x \in M$ and all $t > 0$

$$|\nabla p_t f|(x) \leq e^{-Kt} p_t |\nabla f|(x).$$

(iii) For all $f \in C_c^\infty(M)$ and all $t > 0$

$$\|\nabla p_t f\|_\infty \leq e^{-Kt} \|\nabla f\|_\infty.$$

(iv) For all bounded $f \in C^{\mathrm{Lip}}(M)$ and all $t > 0$

$$\mathrm{Lip}(p_t f) \leq e^{-Kt} \mathrm{Lip}(f).$$

The equivalence of (i) and (ii), perhaps, is one of the most famous general results which relate heat kernels with Ricci curvature. It is due to D. Bakry & M. Emery [BE84], see also [ABC00] and references therein. Property (ii) is successfully used in various applications as a replacement (or definition) of lower Ricci curvature bounds for symmetric Markov semigroups on general state spaces. Our result states that (ii) can be weakened in two respects:
- one can replace the pointwise estimate by an estimate between L^∞-norms;
- one can drop the p_t on the RHS.

Besides being formally weaker than (ii) one other advantage of (iii) is that it is an explicit statement on the smoothing effect of p_t whereas (ii) is implicit (since p_t appears on both sides).

As an easy corollary to the equivalence of the statements (ii) and (iii) one may deduce the well known fact that (ii) is equivalent to the assertion that for all f, x and t as above

$$|\nabla p_t f|(x) \leq e^{-Kt} \left[p_t(|\nabla f|^2)(x) \right]^{1/2}.$$

Property (iv) may be considered as a replacement (or as one possible definition) for lower Ricci curvature bounds for Markov semigroups on metric spaces. For several non-classical examples (including nonlocal generators as well as infinite dimensional or singular finite dimensional state spaces) we refer to [Stu03], [DR02] and [Ren03]. This property turned out to be the key ingredient to prove Lipschitz continuity for harmonic maps between metric spaces in [Stu03].

According to the Kantorovich-Rubinstein duality, property (iv) is equivalent to a contraction property for the heat kernels in terms of the L^1-Wasserstein distance d_1^W. Actually, however, much more can be proven:

–   one obtains contraction in d_θ^W for each $\theta \in [1, \infty]$ and for any initial data;
–   one obtains pathwise contraction for Brownian trajectories.

Corollary 2. *For any smooth complete Riemannian manifold M and any $K \in \mathbb{R}$ the following properties are equivalent:*

(i) $\mathrm{Ric}(M) \geq K$.

(v) For all $x, y \in M$ and all $t > 0$ there exists $\theta \in [1, \infty]$ with

$$d_\theta^W(p_t(x, .), p_t(y, .)) \leq e^{-Kt} \cdot d(x, y).$$

(vi) For all $\theta \in [1, \infty]$, all $\mu, \nu \in \mathcal{P}^\theta(M)$ and all $t > 0$:

$$d_\theta^W(\mu p_t, \nu p_t) \leq e^{-Kt} \cdot d_\theta^W(\mu, \nu).$$

(vii) For all $x_1, x_2 \in M$ there exists a probability space $(\Omega, \mathcal{A}, \mathbb{P})$ and two conservative Brownian motions $(X_1(t))_{t \geq 0}$ and $(X_2(t))_{t \geq 0}$ defined on it, with values in M and starting in x_1 and x_2, respectively, such that for all $t > 0$

$$\mathbb{E}[d(X_1(t), X_2(t))] \leq e^{-Kt/2} \cdot d(x_1, x_2).$$

(viii) There exists a conservative Markov process $(\Omega, \mathcal{A}, \mathbb{P}^x, X(t))_{x \in M \times M, t \geq 0}$ with values in $M \times M$ such that the coordinate processes $(X_1(t))_{t \geq 0}$ and $(X_2(t))_{t \geq 0}$ are Brownian motions on M and such that for all $x = (x_1, x_2) \in M \times M$ and all $t > 0$

$$d(X_1(t), X_2(t)) \leq e^{-Kt/2} \cdot d(x_1, x_2) \qquad \mathbb{P}^x\text{-}a.s.$$

Note that each of the statements (v) and (vi) implicitly includes the conservativity of the heat kernel. Indeed, the finiteness of the Wasserstein distance implies that the measures under consideration must have the same total mass. Thus $p_t(x, M)$ is constant in x, hence also constant in t and therefore equal to 1.

The interpretation of these results is as follows: if we put mass distributions μ and ν on M and if they spread out according to the heat equation then the lower bound for the Ricci curvature of M controls how fast the distances between these distributions may expand (or have to decay) in time.

The second main result in this section deals with transportation inequalities for uniform distributions on spheres and analogous inequalities for uniform distributions on balls. Here the lower Ricci bound is characterized as a control for the increase of the distances if we replace Dirac masses δ_x and δ_y by uniform distributions $\sigma_{r,x}$ and $\sigma_{r,y}$ on spheres around x and y, resp. or if we replace them by uniform distributions $m_{r,x}$ and $m_{r,y}$ on balls around x and y, resp.

Theorem 5. *For any smooth compact Riemannian manifold M and any $K \in \mathbb{R}$ the following properties are equivalent:*

(i) $\mathrm{Ric}(M) \geq K$.
(ix) The normalized surface measure on spheres of radius $\sqrt{2n}\, r$

$$\sigma_{r,x}(A) := \frac{\mathcal{H}^{n-1}(A \cap \partial B_{\sqrt{2n}r}(x))}{\mathcal{H}^{n-1}(\partial B_{\sqrt{2n}r}(x))}, \qquad A \in \mathcal{B}(M)$$

satisfies the asymptotic estimate

$$d_1^W(\sigma_{r,x}, \sigma_{r,y}) \leq \left(1 - Kr^2 + o(r^2)\right) \cdot d(x,y) \tag{6}$$

where the error term is uniform w.r.t. $x, y \in M$.
(x) The normalized Riemannian uniform distribution on balls of radius $\sqrt{2(n+2)}\, r$

$$m_{r,x}(A) := \frac{m(A \cap B_{\sqrt{2(n+2)}r}(x))}{m(B_{\sqrt{2(n+2)}r}(x))}, \qquad A \in \mathcal{B}(M)$$

satisfies the asymptotic estimate

$$d_1^W(m_{r,x}, m_{r,y}) \leq \left(1 - Kr^2 + o(r^2)\right) \cdot d(x,y) \tag{7}$$

where the error term is uniform w.r.t. $x, y \in M$.

The heat kernel on a Riemannian manifold is a fundamental object for analysis, geometry and stochastics. Many properties and precise estimates are known. In most of these results, lower bounds on the Ricci curvature of the underlying manifold play a crucial role. However, for more general spaces, like e.g. metric measure spaces, there is neither a notion of Ricci curvature nor a common notion of bounds for the Ricci curvature (comparable for instance to Alexandrov's notion of bounds for the sectional curvature for metric spaces).

The advantage of the above characterization of Ricci curvature is that it depends only on the basic, robust data: measure and metric. It does not require any heat kernel, any Laplacian or any Brownian motion. It might be used as a guideline in much more general situations.

For instance, let (M, d) be an arbitrary separable metric space equipped with a measure m on its Borel σ-field and assume that (7) holds true (with some number $K \in \mathbb{R}$). Define an operator m_r acting on bounded measurable functions by $m_r f(x) = \int_M f(y) \, m_{r,x}(dy)$. Then by the Arzela-Ascoli theorem there exists a sequence $(l_j)_j \subset \mathbb{N}$ such that

$$p_t f := \lim_{j \to \infty} \left(m_{\sqrt{t/l_j}} \right)^{l_j} f$$

exists (as a uniform limit) for all bounded $f \in \mathcal{C}^{\mathrm{Lip}}(M)$ and it defines a Markov semigroup on M satisfying

$$\mathrm{Lip}(p_t f) \leq e^{-Kt} \mathrm{Lip}(f).$$

For Riemannian manifolds, the invariance principle for Brownian motions implies that this semigroup $(p_t)_t$ is just the usual heat semigroup and thus (x) obviously implies (iv). Analogously, (ix)$\Longrightarrow$ (iv). The converse implications are rather involved and required detailed estimates for the transportation costs. The basic ingredient, however, is an elementary quadrilateral estimate for geodesic parallel transports (and actually this easily explains the final asymptotic formula).

6 Gradient Flows on Metric Spaces and Nonlinear Diffusions

This section is devoted to contraction properties of nonlinear diffusions on $\mathbb{R}^n$ or on a Riemannian manifold M. Following [Ott01] and [OV00], we regard them as gradient flows of appropriate free energy functionals S on $\mathcal{P}^\theta(M)$. Contraction properties for these nonlinear diffusions will be derived from convexity properties of the free energy functional. In particular, we present extensions of the Bakry-Emery criterion to nonlinear equations. For details, proofs and further references we reer to [Stu04].

Given an arbitrary geodesic space (N, d_N), a number $K \in \mathbb{R}$ and a function $S : N \to [-\infty, +\infty]$ we say that S is K-convex iff for each (constant speed, as usual) geodesic $\gamma : [0,1] \to N$ with $S(\gamma_0) < \infty$ and $S(\gamma_1) < \infty$) and for each $t \in [0,1]$:

$$S(\gamma_t) \leq (1-t) \, S(\gamma_0) + t \, S(\gamma_1) - \frac{K}{2} t(1-t) \, d_N^2(\gamma_0, \gamma_1). \tag{8}$$

If S is lower semicontinuous, then it suffices to verify this for all geodesics γ and $t = \frac{1}{2}$.

K-convexity is a local property. The above inequality (8) holds for a given function S and a given geodesic $\gamma : [0,1] \to N$ provided there exists a partition $0 = t_0 < t_1 < \ldots < t_{n+1} = 1$ such that for each $i = 1, \ldots, n$ the geodesic $\gamma : [t_{i-1}, t_{i+1}] \to N$ satisfies (after suitable reparametrization) inequality (8).

A function S is K-convex if and only if for each geodesic $\gamma : [0,1] \to N$ with $S(\gamma_0) < \infty$ and $S(\gamma_1) < \infty$ one has $S(\gamma_t) < \infty$ (for all $t \in [0,1]$) and

$$\liminf_{t \to 0} \frac{1}{t^2} \cdot [S(\gamma_{2t}) - 2S(\gamma_t) + S(\gamma_0)] \geq K \cdot d_N^2(\gamma_0, \gamma_1).$$

Example 7. A smooth function S on a Riemannian manifold (N, d) is K-convex if and only if

$$\text{Hess } S \geq K.$$

Given a function S on a geodesic space (N, d), we say that a map $\sigma : \mathbb{R}_+ \times N \to N$, $(t, x) \mapsto \sigma_t(x)$ is a *gradient flow* for S iff for each $x \in N$, $t \mapsto \sigma(t, x)$ is a curve in N starting in x with $|\partial_t \sigma(t, x)| = -\partial_\sigma S(\sigma)(t, x)$ and $\partial_\sigma S(\sigma)(t_0, x) \leq \partial_\eta S(\eta)(t_0)$ for any t_0 and any other curve η in N with $\sigma(t_0, x) = \eta_{t_0}$.

Proposition 4. *Assume that S is K-convex. Then there exists a unique gradient flow σ for S and it satisfies:*

(i) $d(\sigma(t, x), \sigma(t, y)) \leq e^{-Kt} d(x, y)$ for all $x, y \in N$ and all $t \geq 0$.
(ii) If in addition $K > 0$ and $S > -\infty$ then there exists a unique "ground state" $x_0 \in N$ satisfying $S(x) \geq S(x_0) + \frac{K}{2} d_N^2(x, x_0)$ for all $x \in N$.
(iii) If $K > 0$ and $S(x_0) = 0$ then $-\partial_t S(\sigma(t, x)) \geq 2K \cdot S(\sigma(t, x))$ and thus $S(\sigma(t, x)) \leq e^{-2Kt} S(x)$.

Property (i) is deduced in unpublished papers by Perelman and Petrunin as well as by A. Lytchak (private communication). Properties (ii) and (iii) may be regarded as generalized versions of Talagrand's inequality and Gross' logarithmic Sobolev inequality, resp. This may be seen in Example 8 below where we choose the space N and the function S more specifically.

¿From now on, let N be the space $\mathcal{P}^2(M)$ of probability measures on a smooth complete Riemannian manifold M and let d_N be the L^2-Wasserstein distance d_2^W distance derived from the Riemannian distance $d = d_M$. Following [McC01], K-convex functions on $\mathcal{P}^2(M)$ are also called *displacement K-convex* (to emphasize that it means K-convexity along the geodesics $t \mapsto \gamma_t$ w.r.t. d_2^W and not along the geodesics $t \mapsto (1-t)\gamma_0 + t\gamma_1$ in the linear space of signed measures).

Given an increasing function $U : \mathbb{R} \to \mathbb{R}$ and a lower semicontinuous function $V : M \to \mathbb{R}$ we define the *free energy* $S : \mathcal{P}^2(M) \to [-\infty, \infty]$ by

$$S(\nu) := \int_M U\left(\log \frac{d\nu}{dm}\right) d\nu + \int_M V \, d\nu \tag{9}$$

provided ν is absolutely continuous w.r.t. the Riemannian volume measure m and $\int U_+(\log \frac{d\nu}{dm}) \, d\nu + \int V_+ \, d\nu < \infty$. Otherwise, we define $S(\nu) := +\infty$.

Remark 4. Under minimal assumptions on M, U and V the gradient flow σ for S as above is given by $\sigma(t, \nu)(dx) = \rho(t, x)\, m(dx)$ where the densities ρ solve the nonlinear PDE

$$\partial_t \rho(t, x) = \Delta(\rho U'(\log \rho))(t, x) + \nabla(\rho \cdot \nabla V)(t, x)$$

on $\mathbb{R}_+ \times M$, [OV00], [Vil03].

If we can verify K-convexity of S for some $K > 0$ then this nonlinear diffusion equation has a unique stationary solution and any other solution converges exponentially fast to the stationary solution.

Example 8. The main examples are:

- $U(r) = r, V = 0$ yields the relative entropy $S(\nu) = \int_M \log \frac{d\nu}{dm}\, d\nu$. Its gradient flow is the usual heat equation $\partial_t \rho = \Delta \rho$. More precisely, the densities of the gradient flow are solutions of the heat equation.
- $U(r) = r$ leads to the Fokker-Planck equation

$$\partial_t \rho = \Delta \rho + \nabla(\rho \cdot \nabla V).$$

In this case, an easy calculation shows $S(\sigma(t, \nu)) = \int u^2 \log(u^2) e^{-V} dm$ and $-\partial_t S(\sigma(t, \nu)) = \frac{1}{4} \int |\nabla u|^2 e^{-V} dm$ provided we write $\frac{d\sigma}{dm} = u^2 e^{-V}$. Hence, here we indeed obtain the usual version of the logarithmic Sobolev inequality.
- $U(r) = \frac{1}{a} \exp(ar)$ for a constant $a \neq 0$ and $V = 0$ yields $S(\nu) = \frac{1}{a} \int_M \left(\frac{d\nu}{dm}\right)^a d\nu$. The associated gradient flow is given by the porous medium equation (if $a > 0$) or fast diffusion equation (if $a < 0$)

$$\partial_t \rho = \Delta(\rho^{1+a}).$$

Our main result from [Stu04] yields K-convexity for large classes of energy functionals associated to nonlinear diffusions on Euclidean and Riemannian spaces. As a consequence it yields exponential convergence to equilibrium for the solutions to these equations together with explicit bounds for the rate of convergence.

Theorem 6. *The free energy S from (9) is K-convex if and only if $U''(r) + \frac{1}{n}U'(r) \geq 0$ and*

$$U'(r) \cdot Ric_x(\xi, \xi) + Hess_x V(\xi, \xi) \geq K \cdot |\xi|^2$$

for all $r \in \mathbb{R}$, $x \in M$ and $\xi \in T_x M$.

Applications of this result to heat equation, Fokker-Planck equation and porous medium equation are straightforward.

Corollary 3. *The free energy $S(\nu) = \int_M \log \frac{d\nu}{dm} \, d\nu + \int_M V d\nu$ associated with the Fokker-Planck equation is K-convex if and only if the Bakry-Emery criterion*

$$Ric_x(\xi,\xi) + Hess_x V(\xi,\xi) \geq K \cdot |\xi|^2$$

is satisfied ($\forall x \in M, \forall \xi \in T_x M$).

In particular, the relative entropy $S(\nu) = \int_M \log \frac{d\nu}{dm} \, d\nu$ is a K-convex function on the metric space $\mathcal{P}^2(M)$ if and only if the Ricci curvature of the underlying Riemannian manifold M is bounded from below by K.

Corollary 4. *For any $N > 0$ the free energy $S(\nu) = -N \cdot \int_M \left(\frac{d\nu}{dm}\right)^{-1/N} d\nu$ associated with the fast diffusion equation $\partial_t \rho = \Delta(\rho^{1-1/N})$ is a 0-convex function on the metric space $\mathcal{P}^2(M)$ if and only if the underlying Riemannian manifold M has nonnegative Ricci curvature and dimension $\leq N$.*

Parts of the above corollaries had been obtained in [OV00], [CMS01] and [RS03]. The previous results yields a characterization of the curvature-dimension conditions $CD(K,\infty)$ as well as $CD(0,N)$ of Bakry-Emery in terms of contraction properties of nonlinear diffusions. The general condition $CD(K,N)$ may be characterized in a similar manner:

Theorem 7. *i) For $K > 0$ and $N > 0$ consider the free energy functional*

$$S(\nu) = \int_M \left[\log\left(\frac{d\nu}{dm}\right) - N\left(\frac{d\nu}{dm}\right)^{-1/N}\right] d\nu$$

associated with the nonlinear diffusion equation

$$\partial_t \rho = \Delta(\rho(1 + \rho^{-1/N})).$$

Then S is K-convex if and only if the dimension of the manifold is bounded from above by N and its Ricci curvature is bounded from below by K.

iii) For $K < 0$ and $N > 0$ consider the free energy functional

$$S(\nu) = -N \int_M \log\left[1 + \left(\frac{d\nu}{dm}\right)^{-1/N}\right] d\nu$$

associated with the nonlinear diffusion equation

$$\partial_t \rho = \Delta(\rho(1 + \rho^{1/N})^{-1}).$$

Then S is K-convex if and only if the dimension of the manifold is bounded from above by N and its Ricci curvature is bounded from below by K.

References

[Ale51]   Alexandrov, A.D.: A theorem on triangles in a metric space and some applications (Russian). Trudy Math. Inst. Steklov, **38**, 5–23 (1951)

[ABC00]   Ané, C., Blachère, S., Chafaï, D., Fougères, P., Gentil, I., Malrieu, F., Roberto, C., Scheffer, G.: Sur les inégalités de Sobolev logarithmiques. Société Mathématique de France, Paris (2000)

[Bal95]   Ballmann, W.: Lectures on spaces of nonpositive curvature. DMV Seminar Band **25**, Birkhäuser Verlag, Basel (1995)

[BGS85]   Ballmann, W., Gromov, M., Schroeder, V.: Manifolds of nonpositive curvature. Progress in Mathematics, **61**, Birkhäuser Boston Inc., Boston, MA (1985)

[BE84]   Bakry, D., Emery, M.: Hypercontractivite de semi-groupes de diffusion. C. R. Acad. Sci., Paris, Ser. I, **299**, 775–778 (1984)

[BH99]   Bridson, M.R., Haefliger, A.: Metric spaces of non-positive curvature. Grundlehren der Mathematischen Wissenschaften, **319**. Springer-Verlag, Berlin (1999)

[BBI01]   Burago, D., Burago, Y., Ivanov, S.: A course in metric geometry. Graduate Studies in Mathematics, **33**, American Mathematical Society, Providence, RI (2001)

[Car28]   Cartan, H.: Lecons sur lá geometrie des espaces de Riemann. Gauthiers-Villars, Paris (1928)

[CMS01]   Cordero-Erausquin, D., McCann, R., Schnuckenschläger, M.: A Riemannian interpolation inequality à la Borell, Brascamb and Lieb. Invent. Math., **146**, 219–257 (2001)

[DR02]   Da Prato, G., Röckner, M.: Singular dissipative stochastic equations in Hilbert spaces. Probab. Theory Relat. Fields, **124**, 261–303 (2002)

[Dos49]   Doss, S.: Sur la moyenne d'un élément aléatoire dans un espace distancié. Bull. Sci. Math., **73**, 48–72 (1949)

[Dud89]   Dudley, R.M.: Real analysis and probability. The Wadsworth & Brooks/Cole Mathematics Series. Wadsworth & Brooks/Cole Advanced Books & Software, Pacific Grove, CA. (1989)

[EF01]   Eells, J., Fuglede, B.: Harmonic maps between Riemannian polyhedra. Cambridge Tracts in Mathematics, **142**, Cambridge University Press, Cambridge (2001)

[EM91]   Emery, D., Mokobodzki, G.: Sur le barycentre d'une probabilité dans une variété. Séminaire de Probabilités XXV, 220–233. Lecture Notes in Math. **1485**, Springer, Berlin (1991)

[ESH99]   Es-Sahib, A., Heinich, H.: Barycentre canonique pour un espace métrique à courbure négative. Séminaire de Probabilités, XXXIII, 355–370. Lecture Notes in Math., **1709**, Springer, Berlin (1999)

[Fre48]   Fréchet, M.: Les éléments alétoires de nature quelconque dans un espace distancié, Ann. Inst. H. Poincaré, **10**, 215–310 (1948)

[Gau1809] Carl Friedrich Gauß: Theoria Motus Corporum Celestium (1809)

[Gro81/99] Gromov, M.: Structures métriques pour les variétés Riemanniennes. Rédigé par J. Lafontaine et P. Pansu, Cedic/Fernand Nathan (1981), (1999)

[Her91]   Herer, W.: Espérance mathématique au sens Doss d'une variable aléatoire dans un espace métrique. C. R. Acad. Sci. Paris Sér. I, **302**, 131–134 (1991)

[Jos94]   Jost, J.: Equilibrium maps between metric spaces, Calc. Var. Partial
             Differential Equations, **2**, 173–204 (1994)
[Jos97]   Jost, J.: Nonpositive curvature: geometric and analytic aspects. Lec-
             tures in Mathematics ETH Zürich. Birkhäuser Verlag, Basel (1997)
[Kar77]   Karcher, H.: Riemannian center of mass and mollifier smoothing.
             Comm. Pure Appl. Math., **30**, 509–541 (1977)
[Ken90]   Kendall, W.S.: Probability, convexity, and harmonic maps with small
             images. I. Uniqueness and fine existence. Proc. London Math. Soc.,
             61, 371–406 (1990)
[Ken98]   Kendall, W.S.: From stochastic parallel transport to harmonic maps.
             In: Jost, J., Kendall, W.S., Mosco, U., Röckner, M., Sturm, K.T. (ed)
             New directions in Dirichlet forms. AMS and International Press (1998)
[KS93]   Korevaar, N., Schoen, R.: Sobolev spaces and harmonic maps for met-
             ric space targets. Comm. Anal. Geom., **1**, 561–569 (1993)
[McC01]   McCann, R.: Polar factorization of maps on Riemannian manifolds.
             Geom. Funct. Anal., **11**, 589–608 (2001)
[Ott01]   Otto, F.: The geometry of dissipative evolution equation: the porous
             medium equation. Comm. PDE, **26**, 101–174 (2001)
[OV00]   Otto, F., Villani, C.: Generalization of an inequality by Talagrand and
             links with the logarithmic Sobolev inequality. J. Funct. Anal., **173**,
             361–400 (2000)
[Pic94]   Picard, J.: Barycentres et martingales sur une variéte. Ann. Inst. Henri
             Poincaré, Prob. Stat., **30**, 647–702 (1994)
[RR98]   Rachev, S.T., Rüschendorf, L.: Mass transportation problems. Vol. I.
             Theory. Probability and its Applications. Springer, Berlin Heidelberg
             New York (1998)
[Ren03]   von Renesse, M.-K.: Intrinsic Coupling on Riemannian Manifolds and
             Polyhedra. Preprint, University of Bonn, Bonn (2003)
[RS03]   von Renesse, M.-K., Sturm, K.T.: Transport inequalities, gradient es-
             timates, entropy and Ricci curvature. SFB 611 Preprint, **80**, Univer-
             sity of Bonn, Bonn (2003)
[Str03]   Stroock, D.W.: Markov processes from K. Ito's perspective. Annals
             of Mathematics Studies. Princeton University Press, Princeton, NJ
             (2003)
[Stu98]   Sturm, K.T.: Diffusion processes and heat kernels on metric spaces.
             Ann. Probab., **26**, 1–55 (1998)
[Stu99]   Sturm, K.T.: Metric spaces of lower bounded curvature. Expo. Math.,
             17, 35–47 (1999)
[Stu01]   Sturm, K.T.: Stochastics and analysis on metric spaces. Lecture Notes,
             University of Bonn, Bonn (2001)
[Stu02]   Sturm, K.T.: Nonlinear martingale theory for processes with values in
             metric spaces of nonpositive curvature. Ann. Probab., **30**, 1195–1222
             (2002)
[Stu03]   Sturm, K.T.: A Semigroup Approach to Harmonic Maps. SFB 611
             Preprint, **39**, University of Bonn, Bonn (2003)
[Stu04]   Sturm, K.T.: Nonlinear diffusions as gradient flows: convexity and
             contraction properties. Preprint, University of Bonn, Bonn (2004)
[Vil03]   Villani, C.: Topics in Mass Transportation. Graduate Studies in Math-
             ematics. American Mathematical Society, Providence, RI (2003)

Two Mathematical Approaches
to Stochastic Resonance

Samuel Herrmann[1], Peter Imkeller[2], and Ilya Pavlyukevich[3]

. Université Henri Poincaré Nancy I
  B.P. 239, 54506 Vandoeuvre-lès-Nancy Cedex, France
  herrmann@iecn.u-nancy.fr
. Institut für Mathematik, Humboldt-Universität zu Berlin
  Unter den Linden 6, 10099 Berlin, Germany
  imkeller@mathematik.hu-berlin.de
. Institut für Mathematik, Humboldt-Universität zu Berlin
  Unter den Linden 6, 10099 Berlin, Germany
  pavljuke@mathematik.hu-berlin.de

Summary. We consider a random dynamical system describing the diffusion of a small-noise Brownian particle in a double-well potential with a periodic perturbation of very large period. According to the physics literature, the system is in stochastic resonance if its random trajectories are tuned in an optimal way to the deterministic periodic forcing. The quality of periodic tuning is measured mostly by the amplitudes of the spectral components of the random trajectories corresponding to the forcing frequency. Reduction of the diffusion dynamics in the small noise limit to a Markov chain jumping between its meta-stable states plays an important role.

We study two different measures of tuning quality for stochastic resonance, with special emphasis on their robustness properties when passing to the reduced dynamics of the Markov chains in the small noise limit. The first one is the physicists favourite, spectral power amplification. It is analyzed by means of the spectral properties of the diffusion's infinitesimal generator in a framework where the system switches every half period between two spatially antisymmetric potential states. Surprisingly, resonance properties of diffusion and Markov chain differ due to the crucial significance of small intra-well fluctuations for spectral concepts. To avoid this defect, we design a second measure of tuning quality which is based on the pure transition mechanism between the meta-stable states. It is investigated by refined large deviation methods in the more general framework of smooth periodically varying potentials, and proves to be robust for the passage to the reduced dynamics.

1 Introduction

The physical and mathematical understanding of real world as well as simulation based virtual world phenomena in climate dynamics needs support by stochastically reduced low dimensional climate models. Based on the observation that climate quantities fluctuate on a wide spectrum of spatial or time scales, their heuristic derivation was done in work by Hasselmann around

1975. Their mathematical foundation by means of separation of scales and limit theorems for well mixing deterministic system will remain a challenge.

The scope of reduced stochastic models so far obtained heuristically from this principle range from simple to intermediate complexity models for glacial meta-stability through multi-stable box models of thermohaline circulation to models for the El Niño Southern Oscillation, from linear to delay stochastic oscillators. Their mathematical backbone is non-linear stochastic differential equations with external periodic or internal feedback forcing. Their effective dynamics features phase transitions between meta-stable climate states given by the minima of complex potential functions that describe the non-linear environment. The stochastic analysis of these dynamical climate systems focuses on asymptotic properties such as attractors, bifurcations, hysteresis, stochastic resonance, and Lyapunov stability. It was one of the focal points of the project on the *dynamics of infinite dimensional systems* in the DFG-Schwerpunkt program *Interacting stochastic systems of high complexity*.

This report will concentrate on one particular topic which represents an essential part of the research activity developed in our team in the direction of stochastically reduced climate models. Motivated by periodically forced climate transitions in simple models, we aim at developing the mathematical understanding of the physical paradigm of *stochastic resonance* and spontaneous transitions between meta-stable states. To this end, subjects such as physical quality measures of noise tuning will be studied. Their conceptual output may severely damage the resonance picture due to the eminent role of small stochastic fluctuations in the vicinity of the meta-stable states. It will therefore be compared to contrasting notions of tuning purely based on the transition mechanism.

2 Model Reduction and Stochastic Resonance

The paradigm of *stochastic resonance* (SR) emerged from papers by C. Nicolis [Nic82] and Benzi et al. [BPSV81, BPSV82, BPSV83] which were devoted to the mathematical explanation of the phenomenon of *glacial cycles*. The model they created is based on the following observations. Modern measurement techniques allow to determine concentrations of an oxygen isotope in deep sea core sediments which in turn provide rough estimates of the *global mean temperature* of the earth at the time they were deposited. This way at least seven changes between 'cold' and 'warm' periods were detected during approximately the last 700,000 years. They occur abruptly and with roughly the same period of about 10^5 years. The quoted papers aimed at suggesting a simple mathematical model to account for this deterministic-looking periodicity.

The proposed model just appeals to conservation of radiative energy and supposes that the earth's temperature X satisfies a simple energy-balance equation (for an extended review see [Imk01]), i.e. the instant change of the global temperature is proportional to the difference between *incoming* and *outgoing* radiative energy:

$$c\frac{dX}{dt} = R_{\text{in}} - R_{\text{out}}, \quad c > 0. \tag{1}$$

In the simplest case considered here it is assumed that the total energy flux emitted by the earth is given by the Stefan–Boltzmann law which in fact is valid only for a black body radiator.

The absorbed energy depends on two factors. The *global solar* function describes the flux of the solar energy which reaches the earth. Assuming that the solar activity is a constant, the solar function depends on the distance between the earth and the sun as well as on the inclination of the earth's axis, and due to the gravitational influence of Jupiter exhibits a slow periodic variation of a period of about 10^5 years. The variation is estimated to be 0.1% of the average value.

On the other hand, not all the solar radiation reaching the atmosphere is absorbed: the proportion of absorbed radiation is determined by the earth's albedo which depends *locally* on the earth's average surface temperature T. The simple albedo model used here appears in the papers by Budyko [Bud69] and Sellers [Sel69].

Thus, the right hand side of (1) is a difference of two functions. For appropriate values of parameters the dynamical system (1) has two metastable equilibrium states separated by the unstable state. The lower metastable state is interpreted as describing ice age temparatures whereas the higher one determines warm ages.

This model of climate has major shortcomings and therefore cannot picture reality. Indeed, solutions of (1) converge to one of the metastable states and oscillate with periods of 10^5 years with relatively small amplitudes, due to the small variation of the solar constant. Most importantly, however, the typically observed spontaneous and rapid transitions between 'cold' and 'warm' states are impossible.

To overcome this difficulty C. Nicolis and Benzi et al. added a *noise* term to the energy-balance equation (1) and obtained the following nonlinear SDE for the global temperature. It will be the model equation on which our mathematical approaches of stochastic resonance are based.

On an appropriate probability space $(\Omega, \mathcal{F}, \mathbf{P})$ we consider a one-dimensional diffusion $X^{\varepsilon,T} = (X_t^{\varepsilon,T})_{t\geq 0}$ driven by the stochastic differential equation

$$dX_t^{\varepsilon,T} = -U'(X_t^{\varepsilon,T}, \frac{t}{T})dt + \sqrt{\varepsilon}dW_t, \quad X_0^{\varepsilon,T} = x \in \mathbb{R}, \quad t \geq 0, \tag{2}$$

where W is a standard Brownian motion, and ε a small noise intensity. The potential U is supposed to be double-well in the spatial coordinate. It is temporally periodic with period 1, i.e. $U(\cdot, t) = U(\cdot, t+1)$, for any $t \geq 0$. U' is used to denote the derivative in x. The positive parameter T stand for

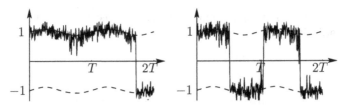

Fig. 1. Sample paths of (2) for small (l.) and 'optimal' (r.) values of ε

the period of the deterministic perturbation. As an example for U one can take $U(x,t) = U_0(x) + ax \sin 2\pi t$, $x \in \mathbb{R}$, $t \geq 0$, with a symmetric potential $U_0 = \frac{x^4}{4} - \frac{x^2}{2}$, $x \in \mathbb{R}$, and a small enough amplitude a to ensure that U does not degenerate to a one-well potential.

Given a large period T, we will be interested in *periodicity properties* of the trajectories of our system (2), in particular to have a mathematically precise concept of how well they are able to follow the deterministic periodic excitation in dependence on the noise intensity ε. It is intuitively clear that if ε is very small, trajectories will almost never be able to leave the well in which they start, and stay close to the starting well's minimum (see Fig.1 (l.)). If ε is very large, the energy of the particle is sufficient to trigger some chaotic changing between the two wells. There will be an intermediate range of small intensities, at which trajectories are more or less close to the deterministic periodic function describing the temporally varying energetically most favorable position in the potential landscape given by the minimum of the deeper of the two wells (see Fig.1 (r.), [Fre00]). The crucial questions to be answered in the sequel are the following. Given T large, for which intensity $\varepsilon(T)$ will the periodicity of the system's trajectories be optimal? And how can we measure the quality of periodicity?

The quasi-deterministic behaviour the trajectories exhibit for small noise raises another question which is very significant since it indicates a route to reduction of complexity relevant for high dimensional systems: is it possible to study the periodic tuning properties of the diffusion by considering instead a simplified two-state Markov chain model, which catches the diffusion's *effective dynamics*?

This approach was extensively studied by physicists. In their pioneering theoretical paper [MW89] McNamara and Wiesenfeld propose a two-state model of stochastic resonance in which the small-noise diffusion in a double-well potential *in adiabatic limit* is replaced by a two-state Markov process. Along with (2) the inhomogeneous Markov chain $Y^{\varepsilon,T}$ is considered, which possesses the 1-periodic infinitesimal generator

$$Q_{\varepsilon,T}(t) = \begin{pmatrix} -\varphi(\frac{t}{T}) & \varphi(\frac{t}{T}) \\ \psi(\frac{t}{T}) & -\psi(\frac{t}{T}) \end{pmatrix} \tag{3}$$

with the infinitesimal transition rates φ and ψ given by time-perturbed *Kramers-Eyring law* [Kra40]

$$\varphi(t) = \frac{1}{2\pi}\sqrt{|U_0''(0)||U_0''(1)|}e^{-\frac{2}{\varepsilon}(\Delta U + a\sin 2\pi t)},$$

$$\psi(t) = \frac{1}{2\pi}\sqrt{|U_0''(0)||U_0''(1)|}e^{-\frac{2}{\varepsilon}(\Delta U - a\sin 2\pi t)}, \quad t \geq 0, \tag{4}$$

where $\Delta U = \frac{1}{4}$ is the height of the potential barrier of the unperturbed potential U_0. Kramers obtained his law heuristically in the autonomous case, i.e. for $a = 0$.

To measure periodicity of the random trajectories of either the diffusion or the Markov chain, we first take the so-called coefficient of *spectral power amplification* (SPA), one of the physicists' favourite characteristics, see e.g. [BPSV83, MW89, DLM+95, GHJM98, ANMS99, WJ98]. It is based on the power spectrum of the average trajectories with respect to the equilibrium of the homogenized Markov processes $(X_{Tt}^{\varepsilon,T}, t \pmod 1)_{t\geq 0}$ resp. $(Y_{Tt}^{\varepsilon,T}, t \pmod 1)_{t\geq 0}$. For the diffusion (2) with equilibrium μ it is defined by

$$\eta^X(\varepsilon, T) = \left| \int_0^1 \mathbf{E}_\mu(X_{Ts}^{\varepsilon,T}) \cdot e^{2\pi i s}\, ds \right|^2. \tag{5}$$

The function η^X depending on noise intensity and the period of time variation of the potential has a clear physical meaning. It describes the amount of energy carried by the averaged path of the diffusion with noise amplitude ε on the frequency $\frac{2\pi}{T}$. The expectation $\mathbf{E}_\mu$ indicates that averages are taken with respect to μ. This will be explained in detail later.

Fig. 2 borrowed from [ANMS99] where Ω corresponds to our $\frac{2\pi}{T}$ and D to the diffusion intensity ε shows that physicists expect a local maximum of the function $\varepsilon \mapsto \eta^X(\varepsilon, \cdot)$. The random paths have their strongest periodic component corresponding to the frequency of the periodic input at the value of ε for which the maximum is taken. In fact, Fig. 2 does not show the SPA

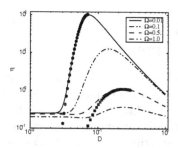

Fig. 2. SPA coefficient as a function of noise amplitude is supposed to have a well pronounced maximum depending on the frequency of periodic perturbation [ANMS99]

coefficient of the diffusion itself, but of its *effective dynamics* given by the two-state Markov chain $Y^{\varepsilon,T}$. It is *a priori* believed in the physical literature that the effective dynamics adequately describes the properties of the diffusion in the small noise limit.

If periodic tuning is measured by SPA, to determine the 'optimal tuning' or *stochastic resonance point* means to find the argument $\varepsilon = \varepsilon(T)$ of a local maximum of $\varepsilon \mapsto \eta^X(\varepsilon, \cdot)$ for large T. In section 3 we address the problem of finding the stochastic resonance point for the diffusion by means of the passage to its effective dynamics in the small noise limit. We shall see that determining the optimal tuning intensity $\varepsilon(T)$ for the Markov chain is a relatively easy task. It turns out, however, that already in the very simple case of a potential hopping every half period between two spatially antisymmetric double-well states with wells of different depths, due to the crucial importance of the diffusion's *inter-well* fluctuations, i.e. small fluctuations in the vicinity of the potential's minima, at low noise, diffusion and Markov chain exhibit different resonance features. In contradiction to physicists' intuition, the SPA notion of resonance is therefore not robust when passing to the reduced model. In section 4 we therefore propose a concept of measuring the quality of periodic tuning which is based on the pure mechanism of transition between the domains of attraction of the potential's local minima, and therefore fails to have this robustness defect.

A different step towards a mathematical understanding of stochastic resonance was made by Berglund and Gentz [BG02a, BG02b, BG02c]. For parameterized deterministic dynamical systems passing through a pitchfork bifurcation point, the relaxation of solutions to stable equilibria are known to happen after well known delays. Berglund and Gentz exploit this observation to derive path-wise estimates for the trajectories of noisy perturbations of these systems. These results are applied to situations in which the parameter moves the system periodically or in hysteresis loops back and forth through bifurcation points, for example in periodically changing double-well potentials. Their papers thus address questions concerning the trajectorial behaviour in the context of stochastic resonance.

3 Periodically Switching Potentials and the Spectral Approach

To catch the essentials of the effect and at the same time to simplify the problem we will work in the first part of this paper with a time-space antisymmetric double well potential switching discontinuously between two states. In the second part we will essentially extend this framework to include continuously varying potentials. In the strip $(x,t) \in \mathbb{R} \times [0,1)$ it is defined by the formula

$$U(x,t) = \begin{cases} U_1(x), & t \in [0, \frac{1}{2}), \\ U_2(x) = U_1(-x), & t \in [\frac{1}{2}, 1). \end{cases} \tag{6}$$

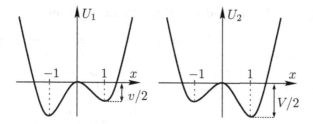

Fig. 3. Time-periodic potential U

It is periodically extended for all times t by the relation $U(\cdot, t) = U(\cdot, t+1)$, see Fig. 3. We assume that the potential has two local minima at ± 1 and a local maximum at 0, that $U_1(-1) = -\frac{V}{2}$, $U_1(1) = -\frac{v}{2}$, $\frac{2}{3} < \frac{v}{V} < 1$, and $U_1(0) = 0$. We also suppose that the extrema of U are not degenerate, i.e. the curvatures at these points do not vanish.

A trajectory of a Brownian particle in this potential is described by the solution of the stochastic differential equation (7).

$$ dX_t^{\varepsilon,T} = -U'(X_t^{\varepsilon,T}, \frac{t}{T})\, dt + \sqrt{\varepsilon}\, dW_t, \quad X_0^{\varepsilon,T} = x \in \mathbb{R}. \tag{7} $$

We aim at finding a resonance intensity $\varepsilon(T)$ for large T which maximizes the SPA coefficient given by (5). The key to the solution of this problem lies in determining the time-dependent invariant density μ of $(X_{Tt}^{\varepsilon,T})_{t\geq 0}$. From now on we follow [Pav02] and [IP02]. Although the diffusion is not time homogeneous, by enlarging its state space we can consider a two-dimensional time homogeneous Markov process $(X_{Tt}^{\varepsilon,T}, t \pmod 1)$ which possesses an invariant law in the usual sense. By definition we identify the time-dependent equilibrium density μ of $(X_{Tt}^{\varepsilon,T})_{t\geq 0}$ with the invariant density of the two-dimensional process. Indeed, with respect to μ and for fixed t, the law of the real random variable $X_{Tt}^{\varepsilon,T}$ has the density $\mu(\cdot, t \pmod 1)$. The invariant density μ is a positive solution of the forward Kolmogorov (Fokker–Planck) equation $A_{\varepsilon,T}^* \mu = 0$, where

$$ A_{\varepsilon,T}^* \cdot = -\frac{1}{T}\frac{\partial}{\partial t}\cdot + \frac{\varepsilon}{2}\frac{\partial^2}{\partial x^2}\cdot + \frac{\partial}{\partial x}\left(\cdot\frac{\partial}{\partial x}U\right) $$

is the formal adjoint of the infinitesimal generator of the two-dimensional diffusion. Moreover, from the time periodicity and time-space antisymmetry of the potential U defined by (6) one concludes that $\mu(x,t) = \mu(-x, t+\frac{1}{2})$ and $\mu(x,t) = \mu(x, t+1)$, $(x,t) \in \mathbb{R} \times \mathbb{R}_+$.

This results in the following boundary-value problem used to determine μ. It is enough to solve the Fokker–Planck equation $A_{\varepsilon,T}^* \mu = 0$ in the strip $(x,t) \in \mathbb{R} \times [0, \frac{1}{2}]$ with boundary condition $\mu(x,0) = \mu(-x, \frac{1}{2})$, $x \in \mathbb{R}$.

3.1 The Spectral Gap and the First Eigenfunction

We have assumed in (6) that the time dependent potential U is a step function of the time variable. In the region $(x, t) \in \mathbb{R} \times (0, \frac{1}{2})$ it is identical to a time independent double well potential U_1, and therefore the Fokker–Planck equation turns into a one-dimensional parabolic PDE

$$\frac{1}{T}\frac{\partial}{\partial t}\mu(x,t) = \frac{\varepsilon}{2}\frac{\partial^2}{\partial x^2}\mu(x,t) + \frac{\partial}{\partial x}\left(\mu(x,t)\frac{\partial}{\partial x}U_1(x)\right). \tag{8}$$

Let L_ε^* denote the second order differential operator appearing on the right hand side of (8).

To determine μ we shall use the Fourier method of separation of variables which consists in expanding the solution of (8) into a Fourier series with respect to the system of eigenfunctions of the operator L_ε^*. It turns out that under the condition that U_1 is smooth and increases at least super-linearly at $\pm\infty$, the operator L_ε^* is essentially self-adjoint in $\mathcal{L}^2(\mathbb{R}, e^{\frac{2U_1}{\varepsilon}} dx)$, its spectrum is discrete and non-positive, and the corresponding eigenspaces are one-dimensional. Denoting by $\|\cdot\|$ and $\langle\cdot,\cdot\rangle$ the norm and the inner product in $\mathcal{L}^2(\mathbb{R}, e^{\frac{2U_1}{\varepsilon}} dx)$ we consider the following formal Floquet type expansion

$$\mu(x,t) = \sum_{k=0}^{\infty} a_k \frac{\Psi_k(x)}{\|\Psi_k\|} e^{-T\lambda_k t}, \quad (x,t) \in \mathbb{R} \times [0, \tfrac{1}{2}], \tag{9}$$

where $\{-\lambda_k, \frac{\Psi_k}{\|\Psi_k\|}\}_{k\geq0}$ is the orthonormal basis corresponding to the spectral decomposition of L_ε^*, where $\lambda_0 < \lambda_1 < \lambda_2 < \cdots$, and the Fourier coefficients a_k are obtained from the boundary condition $\mu(x,0) = \mu(-x, \frac{1}{2}), x \in \mathbb{R}$.

Here is the key observation opening the route towards finding local maxima of the SPA coefficient. The terms in the sum (9) decay in time exponentially fast with rates λ_k, and therefore the terms corresponding to larger eigenvalues contribute less than the ones belonging to the low lying eigenvalues. This underlines their key importance. Fortunately, in the case of a double well potential the following theorem holds.

Theorem 1 ('spectral gap'). *In the limit of small noise, the following asymptotic properties are valid:*

$$\lambda_0 = \lambda_0(\varepsilon) = 0, \quad and \quad \Psi_0 = e^{-\frac{2U_1}{\varepsilon}},$$

$$\lambda_1 = \lambda_1(\varepsilon) = \frac{1}{2\pi}\sqrt{U_1''(1)|U_1''(0)|} \cdot e^{-v/\varepsilon}(1 + \mathcal{O}(\varepsilon)),$$

$$\lambda_2 = \lambda_2(\varepsilon) \geq C > 0 \quad uniformly \ in \ \varepsilon.$$

Moreover, we can provide a very good approximation to the first eigenfunction Ψ_1. Let $-1 < x_- < 0 < x_+ < 1$ such that $U_1(x_-) = -\frac{V}{3}$ and $U_1(x_+) = -\frac{v}{3}$. Fix also some v' such that $\frac{2}{3} < \frac{v'}{v} < 1$. Then the following holds.

Theorem 2. *In the limit of small noise, the first eigenfunction of L_ε^* is found as $\Psi_1 = \Phi_1 e^{-\frac{2U_1}{\varepsilon}}$, where*

$$\max_{x \leq x_-} |\Phi_1 - a(\varepsilon)| \leq e^{-\frac{2V}{3\varepsilon}},$$

$$\max_{x_- \leq x \leq x_+} \left|\Phi_1 - a(\varepsilon) - (1 - a(\varepsilon))\frac{\int_{-1}^{x} e^{\frac{2U(y)}{\varepsilon}} dy}{\int_{-1}^{1} e^{\frac{2U(y)}{\varepsilon}} dy}\right| \leq e^{-\frac{v'}{\varepsilon}},$$

$$\max_{x \geq x_+} |\Phi_1 - 1| \leq e^{-\frac{2v}{3\varepsilon}}$$

with $a(\varepsilon) = -\sqrt{\frac{U_1''(-1)}{U_1''(1)}} e^{-\frac{V-v}{\varepsilon}} (1 + \mathcal{O}(\varepsilon))$.

The result of Theorem 1 plays a crucial role in our analysis. There is a *spectral gap* between the first eigenvalue and the rest of the spectrum. Consequently, only the first two terms of (9) can have an essential contribution to the SPA coefficient η^X.

3.2 Asymptotics of the SPA Coefficient

The following theorem gives the asymptotics of the first two Fourier coefficients a_0 and a_1 in the Floquet type expansion of the previous subsection.

Theorem 3. *We have*

$$a_0 = \|\Psi_0\|,$$

$$a_1 = \frac{\|\Psi_1\|}{\|\Psi_0\|^2} \cdot \frac{\langle \Psi_0(-\cdot), \Psi_1 \rangle}{\|\Psi_1\|^2 - e^{-\frac{1}{2}T\lambda_1} \langle \Psi_1(-\cdot), \Psi_1 \rangle} + r,$$

where r vanishes in the limit of small noise and for $T \geq \exp\{(v + \delta)/\varepsilon\}$, δ being positive and sufficiently small.

Recall the definition (5) of the SPA coefficient. Denote

$$S^X(\varepsilon, T) = \int_0^{\frac{1}{2}} \mathbf{E}_\mu X_{Ts}^{\varepsilon, T} \cdot e^{2\pi i s} \, ds.$$

In these terms we identify $\eta^X = 4|S^X|^2$.

Theorem 4. *Let $T \geq \exp\{(v + \delta)/\varepsilon\}$ for δ positive and sufficiently small. Then the following expansion for S^X holds in the small noise limit $\varepsilon \to 0$*

$$S^X = \frac{1}{\pi i} b_0 + \frac{1}{\pi i - \frac{1}{2}\lambda_1 T} b_1 + r_1$$

where the rest term r_1 tends to zero with ε and the coefficients are given by

$$b_0 = \frac{\int y\, e^{-\frac{2U_1(y)}{\varepsilon}}\, dy}{\int e^{-\frac{2U_1(y)}{\varepsilon}}\, dy},$$

$$b_1 = -\frac{1 + e^{-\frac{1}{2}T\lambda_1}}{2} \cdot \frac{\int y\, \Psi_1(y)\, dy}{\int e^{-\frac{2U_1(y)}{\varepsilon}}\, dy} \cdot \frac{\langle \Psi_0(-\cdot), \Psi_1 \rangle}{\|\Psi_1\|^2 - e^{-\frac{1}{2}T\lambda_1}\langle \Psi_1(-\cdot), \Psi_1 \rangle}.$$

Finally,

$$\eta^X = b_0^2 \frac{4}{\pi^2} \frac{(\lambda_1 T)^2}{4\pi^2 + (\lambda_1 T)^2} + R, \tag{10}$$

where R tends to zero with ε.

Let us now study the *resonance* behaviour of the SPA coefficient η^X, i.e. investigate whether it has a local maximum in ε. We formulate the following Lemma which is obtained by application of Laplace's method of asymptotic expansion of singular integrals, see [Erd56, Olv74] or also [Pav02, IP02].

Lemma 1 ('Laplace's method'). *In the small noise limit, the following holds true:*

$$b_0 = -1 - \frac{1}{4}\frac{U_1^{(3)}(-1)}{U_1''(-1)^2}\varepsilon + \mathcal{O}(\varepsilon^2),$$

$$b_1 = -1 + \mathcal{O}(\varepsilon),$$

and consequently

$$b_0^2 = 1 + \frac{1}{2}\frac{U_1^{(3)}(-1)}{U_1''(-1)^2}\varepsilon + \mathcal{O}(\varepsilon^2), \tag{11}$$

$$(b_0 - b_1)^2 = \mathcal{O}(\varepsilon^2).$$

The following Theorem exhibits the defect of the notion of spectral power amplification for our diffusions in periodically and discontinuously switching potential states.

Theorem 5. *Let us fix δ positive and sufficiently small and $\Delta > v + \delta$. Let also $U_1(x) - 2U_1(-x) < v + V$ for all $x \in \mathbb{R}$ (no strong asymmetry!). Then for $T \to \infty$ and ε satisfying*

$$\frac{v + \delta}{\ln T} \le \varepsilon \le \frac{\Delta}{\ln T}$$

the following asymptotic expansion for the SPA coefficient holds:

$$\eta^X(\varepsilon, T) = \frac{4}{\pi^2}\left(1 + \frac{1}{2}\frac{U_1^{(3)}(-1)}{U_1''(-1)^2}\varepsilon\right) + \mathcal{O}\left(\frac{1}{\ln^2 T}\right).$$

This result has the following surprising consequences.

Corollary 1. *For $T \to \infty$ and $\varepsilon \in [\frac{v+\delta}{\ln T}, \frac{\Delta}{\ln T}]$ the SPA coefficient is a decreasing function of ε if $U_1^{(3)}(-1) < 0$ and an increasing function of ε if $U_1^{(3)}(-1) > 0$.*

Thus, the SPA coefficient as quality measure for tuning shows *no resonance* in a domain above Freidlin's threshold for quasi-deterministic periodicity (see [Fre00]). This contradicts the physical intuition for the 'effective dynamics'. The reason for this surprising phenomenon can only be hidden in the *intra-well* behaviour of the diffusion neglected when passing to the reduced Markov chain. We return to this question later. Let us next study in more detail the 'effective dynamics' of the diffusion (7).

3.3 The 'Effective Dynamics': Two-State Markov Chain

The idea of approximation of diffusions in potential landscapes by appropriate finite state Markov chains in the context of stochastic resonance was suggested by Eckmann and Thomas [ET82], and C. Nicolis [Nic82], and developed by McNamara and Wiesenfeld [MW89]. In this section we follow [Pav02, IP02]. The discrete time case was studied in [IP01].

In order to catch the main features of the spatial *inter-well* transitions of the diffusion (7) we consider the time inhomogeneous Markov chain $Y^{\varepsilon,T}$ living on the diffusion's meta-stable states ± 1. The infinitesimal generator of $Y^{\varepsilon,T}$ is periodic in time and is given by

$$
Q_{\varepsilon,T}(t) =
\begin{cases}
\begin{pmatrix} -\varphi & \varphi \\ \psi & -\psi \end{pmatrix}, & \frac{t}{T} \ (\mathrm{mod}\ 1) \in [0, \tfrac{1}{2}), \\[2mm]
\begin{pmatrix} -\psi & \psi \\ \varphi & -\varphi \end{pmatrix}, & \frac{t}{T} \ (\mathrm{mod}\ 1) \in [\tfrac{1}{2}, 1).
\end{cases}
\tag{12}
$$

The transition rates φ and ψ which are responsible for the similarity of the two processes are chosen to be exponentially small in ε:

$$
\varphi = \frac{1}{2\pi}\sqrt{U_1''(-1)|U_1''(0)|} \cdot e^{-V/\varepsilon} \quad \text{and} \quad \psi = \frac{1}{2\pi}\sqrt{U_1''(1)|U_1''(0)|} \cdot e^{-v/\varepsilon}.
$$

To exponential order they correspond (as they should) to the inverses of the Kramers-Eyring transition times. The invariant measure of $Y_{Tt}^{\varepsilon,T}$ can be obtained as a solution of a forward Kolmogorov equation and is given by

$$
\begin{aligned}
\nu^-(t) &= \frac{\psi}{\varphi+\psi} + \frac{\varphi-\psi}{\varphi+\psi}\frac{e^{-(\varphi+\psi)Tt}}{1+e^{-\frac{1}{2}(\varphi+\psi)Tt}}, \\[2mm]
\nu^+(t) &= \frac{\varphi}{\varphi+\psi} - \frac{\varphi-\psi}{\varphi+\psi}\frac{e^{-(\varphi+\psi)Tt}}{1+e^{-\frac{1}{2}(\varphi+\psi)Tt}}, \quad t \in [0, \tfrac{1}{2}],
\end{aligned}
\tag{13}
$$

and $\nu^\pm(t) = \nu^\mp(t+\tfrac{1}{2})$ for $t \geq 0$.

We define the SPA coefficient η^Y for the Markov chain $Y^{\varepsilon,T}$ analogously to (5). In this much simpler setting given it can be described explicitly.

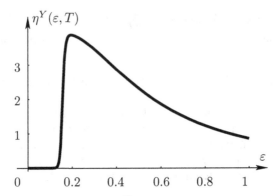

Fig. 4. SPA coefficient η^Y of the two-state Markov chain

Theorem 6. *For all $\varepsilon > 0$ and $T > 0$ the following holds:*

$$\eta^Y(\varepsilon, T) = \frac{4}{\pi^2} \frac{T^2(\varphi - \psi)^2}{4\pi^2 + T^2(\varphi + \psi)^2}. \tag{14}$$

Compare (14) with (10). Since $(\varphi \pm \psi)^2 \approx \lambda_1^2$ in the limit of small ε, the formulae for η^X and η^Y differ only in the 'geometric' pre-factor b_0^2 and the asymptotically negligible remainder term R.

The explicit formula (14) allows to study the local maxima of η^Y as a function of noise intensity for large periods T in great detail (see Fig. 4).

Theorem 7. *In the limit $T \to \infty$ the function $\varepsilon \mapsto \eta^Y(\varepsilon, T)$ has a local maximum at*

$$\varepsilon(T) \approx \frac{v + V}{2} \frac{1}{\ln T},$$

or in the limit $\varepsilon \to 0$ in terms of T

$$T(\varepsilon) \approx \frac{\pi}{\sqrt{2pq}} \sqrt{\frac{v}{V - v}} e^{\frac{V + v}{\varepsilon}}.$$

The 'resonance' behaviours of η^X and η^Y are quite different. Whereas the diffusion's SPA has no extremum for small ε, the Markov chain's *always* has. What can be responsible for this discrepancy? Note that the Markov chain mimics only the *inter-well* dynamics of the diffusion. Thus, the SPA coefficient η^Y measures only the spectral energy contributed by inter-well jumps. On the other hand, η^X also counts the numerous *intra-well* fluctuations of the diffusion the weight of which evidently becomes overwhelming in the small noise limit. These fluctuations have small energy. But since the diffusion spends most of its time near ± 1 the local asymmetries of the potential at these points dominate the picture and destroy optimal tuning.

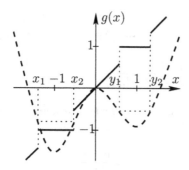

Fig. 5. Function g designed to cut off diffusion's intra-well dynamics

To underpin this heuristics mathematically, let us now make the idea of neglecting the diffusion's intra-well fluctuations precise. For example, we *cut off* those among them which have not enough energy to reach half the height of the potential barrier between the wells. Consider the cut-off function g defined by

$$g(x) = \begin{cases} -1, & x \in [x_1, x_2], \\ 1, & x \in [y_1, y_2], \\ x, & \text{otherwise}, \end{cases}$$

where $x_1 < -1 < x_2 < 0$ and $0 < y_1 < 1 < y_2$ are such that $U_1(x_1) = U_1(x_2) = -\frac{V}{4}$ and $U_1(y_1) = U_1(y_2) = -\frac{v}{4}$, see Fig. 5. Now we study the modified SPA coefficient of a diffusion defined by

$$\widetilde{\eta}^X(\varepsilon, T) = \left| \int_0^1 \mathbf{E}_\mu \left[g(X_{Ts}^{\varepsilon,T}) \right] e^{2\pi i s} \, ds \right|^2.$$

Following the steps of Subsection 3.2 we obtain a formula for $\widetilde{\eta}^X$ which is quite similar to (10) and (14):

$$\widetilde{\eta}^X(\varepsilon, T) = \widetilde{b}_0^2 \frac{4}{\pi^2} \frac{(\lambda_1 T)^2}{4\pi^2 + (\lambda_1 T)^2} + \widetilde{R},$$

where $\widetilde{R}$ is a small remainder term, and

$$\widetilde{b}_0^2 = \left(\frac{\int g(y) e^{-\frac{2U_1(y)}{\varepsilon}} \, dy}{\int e^{-\frac{2U_1(y)}{\varepsilon}} \, dy} \right)^2 = 1 - 4\sqrt{\frac{U_1''(-1)}{U_1''(1)}} e^{-\frac{V-v}{\varepsilon}} (1 + \mathcal{O}(\varepsilon))$$

(compare to (11)).

The modified geometric pre-factor $\widetilde{b}_0^2$ is essentially smaller than its counterpart b_0^2. This has crucial influence on the SPA coefficient $\widetilde{\eta}^X$: in the limit of large period and small noise its behaviour now reminds of η^Y.

Theorem 8. *Let the assumptions of Theorem 5 hold. Then for any $\gamma > 1$ in the limit $T \to \infty$ the function $\varepsilon \mapsto \widetilde{\eta}^X(\varepsilon, T)$ has a local maximum on*

$$\left[\frac{1}{\gamma} \frac{v+V}{2} \frac{1}{\ln T}, \gamma \frac{v+V}{2} \frac{1}{\ln T} \right].$$

In other words, the optimal tuning for the measure of goodness $\widetilde{\eta}^X$ exists and is given approximately by

$$\varepsilon(T) \approx \frac{v+V}{2} \frac{1}{\ln T}.$$

4 Smooth Periodic Potentials and a Robust Resonance Notion

The serious defect of the SPA coefficient in the prediction of the stochastic resonance point in the Markov chain models containing the effective dynamics of complex diffusion models motivates us to look for *robust* notions of quality of periodic tuning. Since the dynamics of the Markov chain only retains the rough mechanism of transitions between the domains of attraction given in the underlying potential landscape, such a notion should only take into account the most important aspects of the *attractor hopping*. Also, as the alternative notions discussed in the preceding section show, the resonance point is by no means a canonical object, independent of the way tuning quality is measured. We think that the methods of advanced large deviations' theory behind the notion to be explained in this section will give it a more natural place, and possibly qualify it as canonical.

At the same time, we essentially generalize the simplified model of time periodic potential considered in the previous section, and lift the study of stochastic resonance to a somewhat more abstract level. The potential function U in the present section will still be supposed to be one-dimensional in space. But its periodic time variation will just be assumed to be continuous, and otherwise quite general. More precisely, we study diffusion processes driven by a Brownian motion of intensity ε given by the stochastic differential equation

$$dX_t^{\varepsilon,T} = -U'\left(X_t^{\varepsilon,T}, \frac{t}{T}\right) dt + \sqrt{\varepsilon}\, dW_t, \quad t \geq 0.$$

The underlying potential landscape (see Fig. 6) is described by a function $U(x,t)$, $x \in \mathbb{R}$, $t \geq 0$, which is periodic in time with period 1, and its temporal variation, by the rescaling with very large T, acts on the diffusion at a very small frequency. U is supposed to have exactly two wells located at ± 1, separated by a saddle at 0. The depth of $U(\cdot, t)$ at ± 1 is given by the 1-periodic depth functions $\frac{1}{2}D_{\pm 1}(t)$ which are assumed to never fall below zero. Let us now look at exponential time scales ρ, related to the natural time scale T

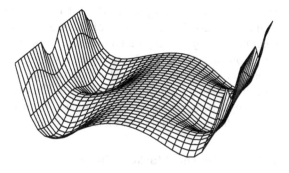

Fig. 6. Potential landscape U

by $T = e^{\rho/\varepsilon}$. In this setting, Freidlin's theory of quasi-deterministic motion indicates that transitions e.g. from the domain of attraction of -1 to the domain of attraction of 1 will occur as soon as D_{-1} gets less than ρ, i.e. at time

$$a_\rho^{\pm 1} = \inf\{t \geq 0 : D_{\pm 1}(t) \leq \rho\}.$$

This triggers periodic behaviour of the diffusion trajectories on long time scales. The modern theory of meta-stability in *time homogeneous diffusion processes* produces the exponential decay rates of transition probabilities between different domains of attraction of a potential landscape together with very sharp multiplicative error estimates, uniformly on compacts in system parameters. Their sharpest forms are presented in some very recent papers by Bovier et al. [BEGK02, BGK02]. We use this powerful machinery in order to obtain very precise estimates of the exponential tails of the laws of the transition times between domains of attraction. To this end, we have to extend the estimates by Bovier et al. [BGK02] to the framework of *time inhomogeneous diffusions*. In the underlying one-dimensional situation, this can be realized by freezing the time dependence of the potential on small time intervals to define lower and upper bound time homogeneous potentials not differing very much from the original one. Comparison theorems are used to control the transition behaviour from above and below through the corresponding time homogeneous diffusions. This allows very precise estimates on the probabilities with which the diffusion at time scale $T = e^{\rho/\varepsilon}$ transits from the domain of attraction of -1 to the domain of attraction of 1 or vice versa within time windows $[(a_\rho^{-1} - h)T, (a_\rho^{-1} + h)T]$ for small $h > 0$. If $\tau_x(X^{\varepsilon,T})$ denotes the transit time to x, it is given by

$$\lim_{\varepsilon \to 0} \varepsilon \ln \left(1 - M(\varepsilon, \rho)\right) = \max_{i = \pm 1} \left\{\rho - D_i(a_\rho^i - h)\right\},$$

with

$$M(\varepsilon, \rho) = \min_{i = \pm 1} P_i(\tau_{-i}(X^{\varepsilon,T}) \in [(a_\rho^i - h)T, (a_\rho^i + h)T]), \quad \varepsilon > 0, \rho \in I_R,$$

and where I_R is the *resonance interval*, i.e. the set of scale parameters for which trivial or chaotic transition behaviour of the trajectories is excluded. The stated convergence is *uniform* in ρ on compact subsets of I_R. This allows us to take $M(\varepsilon, \rho)$ as our measure of periodic tuning, compute the scale $\rho_0(h)$ for which the transition rate is optimal, and define the *stochastic resonance point* as the eventually existing limit of $\rho_0(h)$ as $h \to 0$. This resonance notion has the big advantage of being robust for the passage from the diffusion to the two state Markov chain describing the effective dynamics.

4.1 Transition Times for the Markov Chain

Let us first discuss the effective dynamics modelled by a continuous time two state Markov chain. The states represent the positions of the bottoms of the wells of the double well potential. The transition rates picture the transition mechanism of the diffusion to which we return later. We shall first define the interval of time scales for which transitions are not trivial.

Definition of the Resonance Interval Let us consider the time continuous Markov chain $Y^{\varepsilon,T} = (Y_t^{\varepsilon,T})_{t \geq 0}$ taking values in the state space $\{-1, 1\}$ with initial data $Y_0^{\varepsilon,T} = -1$. Suppose that the infinitesimal generator is given by

$$G_{\varepsilon,T}(t) = \begin{pmatrix} -\varphi(\frac{t}{T}) & \varphi(\frac{t}{T}) \\ \psi(\frac{t}{T}) & -\psi(\frac{t}{T}) \end{pmatrix},$$

where T is an exponentially large time scale (we recall that $T = e^{\rho/\varepsilon}$, $\rho > 0$), ψ and φ are 1-periodic functions describing a rate which just produces the transition dynamics of the diffusion between the potential minima ± 1. Let us recall that, if we consider some time-independent potential U, then the mean transition time between the wells is given by the Kramers-Eyring law. If the diffusion starts in the minimum of one well, the mean exit time is equivalent to $e^{V/\varepsilon}$, where $\frac{V}{2}$ is the height of the barrier separating the two minima of the potential. Consequently the transition rate should be proportional to $e^{-V/\varepsilon}$.

In the framework we now consider the depth of the wells depends continuously on time. In this situation it is natural to postulate the following periodic infinitesimal probabilities

$$\varphi(t) = e^{-D_{-1}(t)/\varepsilon}, \quad t \geq 0. \tag{15}$$

Let us assume that $D_1(t) = D_{-1}(t + \alpha)$, $t \geq 0$, with phase shift $\alpha \in (0, 1)$ and

- all local extrema of $D_{\pm 1}(\cdot)$ are global;
- the functions $D_{\pm 1}(\cdot)$ are strictly monotonous between the extrema.

Hence $\psi(t) = \varphi(t + \alpha)$, $t \geq 0$, and

$$\psi(t) = e^{-D_1(t)/\varepsilon}, \quad t \geq 0. \tag{16}$$

Let us define S_{-1} to be the normalized time of the first jump from the state -1 to 1, i.e. $S_{-1} = \inf\{t \geq 0 : Y_{tT}^{\varepsilon,T} = 1\}$ Analogously, S_1 will be the time of first jump from state 1 to -1, starting with $Y_0 = 1$. We are especially interested in the behaviour of S as T becomes very large, that is as $\varepsilon \to 0$. In fact we get the following dichotomy of possible behaviour:

- If $\rho > \inf_{t \geq 0} D_{-1}(t)$, the law of S_{-1} tends to the Dirac measure in the point a_ρ^{-1} given by

$$a_\rho^{\pm 1} = \inf\{t \geq 0 : D_{\pm 1}(t) \leq \rho\}. \tag{17}$$

- If $\rho \leq \inf_{t \geq 0} D_{-1}(t)$, then the probability measure of S_{-1} tends weakly to the null measure.

It suffices to replace D_{-1} by D_1 and a_ρ^{-1} by a_ρ^1 to obtain similar results for S_1.

This leads to the following interpretation:

If $\rho \geq D_{-1}(0)$, that is, if the time scale T is very large, then on this exponential scale, the asymptotic behaviour of the Markov chain is characterized by an instantaneous jump, i.e. $a_\rho^{-1} = 0$. This just means that a clock ticking in units of T will record a jump of the process as instantaneous, since it occurs on a smaller scale.

In case $\rho < \inf_{t \geq 0} D_{-1}(t)$, the time scale T is too small compared to the transition rates. Consequently no transitions will be observed, and the process never jumps on this scale.

In the last case $D_{-1}(0) > \rho > \inf_{t \geq 0} D_{-1}(t)$. So the infinitesimal probability at time 0 is too small to allow any transition, and the Markov chain will have to wait until this probability is large enough to allow for jumps, that is approximatively $a_\rho T$. This case is the only interesting case, in the sense that the chain stays for some time in the starting state before it jumps to the other one.

To observe stochastic resonance we obviously need to study both transitions from -1 to 1 and vice versa. So we define some interval I_R called *interval of resonance* (see Fig. 7) which is to contain those exponential scales in which the process on the one hand asymptotically cannot always stay in the same state with positive probability, and on the other hand asymptotically cannot jump instantaneously from one state to the other.

$$I_R = (\max_{i=\pm 1} \inf_{t \geq 0} D_i(t), \inf_{t \geq 0} \max_{i=\pm 1} D_i(t)).$$

Optimal Tuning for the Markov Chain Let us now assume that we are in the range of non-trivial jumping, that is $\rho \in I_R$. We next determine an optimal tuning rate or stochastic resonance point. It will be based on the density of the first jump, in particular the intensity of its peak, which we propose as a new measure of quality of tuning. For $h > 0$ we shall compare the probabilities with which the first transition takes place within the window

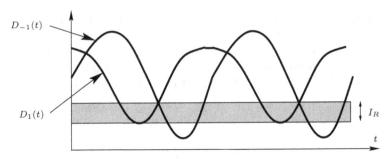

Fig. 7. Resonance interval I_R:

of exponential length $[(a^i_\rho - h)T, (a^i_\rho + h)T]$, $i = \pm 1$, for different ρ, maximize this quantity in ρ and finally take the window length to 0. More formally, for $h > 0$ small enough define

$$N(\varepsilon, \rho) = \min_{i=\pm 1} \mathbf{P}_i(S_i \in [(a^i_\rho - h)T, (a^i_\rho + h)T]), \quad \varepsilon > 0, \ \rho \in I_R, \qquad (18)$$

and call it *transition probability for a time window of width h* for the Markov chain. The optimal parameter ρ_0 will tell us at which time scale it is most likely to see trajectories of the chain with first jump in the corresponding window, and further jumps in accordingly displaced windows. In particular, it will tell us at which scale periodic trajectories of just this period are most probable. Since the probability density of the first transition times from one state to the other is well known, for example the density of S_{-1} equals

$$p(t) = \varphi(t)e^{-\int_0^t \varphi(s)ds},$$

we can compute an explicit expression for $N(\varepsilon, \rho)$. The optimal time scale will be determined by a combination of a large deviations result concerning the first jump of the Markov chain parameterized by the logarithmic scale ρ of time, and a maximization problem for the large deviation rates in ρ to which the transition probabilities converge uniformly.

Using Laplace's method to estimate the singular integrals appearing as $\varepsilon \to 0$, we obtain the required asymptotic result.

Theorem 9. *Let Γ be a compact subset of I_R, $h_0 < \max\{a^{-1}_\rho, \frac{T}{2} - a^{-1}_\rho\}$. Then for $0 < h \le h_0$*

$$\lim_{\varepsilon \to 0} \varepsilon \ln(1 - N(\varepsilon, \rho)) = \max_{i=\pm 1}\left\{\rho - D_i(a^i_\rho - h)\right\} \qquad (19)$$

uniformly for $\rho \in \Gamma$.

Since the convergence is uniform in ρ, it suffices to minimize the left hand side of (19) to obtain an optimal tuning point. For h small the eventually existing global minimizer $\rho_R(h)$ of

$$I_R \ni \rho \mapsto \max_{i=\pm 1} \{\rho - D_i(a^i_\lambda - h)\}$$

is a good candidate for our resonance point. But it still depends on h. To get rid of this dependence, we shall consider the limit of $\lambda_R(h)$ as $h \to 0$.

Definition 1. *Suppose that*

$$I_R \ni \rho \mapsto \max_{i=\pm 1} \{\rho - D_i(a^i_\rho - h)\}$$

possesses a global minimum $\rho_R(h)$. Suppose further that

$$\rho_R = \lim_{h \to 0} \rho_R(h)$$

exists in I_R. We call ρ_R the stochastic resonance point *of the Markov chain $Y^{\varepsilon,T}$ with time periodic infinitesimal generator $G_{\varepsilon,T}$.*

In fact the stochastic resonance point exists if one of the depth functions, and therefore both, due to the phase lag, has a unique point of maximal decrease in the interval in which it is strictly decreasing.

   Example: In fact all the results presented before, in the case of a time dependent potential U with meta-stable states at ± 1 also hold true if the meta-stable states are allowed to move periodically but stay away from the saddle 0. Then the state -1 of the Markov chain represents the left meta-stable state and 1 represents the right one. We shall mention one classical example in stochastic resonance (see, for instance [GHJM98]) which is the over-damped motion of a Brownian particle in the potential

$$2U(x,t) = V(x) + Ax\cos(2\pi t),$$

where V denotes a reflection-symmetric potential with two wells located at ± 1. In this particular case, for $0 < A < V(0) - V(-1)$,

$$D_{\pm 1}(t) = V(0) - V(-1) \pm A\cos(2\pi t).$$

Hence the phase lag α is equal to π and the resonance interval is

$$I_R = (V(0) - V(-1) - A, V(0) - V(-1)).$$

Let $h > 0$ small enough, then the logarithmic time scale which asymptotically optimizes the quality measure $N(\varepsilon, \rho)$ is given by

$$\rho_R(h) = V(0) - V(-1) - A\sin(\pi h).$$

In order to obtain the resonance point, we just let h tend to zero, to obtain $\rho_R = \lim_{h \to 0} \rho_R(h) = V(0) - V(-1)$, that is the average depth of the time periodic potential U. In this particular case, it is obvious that the resonance point coincides with the point of maximal decrease of the depth functions D_{-1} and D_1. This example is treated in detail in [HI02].

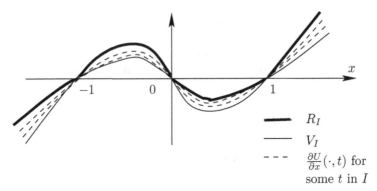

$$R_I$$
$$V_I$$
$$\frac{\partial U}{\partial x}(\cdot, t) \text{ for some } t \text{ in } I$$

Fig. 8. Definition of V_I and R_I

4.2 Transition Times for the Diffusion and Robustness

As seen in the preceding subsection, for the effective dynamics we obtain both simple and explicit results. Now we shall show how our measure of quality based purely on the jumps for the two-state Markov chain can be extended to the diffusion case. We just have to generalize the notion of jumps to the transition times between the two domains of attraction of the potential landscape, i.e. the two wells. The accordingly generalized measure of quality of periodic tuning possesses the desired property of being robust. The analogous notion of interval of resonance will then be presented in the following subsection. In our presentation we follow [HI].

Resonance Interval for Diffusions Recall that the underlying potential is described by a function $U(x, t)$, $x \in \mathbb{R}$, $t \geq 0$, such that $U'(\cdot, \cdot)$ is both continuous in time and space. The local minima are located at ± 1 and the saddle point at 0, independently of time. Our main concern will be the asymptotics of the transition times from the domain of attraction $(-\infty, 0)$ of -1 to the domain of attraction $(0, \infty)$ associated with 1 of the time inhomogeneous diffusion in the small noise limit $\varepsilon \to 0$. More precisely we will be interested in describing the exponential transition rate from the domain of attraction of -1 to the domain of attraction of 1. Our potential not being time homogeneous, we shall make use of comparison arguments with diffusions possessing time independent potentials in order to perform a careful reduction of the inhomogeneous exit problem to the homogeneous one, and use the asymptotic results well known for this particular case. This will be achieved by freezing the driving force derived from the potential on small time intervals on the minimal or maximal level it takes there. To be more precise, for each interval $I \subset \mathbb{R}_+$ let

$$V_I(x) = \sup_{t \in I} \frac{\partial U}{\partial x}(t, x) \quad \text{and} \quad R_I(x) = \inf_{t \in I} \frac{\partial U}{\partial x}(t, x). \tag{20}$$

The regularity conditions valid for U imply that V and R are continuous functions. Moreover $V_I(-1) = R_I(-1) = 0$, see Fig. 8. If $I = [a, b]$, we denote by $\overline{X}^{\varepsilon,I}$ the solution of the SDE on $\mathbb{R}_+$

$$\begin{cases} d\overline{X}_t^{\varepsilon,I} = -R_I(\overline{X}_t^{\varepsilon,I})\, dt + \sqrt{\varepsilon}\, dW_t, \\ \overline{X}_0^{\varepsilon,I} = X_{aT}^{\varepsilon,T}. \end{cases} \tag{21}$$

$\underline{X}^{\varepsilon,I}$ is defined in the same way replacing R_I by V_I. These two *time homogeneous* diffusions are used to control the *time inhomogeneous* diffusion $X^{\varepsilon,T}$ as long as time runs in the interval I. In fact, we have P-a.s.

$$\underline{X}_{tT}^{\varepsilon,I} \leq X_{(t+a)T}^{\varepsilon,T} \leq \overline{X}_{tT}^{\varepsilon,I}, \quad t \in [0, b-a].$$

Hence in order to study the time the diffusion needs to reach 1 starting in the left well, we shall consider the diffusion on one period. This time interval can be decomposed into finitely many small time intervals I_n, $0 \leq n \leq n_0$. We shall then freeze the potential on I_n and analyze if the the diffusions $\underline{X}^{\varepsilon,I_n}$ and $\overline{X}^{\varepsilon,I_n}$ have enough time in I_n to reach the top of the barrier between the two wells and, consequently on the same scale reach 1, the bottom of the right well. In other words we need to get information on the exit problem for the homogeneous diffusions $\underline{X}^{\varepsilon,I}$ and $\overline{X}^{\varepsilon,I}$.

We shall refer to the most recent and advanced development of sharp estimates for transition times presented in Bovier et al. [BEGK02] and [BGK02]. They are valid far beyond our modest framework, and we just present the results we will use here. For this purpose, suppose that $U_1(\cdot)$ is a purely space dependent C^2 potential function of the shape similar to those on Fig. 3. It possesses only ± 1 as local minima, separated by the saddle 0. Suppose that the curvature of U_1 at -1 is strictly positive, i.e. $U_1''(-1) > 0$. As for ultra- or hypercontractivity type properties for U_1, we shall assume that it has exponentially tight level sets, i.e. there is $M_0 > 0$ such that for any $M \geq M_0$ there exists a constant $C(M)$ such that for $\varepsilon \leq 1$

$$\int_{\{y:U_1(z)\geq M\}} e^{-2U_1(z)/\varepsilon}\, dz < C(M)e^{-M/\varepsilon}. \tag{22}$$

We shall concentrate in this situation on an exit of the domain of attraction of the meta-stable point -1 for the diffusion associated with the SDE

$$\begin{cases} dX_t^\varepsilon = -U_1'(X_t^\varepsilon)\, dt + \sqrt{\varepsilon}\, dW_t, \\ X_0^\varepsilon = x < 0. \end{cases}$$

We are interested in the asymptotics of the first time X^ε reaches 1:

$$\tau_1(X^\varepsilon) = \inf\{t > 0 : X_t^\varepsilon = 1\}.$$

Then we obtain the following result.

Theorem 10. *Let $\lambda(\varepsilon)$ denote the principal eigenvalue of the linear operator*

$$L_\varepsilon u = \frac{\varepsilon}{2}u'' - U_1'u', \quad u \in \mathcal{L}^2((-\infty, 1], e^{-2U_1/\varepsilon}dx)$$

with Dirichlet boundary conditions at 1. Then for every compact $K \subseteq (-\infty, 0)$ there is a constant $c > 0$ such that

$$\mathbf{P}_x(\tau_1(X^\varepsilon) > t) = e^{-\lambda(\varepsilon)t}(1 + \mathcal{O}_K(e^{-c/\varepsilon})), \tag{23}$$

where $\mathcal{O}_K$ denotes an error term which is uniform in $x \in K$, $t \geq 0$. Moreover, for the asymptotic behaviour of the eigenvalue $\lambda(\varepsilon)$ the following holds

$$\lambda(\varepsilon)\mathbf{E}_x[\tau_1(X^\varepsilon)] \to 1 \quad \text{uniformly on compacts } K \subseteq (-\infty, 0) \text{ as } \varepsilon \to 0. \tag{24}$$

Large deviations' theory reveals the asymptotic behaviour of the principal eigenvalue:

$$\lim_{\varepsilon \to 0} \varepsilon \ln \lambda(\varepsilon) = -2(U_1(0) - U_1(-1)).$$

This allows us to deduce that the mean hitting time $\mathbf{E}_x[\tau_1(X^\varepsilon)]$ is equivalent to $e^{\frac{2}{\varepsilon}(U_1(0)-U_1(-1))}$ as $\varepsilon \to 0$. Here $U_1(0) - U_1(-1)$ is the depth of the starting well. Moreover, by Theorem 10, the normalized hitting time $\frac{\tau_1(X^\varepsilon)}{\mathbf{E}_x[\tau_1(X^\varepsilon)]}$ converges in law to an exponential random variable with mean 1 as $\varepsilon \to 0$.

These results are very precise. They describe the asymptotic time of the barrier crossing and at the same time give an estimation of the probability to cross the barrier in a small time window around this asymptotic deterministic time. We can apply them to the 'frozen' potential $U(\cdot, \cdot)$ on the small time intervals I_n. We thereby assume for simplicity that the frozen potentials are regular of order $\mathcal{C}^2$. Let us choose $n \geq 0$ and set $I_n = [r_n, r_{n+1}]$. We assume that $X^{\varepsilon,T}$ has not reached the top of the barrier before $r_n T$ and study what happens during the time interval $[r_n T, r_{n+1}T]$. We have already seen that $X^{\varepsilon,T}$ is controlled by both $\underline{X}^{\varepsilon,I_n}$ and $\overline{X}^{\varepsilon,I_n}$. On the one hand, it suffices to prove that $\underline{X}^{\varepsilon,I_n}$ reaches 1 before $r_{n+1}T$ in order to get $\tau_1(X^{\varepsilon,T}) \leq r_{n+1}T$. On the other hand, if we get that $\overline{X}^{\varepsilon,I_n}$ does not hit 1 then so does $X^{\varepsilon,T}$. As $\varepsilon \to 0$, Theorem 10 tells us, for example, that the probability that $\underline{X}^{\varepsilon,I_n}$ reaches 1 before $r_{n+1}T$ is close to 1 if the depth of the left well is smaller than $\lim_{\varepsilon \to 0} \varepsilon \ln(r_{n+1} - r_n)T = \rho$. Indeed we get $\lim_{\varepsilon \to 0}(r_{n+1} - r_n)\lambda(\varepsilon)T = +\infty$ which implies by (23) that

$$\lim_{\varepsilon \to 0} \mathbf{P}_x(\tau_1(\underline{X}^{\varepsilon,I_n}) > (r_{n+1} - r_n)T) = 0.$$

The statements depend weakly on the depth of the well of the potential associated with $\underline{X}^{\varepsilon,I_n}$ and $\overline{X}^{\varepsilon,I_n}$. Since $\frac{\partial U}{\partial x}$ is continuous both in x and t, if we choose the length of all intervals I_n small enough then the well depth functions associated with the two time homogeneous diffusions are equivalent to $D_{-1}(r_n)$, the depth of the left well of the landscape U. Hence the diffusion

$X_{tT}^{\varepsilon,T}$ reaches 1 asymptotically as soon as the depth $D_{-1}(t)$ goes below the level ρ. This means

$$\lim_{\varepsilon \to 0} \frac{\tau_1(X^{\varepsilon,T})}{T} = a_\rho^{-1},$$

where a_ρ^{-1} was defined in (17).

Knowing the asymptotics of the time at which the diffusion reaches the barrier separating the two wells in order to hit 1 puts us again in a position in which we can discuss a resonance interval as for the reduced model. We obtain the same interval

$$I_R = (\max_{i=\pm 1} \inf_{t \geq 0} D_i(t), \inf_{t \geq 0} \max_{i=\pm 1} D_i(t)).$$

Optimal Tuning for the Diffusion and Robustness The comparison between time inhomogeneous and homogeneous potentials and the asymptotic result 10 enable us to proceed to the completion of our approach of stochastic resonance for diffusions. We have very precise estimates on the probabilities with which the diffusion at time scale $T = e^{\rho/\varepsilon}$ transits from the domain of attraction of -1 to the domain of attraction of 1 and vice versa within the time windows $[(a_\rho^i - h)T, (a_\rho^i + h)T]$ for small $h > 0$. On their basis we may define a measure of quality of tuning for the diffusion which corresponds to (18):

$$M(\varepsilon, \rho) = \min_{i=\pm 1} \mathbf{P}_i(\tau_{-i}(X^{\varepsilon,T}) \in [(a_\rho^i - h)T, (a_\rho^i + h)T]), \quad \varepsilon > 0, \rho \in I_R. \quad (25)$$

We may now state our main result on uniform transition rates.

Theorem 11. *Let Γ be a compact subset of I_R, $h_0 > 0$ small enough. Then*

$$\lim_{\varepsilon \to 0} \varepsilon \ln(1 - M(\varepsilon, \rho)) = \max_{i=\pm 1} \left\{ \rho - D_i(a_\rho^i - h) \right\} \quad (26)$$

uniformly for $\rho \in \Gamma$.

The stated convergence is *uniform* in ρ on compact subsets of I_R. This allows us to take $M(\varepsilon, \rho)$ as our measure of periodic tuning, compute the scale $\rho_0(h)$ for which the transition rate is optimal, and define the *stochastic resonance point* as the eventually existing limit of $\rho_0(h)$ as $h \to 0$. This notion of quality has the big advantage of being robust for the passage from the two state Markov chain to the diffusion. So the following final robustness result holds true.

Theorem 12. *The resonance points of the diffusion $X^{\varepsilon,T}$ with periodic potential U and of the Markov chain $Y^{\varepsilon,T}$ with exponential transition rate functions $D_{\pm 1}$ coincide.*

References

[ANMS99]   V. S. Anishchenko, A. B. Neiman, F. Moss, and L. Schimansky-Geier. Stochastic resonance: noise-enhanced order. *Physics–Uspekhi*, 42(1):7–36, 1999.

[BEGK02]   A. Bovier, M. Eckhoff, V. Gayrard, and M. Klein. Metastability in reversible diffusion processes I. Sharp asymptotics for capacities and exit times. Preprint No. 767, Weierstraß-Institut für angewandte Analysis und Stochastik (WIAS), Berlin, 2002. To appear in J. Eur. Math. Soc.

[BG02a]   N. Berglund and B. Gentz. A sample-paths approach to noise-induced synchronization: Stochastic resonance in a double-well potential. *Ann. Appl. Probab.*, 12(4):1419–1470, 2002.

[BG02b]   N. Berglund and B. Gentz. Beyond the Fokker-Planck equation: Pathwise control of noisy bistable systems. *J. Phys. A, Math. Gen.*, 35(9):2057–2091, 2002.

[BG02c]   N. Berglund and B. Gentz. Pathwise description of dynamic pitchfork bifurcations with additive noise. *Probab. Theory Relat. Fields*, 122(3):341–388, 2002.

[BGK02]   A. Bovier, V. Gayrard, and M. Klein. Metastability in reversible diffusion processes II. Precise asymptotics for small eigenvalues. Preprint No. 768, Weierstraß-Institut für angewandte Analysis und Stochastik (WIAS), Berlin, 2002. To appear in J. Eur. Math. Soc.

[BPSV81]   R. Benzi, G. Parisi, A. Sutera, and A. Vulpiani. The mechanism of stochastic resonance. *J. Phys. A*, 14:453–457, 1981.

[BPSV82]   R. Benzi, G. Parisi, A. Sutera, and A. Vulpiani. Stochastic resonance in climatic changes. *Tellus*, 34:10–16, 1982.

[BPSV83]   R. Benzi, G. Parisi, A. Sutera, and A. Vulpiani. A theory of stochastic resonance in climatic change. *SIAM J. Appl. Math.*, 43:563–578, 1983.

[Bud69]   M. I. Budyko. The effect of solar radiation variations on the climate of the earth. *Tellus*, 21:611–619, 1969.

[DLM* 95]   M. I. Dykman, D. G. Luchinskii, R. Mannella, P. V. E. McClintock, N. D. Stein, and N. G. Stocks. Stochastic resonance in perspective. *Nuovo Cimento D*, 17:661–683, 1995.

[Erd56]   A. Erdélyi. *Asymptotic expansions.* Dover Publications, Inc., New York, 1956.

[ET82]   J.-P. Eckmann and L. E. Thomas. Remarks on stochastic resonance. *J. Phys. A*, 15:261–266, 1982.

[Fre00]   M. I. Freidlin. Quasi-deterministic approximation, metastability and stochastic resonance. *Physica D*, 137(3-4):333–352, 2000.

[GHJM98]   L. Gammaitoni, P. Hänggi, P. Jung, and F. Marchesoni. Stochastic resonance. *Reviews of Modern Physics*, 70:223–287, January 1998.

[HI]   S. Herrmann and P. Imkeller. The exit problem for diffusions with time periodic drift and stochastic resonance. To appear in Ann. Appl. Probab.

[HI02]   S. Herrmann and P. Imkeller. Barrier crossings characterize stochastic resonance. *Stochastics and Dynamics*, 2(3):413–436, 2002.

[Imk01]   P. Imkeller. Energy balance models — viewed from stochastic dynamics. In Imkeller, P. et al., editors, *Stochastic climate models. Proceedings of a workshop, Chorin, Germany, Summer 1999.*, volume 49 of *Prog. Probab.*, pages 213–240, Basel, 2001. Birkhäuser.

[IP01]   P. Imkeller and I. Pavlyukevich. Stochastic resonance in two-state
          Markov chains. *Arch. Math.*, 77(1):107–115, 2001.
[IP02]   P. Imkeller and I. Pavlyukevich. Model reduction and stochastic reso-
          nance. *Stochastics and Dynamics*, 2(4):463–506, 2002.
[Kra40]   H. A. Kramers. Brownian motion in a field of force and the diffusion
          model of chemical reactions. *Physica*, 7:284–304, 1940.
[MW89]   B. McNamara and K. Wiesenfeld. Theory of stochastic resonance.
          Physical Review A (General Physics), 39:4854–4869, May 1989.
[Nic82]   C. Nicolis. Stochastic aspects of climatic transitions — responses to
          periodic forcing. *Tellus*, 34:1–9, 1982.
[Olv74]   F. W. J. Olver. *Asymptotics and special functions*. Computer Science
          and Applied Mathematics. Academic Press, a subsidiary of Harcourt
          Brace Jovanovich, Publishers, New York - London, 1974.
[Pav02]   I. E. Pavlyukevich. *Stochastic Resonance*. PhD thesis, Humboldt-
          Universität, Berlin, 2002. Logos–Verlag, ISBN 3-89722-960-9.
[Sel69]   W. B. Sellers. A global climate model based on the energy balance of
          the earth-atmosphere system. *J. Appl. Meteor.*, 8:301–320, 1969.
[WJ98]   K. Wiesenfeld and F. Jaramillo. Minireview of stocastic resonance.
          Chaos, 8:539–548, September 1998.

Continuity Properties of Inertial Manifolds for Stochastic Retarded Semilinear Parabolic Equations

Igor Chueshov[1], Michael Scheutzow[2], and Björn Schmalfuß[3]

[.] Department of Mechanics and Mathematics, Kharkov University, 4 Svobody Square, 310077 Kharkov, Ukraine
  chueshov@univer.kharkov.ua
[.] Technische Universität Berlin, Fakultät II - Mathematik und Naturwissenschaften, Institut für Mathematik, Sekr. MA 7-5, Str. des 17 Juni 136, 10623 Berlin, Germany
  ms@math.tu-berlin.de
[.] Institut für Mathematik, Universität Paderborn, Warburger Str. 100, 33095 Paderborn, Germany
  bjoern.schmalfuss@math.upb.de

Summary. We study continuity properties of inertial manifolds for a class of random dynamical systems generated by retarded semilinear parabolic equations subjected to additive white noise. We focus on two cases: (i) the delay time tends to zero and (ii) the intensity of the noise becomes small.

Key words: inertial manifold, stochastic PDE with delay, perfect cocycle, inertial form, stochastic convolution.

Introduction

One of the approaches to the study of the long-time behaviour of infinite dimensional dynamical systems is based on the concept of an *inertial manifold* (IM) (see, e.g. (Temam, 1988), (Constantin et al., 1989), (Chueshov, 1992, 1999) and the references therein). These manifolds are finite dimensional invariant surfaces which attract exponentially all solutions. The long-time dynamics restricted to the IM reduces to a finite system of ordinary differential equations, which is called an *inertial form* (IF). Inertial manifold theory has been developed and widely studied for deterministic and stochastic systems by many authors (see, e.g. the literature in the references of this paper).

A different (and older) concept used in the study of the long-time behaviour of a finite or infinite dimensional dynamical system is that of an *attractor*. The corresponding concept for a *random dynamical system* – a *random attractor* – was introduced independently in (Crauel and Flandoli, 1994) and (Schmalfuß, 1992). An attractor is a compact subset which is invariant under the dynamical system. If it exists, then it is (typically) contained in any inertial manifold (we will make precise statements below). The disadvantage of attractors is that they are often either trivial (i.e. they consist of one point only; this is typically true in the random case when the noise is nondegenerate) or that they are hard to describe explicitly. Inertial manifolds also do not always exist but if they do, then they are often easier to deal with than attractors. In addition, the approach provides some flexibility in choosing the dimension of the manifold: by increasing the dimension it is possible to improve the rate of exponential convergence of solutions to the manifold.

In this paper, we study continuity properties with respect to the retardation time and the intensity of the noise of stochastic IMs for a class of retarded semilinear parabolic equations subjected to additive white noise. These IMs were constructed in (Chueshov and Scheutzow, 2001). They are generated by stationary nonanticipating random processes and the corresponding IF is a finite dimensional (non-retarded) stochastic Itô differential equation with a random stationary drift. The approach in (Chueshov and Scheutzow, 2001) is based on the well-known method of integral equations due to Lyapunov and Perron in the form presented in (Boutet de Monvel et al., 1998) for deterministic systems with delay and in (Chueshov, 1995) for stochastic (non-retarded) equations. As in (Boutet de Monvel et al., 1998) we need a suitable spectral gap condition and also a restriction on the delay time. Another approach to the construction of an inertial manifold is the *graph transform* (Hadamard) method. For equations without retardation, this method can be found in (Constantin et al., 1989) in the deterministic case and in (Schmalfuß, 1998), for stochastic systems. Its stochastic version involves the study of a random dynamical system arising on graphs and general results on random fixed points from (Schmalfuß, 1996, 1998). A similar method was also used in (Girya and Chueshov, 1995). In this paper we use the Lyapunov–Perron method. The main reason is that this approach allows us to refer to calculations which were made in (Chueshov and Scheutzow, 2001).

The paper is organized as follows: in Section 1 we provide some notation and present background material about retarded stochastic semilinear parabolic equations and on the generation of a filtered random dynamical system by parabolic retarded equations. We also establish additional properties of infinite dimensional Ornstein-Uhlenbeck processes which may be of general interest (see Proposition 1.6). In Section 2 we describe briefly the construction of inertial manifolds and explain their basic properties. Sections 1 and 2 (except Proposition 1.6 and the final statement in Lemma 2.1) are essentially a condensed form of (Chueshov and Scheutzow, 2001) and detailed proofs can be found there. In Section 3 we study the limit of zero delay time. Section 4 is devoted to the case where the noise intensity goes to zero.

1 Preliminaries

Let H be an infinite dimensional separable real Hilbert space. The main object of the paper is the following retarded differential equation for the H-valued stochastic process $u(t), t \geq \sigma$ excited by a white noise process $\dot{W}(t)$ in H:

$$\frac{du}{dt} + Au = B(u_t, \vartheta(t)\omega) + \dot{W}(t), \text{ for } t \geq \sigma,$$

$$u_t(\theta) := u(t + \theta), \ \theta \in [-r, 0], \ r > 0.$$

We will specify all assumptions on A, B, Ω, ϑ and W below, and explain what we mean by a solution to the equation.

Let A be a positive operator on H with discrete spectrum: there exists an orthonormal basis $\{e_k\}$ of H such that

$$Ae_k = \lambda_k e_k, \quad \text{with } 0 < \lambda_1 \leq \lambda_2 \leq \dots, \quad \lim_{k \to \infty} \lambda_k = \infty.$$

For $\alpha \geq 0$ we denote by $C([a, b]; D(A^\alpha))$ the space of strongly continuous functions on an interval $[a, b]$ with values in $D(A^\alpha)$. For $r > 0$, we denote for short $C_\alpha = C([-r, 0]; D(A^\alpha))$. C_α is a Banach space with norm

$$|v|_{C_\alpha} \equiv \sup\{\| A^\alpha v(\theta) \|: \theta \in [-r, 0]\}.$$

Here and below, $\| \cdot \|$ and $(\cdot, \cdot)$ denote the norm and the inner product in H. We also denote by $\mathcal{B}(X)$ the σ-algebra of Borel sets in the topological space X.

We suppose that $(\vartheta, \mathbf{F}) = (\Omega, \mathcal{F}, \{\mathcal{F}_t, t \in \mathbb{R}\}, \mathbb{P}, \{\vartheta(t), t \in \mathbb{R}\})$ is a filtered metric dynamical system in the sense of the following definitions. For future reference we provide the definition of a (filtered) random dynamical system (see also (Arnold, 1998)). We will denote by $\mathbb{R}_+$ the set of all non-negative elements from $\mathbb{R}$.

Definition 1.1 Let X be a topological space. *A random dynamical system* (RDS) with time $\mathbb{R}_+$ and state space X is a pair (ϑ, ϕ) consisting of the following two objects:

1. A metric dynamical system (MDS) $\vartheta \equiv (\Omega, \mathcal{F}, \mathbb{P}, \{\vartheta(t), t \in \mathbb{R}\})$, i.e. a probability space $(\Omega, \mathcal{F}, \mathbb{P})$ with a family of measure preserving transformations $\{\vartheta(t) : \Omega \mapsto \Omega, t \in \mathbb{R}\}$ such that
   a) $\vartheta(0) = \text{id}, \quad \vartheta(t) \circ \vartheta(s) = \vartheta(t + s) \quad \text{for all } t, s \in \mathbb{R};$
   b) the map $(t, \omega) \mapsto \vartheta(t)\omega$ is measurable and $\vartheta(t)\mathbb{P} = \mathbb{P}$ for all $t \in \mathbb{R}$.

2. A (perfect) cocycle ϕ over ϑ of continuous mappings of X with one-sided time $\mathbb{R}_+$, i.e. a measurable mapping

$$\phi : \mathbb{R}_+ \times \Omega \times X \mapsto X, \quad (t, \omega, x) \mapsto \phi(t, \omega, x)$$

such that
   a) the mapping $\phi(\cdot, \omega, \cdot) : (t, x) \mapsto \phi(t, \omega, x)$ is continuous for all $\omega \in \Omega$
      and
   b) it satisfies the cocycle property:

$$\phi(0, \omega, \cdot) = \mathrm{id}, \quad \phi(t + s, \omega, \cdot) = \phi(t, \vartheta(s)\omega, \phi(s, \omega, \cdot))$$

   for all $t, s \geq 0$ and $\omega \in \Omega$.

Definition 1.2 Let ϑ be an MDS, $\overline{\mathcal{F}}$ the $\mathbb{P}$-completion of $\mathcal{F}$ and $\mathbf{F} = \{\mathcal{F}_t, t \in \mathbb{R}\}$ a family of sub-σ-algebras of $\overline{\mathcal{F}}$ such that

1. $\mathcal{F}_s \subseteq \mathcal{F}_t, \quad s < t$;
2. $\mathcal{F}_s = \bigcap_{h>0} \mathcal{F}_{s+h}, \quad s \in \mathbb{R}$, i.e., the filtration $\mathbf{F}$ is right-continuous;
3. $\mathcal{F}_s$ contains all sets in $\overline{\mathcal{F}}$ of $\mathbb{P}$-measure 0, $s \in \mathbb{R}$;
4. $\vartheta(s)$ is $(\mathcal{F}_{t+s}, \mathcal{F}_t)$-measurable for all $s, t \in \mathbb{R}$.

Then $(\vartheta, \mathbf{F})$ is called a *filtered metric dynamical system* (FMDS). If – in addition – (ϑ, ϕ) is an RDS such that $\phi(t, \cdot, x)$ is $(\mathcal{F}_t, \mathcal{B}(X))$-measurable for every $t \geq 0$ and $x \in X$, then $(\vartheta, \mathbf{F}, \phi)$ is called a *filtered random dynamical system* (FRDS).

We recall that an X-valued stochastic process $Y(t), t \in T \subseteq \mathbb{R}$ is called *adapted* with respect to the filtration $\mathbf{F}$ if $Y(t)$ is $(\mathcal{F}_t, \mathcal{B}(X))$- measurable for every $t \in T$. Therefore $(\vartheta, \mathbf{F}, \phi)$ is an FRDS if and only if (ϑ, ϕ) is an RDS, $(\vartheta, \mathbf{F})$ is an FMDS and $\phi(\cdot, \cdot, x)$ is adapted to $\mathbf{F}$ for every $x \in X$.

Now we fix $\alpha \in [0, 1[$ and $M > 0$ and let B be a (nonlinear) $(\mathcal{B}(C_\alpha) \otimes \mathcal{F}_0, \mathcal{B}(H))$-measurable map from $C_\alpha \times \Omega$ to H which satisfies

$$\| B(v_1, \omega) - B(v_2, \omega) \| \leq M \max_{\theta \in [-r, 0]} \| A^\alpha(v_1(\theta) - v_2(\theta)) \|, \quad \omega \in \Omega, \quad (1)$$

for all $v_1, v_2 \in C_\alpha$. We also assume that $b_0 := \sup_{\omega \in \Omega} \|B(0, \omega)\| < \infty$.

Finally we assume that $W(t), t \in \mathbb{R}$, is a two-sided H-valued Wiener process with covariance operator $K = K^* \geq 0$ such that $\mathrm{tr}\,(K \cdot A^{2(\alpha+\varepsilon)-1}) < \infty$ for some $\varepsilon > 0$. This condition will ensure the existence of a continuous modification of the solution. For definitions and properties of such processes see (Da Prato and Zabczyk, 1992). If $\mathrm{tr}\,K < \infty$, then W almost surely has strongly continuous trajectories in H. In all cases there exists a Hilbert space $\overline{H}$ such that W has strongly continuous paths in $\overline{H}$. In addition to the distributional properties of W we will assume the following:

(a) $W(t) - W(u)$ is $(\mathcal{F}_t, \mathcal{B}(\overline{H}))$-measurable, $t \in \mathbb{R}, u \leq t$;

(b) $W(t) - W(s)$ is independent of $\mathcal{F}_s$, $s < t$;

(c) $W(t + s, \omega) - W(s, \omega) = W(t, \vartheta(s)\omega)$, $s, t \in \mathbb{R}, \omega \in \Omega$.

Property (c) is called *helix property* or *additive cocycle property*.

We remark that given H, $0 \leq \alpha < 1$ and K as above it is always possible to find an FMDS and a Wiener process W with covariance operator K which enjoys properties (a)-(c) above e.g. by choosing $\Omega = C_0(\mathbb{R}, \overline{H})$, the space of continuous $\overline{H}$-valued functions which are 0 at 0, $\mathcal{F}$ the σ-algebra generated by the evaluations, $\mathbb{P}$ the law of W, $\vartheta(t)(\omega) = \omega(t + \cdot) - \omega(t)$ and $\mathcal{F}_t$ the completed σ-algebra generated by $\omega(u) - \omega(v), v \leq u \leq t$ (Arnold and Scheutzow, 1995). However a richer space may be needed if B is not just a function of $W(u), u \leq 0$.

As mentioned in the beginning we will study the following stochastic functional differential equation excited by the white noise process $\dot{W}(t)$ in H:

$$\frac{du}{dt} + Au = B(u_t, \vartheta(t)\omega) + \dot{W}(t), \quad \text{for } t > \sigma, \tag{2}$$

with initial data $u_\sigma \in C_\alpha$ at time $\sigma \in \mathbb{R}$:

$$u(\sigma + \theta) = u_\sigma(\theta) \quad \text{for } \theta \in [-r, 0]. \tag{3}$$

Here $u_t = u_t(\theta)$ denotes the element of C_α such that $u_t(\theta) = u(t + \theta)$ for $t + \theta > \sigma$ and $u_t(\theta) = u_\sigma(t - \sigma + \theta)$ for $\sigma - r \leq t + \theta \leq \sigma$. We rewrite (2) and (3) in the integral form

$$u(t, \omega) = e^{-(t-\sigma)A} u_\sigma(0) + \int_\sigma^t e^{-(t-\tau)A} B(u_\tau, \vartheta(\tau)\omega) \, d\tau + \eta(t, \sigma, \omega), \tag{4}$$

where

$$\eta(t, \sigma) = \int_\sigma^t e^{-(t-\tau)A} dW(\tau). \tag{5}$$

A solution of (4) which belongs to $C([\sigma - r, \infty); D(A^\alpha))$ for almost every ω is usually referred to as a *mild solution* of (2) and (3). The integral in (5) exists as an operator stochastic integral (see e.g.(Kuo, 1975)) and it is an H-valued Gaussian process such that $\mathbb{E}\eta(t, s) = 0$ and

$$\mathbb{E} \parallel A^\alpha R\eta(t, s) \parallel^2 = 1/2 \quad \text{tr} \, (KR^2 A^{2\alpha - 1}(1 - e^{-2(t-s)A})),$$

$$(\mathbb{E} \parallel A^\alpha R\eta(t, s) \parallel^l)^{1/l} \leq c_l (\mathbb{E} \parallel A^\alpha R\eta(t, s) \parallel^2)^{1/2}, \quad l \geq 1 \tag{6}$$

for any bounded operator R commuting with A and for any $-\infty \leq s < t < \infty$, where c_l is independent of R (below we use these formulas with R an eigen-projector of A). Here and below, $\mathbb{E}$ is the mean value on $(\Omega, \mathcal{F}, \mathbb{P})$. We refer the reader to (Da Prato and Zabczyk, 1992) or (Da Prato and Zabczyk, 1996) for these and further properties of η and more general *stochastic convolutions*.

Solutions of (4) generate an FRDS $(\vartheta, \mathbf{F}, \phi)$ if we define $\phi(t, \omega, u_0) := u(t, \omega)$, where $u(t, \omega)$ solves (4) with $\sigma = 0$. For this to be true we have to pick a particular modification $\overline{\eta}$ of the process η which we will construct in the following proposition. This procedure is usually referred to as *perfection*. Perfection theorems have been shown in various different cases e.g. (Arnold and Scheutzow, 1995), (Scheutzow, 1996), (Kager and Scheutzow, 1997) and (Sharpe, 1988). In our case the main task is the perfection of the stationary process $\eta(t, -\infty, \omega)$. For future reference we pick $\overline{\eta}$ in such a way that it also satisfies other desirable properties for *all* ω like the *temperedness* property (vi) below.

Let

$$\Delta := \{(t, s) : -\infty \le s \le t < \infty, t > -\infty\}.$$

Proposition 1.3 *Let* $\eta(t, s)$ *be defined as in (5),* $(t, s) \in \Delta$. *There exists a function* $\overline{\eta} : \Delta \times \Omega \to H$ *such that*

(i) $\mathbb{P}(\{\omega : \overline{\eta}(t, s, \omega) = \eta(t, s, \omega)\}) = 1, \quad (t, s) \in \Delta$;
(ii) $\overline{\eta}(t, s, \omega) = \overline{\eta}(t + \tau, s + \tau, \vartheta(-\tau)\omega), \quad (t, s) \in \Delta, \tau \in \mathbb{R}, \omega \in \Omega$;
(iii) $(t, s) \mapsto \overline{\eta}(t, s, \omega)$ *is continuous from* Δ *to* $D(A^\alpha), \quad \omega \in \Omega$;
(iv) $(t, s, \omega) \mapsto \overline{\eta}(t, s, \omega)$ *is measurable as a map from* $\Delta \times \Omega$ *to* $D(A^\alpha)$;
(v) $\overline{\eta}(t, s, \omega) = \overline{\eta}(t, \tau, \omega) - e^{-(t-s)A}\overline{\eta}(s, \tau, \omega), \quad -\infty \le \tau < s \le t, \omega \in \Omega$;
(vi) $\sup_{t \in \mathbb{R}}\{\| A^\alpha \overline{\eta}(t, -\infty, \omega) \| \, e^{-\beta|t|}\} < \infty$ *for all* $\beta > 0, \omega \in \Omega$.

For the proof we refer the reader to (Chueshov and Scheutzow, 2001)

From now on we will assume that η in (5) has all the properties stated for $\overline{\eta}$ in Proposition 1.3.

The following theorem states that problem (2) generates an RDS in the space C_α (see (Chueshov and Scheutzow, 2001) for the proof).

Theorem 1.4 *For every* $\omega \in \Omega, \sigma \in \mathbb{R}$ *and* $u_\sigma \in C_\alpha$ *equation (4) has a unique solution* $u(t, \sigma, \omega, u_\sigma), t \ge \sigma - r$ *where* $u(t, \sigma, \omega, u_\sigma) := u_\sigma(t - \sigma)$ *in case* $\sigma - r \le t \le \sigma$. *For every* $\omega \in \Omega$ *and* $\sigma \in \mathbb{R}$, u *is jointly continuous as a function from* $[\sigma - r, \infty) \times C_\alpha$ *to* $D(A^\alpha)$. *For every* $\sigma \in \mathbb{R}, \bar{u} \in C_\alpha$ *the process* $u(\cdot, \sigma, \cdot, \bar{u})$ *is adapted to the filtration* $\mathbf{F}$.

Define the map $\phi : \mathbb{R}_+ \times \Omega \times C_\alpha \to C_\alpha$ *by* $\phi(t, \omega, \bar{u})(\theta) := u(t + \theta, 0, \omega, \bar{u})$. *Then*

(i) $(\vartheta, \mathbf{F}, \phi)$ *is an FRDS;*
(ii) $u(t, \omega) := \phi(t - \sigma, \vartheta(\sigma)\omega, u_\sigma)(0), t \ge \sigma$ *solves (4) for every* $\omega \in \Omega, \sigma \in \mathbb{R}$ *and* $u_\sigma \in C_\alpha$.

We now introduce some more notation. We fix an integer $N \ge 1$ and denote by $P = P_N$ the orthogonal projector onto the space spanned by the first N eigenvectors of A. Let $Q = I - P$. We also define the N-dimensional projector $\hat{P} = \hat{P}_N$ in C_α by

$$\hat{P}v = (\hat{P}v)(\theta) = \sum_{k=1}^{N} e^{-\lambda_k \theta}(v(0), e_k)e_k \equiv e^{-A\theta}\hat{P}v(0), \tag{7}$$

where $-r \le \theta \le 0$ and $v = v(\theta)$ is an element of C_α. We note that $\hat{P}$ is the spectral projector of the infinitesimal generator of the linear semigroup in C_α generated by the solution of (2) with $B \equiv 0$ and $W \equiv 0$. This observation is important for the construction of inertial manifolds for retarded systems (see (Boutet de Monvel et al., 1998) or (Chueshov and Scheutzow, 2001) for details).

Later on we will suppose that for $N \in \mathbb{N}$ and $\mu > 0$ the following *spectral gap* condition is satisfied:

$$\lambda_{N+1} - \lambda_N \ge \mu(\lambda_N^\alpha + \lambda_{N+1}^\alpha + k(\lambda_{N+1} - \lambda_N)^\alpha). \tag{8}$$

Here and below we have set:

$$k = \alpha^\alpha \int_0^\infty \xi^{-\alpha}e^{-\xi}d\xi \quad \text{if} \quad \alpha \in (0,1) \quad \text{and} \quad k = 0 \quad \text{if} \quad \alpha = 0. \tag{9}$$

Note that we need $\alpha < 1$ here! We will also use the notation:

$$\gamma = \lambda_N + \mu\lambda_N^\alpha, \quad \delta = \frac{2}{\mu}Me^{\gamma r}, \tag{10}$$

where M is defined in (1).

The following elementary estimates will be useful in the following sections (see, e.g. (Temam, 1988), (Foias et al., 1988) (Chueshov, 1992) and (Chueshov, 1999)).

Lemma 1.5 (i)   *We have the following estimates*

$$\| A^\alpha e^{-tA}P \| \le \lambda_N^\alpha \cdot e^{\lambda_N |t|}, \quad t \in \mathbb{R}, \ \alpha \ge 0,$$

$$\| A^\alpha e^{-tA}Q \| \le [(\alpha/t)^\alpha + \lambda_{N+1}^\alpha] \cdot e^{-\lambda_{N+1}t}, \quad t > 0, \ \alpha > 0,$$

and

$$\| A^\alpha e^{-sA} \| \le \max_{\lambda>0} |\lambda^\alpha e^{-s\lambda}| = \left(\frac{\alpha}{es}\right)^\alpha, \quad s > 0, \ \alpha > 0. \tag{11}$$

(ii)   *For all $t \le s$ and $\alpha \in [0, 1[$ we have*

$$\int_t^s \| A^\alpha e^{-(t-\tau)A}P \| e^{-\gamma(\tau-s)} \, d\tau \le \frac{\lambda_N^\alpha}{\gamma-\lambda_N} \cdot (e^{-\gamma(t-s)} - e^{-\lambda_N(t-s)}),$$

$$\int_{-\infty}^t \| A^\alpha e^{-(t-\tau)A}Q \| e^{-\gamma\tau} \, d\tau \le \frac{k(\lambda_{N+1}-\gamma)^\alpha + \lambda_{N+1}^\alpha}{\lambda_{N+1}-\gamma} \cdot e^{-\gamma t}.$$

$$\tag{12}$$

(iii) *For any $s > 0$ and $0 < \beta \leq 1$ we have*

$$\| A^{-\beta}(1 - e^{-sA}) \| \leq \sup_{\lambda > 0} \left\{ \frac{1 - e^{-s\lambda}}{\lambda^\beta} \right\} \leq \sup_{\lambda > 0} \left\{ \frac{1 \wedge (s\lambda)}{\lambda^\beta} \right\} = s^\beta. \quad (13)$$

To study continuity properties of inertial manifolds for problem (2) we need to show the following integrability properties for the random process

$$(t, s, \omega) \mapsto \eta(t, s, \omega), \quad (t, s) \in \Delta.$$

Proposition 1.6 *Let $\eta(t, s, \omega)$ be given by (5). Then for any $\varrho > 0$, $\sigma \geq 0$, $0 < \varepsilon \leq 1/2$ and $l \geq 1$ there exists a constant $C_1 = C_1(\varrho, \varepsilon, l) > 0$ such that*

$$\mathbb{E} \left(\sup_{t \leq 0} \left\{ e^{\varrho t} \| A^\sigma \eta(t, -\infty) \| \right\} \right)^l \leq C_1 \cdot q_{\sigma + \varepsilon}(K)^{l/2} \quad (14)$$

provided $q_{\sigma + \varepsilon}(K) := \mathrm{tr}\left(K A^{2(\sigma + \varepsilon) - 1} \right) < \infty$, where K is the covariance operator of the Wiener process W. Furthermore, for each $r_0 > 0$ we have that

$$\mathbb{E} \left(\sup_{t \leq 0} \left\{ e^{\varrho t} \sup_{\theta \in [-r, 0]} \| A^\sigma \eta(t, t + \theta) \| \right\} \right)^l \leq C_2 \cdot q_{\sigma + \varepsilon}(K)^{l/2} \cdot r^{l\kappa\varepsilon} \quad (15)$$

for any $\kappa \in]0, 1[$ and $r \in (0, r_0]$, where $C_2 = C_2(\varrho, \varepsilon, l, \kappa, r_0)$ is a positive constant. Finally, there exists a random variable $\hat{C}_{\sigma, \rho, \varepsilon, \kappa, r_0}(\omega)$, such that

$$\sup_{t \leq 0} \left\{ e^{\varrho t} \sup_{\theta \in [-r, 0]} \| A^\sigma \eta(t, t + \theta, \omega) \| \right\} \leq \hat{C}_{\sigma, \rho, \varepsilon, \kappa, r_0}(\omega) r^{\kappa\varepsilon} \quad (16)$$

for any $r \leq r_0$.

Proof. The fact that $(t, \omega) \mapsto \eta(t, -\infty)$ is a stationary process implies that

$$\mathbb{E} \left(\sup_{t \leq 0} \left\{ e^{\varrho t} \| A^\sigma \eta(t, -\infty) \| \right\} \right)^l = \mathbb{E} \sup_{t \leq 0} \left\{ e^{\varrho l t} \| A^\sigma \eta(t, -\infty) \|^l \right\}$$

$$\leq \sum_{k=0}^{\infty} e^{-\varrho l k} \mathbb{E} \sup_{-t \in [k, k+1]} \| A^\sigma \eta(t, -\infty) \|^l$$

$$= \sum_{k=0}^{\infty} e^{-\varrho l k} \mathbb{E} \sup_{-t \in [0, 1]} \| A^\sigma \eta(t, -\infty) \|^l$$

$$= \frac{1}{1 - e^{-\varrho l}} \mathbb{E} \sup_{t \in [0, 1]} \| A^\sigma \eta(t, -\infty) \|^l.$$

To estimate the last term we will use Kolmogorov's theorem (see (Kunita, 1990), Theorem 1.4.1). Using Proposition 1.3(v), we have for $t > \tau$

$$\|A^\sigma(\eta(t, -\infty) - \eta(\tau, -\infty))\| \le \|A^\sigma \eta(t, \tau)\|$$
$$+ \|A^\sigma(1 - e^{-(t-\tau)A})\eta(\tau, -\infty)\|.$$

According to (6) and (13) we have that

$$\mathbb{E}\|A^\sigma \eta(t, \tau)\|^l \le C \cdot \left(q_{\sigma+\varepsilon}(K) \cdot |t - \tau|^{2\varepsilon}\right)^{l/2}.$$

for some appropriate constant $C = C_{\varepsilon,l} > 0$. Similarly,

$$\mathbb{E}\|A^\sigma(1 - e^{-(t-\tau)A})\eta(\tau, -\infty)\|^l$$
$$\le \|A^{-\varepsilon}(1 - e^{-(t-\tau)A})\|^l \cdot \mathbb{E}\|A^{\sigma+\varepsilon}\eta(\tau, -\infty)\|^l$$
$$\le C \cdot |t - \tau|^{\varepsilon l} \cdot (q_{\sigma+\varepsilon}(K))^{l/2}.$$

Therefore there exists a constant $c_{l,\varepsilon} > 0$ such that

$$\mathbb{E}\|A^\sigma(\eta(t, -\infty) - \eta(\tau, -\infty))\|^l \le c_{l,\varepsilon} q_{\sigma+\varepsilon}(K)^{l/2}|t - \tau|^{\varepsilon l}.$$

Now, Kolmogorov's theorem implies that for $\varepsilon l > 1$ we have

$$\mathbb{E} \sup_{0 \le t, \tau \le r_0 + 1} \|A^\sigma(\eta(t, -\infty) - \eta(\tau, -\infty))\|^l \le \tilde{k}_{l,\varepsilon,r_0} q_{\sigma+\varepsilon}(K)^{l/2} \tag{17}$$

and

$$\mathbb{E}\left[\sup_{\substack{0 \le t, \tau \le r_0 + 1, \\ |t-\tau| \le r}} \|A^\sigma(\eta(t, -\infty) - \eta(\tau, -\infty))\|^l \right] \le \tilde{k}_{l,\varepsilon,\kappa,r_0} q_{\sigma+\varepsilon}(K)^{l/2} r^{\kappa \varepsilon l} \tag{18}$$

for any $r \in\,]0, r_0]$, $\kappa \in\,]0, 1[$, some $\tilde{k}_{l,\varepsilon,\kappa,r_0}$ and l sufficiently large. Observe that both (17) and (18) are in fact valid for arbitrary $0 < \varepsilon \le 1/2$, $0 < \kappa < 1$ and $l \ge 1$ by Hölder's inequality. Using (6), we have

$$\mathbb{E}\|A^\sigma \eta(0, -\infty)\|^l \le C_l \cdot \|A^{-\varepsilon}\| \cdot q_{\sigma+\varepsilon}(K)^{l/2} \le C_{\varepsilon,l} \cdot q_{\sigma+\varepsilon}(K)^{l/2}. \tag{19}$$

Hence, by (17) and (19) we obtain

$$\mathbb{E} \sup_{0 \le t \le 1} \|A^\sigma \eta(t, -\infty)\|^l \le \hat{k}_{l,\varepsilon} q_{\sigma+\varepsilon}(K)^{l/2} \tag{20}$$

for some $\hat{k}_{l,\varepsilon}$. This implies that

$$\mathbb{E} \left(\sup_{t \le 0} \{e^{\varrho t} \|A^\sigma \eta(t, -\infty)\|\} \right)^l \le \frac{\hat{k}_{l,\varepsilon}}{1 - e^{-\varrho l}} \cdot q_{\sigma+\varepsilon}(K)^{l/2}$$
$$\equiv k_{l,\varepsilon,\varrho} \, q_{\sigma+\varepsilon}(K)^{l/2},$$

i.e. we have proved (14) (of course this relation also holds for larger values of ε but we need $0 < \varepsilon \le 1/2$ for (15)).

To prove (15), we use a similar argument. Let

$$\xi \equiv \xi(r, \varrho, \sigma; \omega) = \sup_{t \leq 0} \left\{ e^{\varrho t} \sup_{\theta \in [-r,0]} \|A^{\sigma} \eta(t, t + \theta)\| \right\}.$$

As above, using Proposition 1.3(ii) we have that

$$\mathbb{E}\xi^l \leq \sum_{k=0}^{\infty} e^{-\varrho l k} \mathbb{E} \sup_{-t \in [k,k+1]} \sup_{\theta \in [-r,0]} \|A^{\sigma} \eta(t, t + \theta)\|^l \tag{21}$$

$$= \frac{1}{1-e^{-\varrho l}} \cdot \mathbb{E} \left[\sup_{t \in [0,1], \, \theta \in [-r,0]} \|A^{\sigma} \eta(t, t + \theta)\|^l \right].$$

It follows from Proposition 1.3(v) that

$$\eta(t, t + \theta) = \eta(t, -\infty) - e^{\theta A} \eta(t + \theta, -\infty)$$

$$= \left(1 - e^{\theta A}\right) \eta(t, -\infty) + e^{\theta A} \left(\eta(t, -\infty) - \eta(t + \theta, -\infty)\right). \tag{22}$$

As above, using (13) we have

$$\|A^{\sigma} \left(1 - e^{\theta A}\right) \eta(t, -\infty)\| \leq |\theta|^{\varepsilon \kappa} \|A^{\sigma + \varepsilon \kappa} \eta(t, -\infty)\| \tag{23}$$

for every $0 < \kappa < 1$. Therefore from (20) (with rescaled σ and ε) we obtain that

$$\mathbb{E} \left[\sup_{\substack{t \in [0,1] \\ \theta \in [-r,0]}} \|A^{\sigma} \left(1 - e^{\theta A}\right) \eta(t, -\infty)\|^l \right] \leq C_{l,\varepsilon,\kappa} r^{\varepsilon \kappa l} q_{\sigma + \varepsilon}(K)^{l/2} \tag{24}$$

for arbitrary $0 < \varepsilon \leq 1/2$, $0 < \kappa < 1$ and $l \geq 1$.

It follows from (18) that

$$\mathbb{E} \left[\sup_{\substack{t \in [0,1] \\ \theta \in [-r,0]}} \|A^{\sigma} e^{\theta A} \left(\eta(t, -\infty) - \eta(t + \theta, -\infty)\right)\|^l \right] \leq C r^{\kappa \varepsilon l} q_{\sigma + \varepsilon}(K)^{l/2}, \tag{25}$$

for any $r \in]0, r_0]$, where $C > 0$ depends on l, ε, κ and r_0. Now (21), (22), (23), (24) and (25) imply (15).

To show the final part of the proposition we fix $\kappa \in (0, 1)$ and let $\hat{\kappa} \in (0, \kappa)$. For ξ defined as above and arbitrary $c > 0$, Chebychev's inequality implies that

$$\mathbb{P}(\xi(r, \omega) \geq c r^{\hat{\kappa} \varepsilon}) \leq \frac{\mathbb{E}\xi(r)}{c r^{\hat{\kappa} \varepsilon}} \leq C_2 q_{\sigma + \varepsilon}(K)^{1/2} c^{-1} r^{(\kappa - \hat{\kappa}) \varepsilon}.$$

Using the Borel-Cantelli Lemma, we obtain that there exists some $n_0 = n_0(\omega)$ such that $\xi(r, \omega) \geq c r^{\hat{\kappa} \varepsilon}$ for all $r = 2^{-n}$ satisfying $n \geq n_0$. Now the fact that $r \mapsto \xi(r, \omega)$ is nondecreasing implies (16) and the proof of Proposition 1.6 is complete. $\qquad \square$

2 Construction of inertial manifolds

In this section, we describe briefly the construction of an inertial manifold for problem (2) relying on a version of the Lyapunov-Perron method presented in (Boutet de Monvel et al., 1998) in the deterministic retarded setting and in (Chueshov and Scheutzow, 2001) in the stochastic retarded setting (see, (Chueshov and Girya, 1994, 1995), (Chueshov, 1999) for further cases).

Fix $\alpha \in [0,1[, N \in \mathbb{N}, \mu > 0$ and define γ as in (10) and the orthoprojectors P and Q as before. We continue to assume that $\eta(s,t)$ is defined by (5) and is assumed to enjoy all properties stated in Proposition 1.3 for $\bar{\eta}$. For $s \in \mathbb{R}$ let $\mathcal{C}_s$ be the Banach space of strongly continuous functions $v(t)$ on the semiaxis $\mathbb{R}_-(s) =] - \infty, s]$ with values in $D(A^\alpha)$ such that

$$|v|_s \equiv \sup_{t \in \mathbb{R}_-(s)} \{e^{\gamma(t-s)} \parallel A^\alpha v(t) \parallel\} < \infty,$$

where $s \in \mathbb{R}$. According to the standard procedure (see (Chow and Lu, 1988), (Boutet de Monvel et al., 1998), (Chueshov, 1999) and ((Chueshov and Scheutzow, 2001), for instance) used to construct an inertial manifold, we first need to solve the following integral equation in the space $\mathcal{C}_s$:

$$v(t) = \mathcal{B}_p^{s,\omega}[v](t), \quad -\infty < t \le s, \tag{26}$$

where

$$\mathcal{B}_p^{s,\omega}[v](t) = e^{-(t-s)A}p - \int_t^s e^{-(t-\tau)A}PB(v_\tau, \vartheta(\tau)\omega) \, d\tau$$

$$+ \int_{-\infty}^t e^{-(t-\tau)A}QB(v_\tau, \vartheta(\tau)\omega) \, d\tau - e^{-(t-s)A}P\eta(s,t,\omega) + Q\eta(t, -\infty, \omega),$$

$t \le s$, $p \in PH$ and $v_\tau \in \mathcal{C}_\alpha$ defined by $v_\tau(\theta) = v(\tau + \theta)$ for $-r \le \theta \le 0$. It is clear that the solution $v(t)$ of (26) is a particular solution of equation (4) on $] - \infty, s]$.

The following lemma provides conditions which guarantee the solvability of equation (26).

Lemma 2.1 *Assume that there exists some $\mu > 0$ for which the spectral gap condition (8) is satisfied and for which $\delta < 1$ (see (10)). Then the mapping $\mathcal{B}_p^{s,\omega}[\cdot]$ is a contraction with contraction constant δ and equation (26) has a unique solution $v(t) = v(t; s, p, \omega) \in \mathcal{C}_s$. $v(t)$ is $\mathcal{F}_s$-measurable for every $t \le s$. Moreover the relation*

$$v(t; s, p, \omega) = v(t - s; 0, p, \vartheta(s)\omega), \quad t \le s, \ p \in PH.$$

holds and we have the estimates

$$|v(\cdot; s, p_1) - v(\cdot; s, p_2)|_s \le (1 - \delta)^{-1} \parallel A^\alpha(p_1 - p_2) \parallel, \quad p_1, p_2 \in PH,$$

and

$$|v(\cdot; s, p)|_s \leq (1-\delta)^{-1} \left(\| A^\alpha p \| + \frac{2b_0}{\mu} + 2|\eta(\cdot, -\infty)|_s \right), \quad p \in PH, \ s \in \mathbb{R},$$

(27)

where $b_0 = \sup_{\omega \in \Omega} \| B(0, \omega) \|$.

Proof. The proof is based on the standard fixed point argument for the map $\mathcal{B}_p^{s,\omega}$ in $\mathcal{C}_s$. All statements of the lemma, except relation (27) were proved in (Chueshov and Scheutzow, 2001). To obtain estimate (27) we note that $v = v(t; s, p)$ satisfies

$$|v|_s \leq |\mathcal{B}_p^s[v] - \mathcal{B}_p^s[0]|_s + |\mathcal{B}_p^s[0]|_s \leq \delta|v|_s + |\mathcal{B}_p^s[0]|_s.$$

Further, we have that

$$\| A^\alpha \mathcal{B}_p^s[0](t) \| \leq \delta(s, t) \cdot b_0$$
$$+ \left\| A^\alpha \left(e^{-(t-s)A} p - e^{-(t-s)A} P\eta(s, t) + Q\eta(t, -\infty) \right) \right\|$$

for $t \leq s$, where

$$\delta(s, t) = \int_t^s \| A^\alpha e^{-(t-\tau)A} P \| e^{\gamma(s-\tau)} \, d\tau + \int_{-\infty}^t \| A^\alpha e^{-(t-\tau)A} Q \| e^{\gamma(s-\tau)} \, d\tau.$$

It follows from (12) and the spectral gap condition (8) that $\delta(s, t) \leq \frac{2}{\mu} e^{\gamma(s-t)}$. From Proposition 1.3(v) we have that

$$-e^{(s-t)A} P\eta(s, t) + Q\eta(t, -\infty) = -e^{(s-t)A} P\eta(s, -\infty) + \eta(t, -\infty). \quad (28)$$

Therefore,

$$\| A^\alpha \mathcal{B}_p^s[0](t) \| \leq \| A^\alpha \eta(t, -\infty) \|$$
$$+ e^{\gamma(s-t)} \left(\frac{2b_0}{\mu} + \| A^\alpha p \| + \| A^\alpha P\eta(s, -\infty) \| \right).$$

This implies (27). □

Following the approach presented in (Chueshov and Scheutzow, 2001), we introduce a family $\{ \mathcal{M}_s(\omega) \}$ of surfaces in C_α in the following way. Let $\hat{P}$ be given by (7) and $\Psi_s(\omega, \cdot)$ be the map from PH to $(1 - \hat{P})C_\alpha$ defined by the formula

$$\Psi_s(\omega, p, \theta) = \int_{-\infty}^{s+\theta} e^{-(s+\theta-\tau)A} QB(v_\tau, \vartheta(\tau)\omega) \, d\tau$$
$$- \int_{s+\theta}^s e^{-(s+\theta-\tau)A} PB(v_\tau, \vartheta(\tau)\omega) \, d\tau$$
$$- e^{-\theta A} P\eta(s, s+\theta, \omega) + Q\eta(s+\theta, -\infty, \omega)$$
$$\equiv v(s+\theta; s, p, \omega) - e^{-A\theta} p \equiv v(\theta; 0, p, \vartheta(s)\omega) - e^{-A\theta} p, \quad (29)$$

where $v(t) \equiv v(t; s, p, \omega)$ solves (26). We set

$$\mathcal{M}_s(\omega) = \{\hat{p}(\cdot) + \Psi_s(\omega, \hat{p}(0), \cdot) : \hat{p} \in \hat{P}C_\alpha\} \subset C_\alpha. \tag{30}$$

We have the following result on the existence of an inertial manifold $\mathcal{M}(\omega)$ for problem (4) (we refer to (Chueshov and Scheutzow, 2001) for the proof).

Theorem 2.2 *Assume that the spectral gap condition (8) is satisfied, and that $\delta < 1$ (see (10)). Then the random map Ψ_s from PH into $(1 - \hat{P})C_\alpha$ defined by (29) is $\mathcal{F}_s$-measurable and possesses the following properties:*

(i) *it satisfies the Lipschitz condition:*

$$\| A^\alpha(\Psi_s(\omega, p_1, \theta) - \Psi_s(\omega, p_2, \theta)) \| \le \frac{\delta \cdot \exp\{-\gamma\theta\}}{1 - \delta} \cdot \| A^\alpha(p_1 - p_2) \|$$

for all $p_1, p_2 \in PH$ and $s \in \mathbb{R}$;

(ii) *$\Psi_t(\omega, p, \cdot) = \Psi_0(\vartheta(t)\omega, p, \cdot)$. In particular $\Psi_t(\omega, p, \cdot)$ is a stationary process with values in C_α for fixed $p \in PH$;*

(iii) *the family of N-dimensional random Lipschitz manifolds $\{\mathcal{M}_s(\omega)\}$ given by (30) is invariant under the cocycle ϕ, i.e. $\phi(t - s, \vartheta(s)\omega)\mathcal{M}_s(\omega) \subseteq \mathcal{M}_t(\omega)$ for $t \ge s$. Defining $\mathcal{M} = \mathcal{M}_0$ we have*

$$\mathcal{M}_s(\omega) = \mathcal{M}(\vartheta(s)\omega), \quad s \in \mathbb{R}, \omega \in \Omega.$$

(iv) *If we additionally assume that $\delta \le 1/2$, then for every $u_0 \in C_\alpha$ there exists an induced trajectory i.e. there exists a random variable $u_0^*(\omega) \in \mathcal{M}(\omega)$ such that*

$$|\phi(t, \omega, u_0) - \phi(t, \omega, u_0^*)|_{C_\alpha} \le 4e^{-\gamma(t-r)}|(1 - \hat{P})u_0 - \Psi_0(\omega, Pu_0(0))|_{C_\alpha}$$

for all $\omega \in \Omega$. In particular this means that $\mathcal{M}$ has the asymptotic completeness property.

Remark 2.3 1. Theorem 2.2 and the structure of the projector $\hat{P}$ (see (7)) allow us to conclude that any solution $u_t = u_t(\theta)$ lying on the manifold $\mathcal{M}_t$ is of the form

$$u_t(\theta) = \Pi_t(\omega, p(t, \omega))(\theta) \equiv e^{-A\theta}p(t, \omega) + \Psi_t(\omega, p(t, \omega), \theta)$$
$$= v(t + \theta; t, p(t, \omega), \omega),$$

where $p(t)$ is a function with values in the N-dimensional space PH satisfying the equation

$$\frac{dp}{dt} + Ap = PB(\Pi_t(\omega, p(t)), \vartheta(t)\omega) + P\dot{W}(t), \quad t > \sigma, \quad p(\sigma) = Pu_\sigma(0). \tag{31}$$

Theorem 2.2(iv) shows that the long-time behaviour of the RDS (ϑ, ϕ) is completely described by the *inertial form* (31). This inertial form (IF) is a finite-dimensional ordinary Itô differential equation without retardation. It was shown in (Chueshov and Scheutzow, 2001) that the IF (31) generates itself an FRDS. We point out however that (31) does *not* generate a Markov process on PH even if B is non-random (except for a few highly degenerate cases).

2. The initial condition u_0^* of the induced trajectory is – in general – neither $\mathcal{F}_0$-measurable nor measurable with respect to the σ-algebra generated by $W(u), u \geq 0$.

3. If $\operatorname{esssup}_\omega \sup_v \|B(v, \omega)\| < \infty$, then one can show (Chueshov and Scheutzow, 2001) that the RDS (ϑ, ϕ) associated with problem (4) has a (global) pullback attractor in the sense of the theory of RDS which is a compact subset of the inertial manifold $\mathcal{M}(\omega)$. For the definition of a (random) attractor, (Arnold, 1998, p. 483), (Crauel and Flandoli, 1994) or (Schmalfuß, 1992, 1997).

We also note that the Lyapunov-Perron method presented above can be used for the construction of random fixed points or an *invariant Dirac measure* of (ϑ, ϕ) in the sense of RDS. For the definition see, e.g. (Arnold, 1998). If we formally suppose $P \equiv 0$ in (26), then we obtain the equation

$$v(t, \omega) = \int_{-\infty}^t e^{-(t-\tau)A} B(v_\tau, \vartheta(\tau)\omega) \, d\tau + \eta(t, -\infty, \omega). \tag{32}$$

The following assertion was proved in (Chueshov and Scheutzow, 2001).

Theorem 2.4 *Assume that the Lipschitz constant M in (1) satisfies*

$$M(1 + k) < \lambda_1^{1-\alpha},$$

where k is defined in (9). Then for any retardation time $r > 0$ and for any parameter $s \in \mathbb{R}$ the equation (32) has a unique $\mathcal{F}_t$-measurable solution $v(t, \omega)$ in the space $\mathcal{C}_s$ with some positive constant $\gamma = \gamma(r)$. This solution is a stationary process and $v(t, \omega) = v(0, \vartheta(t)\omega) \equiv v(\vartheta(t)\omega)$ for all $t \in \mathbb{R}$. Moreover this solution $v(t, \omega)$ is a global exponential attractor for the system (4) in the following sense:

$$\|A^\alpha(u(t, \omega) - v(\vartheta(t)\omega))\| \leq C \cdot e^{-\gamma t} \cdot |u_0(\omega) - v_0(\omega)|_{C_\alpha}, \quad t > 0,$$

for any solution $u(t, \omega)$ to (4) with $u(\theta, \omega) = u_0(\theta, \omega)$, $\theta \in [-r, 0]$. Here $v_0(\omega) \in C_\alpha$ is defined by $v_0(\theta, \omega) = v(\vartheta(\theta)\omega)$ and C is a non-random constant depending on λ_1, M and r.

In conclusion of this section we recall two examples from (Chueshov and Scheutzow, 2001).

Example 2.5 Theorems 2.2 and 2.4 can be applied to the problem

$$\partial_t u = \nu \partial_{xx} u + \int_{-r}^0 f(\theta, x, u(t + \theta, x))\zeta(d\theta) + \dot{W}(t, x) \tag{33}$$

on the interval $(0,1)$ with the Dirichlet boundary conditions

$$u|_{x=0} = u|_{x=1} = 0 \tag{34}$$

and initial data

$$u|_{t\in[-r,0]} = u_0(t,x). \tag{35}$$

Here ν and r are positive numbers; $\dot{W}(t,x)$ is white noise with identity correlation operator, $f(\theta,x,y)$ is a continuous function on the set $[-r,0]\times[0,1]\times\mathbb{R}$ such that

$$|f(\theta,x,y_1) - f(\theta,x,y_2)| \leq L|y_1 - y_2|$$

uniformly in θ and x; $\zeta(d\theta)$ is a finite signed measure on $[-r,0]$ such that $\mathrm{Var}\{\zeta\} = \int_{-r}^{0} |\zeta(d\theta)| = 1$. For this problem we have $H = L^2(0,1)$ and $A = -\nu\partial_{xx}$ with Dirichlet boundary conditions. The corresponding eigenvalues are $\lambda_k = \nu\pi^2 k^2$, $k \geq 1$. We can apply Theorem 2.4 with $\alpha = 0$ and $M = L$. It shows that under the condition $L/\nu < \pi^2$ there exists a stationary exponentially attractive process. If $L/\nu \geq \pi^2$, then choosing $\mu = 3L$ we obtain the conditions

$$\nu\pi^2(2N+1) \geq 6L, \quad \exp\{r(\nu\pi^2 N^2 + 3L)\} < \frac{3}{2}$$

on N and r under which we can apply Theorem 2.2 and prove the existence of an N-dimensional invariant surface in the space $\mathcal{C} = C([-r,0]; L^2(0,1))$. If we let $\mu = 5L$, then the conditions

$$\nu\pi^2(2N+1) \geq 10L, \quad \exp\{r(\nu\pi^2 N^2 + 5L)\} < \frac{5}{4}$$

allow us to prove the existence of an N-dimensional invariant exponentially complete manifold in $\mathcal{C}$ for the RDS generated by (33).

Example 2.6 The results mentioned above can also be applied to the equation:

$$\partial_t u = \nu\partial_{xx}u - u + f(\int_{-r}^{0} g(u(x,t+\theta))d\theta) + \dot{W}(t,x), \quad x \in (0,1), \ t > 0;$$

$$\partial_x u|_{x=0} = 0, \quad \partial_x u|_{x=1} = 0, \quad u|_{t\in[-r,0]} = u_0(t,x),$$

$$\tag{36}$$

where $\dot{W}(t,x)$ is white noise with identity correlation operator and $f(u)$ and $g(u)$ are continuous functions on $\mathbb{R}$ such that

$$|f(u)| \leq L_0, \quad |f(u_1) - f(u_2)| \leq L_1|u_1 - u_2|$$

and

$$|g(u_1) - g(u_2)| \le L_2 |u_1 - u_2|.$$

For this problem we have $H = L^2(0,1)$ and $A = -\nu\partial_{xx}+1$ with Neumann boundary conditions. The corresponding eigenvalues are $\lambda_k = \nu\pi^2(k-1)^2+1$, $k \ge 1$. It is easy to see that for the nonlinear operator

$$B(u) = f(\int_{-r}^{0} g(u(x,t+\theta))d\theta), \quad u \in \mathcal{C} = C([-r,0]; L^2(0,1))$$

we have the estimate

$$\|B(u_1) - B(u_2)\|_{L^2} \le r \cdot L_1 \cdot L_2 \cdot \max_{\theta \in [-r,0]} \|u_1(\theta) - u_2(\theta)\|_{L^2}$$

Therefore we can apply Theorem 2.4 with $\alpha = 0$ and $M = r \cdot L_1 \cdot L_2$. Thus, under the condition $r \cdot L_1 \cdot L_2 < 1$, we have the existence of a stationary exponentially attractive process for problem (36). As in Example 2.5, it is easy to state conditions on N and r that guarantee the existence of an N-dimensional invariant exponentially complete manifold in $\mathcal{C}$.

3 Dependence of inertial manifolds on the delay time

In this section we will study the dependence of the inertial manifold on the delay time $r \in [0, r_0]$ for some fixed $r_0 > 0$. We will often use the notation $\mathcal{C}_{\alpha,r}$ instead of $\mathcal{C}_\alpha$ to emphasize the dependence on r. Fix $M \ge 0$. We consider nonlinear operators B^r satisfying (1) for $\mathcal{C}_\alpha \equiv \mathcal{C}_{\alpha,r}$ with a Lipschitz constant M; (2) with $B = B^r$ then yields a random dynamical system ϕ^r on the phase space $\mathcal{C}_{\alpha,r}$ (see Section 2). We know from Theorem 2.2 that under appropriate assumptions the random dynamical systems ϕ^r possess inertial manifolds $\mathcal{M}_s^r = \mathcal{M}^s(\vartheta(s)\omega)$ having all the properties of that theorem. These manifolds have graphs $\Psi^r(\vartheta(s)\omega, p, \theta)$ for $\theta \in [-r,0]$.

In addition to r_0 and M we fix nonnegative numbers M_0, M_1 and M_2. We assume that the nonlinear function $B^r(v,\omega)$ consists of two parts: a non-retarded (B_0^r) and a retarded one (B_1^r), i.e. B^r is a (nonlinear) map from $\mathcal{C}_{\alpha,r}$ to H of the form

$$v \mapsto B^r(v,\omega) = B_0^r(v(0),\omega) + B_1^r(v,\omega), \tag{37}$$

where $B_0^r(\cdot,\omega)$ and $B_1^r(\cdot,\omega)$ are maps from $D(A^\alpha)$, resp. $\mathcal{C}_{\alpha,r}$ to H such that

$$\| B_0^r(w_1,\omega) - B_0^r(w_2,\omega) \| \le M_0 \| A^\alpha(w_1 - w_2) \|, \quad \text{for } w_1, w_2 \in D(A^\alpha) \tag{38}$$

and

$$\| B_1^r(v_1,\omega) - B_1^r(v_2,\omega) \| \le M_1 |v_1 - v_2|_{\mathcal{C}_{\alpha,r}}, \quad \text{for } v_1, v_2 \in \mathcal{C}_{\alpha,r}. \tag{39}$$

We will assume that B_1^r is a *purely retarded* function in the sense that $B_1^r(v, \omega) = 0$ for any function v which is constant on $[-r, 0]$. Observe that B^r can always be decomposed (uniquely) into the sum of a non-retarded and a purely retarded function by defining $B_0^r(y, \omega) := B^r(y\mathbf{1}, \omega)$, where $y \in D(A^\alpha)$ and $\mathbf{1}$ is the function which equals 1 on $[-r, 0]$. Therefore we can (and will) assume that the inequalities $M_0 + M_1 \geq M \geq M_0$ and $M_1 \leq M + M_0$ hold. Further, we will assume that we have $\sup_\omega \|B_0^r(0, \omega)\| \leq M_2$.

Let us also suppose that the spectral gap condition (8) is satisfied for some $\mu > 0$ such that

$$\delta := \frac{2Me^{\gamma r_0}}{\mu} < 1. \tag{40}$$

We consider problem (2) as a perturbation of the non-retarded problem:

$$\frac{du}{dt} + Au = B_0^r(u, \vartheta(t)\omega) + \dot{W}(t), \quad t > \sigma, \quad u|_{t=\sigma} = u_\sigma. \tag{41}$$

This equation generates a random dynamical system with phase space $D(A^\alpha)$. Theorem 2.2 also applies to non-retarded systems of the form (41). Therefore (41) possesses an inertial manifold $\mathcal{M}^0(\omega)$. In this case the corresponding function $\Psi^0(\omega, p)$ does not depend on an additional parameter θ. We will denote all objects corresponding to the non-retarded problem by an additional upper index 0 and those corresponding to a retarded problem with delay r by an upper index r. We will prove the following assertion on the proximity of the inertial manifolds for the retarded and non-retarded problems.

Theorem 3.1 *Fix r_0, M, M_0, M_1, and M_2 as above. Assume that the spectral gap condition (8) is satisfied for some $\mu > 0$ and that δ, defined in (40), is less than 1. Let B^r be a nonlinear function as above with delay time $r \leq r_0$ satisfying all properties above. Let $0 < \beta < 1 - \alpha$ and $\varepsilon > 0$ and assume that*

$$q_{\alpha+\beta+\varepsilon}(K) := \operatorname{tr}KA^{2(\alpha+\beta+\varepsilon)-1} < \infty. \tag{42}$$

Further, let $\kappa \in (0, (\beta + \varepsilon) \wedge \frac{1}{2})$. Then

$$\sup_{-r \leq \theta \leq 0} \|A^\alpha(\Psi^0(\vartheta(s)\omega, p) - \Psi^r(\vartheta(s)\omega, p, \theta))\|$$

$$\leq C_1 r^\beta \left[\|A^\alpha p\| + \xi_1(\vartheta(s)\omega)\right] + C_2 \xi_2(\vartheta(s)\omega)$$

$$\leq C_1 r^\beta \left[\|A^\alpha p\| + \xi_1(\vartheta(s)\omega)\right] + C_2 \hat{C}_{\alpha,\gamma,(\beta+\varepsilon)\wedge\frac{1}{2},\tilde\kappa,r_0}(\vartheta(s)\omega) r^\kappa$$

for every $s \in \mathbb{R}$, where the constants C_1 and C_2 only depend on A, N, α, β, ε, κ, μ, γ, r_0, M, M_0, M_1 and M_2 (but not on r, p, ω, and s), $\hat{C}$ is the random variable in (16) and $\tilde\kappa := \frac{\kappa}{(\beta+\varepsilon)\wedge 1/2}$. The random variables $\xi_1(\omega)$ and $\xi_2(\omega)$ have the form

$$\xi_1(\omega) = \sup_{t \leq 0} \left\{ e^{\gamma t} \|A^{\alpha+\beta}\eta(t, -\infty, \omega)\| \right\} + 1 \tag{43}$$

and

$$\xi_2(\omega) = \sup_{t \leq 0} \left\{ e^{\gamma t} \sup_{\theta \in [-r,0]} \| A^\alpha \eta(t, t+\theta, \omega) \| \right\}. \tag{44}$$

Proof. It suffices to prove the theorem for $s = 0$. Let $v^r \equiv v^r(t; 0, p, \omega) \in \mathcal{C}_0$ be the fixed point of the operator equation

$$v^r = \mathcal{B}_{p,r}^{0,\omega}[v^r] \tag{45}$$

where we replace B by B^r in (26). In addition, we denote by

$$v^0 \equiv v^0(t; 0, p, \omega) \in \mathcal{C}_0 \equiv \left\{ v : |v|_0 \equiv \sup_{t \leq 0} \{ e^{\gamma t} \, \| A^\alpha v(t) \| \} < \infty \right\}$$

the fixed point of the equation

$$v^0 = \mathcal{B}_{p,0}^{0,\omega}[v^0] \tag{46}$$

given by (26) if we replace B by B_0^r.

By (40) the contraction constants for (45), (46) are bounded by $\delta < 1$ uniformly for $r \in [0, r_0]$ and $\omega \in \Omega$, $p \in PH$. We obtain by (Zeidler, 1985), Theorem 1.A.c. and (12) that

$$|v^r - v^0|_0 \leq \frac{1}{1-\delta} |\mathcal{B}_{p,r}^{0,\omega}(v^0) - \mathcal{B}_{p,0}^{0,\omega}(v^0)|_0$$

$$\leq \frac{2}{\mu(1-\delta)} \sup_{\tau \leq 0} \{ e^{\gamma \tau} \| A^\alpha (B^r(v_\tau^0, \vartheta(\tau)\omega) - B_0^r(v^0(\tau), \vartheta(\tau)\omega)) \| \}$$

$$= \frac{2}{\mu(1-\delta)} \sup_{\tau \leq 0} \{ e^{\gamma \tau} \| A^\alpha B_1^r(v_\tau^0, \vartheta(\tau)\omega) \| \}$$

$$= \frac{2}{\mu(1-\delta)} \sup_{\tau \leq 0} \{ e^{\gamma \tau} \| A^\alpha (B_1^r(v_\tau^0, \vartheta(\tau)\omega) - B_1^r(v^0(\tau)\mathbf{1}, \vartheta(\tau)\omega)) \| \}$$

$$\leq \frac{2M_1}{\mu(1-\delta)} \sup_{\tau \leq 0} \{ e^{\gamma \tau} \sup_{\theta \in [-r,0]} \| A^\alpha(v^0(\tau+\theta) - v^0(\tau)) \| \}.$$

We have used that B_1^r is purely retarded.

Therefore,

$$\sup_{\theta \in [-r,0]} \| A^\alpha (\Psi^0(\omega, p) - \Psi^r(\omega, p, \theta)) \|$$

$$= \| A^\alpha (v^0(0; 0, p, \omega) - p - v^r(\theta; 0, p, \omega) + e^{-\theta A} p) \|$$

$$\leq e^{\gamma r_0} |v^0 - v^r|_0 + \sup_{\theta \in [-r,0]} \| A^\alpha(v^0(0) - v^0(\theta)) \| + \lambda_N r \| A^\alpha p \| \tag{47}$$

$$\leq \frac{4M_1 e^{\gamma r_0}}{\mu(1-\delta)} \sup_{\tau \leq 0} \{ e^{\gamma \tau} \sup_{\theta \in [-r,0]} \| A^\alpha(v^0(\tau+\theta) - v^0(\tau)) \| \}$$

$$+ \lambda_N r \| A^\alpha p \|.$$

The conclusion of the theorem now follows from Lemma 3.3 below. $\qquad \square$

Corollary 3.2 *Suppose that the assumptions of Theorem 3.1 are satisfied. Then the following estimate holds*

$$\left(\mathbb{E}\sup_{-r\leq\theta\leq0}\|A^\alpha(\Psi^0(\omega,p)-\Psi^r(\omega,p,\theta))\|^l\right)^{\frac{1}{l}}$$

$$\leq C_1 r^\beta\left[\|A^\alpha p\|+(\mathbb{E}\xi_1(\omega)^l)^{\frac{1}{l}}\right]+C_2(\mathbb{E}\xi_2(\omega)^l)^{\frac{1}{l}}$$

$$\leq \tilde{C}_1 r^\beta(\|A^\alpha p\|+1)+\tilde{C}_2 r^\kappa.$$

This corollary follows easily from the previous theorem and Proposition 1.6.

To obtain the conclusion of Theorem 3.1 we prove the following lemma.

Lemma 3.3 *Suppose that the assumptions of Theorem 3.1 are satisfied. Then we have*

$$\sup_{\tau\leq0}\{e^{\gamma\tau}\sup_{\theta\in[-r,0]}\|A^\alpha(v^0(\tau+\theta)-v^0(\tau))\|\}$$

$$\leq \bar{C}_1 r^\beta\left[\|A^\alpha p\|+\xi_1(\omega)\right]+\xi_2(\omega) \tag{48}$$

$$\leq \bar{C}_1 r^\beta\left[\|A^\alpha p\|+\xi_1(\omega)\right]+\hat{C}_{\alpha,\gamma,(\beta+\varepsilon)\wedge\frac{1}{2},\tilde{\kappa},r_0}(\omega)r^\kappa,$$

where the constant $\bar{C}_1$ is independent of r and p and the random variables $\xi_1(\omega)$ and $\xi_2(\omega)$ are defined by (43) and (44).

Proof. Let $t\leq0$. The fixed point v^0 of (46) fulfills the equation

$$v^0(t)=e^{\theta A}v^0(t+\theta)+\int_{t+\theta}^t e^{-(t-\tau)A}B_0^r(v^0(\tau))\,d\tau+\eta(t,t+\theta).$$

Consequently, we have the estimate

$$\|A^\alpha(v^0(t)-v^0(t+\theta))\|\leq\|A^{-\beta}(e^{\theta A}-1)\|\cdot\|A^{\alpha+\beta}v^0(t+\theta)\|$$

$$+\int_{t+\theta}^t\|A^\alpha e^{-(t-\tau)A}\|\cdot\|B_0^r(v^0(\tau))\|\,d\tau+\|A^\alpha\eta(t,t+\theta)\|.$$

Therefore, using the estimate

$$\|B_0^r(v^0(\tau))\|\leq\|B_0^r(0)\|+M_0 e^{-\gamma\tau}|v^0|_0 \tag{49}$$

and relations (11) and (13), we get

$$\|A^\alpha(v^0(t)-v^0(t+\theta))\|\leq|\theta|^\beta\cdot\|A^{\alpha+\beta}v^0(t+\theta)\|+\|A^\alpha\eta(t,t+\theta)\|$$

$$+C(\alpha)|\theta|^{1-\alpha}(M_2+M_0 e^{-\gamma(t-r_0)}\cdot|v^0|_0), \tag{50}$$

where $C(\alpha)=(1-\alpha)^{-1}\left(\frac{\alpha}{e}\right)^\alpha$ for $0<\alpha<1$ and $C(\alpha)=1$ for $\alpha=0$. Using (26) with $s=0$ we obtain

$$\|A^{\alpha+\beta}v^0(t)\| \le e^{-\lambda_N t}\|A^{\alpha+\beta}p\| + \int_t^0 \|A^{\alpha+\beta}e^{-(t-\tau)A}P\| \cdot \|B_0^r(v^0(\tau))\| \, d\tau$$

$$+ \int_{-\infty}^t \|A^{\alpha+\beta}e^{-(t-\tau)A}Q\|\|B_0^r(v^0(\tau))\| \, d\tau$$

$$+ \|A^{\alpha+\beta}\left(-e^{-tA}P\eta(0,t) + Q\eta(t,-\infty)\right)\|.$$

Therefore, using (49), (12) and (28), we get

$$e^{\gamma t}\|A^{\alpha+\beta}v^0(t)\| \le \|A^{\alpha+\beta}p\| + C(\alpha,\beta)\left(1 + |v^0|_0\right)$$
$$+ \|A^{\alpha+\beta}\eta(0,-\infty)\| + e^{\gamma t}\|A^{\alpha+\beta}\eta(t,-\infty)\|,$$

where the constant $C(\alpha,\beta)$ also depends on M_0, M_2. Thus

$$e^{\gamma t}\|A^{\alpha+\beta}v^0(t)\| \le \lambda_N^\beta \|A^\alpha p\| + C(\alpha,\beta) \cdot (1 + |v^0|_0) + 2\xi_1 \tag{51}$$

for all $t \le 0$. Since $|\eta(\cdot,-\infty)|_0 \le \lambda_1^{-\beta}\xi_1$, it follows from (27) that

$$|v^0|_0 \le (1-\delta)^{-1}\left(\|A^\alpha p\| + \frac{2M_0}{\mu} + \frac{2}{\lambda_1^\beta}\xi_1\right), \quad p \in PH. \tag{52}$$

Consequently, using (16), (50), (51) and (52) we obtain (48).   □

Example 3.4 Theorem 3.1 and Corollary 3.2 can be applied to problem (33)–(35). One can show that for r small enough, an inertial manifold for this system is close to an inertial manifold for the non-retarded stochastic equation

$$\frac{\partial u}{\partial t} = \nu \frac{\partial^2 u}{\partial^2 x} + f_r(x,u) + \dot{W}, \quad x \in (0,\pi), \quad t > 0;$$

$$u|_{x=0} = u|_{x=\pi} = 0, \quad u|_{t=0} = u_0(x),$$

where

$$f_r(x,u) = \int_{-r}^0 f(\theta, x, u(t,x))\zeta(d\theta).$$

In this case $B_0^r(u,\omega)$ is the mapping given by

$$u(x) \mapsto f_r(x,u(x)), \quad u \in L^2(0,1),$$

and $B_1^r(v,\omega)$ is given by

$$B_1^r(v,\omega) = \int_{-r}^0 f(\theta, x, v(\theta, x))\zeta(d\theta) - f_r(x, v(0,x))$$

for $v \in C\left([0,1], L^2(0,1)\right)$.

4 Dependence of the inertial manifold on the intensity of the noise

We now consider solutions of equation (2) as functions of the covariance operator K and denote the dependence of $\mathcal{M}$, Ψ, η etc. on K by an additional upper index. If K is zero (i.e. there is no additive white noise term), then the index is zero. We drop the lower index of Ψ in case it is 0. Under the assumptions of Theorem 2.2 the random dynamical system generated by (2) has an inertial manifold $\mathcal{M}^K$ with graph Ψ^K:

$$\mathcal{M}^K = \{\hat{p}(\cdot) + \Psi^K(\omega, \hat{p}(0), \cdot) : \hat{p} \in \hat{P}C_\alpha\}.$$

In this section we will show that the graph Ψ^K of $\mathcal{M}^K$ tends to Ψ^0 uniformly in L^l for any $l \geq 1$ as

$$q(K) \equiv \operatorname{tr}(K\,A^{2(\alpha+\varepsilon)-1}) \to 0,$$

where $\varepsilon \in (0,1)$ is fixed. The main (and only) result of this section in the following assertion.

Theorem 4.1 *Let $\Psi^K(\cdot)$ be the graph of the inertial manifold $\mathcal{M}^K$. Then*

$$\sup_{p \in PH} \sup_{\theta \in [-r,0]} \|A^\alpha(\Psi^K(\omega, p, \theta) - \Psi^0(\omega, p, \theta))\| \leq \frac{e^{\gamma r}}{1-\delta} \cdot \Delta^K(\omega),$$

where

$$\Delta^K(\omega) = \|A^\alpha \eta^K(0, -\infty, \omega)\| + |\eta^K(\cdot, -\infty, \omega)|_0 \leq 2|\eta^K(\cdot, -\infty, \omega)|_0,$$

and for each $l \geq 1$ there exists a constant $c_{l,\varepsilon}$ which does not depend on K such that

$$\mathbb{E}\left[\sup_{p \in PH} \sup_{\theta \in [-r,0]} \|A^\alpha(\Psi^K(p, \theta) - \Psi^0(p, \theta))\|\right]^l \leq c_{l,\varepsilon} q(K)^{l/2}.$$

Proof. In the following we denote the operator introduced in (26) for $s = 0$ and its fixed point by $\mathcal{B}_p^{0,\omega,K}$ and $v^K(\cdot; 0, p, \omega) \in C_0$, $p \in PH$ respectively.

We have from representation (29) that

$$\Psi^K(\omega, p, \theta) - \Psi^0(p, \theta, \omega) = v^K(\theta; 0, p, \omega) - v^0(\theta; 0, p, \omega).$$

Therefore

$$\sup_{\theta \in [-r,0]} \|A^\alpha(\Psi^K(p, \theta) - \Psi^0(p, \theta))\| \leq e^{\gamma r} |v^K(\cdot; 0, p) - v^0(\cdot; 0, p)|_0. \tag{53}$$

Using relation (28), we get

$$|v^K(\cdot;0,p) - v^0(\cdot;0,p)|_0 = |\mathcal{B}_p^{0,K}(v^K) - \mathcal{B}_p^{0,0}(v^0)|_0$$
$$\leq \delta|v^K(\cdot;0,p) - v^0(\cdot;0,p)|_0 + |-e^{-\cdot A}P\eta^K(0,\cdot) + Q\eta^K(\cdot,-\infty)|_0$$
$$= \delta|v^K(\cdot;0,p) - v^0(\cdot;0,p)|_0$$
$$+ \sup_{t\leq 0}\left\{e^{\gamma t}\| - A^\alpha e^{-tA}P\eta^K(0,-\infty) + A^\alpha\eta^K(t,-\infty)\|\right\}.$$

Therefore,

$$|v^K(\cdot;0,p,\omega) - v^0(\cdot;0,p,\omega)|_0 \leq (1-\delta)^{-1}\Delta^K(\omega).$$

Consequently, the first assertion of the theorem follows from (53). It is clear that $\Delta^K(\omega) \leq 2|\eta^K(\cdot,-\infty)|_0$. Therefore, the second assertion follows from the first one together with relation (14) in Proposition 1.6. $\square$

References

Arnold, L. (1998). *Random Dynamical Systems.* Springer, Berlin-Heidelberg-New York.

Arnold, L. and Scheutzow, M. (1995). Perfect cocycles through stochastic differential equations. *Prob. Theory Related Fields*, 101:65–88.

Bensoussan, A. and Flandoli, F. (1995). Stochastic inertial manifolds. *Stoch. Stoch. Rep.*, 53:13–39.

Boutet de Monvel, L., Chueshov, I., and Rezounenko, A. (1998). Inertial manifolds for retarded semilinear parabolic equations. *Nonlinear Anal.*, 34:907–925.

Chow, S.-N. and Lu, K. (1988). Invariant manifolds for flows in Banach spaces. *J. Differential Equations*, 74:285–317.

Chueshov, I. (1992). *Introduction to the theory of inertial manifolds.* Kharkov Univ. Press, Kharkov.

Chueshov, I. (1995). Approximate inertial manifolds of exponential order for semilinear parabolic equations subjected to additive white noise. *J. Dynam. Differential Equations*, 7:42–45.

Chueshov, I. (1999). *Introduction to the Theory of Infinite-Dimensional Dissipative Systems.* Acta, Kharkov. in Russian; English translation: Acta, Kharkov, 2002, see also http://www.emis.de/monographs/Chueshov/.

Chueshov, I. and Girya, T. (1994). Inertial manifolds for stochastic dissipative dynamical systems. *Dokl. Acad. Sci. Ukraine*, 7:42–45.

Chueshov, I. and Girya, T. (1995). Inertial manifolds and forms for semilinear parabolic equations subjected to additive white noise. *Lett. Math. Phys.*, 34:69–76.

Chueshov, I. and Scheutzow, M. (2001). Inertial manifolds and forms for stochastically perturbed retarded semilinear parabolic equations. *J. Dynam. Differential Equations*, 13:355–380.

Constantin, P., Foias, C., Nicolaenko, B., and Temam, R. (1989). *Integral Manifolds and Inertial Manifolds for Dissipative Partial Differential Equations*. Springer, Berlin-Heidelberg-New York.

Crauel, H. and Flandoli, F. (1994). Attractors for random dynamical systems. *Prob. Theory Related Fields*, 100:365–393.

Da Prato, G. and Debussche, A. (1996). Construction of stochastic inertial manifolds using backward integration. *Stoch. Stoch. Rep.*, 59:305–324.

Da Prato, G. and Zabczyk, J. (1992). *Stochastic Equations in Infinite Dimensions*. Cambridge Univ. Press, Cambridge.

Da Prato, G. and Zabczyk, J. (1996). *Ergodicity for Infinite Dimensional Systems*. Cambridge Univ. Press, Cambridge.

Duan, J., Lu, K., and Schmalfuß, B. (2003). Invariant manifolds for stochastic partial differential equations. *Ann. Prob.*, 31:2109–2135.

Foias, C., Sell, G., and Temam, R. (1988). Inertial manifolds for nonlinear evolutionary equations. *J. Differential Equations*, 73:309–353.

Girya, T. and Chueshov, I. (1995). Inertial manifolds and stationary measures for stochastically perturbed dissipative dynamical systems. *Sb. Math.*, 186:29–45.

Kager, G. and Scheutzow, M. (1997). Generation of one-sided random dynamical systems by stochastic differential equations. *Electron. J. Prob.*, 2(8):1–17.

Kunita, H. (1990). *Stochastic Flows and Stochastic Differential Equations*. Cambridge Univ. Press, Cambridge.

Kuo, H.-H. (1975). *Gaussian Measures in Banach Spaces*. Springer, Berlin-Heidelberg-New York.

Scheutzow, M. (1996). On the perfection of crude cocycles. *Random Comp. Dynamics*, 4:257–268.

Schmalfuß, B. (1992). Backward cocycles and attractors of stochastic differential equations. In Reitmann, V., Riedrich, T., and Koksch, N., editors, *International Seminar on Applied Mathematics–Nonlinear Dynamics: Attractor Approximation and Global Behaviour*, pages 185–192.

Schmalfuß, B. (1996). A random fixed point theorem based on Lyapunov exponents. *Random Comp. Dynamics*, 4(4):257–268.

Schmalfuß, B. (1997). The random attractor of the stochastic Lorenz system. *Z. Angew. Math. Phys.*, 48:951–975.

Schmalfuß, B. (1998). A random fixed point theorem and the random graph transformation. *J. Math. Anal. Appl.*, 225(1):91–113.

Sharpe, M. (1988). *General Theory of Markov Processes*. Academic Press, Boston.

Temam, R. (1988). *Infinite Dimensional Dynamical Systems in Mechanics and Physics*. Springer, Berlin-Heidelberg-New York.

Zeidler, E. (1985). *Nonlinear Functional Analysis and its Applications*, volume 1. Springer, Berlin-Heidelberg-New York.

The Random Walk Representation
for Interacting Diffusion Processes

Jean-Dominique Deuschel

Fakultät II - Mathematik und Naturwissenschaften, Institut für Mathematik,
Technische Universität Berlin, Straße des 17. Juni 136, D-10623 Berlin
Tel. (0049 30) 314-25193, Fax: (0049 30) 314-21695
deuschel@math.tu-berlin.de

Summary. We investigate a system of lcal interacting diffusion processes with attractive interaction. We show how the random walk representation can be used to express the gradient of the semigroup and to estimates for the time-space correlations. In particular we can answer questions dealing with localization, convergence rates to equilibrium and aging properies of the system.

Introduction

One of the main theme of this "Schwerpunkt" deals with interacting particles systems. In this note we will be considering a wide class of locally interacting diffusion processes. These models are presented else where in this Schwerpunkt program. Depending on the context, the process describes in [25] the intensity of a biological population, such as the Flemming-Viot model, or in [24], motivated from magneto-hydrodynamics, biological evolution processes and spatial branching processes, the parabolic Anderson model, and in [1] the stochastic evolution of quantum anharmonic crystals related to Euclidean Gibbs measures. One of the main interest deals with the long time asymptotic of the system such as the existence of invariant measure and convergence rate to equilibrium. These models are all attractive and positively correlated. We will be particularly interested in the massless or strongly interactive case. Here the standard perturbative methods do not apply, and our primary aim is to present a method: *the random walk representation*. This technique allows to express the gradient of the semigroup in terms of the Green function of a random walk in a dynamical random environment, cf. [22]. One can then obtain via martingale methods, estimates for the time-space correlations and use them to understand how the dimension of lattice affects both convergence to equilibrium and the aging properties of the system, cf. [2].

The rest of the note is as follows: in Section 1 we present the random walk representation. Section 2 shows how this representation can be applied to estimate the space-time correlations. Section 3 deals with various examples, such as the Landau-Ginzburg model, the super-random walk, the stepping stone model and the Parabolic Anderson model.

A concluding remark: since our objective is to present a useful technique, rather than survey the different models, we have tried here to give a hint of the proofs without being to technical.

1 The Random Walk Representation

The object of this section is the random walk representation, that is, we show how the derivative of the semigroup can be expressed with the help of the random walk. The result is a simple consequence of commutation of the generator of the process and the gradient operator. We also show in this section how correlations and exponential moments can be expressed with the help of the gradient of the semigroup.

Let $\Lambda \subseteq \mathbb{Z}^d$ and $I \subseteq \mathbb{R}$ be an interval. We consider a special class of interacting stochastic differential equations $\phi_t = (\phi_t(x))_{x \in \Lambda} \in I^\Lambda$ of the form

$$d\phi_t(x) = b_x(\phi_t)\,dt + \sigma_x(\phi_t(x))dw_t(x), \qquad x \in \Lambda, \tag{1.1}$$

where $(w_t(x))_{t \geq 0}, x \in \Lambda$ is a family of independent Wiener processes.

We make the following assumptions on the drift and diffusion coefficient: for each $x \in \Lambda$ there exists a finite set $N(x) \subset \Lambda$ with bounded diameter: $\mathrm{diam}(N(x)) = \max_{y,z \in N(x)} |y - z| \leq R$ such that $b_x \in C_b^1(I^{N(x)})^1$. Moreover $a_x \equiv \sigma_x^2 \in C_b^1(I)$ with $a_x > 0$ in the interior of I and $a_x \equiv 0$ on the boundary of I. In case of unbounded I and Λ, we introduce the set of tempered configurations

$$E_r = \{\phi \in I^\Lambda : \sum_{x \in \Lambda} \phi(x)^2 e^{-r|x|} < \infty\}$$

for some $r > 0$. Then for each $\phi_0 \in E_r$, the above SDE has a unique strong solution $\phi_t \in E_r, \forall t \geq 0$, cf. [28].

Let $\{P_t, t \geq 0\}$ be the corresponding Markov semigroup associated with the generator L

$$Lf(\phi) = \sum_y \left(b_y(\phi)\partial_y f(\phi) + \frac{1}{2}a_y(\phi(y))\partial_y^2 f(\phi) \right), \quad f \in C_b^2(E_r).$$

The object of this note is to express for $F \in C_b^1(E_r)$

$$\partial_x \mathbb{E}_\phi[F(\phi_t)] = \partial_x P_t F(\phi) = \frac{\partial}{\partial \phi(x)} P_t F(\phi), \qquad x \in \Lambda.$$

Our main assumption is the attractivity of interaction:

$$\partial_y b_x \geq 0, \qquad y \in N(x) \setminus \{x\}. \tag{1.2}$$

. This is the space of continuously differentiable functions with bounded derivatives

We set

$$u(\phi, x) = \partial_x b_x(\phi) - \sum_{y \neq x} \partial_y b_x(\phi).$$

The case

$$u(\phi, x) \leq -m \quad \text{for some } m > 0, \tag{1.3}$$

corresponds to the *massive case*. When $u(\phi, x) \equiv 0$, we have a *massless case*. A typical massless situation is when the interaction is of *gradient type*. More precisely let $\Lambda_* \equiv \{b = (b_1, b_2) : b \in \Lambda, \text{with} |b_2 - b_1| = 1\}$ be the set of oriented nearest neighbor bonds of Λ and introduce the discrete gradient

$$\nabla \phi(b) = \phi(b_2) - \phi(b_1), \quad b \in \Lambda_*.$$

Then drifts of gradient form $b_x(\phi) = \tilde{b}_x(\nabla \phi)$, imply $u(\phi, x) \equiv 0$. If the initial distribution is translation invariant with finite first moment: $\langle \phi(x) \rangle_\nu \equiv \int \phi(x) d\nu = m$, $\quad \forall x \in \Lambda$ then the first moment is preserved:

$$\mathbb{E}_\nu[\phi_t(x)] = m \quad \forall t \geq 0, x \in \Lambda. \tag{1.4}$$

In the special case of constant diffusion coefficients:

$$a_x(\phi(x)) \equiv \alpha_x, \quad \text{for some } \alpha_x > 0, \tag{1.5}$$

then the gradient process $(\nabla \phi_t)_{t \geq 0}$ is itself a diffusion, solution of the SDE

$$d\nabla \phi_t(b) = \nabla(\tilde{b}.(\nabla \phi_t))(b) \, dt + \sqrt{\alpha_{b_2}} dw_t(b_2) - \sqrt{\alpha_{b_1}} dw_t(b_1), \quad b = (b_1, b_2) \in \Lambda_*.$$

In order to introduce the random walk representation let us consider the process $(\hat{\phi}_t, \xi_t)_{t \geq 0}$ on $E_r \times \Lambda$ generated by

$$\hat{L}F(\phi, x) = \sum_y \left([b_y(\phi) + \frac{1}{2}\delta_x(y)a'_x(\phi(x))] \partial_y F(\phi, x) + \frac{1}{2}a_y(\phi_y)\partial_y^2 F(\phi, x) \right)$$

$$+ \sum_{y \in N(x) \backslash \{x\}} \partial_y b_x(\phi)(F(\phi, y) - F(\phi, x)).$$

This is the generator if a diffusion $(\hat{\phi}_t)_{t \geq 0}$ and a random walk on $(\xi_t)_{t \geq 0}$. At time t, the random walk jumps from $\xi_t = x$ to $y \in N(x) \backslash \{x\}$ at rate $\partial_y b_x(\hat{\phi}_t)$, while the diffusion $(\hat{\phi}_t)_{t \geq 0}$ solves the SDE

$$d\hat{\phi}_t(x) = b_x(\hat{\phi}_t) + \frac{1}{2}1_x(\xi_t)a'_x(\hat{\phi}_t(x))) \, dt + \sigma_x(\hat{\phi}_t(x))dw_t(x), \quad x \in \Lambda. \tag{1.6}$$

We write $\mathbb{P}_{\phi, x}$ for the law of the process starting at (ϕ, x). Our main result is the following expression of the derivative of the semigroup:

Theorem 1.1. *Assume (1.2). Let $f \in C_b^1(E_r)$, then*

$$\partial_x P_t f(\phi) = \partial_x \mathbb{E}_\phi[f(\phi_t)] = \sum_y \mathbb{E}_{\phi, x}[\partial_y f(\hat{\phi}_t)1_y(\xi_t) \exp(\int_0^t u(\hat{\phi}_s, \xi_s) \, ds)].$$

$$\tag{1.7}$$

Proof. The major difficulty is to show that for given $f \in C_b^1(E_r)$, the function $v_t = P_t f \in C_b^3(E_r)$, cf. [12]. We skip this part and just explain why (1.7) is true: the functions $z_t(\phi, x) \equiv \partial_x v_t(\phi), x \in \Lambda$ solve the PDE

$$\partial_t z_t(\phi, x) = \partial_x(L v_t(\phi)) = \partial_x \sum_y \left(b_y(\phi) \partial_y v_t(\phi) + \frac{1}{2} a_y(\phi(y)) \partial_y^2 v_t(\phi) \right)$$

$$= \sum_y \left(b_y(\phi) \partial_y \partial_x v_t(\phi) + \frac{1}{2} a_y(\phi(y)) \partial_y^2 \partial_x v_t(\phi) \right.$$

$$\left. + \partial_x b_y(\phi) \partial_y v_t(\phi) + \frac{1}{2} \delta_x(y) a_x'(\phi(x)) \partial_x \partial_y v_t(\phi) \right)$$

$$= \hat{L} z_t(\phi, x) + u(\phi, x) z_t(\phi, x)$$

with $z_0(\phi, x) = \partial_x f(\phi)$. Now the result follows from the Feynman-Kac formula. $\square$

Note that in general both random walk and diffusion are dependent of each others, however two special cases are of special interest: The first one is when the diffusion coefficients are constant, cf. (1.5). In this case of course $a_x' \equiv 0$ and therefore the diffusion $(\hat{\phi}_t = \phi_t)_{t \geq 0}$ is independent of the random walk $(\xi_t)_{t \geq 0}$. The next case is when the drift is linear

$$b_x(\phi) = \sum_{y \in N(x)} q(x, y) \phi(y) \tag{1.8}$$

with $q(x, y) \geq 0, x \neq y$, then $\partial_y b_x(\phi) = q(x, y)$ is constant and the random walk $(\xi_t)_{t \geq 0}$ is independent of the diffusion $(\hat{\phi}_t)_{t \geq 0}$. In the massless case we simply have

$$b_x(\phi) = \sum_{y \in N(x) \setminus \{x\}} q(x, y) (\phi(y) - \phi(x)). \tag{1.9}$$

In order to understand how the above can be applied let us introduce the bilinear form Γ on $C_b^2(E_r)$:

$$\Gamma(f, g) = L(f \cdot g) - f \cdot Lg - g \cdot Lf = \sum_x a_x \cdot \partial_x f \cdot \partial_x g.$$

Also for given $\Phi \in C_b^2(\mathbb{R}^d)$ and $f = (f_1, ..., f_d)$ with $f_k \in C_b^2(E_r)$ set

$$\Gamma_\Phi(f) = L(\Phi \circ f) - \sum_{k=1}^d (\partial_{x_k} \Phi) \circ f \cdot Lf_k = \frac{1}{2} \sum_{j,k=1}^d (\partial_{x_j} \partial_{x_k} \Phi) \circ f \cdot \Gamma(f_j, f_k). \tag{1.10}$$

By definition we simply have $\Gamma_\Phi = \Gamma$ for $\Phi(f_1, f_2) = f_1 \cdot f_2$.

Lemma 1.2. *We have*

$$P_t(\Phi \circ f) = \Phi \circ (P_t f) - \int_0^t P_{t-s}(\Gamma_\Phi(P_s f)) \, ds. \qquad (1.11)$$

If μ is an invariant measure such that

$$\lim_{t \to \infty} \langle \Phi \circ (P_t f) \rangle_\mu = \Phi \circ (\langle f \rangle_\mu), \qquad (1.12)$$

then

$$\langle \Phi \circ (P_t f) \rangle_\mu = \Phi \circ (\langle f \rangle_\mu) + \int_t^\infty \langle \Gamma_\Phi(P_s f) \rangle_\mu \, ds. \qquad (1.13)$$

Proof. In order to show (1.11) just note that

$$\partial_t [\Phi \circ (P_t f)] = L(\Phi \circ (P_t f)) - \Gamma_\Phi(P_t f).$$

Next if μ is invariant we have for any $T > 0$ integrating both sides of (1.11)

$$\langle \Phi \circ f \rangle_\mu = \langle \Phi \circ (P_T f) \rangle_\mu - \int_0^T \langle \Gamma_\Phi(P_s f) \rangle_\mu \, ds.$$

In view of (1.12), letting $T \to \infty$ yields (1.13). $\qquad \square$

The condition (1.12) is usually verified when μ is an extremal invariant measure. For simplicity we will say that μ is *extremal* if (1.12) holds for $\Phi(f_1, f_2) = f_1 \cdot f_2$. The next proposition shows how covariances and exponential moments can be expressed using the operator Γ:

Proposition 1.3. Let $f, g \in C_b^1(E_r)$ then

$$cov_\mu(f(\phi_s), g(\phi_t)) = cov_\mu(P_s f, P_t g) + \int_0^s \mathbb{E}_\mu[\Gamma(P_u f, P_{t-s+u} g)(\phi_{s-u})] \, du \qquad (1.14)$$

and for bounded f

$$\mathbb{E}_\mu[e^{f(\phi(t))}] = \langle e^{P_t f(\phi)} \rangle_\mu + \frac{1}{2} \int_0^t \mathbb{E}_\mu[(e^{P_s f} \Gamma(P_s f, P_s f))(\phi_{t-s})] \, ds. \qquad (1.15)$$

Let μ be extremal, then

$$cov_\mu(f(\phi_0), g(\phi_t)) = \int_0^\infty \langle \Gamma(P_u f, P_{t+u} g) \rangle_\mu \, du \qquad (1.16)$$

Proof. Note first that

$$cov_\mu(f(\phi_s), g(\phi_t)) = cov_\mu(P_s f, P_t g) + \mathbb{E}_\mu[cov(f(\phi_s), g(\phi_t)|\phi_0)]$$

where

$$cov(f(\phi_s), g(\phi_t)|\phi_0) = P_s(f \cdot P_{t-s} g)(\phi_0) - P_s f(\phi_0) \cdot P_t g(\phi_0)$$

Now (1.14) follows from (1.11) with using $\Phi(f, g) = f \cdot g$ and integrating both sides. Next let $\Phi(f) = e^f$, then in view of (1.10) $\Gamma_\Phi(f) = \frac{1}{2} e^f \Gamma(f, f)$, thus (1.15) is a simple application of (1.11) again. $\qquad \square$

2 Estimates for the Correlations

In this section we present some applications of the random walk representation based on estimates of the Green function of the random walk. In particular we show in the massless case, that localization takes place in higher lattice dimensions $d \geq 3$, whereas we have delocalization in $d = 1, 2$. Also the system exhibits aging in lower dimensions. We also derive algebraic L^2 rates of convergence for $d \geq 3$. Finally we show that the dynamic of the discrete gradient converges faster to equilibrium.

The first thing to notice is the triviality of the positive mass case, cf. (1.3). In this situation if we apply (1.7) and Proposition 1.3, we get exponentially decaying correlations both in time and space variables.

In the massless case, the situation is subtler. If we want to apply the above Proposition, we need some estimates in the random walk representation. More precisely, for given trajectory of the diffusion $(\hat{\phi})_{t \geq 0}$, set

$$p(s, x; t, y) = \mathbb{P}(\xi_t = y \,|\, \xi_s = x, \hat{\phi}_u, s \leq u \leq t).$$

Our main assumption is the existence of constant c_1, c_2, c_2, c_4 [2] such that

$$p^*_{c_1 t}((x - y)c_2) \leq p(s, x; s + t, y) \leq p^*_{c_3 t}((x - y)c_4), \quad x, y \in \mathbb{Z}^d, \qquad (2.1)$$

where $p^*_t(x, y) = p^*_t(x - y)$ is the transition function of the simple random walk on $\mathbb{Z}^d$. This is trivially the case for constant jump coefficients (1.8) when the matrix q is irreducible, but such an estimate also applies for symmetric uniformly elliptic rates:

$$0 < c_- \leq \partial_y b_x(\phi) = \partial_x b_y(\phi) \leq c_+, \quad \forall y \in N(x) \setminus \{x\}, \qquad (2.2)$$

cf. [7], [30].

Unless otherwise stated we will always assume the estimate (2.1). Moreover ν will be an initial i.i.d. distribution with finite variance. Our next result is the basic tool for the estimate of covariances

Proposition 2.1. *Assume that*

$$A_1(t) \leq \mathbb{E}_\nu[a_x(\phi_t(x))] \leq A_2(t). \qquad (2.3)$$

If $u(\phi, x) \leq 0$, then for all $f, g \in C^1_b(E_r)$,

$$var_\nu(f(\phi_s)) \leq c_1 \int_0^s A_2(u)(1 \vee u)^{-d/2} du \qquad (2.4)$$

and

[.] in what follows $c_., c_., \ldots$ will denote positive constants which may differ from lines to lines, depending on the context, they may depend on the test functions f, g and the initial measure ν, but not on the time variables t, s!

$$cov_\nu(f(\phi_s), g(\phi_{s+t})) \leq c_2 \int_0^s A_2(u)(1 \vee (t+u))^{-d/2} \, du \qquad (2.5)$$

Next assume $u(\phi, x) \geq 0$ then the corresponding lower bound holds with A_1 for monotone

$$f, g \in C_b^{1,+}(E_r) = \{f \in C_b^1(E_r) : \|\partial_x f\|_{\inf} = \inf_\phi \partial_x f(\phi) \geq 0, \sum_x \|\partial_x f\|_{\inf} > 0\}.$$

Finally assume that ν has Gaussian tail:

$$\log \langle \exp(L(\phi(x) - \langle \phi(x) \rangle_\nu))) \rangle_\nu \leq c_3 L^2, \qquad (2.6)$$

$u(\phi, x) \leq 0$ and that the diffusion coefficients are bounded: $a_x(\phi) \leq c_4$, then for $f \in C_b^1(E_r)$ we have

$$\mathbb{P}_\nu(|f(\phi_t) - \langle P_t f \rangle_\nu| \geq L) \leq c_5 \exp(-\frac{L^2}{\kappa(t)}), \qquad (2.7)$$

where $\kappa(t) \geq c \int_0^t (1 \vee u)^{-d/2} du$.

Proof. First note that for $f \in C_b^1(E_r)$ in case $u(\phi, x) \leq 0$

$$|\partial_x P_t f(\phi)| \leq \sum_y p_{c_3 t}^*((x-y)c_4)\|\partial_y f\|_\infty, \qquad (2.8)$$

and in case $u(\phi, x) \geq 0$ for monotone $f \in C_b^{1,+}(E_r)$

$$\partial_x P_t f(\phi) \geq \sum_y p_{c_1 t}^*((x-y)c_2)\|\partial_y f\|_{\inf}. \qquad (2.9)$$

Next we have for the transition function of the simple random walk the standard estimate:

$$\frac{c_1}{(1 \vee t)^{d/2}} \exp(-|x-y|^2/c_2 t) \leq p_t^*(x-y) \leq \frac{c_3}{(1 \vee t)^{d/2}} \exp(-|x-y|^2/c_4 t)$$

when $t \geq |x|$. Also if ν is i.i.d. with finite variance,

$$|cov_\nu(f, g)| \leq c_6 \sum_x \|\partial_x f\|_\infty \|\partial_x g\|_\infty, \quad f, g \in C_b^1(E_r).$$

Now the result follows from the above estimates and the preceding Proposition. Finally in order to prove the exponential estimate, we note that in view of (1.14), (2.8) and (2.6)

$$\log \mathbb{E}_\nu[\exp(L(f(\phi_t) - \langle P_t f \rangle_\nu))] \leq c_5 L^2 \int_0^t (1 \vee u)^{-d/2} du, \quad L \geq 0.$$

$\square$

The first application of the above estimates deals with localization and delocalization:

Proposition 2.2. *Let $u(\phi, x) \geq 0$ and $A_1(t) \geq c_1$, then we have delocalization in $d = 1, 2$ with*

$$var_\nu(f(\phi_t)) \approx \begin{cases} \log t & d = 2 \\ t^{1/2} & d = 1 \end{cases}, \qquad \forall f \in C_b^{1,+}(E_r), t \to \infty, \qquad (2.10)$$

On the other hand if $u(\phi, x) \leq 0$, $A_2(t) \leq c_2$ and

$$\mathbb{E}_\nu[\phi_t(x)] \leq c_3, \qquad (2.11)$$

then we have localization in $d \geq 3$ with

$$var_\nu(f(\phi_t)) \approx 1 \qquad \forall f \in C_b^{1,+}(E_r). \qquad (2.12)$$

Note that an estimate of the variance alone is not sufficient for localization, cf. entropic repulsion below, since one also needs a control of the mean (2.11). However in the gradient interacting case, in view of (1.4), localization is always established for $d \geq 3$ once $A_2(t) \leq c_2$ is verified!

The next application deals with the aging phenomena. For $f, g \in C_b^{1,+}(E_r)$, set

$$\rho(s, t) = cor_\nu(f(\phi_s), g(\phi_{s+t})) \equiv \frac{cov_\nu(f(\phi_s), g(\phi_{s+t}))}{\sqrt{var_\nu(f(\phi_s))}\sqrt{var_\nu(g(\phi_{s+t}))}}.$$

Typically $\rho(s, t) \to 0$ as $t \to \infty$ uniformly in $s > 0$. We say in this case that no aging takes place. On the other hand if the time variable s has an influence on the convergence of $\rho(s, t)$ as $t \to \infty$, then we speak of aging for this model, cf. [2], [8].

Proposition 2.3. *Consider the massless case $u(\phi, x) \equiv 0$ and suppose that*

$$c_1 \leq A_1(t) \leq A_2(t) \leq c_2. \qquad (2.13)$$

Then for all $f, g \in C_b^{1,+}(E_r)$, we have no aging in $d \geq 3$. For $d = 2$, we have aging at logarithmic scale with

$$\rho(s, t) \approx 1 - \frac{\log t}{\log s}, \qquad s, t \to \infty, t \leq s. \qquad (2.14)$$

In case $d = 1$ we have aging at linear scale with

$$\rho(s, t) \approx \frac{(1 + t/s)^{1/2} - (t/s)^{1/2}}{(1 + t/s)^{1/4}}, \qquad s, t \to \infty. \qquad (2.15)$$

Proof. We simply note that for monotone $f, g \in C_b^{1,+}(E_r)$ we have

$$\text{cov}_\nu(f(\phi_s), g(\phi_{s+t})) \approx \begin{cases} (1 \vee t)^{-d/2+1} - (1 \vee (t+s))^{-d/2+1} & d \geq 3 \\ \log(1 \vee (t+s)) - \log(1 \vee s), & d = 2 \\ (1 \vee (t+s))^{1/2} - (1 \vee s)^{1/2}, & d = 1. \end{cases}$$

Now the result follows from this and the variance estimates (2.10) and (2.12).

$\square$

The next results deals with the correlations and convergence to equilibrium.

Proposition 2.4. *Let $d \geq 3$ and μ be an extremal invariant distribution such that*

$$\langle a_x(\phi(x)) \rangle_\mu \leq c_1. \tag{2.16}$$

Consider the massless case then, for all $f, g \in C_b^1(E_r)$

$$\left| \text{cov}_\mu(f(\phi_0), g(\phi_t)) \right| \leq \sum_{x,y} \|\partial_x f\|_\infty \frac{c_2}{\left((1 \vee t)^{1/2} \vee \|x - y\|\right)^{d-2}} \|\partial_y g\|_\infty.$$

In particular we have the following algebraic $L^2(\mu)$ rate of convergence

$$\|P_t f - \langle f \rangle_\mu\|_{L^2(\mu)} \leq c_3 (1 \vee t)^{-d/4+1/2} \sum_x \|\partial_x f\|_\infty.$$

Moreover the corresponding lower bounds holds for monotone $f, g \in C_b^{1,+}(E_r)$.

Proof. This is a simple consequence of (1.16) and the estimates (2.8) and (2.9).

$\square$

Consider now the gradient interacting case with preserved mean, cf. (1.4). Then functions of the discrete gradient, which in a sense stay orthogonal to the preserved mean, converge faster to equilibrium. More precisely let $f \in C_b^1(E_r)$ be of the form $f(\phi) = \tilde{f}(\nabla\phi)$ and write $\partial_b \tilde{f} \equiv \frac{\partial \tilde{f}}{\partial \nabla \phi(b)}$, then using (1.7) and summation by part

$$\partial_x \mathbb{E}_\phi[\tilde{f}(\nabla\phi_t)] = \sum_y \mathbb{E}_{\phi,x}[\partial_y \tilde{f}(\nabla\hat{\phi}_t) 1_y(\xi_t)]$$

$$= \sum_y \mathbb{E}_\phi[\partial_y \tilde{f}(\nabla\hat{\phi}_t) p(0, x; t, y)]$$

$$= \sum_b \mathbb{E}_\phi[\partial_b \tilde{f}(\nabla\hat{\phi}_t) \nabla(p(0, x; t, \cdot))(b)].$$

In case of linear drift (1.9), that is of constant irreducible jumps rate for the random walk, we have the following estimate for the gradient of the kernel:

$$|\nabla(p(0,x;t,\cdot))(b)| \le \frac{p^*_{c_3 t}((x-b_1)c_4)}{(1 \vee t)^{1/2}}. \tag{2.17}$$

In the random case the situation is more complicated since the one cannot expect such a result for fixed diffusion $(\hat{\phi}_t)_{t \ge 0}$. However, starting form translation invariant measure ν, one can prove, for constant diffusivity (1.3) and uniformly ellipticity (2.2), the following annealed or integrated estimate, cf. [7]:

$$\mathbb{E}_\nu[|\nabla(p(0,x;t,\cdot))(b)|^2]^{1/2} \le \frac{p^*_{c_3 t}((x-b_1)c_4)}{(1 \vee t)^{1/2}}. \tag{2.18}$$

In particular, if we either have constant coefficients (1.9) and $\mathbb{E}_\nu[a_x(\phi_t(x))] \le c_1$, or symmetric elliptic jumps rates (2.2) and constant diffusive coefficients (1.3), then using the above we see that the gradient process remains localized in any lattice dimensions since

$$\text{var}_\nu(|\nabla\phi_t(b)|^2) \le c_4 \int_0^t (1 \vee u)^{-d/2-1} \, du \approx 1.$$

Also in this case we have no aging in any dimension!

Let $d \ge 3$ and μ be an invariant distribution for $(\phi_t)_{t \ge 0}$, we then write $\tilde{\mu} \equiv \mu \circ \nabla^{-1}$ for the corresponding invariant measure of the gradients. In case of constant diffusive coefficients (1.3) $(\nabla\phi_t)_{t \ge 0}$ is itself a Markov process. Using the localization result, we see that invariant measures $\tilde{\mu}$ exist for the gradient process even in lower lattice dimensions $d = 1, 2$. The estimates (2.17) and (2.18) then yields the following faster rate to equilibrium for functions of the gradient:

Proposition 2.5. *Let $\tilde{\mu}$ be an extremal invariant distribution for the gradient process, where we assume either, $d \ge 3$, (1.9) and $\tilde{\mu} = \mu \circ \nabla^{-1}$ for some invariant μ with $\langle a_x(\phi(x)) \rangle_\mu \le c_1$, or $d \ge 1$, (2.2) and constant a_x, then for each $\tilde{f}, \tilde{g} \in C_b^1(\tilde{E}_r)$*

$$\left| cov_{\tilde{\mu}}(\tilde{f}(\nabla\phi_0), \tilde{g}(\nabla\phi_t)) \right| \le \sum_{b,b'} \|\partial_b \tilde{f}\|_\infty \frac{c_2}{\left((1 \vee t)^{1/2} \vee \|b_1 - b_1'\|\right)^d} \|\partial_{b'} \tilde{g}\|_\infty.$$

In particular we have the following algebraic $L^2(\tilde{\mu})$ rate of convergence

$$\|P_t \tilde{f} - \langle \tilde{f} \rangle_{\tilde{\mu}}\|_{L^2(\tilde{\mu})} \le c_3 (1 \vee t)^{-d/4} \sum_b \|\partial_b \tilde{f}\|_\infty.$$

3 Examples

We present here various models where the results of the previous section apply. We first look at the Landau-Ginzburg model, including a discussion on entropic repulsion and wetting transition. Next we consider linear interacting models such as the super-random walk, the stepping stone model and the parabolic Anderson model.

The Landau Ginzburg model

The first example deals with the Landau Ginzburg model. This is a model for an interface with a gradient interaction. This model has attracted a lot of attention, see eg. the review articles [20] or [16]. Also the hydrodynamic limes of such a system has been derived in several papers including study of the corresponding large deviations and fluctuations, cf. [17], [21].

In this case the diffusion coefficient is constant: $a_x \equiv 2, x \in \mathbb{Z}^d$. Next let $V \in C_b^2(\mathbb{R})$ be even and strictly convex:

$$c_- \leq V'' \leq c_+ \qquad (3.1)$$

for some $0 < c_- \leq c_+ < \infty$. The drift is a function of the discrete gradient:

$$b_x(\phi) = -\sum_{e:|e|=1} V'(\phi(x) - \phi(x+e)) = \sum_{b:b_1=x} V'(\nabla\phi(b)) = \frac{1}{2}\mathrm{div}(V'(\nabla\phi))(x),$$
$$(3.2)$$

where div is the discrete divergence:

$$\mathrm{div}(\alpha)(x) = \sum_{b:b_2=x} \alpha(b) - \sum_{b:b_1=x} \alpha(b).$$

This yields the SDE

$$d\phi_t(x) = \mathrm{div}(V'(\nabla\phi_t))(x)dt + \sqrt{2}dw_t(x), \quad x \in \Lambda. \qquad (3.3)$$

The function V plays the role of an elastic force or local surface tension. The formal Hamiltonian then gives the total energy or surface tension

$$H(\phi) = \frac{1}{2}\sum_{b\in\mathbb{Z}_*^d} V(\nabla\phi(b)),$$

so that the drift can be expressed as partial derivative of H: $b_x(\phi) = -\partial_x H(\phi)$. This implies that the formal Gibbs measure

$$\mu_\Lambda(d\phi) = \frac{1}{Z_\Lambda}\exp(-H(\phi))\prod_{x\in\Lambda} d\phi(x) \qquad (3.4)$$

is a reversible equilibrium for the dynamic. Of course (3.4) is just formal and makes no sense for infinite Λ. In fact since the Hamiltonian depends only on the discrete gradients, a classical results show that no infinite Gibbs measure exists on the whole $\mathbb{Z}^d$ in lower dimensions $d = 1, 2$. In higher dimensions $d \geq 3$, one can use the Brascamp-Lieb inequality to show for each $m \in \mathbb{R}$ the existence of a unique extremal translation invariant Gibbs measure μ_m such that

$$\langle\phi(x)\rangle_{\mu_m} = m, \qquad \forall x \in \mathbb{Z}^d.$$

A special case is for quadratic $V^*(\nabla\phi(b)) = \frac{1}{2}|\nabla\phi(b)|^2$ with linear drift

$$b_x^*(\phi) = \sum_{e:|e|=1} (\phi(x+e) - \phi(x)) = \Delta\phi(x)$$

the discrete Laplacian. The corresponding process is then an Ornstein Uhlenbeck process, and the extremal Gibbs μ_m^* is a Gaussian measure, also called *harmonic crystal*, while the general convex is known as *anharmonic crystal*, cf. [1]. Turning now to the random walk representation which was first introduced by [22], see also [27], we first note that, since V is even, we have in view of (3.1) uniformly elliptic symmetric jumps rate for the random walk $(\xi_t)_{t\geq 0}$:

$$\partial_y b_x(\phi_t) = V''(\phi_t(x) - \phi_t(y)) = \partial_x b_y(\phi_t), \quad |x-y| = 1.$$

In particular the classical theory of uniformly elliptic symmetric random walks applies, and the bound (2.1) is verified, cf. [7]. Thus we see that all of the results of the preceding section, including the faster decay for functions of the gradients, apply.

We can add a self potential $U \in C_b^2(\mathbb{R})$ to the Hamiltonian:

$$H(\phi) = \frac{1}{2}\sum_{b\in\mathbb{Z}_*^d} V(\nabla\phi(b)) + \sum_x U(\phi(x))$$

so that the SDE becomes

$$d\phi_t(x) = \left[\operatorname{div}(V'(\nabla\phi_t))(x) - U'(\phi_t(x))\right]dt + \sqrt{2}dw_t(x), \quad x \in \mathbb{Z}^d. \quad (3.5)$$

Now if U is convex, then $u(\phi, x) = -U''(\phi(x)) \leq 0$, so that the bound (2.7) applies also in this case. Taking now $U_\delta(\phi(x)) = \frac{1}{\delta}\phi(x)^2 1_{\phi(x)\leq 0}$ and letting $\delta \to 0$, we get the SDE with *reflection* at the wall $\{\phi(x) = 0, x \in \mathbb{Z}^d\}$:

$$d\phi_t(x) = \operatorname{div}(V'(\nabla\phi_t))(x)dt + d\ell_t(x) + \sqrt{2}dw_t(x), \quad x \in \Lambda, \quad (3.6)$$

where $(\ell_t(x))_{t\geq 0}$ is the local time of $(\phi_t(x))_{t\geq 0}$ at 0, cf. [13], [18]. We have here the following random walk representation, cf. [15]:

Proposition 3.1. *Let $(\phi_t)_{t\geq 0}$ be the solution of the SDE (3.6), then for each* $f \in C_b^1(E_r)$

$$\partial_x P_t(f)(\phi) = \mathbb{E}_{\phi,x}\Big[\sum_y \partial_y f(\phi_t)1_y(\xi_t)1_{\tau<t}\Big] \quad (3.7)$$

where $\tau = \inf\{t \geq 0 : \phi_t(\xi_t) = 0\}$.

It is clear from (3.7), that the bound (2.7) also holds. However although we have Gaussian concentration around the mean, the random walk representation gives no clue for the mean itself $\mathbb{E}_\nu[\phi_t(x)]$. In fact one can show that starting with i.i.d. ν of finite variance, then, as $t \to \infty$

$$\mathbb{E}_\nu[\phi_t(x)] \approx \begin{cases} t^{1/2}, & d = 1 \\ \log(t), & d = 2 \\ \sqrt{\log(t)}, & d \geq 3, \end{cases}$$

cf. [13]. In particular this indicates that we have delocalization in any dimension. In fact looking at the corresponding Gibbs state on a finite box $V_N = [-N, N]^d \cap \mathbb{Z}^d$ with boundary 0-condition

$$\mu_N^+(d\phi) = \frac{1}{Z_N^+} \exp(-H(\phi)) 1_{\Omega^+} \prod_{x \in V_N} d\phi(x) \prod_{x \notin V_N} \delta_0(d\phi(x)),$$

where $\Omega^+ = \{\phi : \phi(x) \geq 0, x \in \mathbb{Z}^d\}$, then the random fields undergoes an entropic repulsion with

$$\langle \phi(x) \rangle_{\mu_N^+} \approx \begin{cases} N^{1/2}, & d = 1 \\ \log(N), & d = 2 \\ \sqrt{\log(N)}, & d \geq 3, \end{cases}$$

cf. [3], [10], which is consistent with the above dynamical entropic repulsion.

The situation is more intricate in presence of both *repulsion* and *weak pinning*, corresponding to a formal Gibbs measure of the form

$$\mu_\Lambda^{+,J}(d\phi) = \frac{1}{Z_\Lambda^{+,J}} \exp(-H(\phi)) 1_{\Omega^+} \prod_{x \in \Lambda} (d\phi(x) + e^J \delta_0(\phi(x))),$$

where $J \in \mathbb{R}$ is a pinning parameter. This is the formal reversible equilibrium of the SDE

$$d\phi_t(x) = \mathrm{div}(V'(\nabla\phi_t))(x) 1_{(0,\infty)}(\phi_t(x)) dt + d\ell_t(x) + \sqrt{2} 1_{(0,\infty)}(\phi_t(x)) dw_t(x),$$

where $(\ell_t(x))_{t \geq 0}$ satisfies the sticky condition:

$$\int_0^\infty \phi_t(x) \, d\ell_t(x) = 0, \qquad e^J \ell_t(x) = \int_0^t 1_{\{0\}}(\phi_s(x)) \, ds$$

cf. [16]. Without repulsion, weak pinning alone suffices for localization in any dimensions, cf. [14]. Moreover the unique Gibbs measure μ^J has exponential decaying correlations, cf. [26]. However when both effect compete, we have a wetting transition in lower dimensions $d = 1, 2$ with delocalization for weak pinning $J \leq J_0$ and localization for strong pinning $J > J_0$, cf. [5]. In higher dimensions $d \geq 3$, pinning always wins and we have localization for any $J \in \mathbb{R}$, cf. [4]. So far, these results were obtained only for the Gibbs measure and virtually, nothing is known for the dynamic in presence of pinning. One of the difficulty is that weak pinning cannot be approximated by a convex self-potential U so that the random walk representation gives little information in this case.

However in the localized regime one expects an exponential convergence to equilibrium which is predicted by the exponential decay of the space correlations. Only in the one dimensional case where both fluctuations and repulsion are of the same size, one can see that starting from equilibrium in a box of size N, the properly rescaled space-time process converges as $N \to \infty$ to reflected partial stochastic differential equation of the Nualard-Pardoux type, cf. [18], [30], [12].

Linear Drifts

From now on we assume that the drift is linear cf. (1.3), with $\{q(x, y)\}$ being the generator of a symmetric random walk on $\mathbb{Z}^d$, that is

$$q(x, y) = q(y, x) = q(0, y - x), \qquad q(x, x) = - \sum_{y \neq x} q(x, y)$$

with a SDE of the form

$$d\phi_t(x) = \sum_{y \in N(x)} q(x, y)[\phi_t(y) - \phi_t(x)]dt + \sigma_x(\phi_t(x))dw_t(x), \quad x \in \Lambda.$$

This is again a drift of gradient form therefore the mean is preserved, cf. (2.13). Here the random walk is trivial since it does not depend on $(\hat{\phi}_t)_{t \geq 0}$. In particular the estimate (2.1) holds. Thus if the diffusion coefficient is bounded above and below

$$c_1 \leq a_x \leq c_2$$

then the results of aging and convergence to equilibrium apply. The situation is more complicated, when a_x is unbounded or degenerate. Here a general theory is not applicable, so we just look at different situations.

Super-random walk

In this case $I = [0, \infty)$ and $a_x(\phi(x)) = \alpha^2 \phi(x)$, so that if we start with i.i.d. ν, then

$$\mathbb{E}_\nu[a_x(\phi_t(x))] = \alpha^2 \mathbb{E}_\nu[\phi_t(x)] = \alpha^2 \langle \phi(x) \rangle_\nu$$

remains constant. Thus we see from Prop. 2.2 that we have localization in $d \geq 3$, whereas for $d = 1, 2$, $\delta_0{}^3$ is the only equilibrium. Note that both the aging result of Prop. 2.3 and Prop. 2.4 hold in this case.

$^\cdot$ $\mathbf{0}$ is the configuration with $\phi(x) \equiv 0, \forall x \in \mathbb{Z}^d$

Stepping Stone model

We consider a bounded interval, for simplicity $I = [0,1]$ with $a_x(\phi(x)) \equiv a(\phi(x))$ independent of x and $a(0) = a(1)$. The standard case is the stepping stone model with $a(\phi(x)) = \phi(x)(1-\phi(x))$. Of course in case of compact I, the localization problem is not relevant. For $d = 1, 2$ then the only equilibrium for this model convex combinations of the trivial δ_0 or δ_1. In higher dimensions $d \geq 3$ we have as above for each $m \in (0,1)$ unique extremal invariant μ_m with mean m, [6]. In particular we can apply Prop. 2.4 in this case.

3.1 Parabolic Anderson model

Here $I = \mathbb{R}$ and the diffusion coefficient is quadratic: $a_x(\phi(x)) = \alpha\phi(x)^2$. Assuming an initial i.i.d. ν with finite variance, one sees that

$$v_t(x - y) = \mathbb{E}_\nu[\phi_t(x) \cdot \phi_t(y)]$$

satisfies

$$\partial_t v_t(x) = 2\sum_y q(x,y)\big(v_t(y) - v_t(x)\big) + \alpha^2 1_0(x)v_t(x)$$

with $v_0(x) = 1_0(x)\langle\phi(0)^2\rangle_\nu + \langle\phi(0)^2\rangle_\nu$. Thus using the Feynman-Kac formula

$$\mathbb{E}_\nu\left[\phi^2(0)\right] = \mathbb{E}_0\left[(\sigma^2 1_0(\xi_t) + m^2)\exp(\alpha^2\int_0^t 1_0(\xi_s)\,ds)\right]$$

where $(\xi_s)_{s\geq 0}$ is the random walk starting at x generated by the matrix $2q$. In particular in recurrent dimensions we have delocalization. In case $d \geq 3$, there exists $\alpha_0 > 0$ so that we have extremal invariant μ_m with finite variance for $\alpha < \alpha_0$, whereas we have delocalization for $\alpha > \alpha_0$, cf. [23]. Now Prop. 2.4 apply in case $\alpha < \alpha_0$.

References

1. S. Albeverio, Y. Kondratiev, M. Röckner and T. Pasurek, *Euclidean Gibbs measures of quatum crystals: existence, uniqueness and a priori estimates* here (2004).
2. G. Ben Arous, A. Dembo and A. Guionnet, *Aging of spherical spin glasses*, Prob. Theor. Rel. Fields, **121** (2001), 1–67.
3. E. Bolthausen, J.-D. Deuschel and O. Zeitouni *Entropic repulsion of the lattice free field* Comm. Math. Phys., 170 (1995), 417–443.
4. E. Bolthausen, J.-D. Deuschel and O. Zeitouni *Absence of a wetting transition for a pinned harmonic crystal in dimensions three and larger*, J. Math. Phys. **41** (2001).

5. P. Caputo and I. Velenik *A note on wetting transition for gradient fields* Stoch. Proc. Appl. **87** (2000), 107–113.

6. J.T. Cox, A. Greven, and T. Shiga, *Finite and infinite systems of interacting diffusions*, Probab. Rel. Fields, **103** (1994), 165-197.

7. T. Delmotte and J.–D. Deuschel, *On estimating the derivatives of symmetric diffusions in stationary random environment with applications to $\nabla \phi$ interface models*, to appear in PTRF 2004.

8. A. Dembo and J.–D. Deuschel, *Algebraic L. decay of attractive critical processes on the lattice*, Ann. of Prob. **22** (1994), 264–283.

9. J.–D. Deuschel, *Aging and the fluctuation dissipation theorem*, Preprint (2004)

10. J.–D. Deuschel, and G. Giacomin, *Entropic repulsion for massless free fields*, Stoch. Proc. Appl. **89** (2000), 333-354.

11. J.–D. Deuschel, G. Giacomin and D. Ioffe, *Large Deviations and Concentration Properties for $\nabla \phi$ Interface Models*, Prob. Theor. Rel. Fields **117** (2000), 49–111.

12. J.–D. Deuschel, G. Giacomin, and L. Zambotti, *Scaling limits of wetting models in $(1 + 1)$- dimensions* Preprint 2004.

13. J.–D. Deuschel and T. Nishikawa *The dynamic of entropic repulsion* , Preprint (2003).

14. J.–D. Deuschel and I. Velenik, *Non-Gaussian surfaces pinned by weak potentials*, Prob. Theor. Rel. Fields **116** (2000), 359–377.

15. J.–D. Deuschel, L. Zambotti, *Bismut-Elworthy formula and Random Walk representation for SDEs with reflection* Preprint (2003).

16. T. Funaki *Stochastic Interface Models* Lecture Notes St Flour (2003).

17. T. Funaki and T. Nishikawa, *Large deviations for the Ginzburg-Landau $\nabla \phi$ interface model* Prob. Theor. Rel. Fields **120**, (2001), 535–568.

18. T. Funaki and S. Olla, *Fluctuations for $\nabla \phi$ interface model on a wall* Stoch. Proc. Appl. **94**, (2001), 1–27.

19. T. Funaki and H. Spohn, *Motion by mean curvature from the Ginzburg-Landau $\nabla \phi$ interface model* Comm. Math. Phys. **185**, (1997), 1–36.

20. G. Giacomin *Aspects of statistical mechanics of random surfaces* Notes given at IHP (2002).

21. G. Giacomin, S. Olla and H. Spohn, *Equilibrium fluctuations for $\nabla \phi$ interface model*, Ann. Probab. **29** (2001), 1138–1172.

22. B. Helffer and J. Sjöstrand, *On the correlation for Kac–like models in the convex case*, J. Stat. Phys. **74**(1/2) (1994), 349–409.

23. F. den Hollander and J. Gärtner *Intermittency in a catalytic random medium*, Preprint (2003).

24. J. Gärtner and W. König, *Intermittency in the parabolic Anderson model* here (2004).

25. A. Greven, *Renormalization and universality for multitype population models*, here (2004).

26. D. Ioffe and I. Velenik, *A note on the decay of correlations under δ-pinning*, Prob. Theor. Rel. Fields **116** (2000), 379–389.

27. A. Naddaf and T. Spencer, *On homogenization and scaling limit of some gradient perturbation of a massless free field*, Comm. Math. Phys. **183** (1997), 55-84.

28. T. Shiga and A. Shimizu, *Infinite-dimensional stochastic differential equations and their applications*, J. Math. Kyoto Univ. **20** (1980), 395–416.

29. D.W. Stroock and W. Zheng, *Markov chain approximations to symmetric diffusions*, Ann. Inst. H. Poincaré **33** (1997), 619–649.
30. L. Zambotti, *Fluctuations for the $\nabla\phi$ interface model*, Preprint (2003).

Part IV

Applications of Stochastic Analysis in Finance, Engineering and Algorithms

On Worst-Case Investment with Applications in Finance and Insurance Mathematics

Ralf Korn and Olaf Menkens

Fachbereich Mathematik, Universität Kaiserslautern, 67653 Kaiserslautern

Summary. We review recent results on the new concept of worst-case portfolio optimization, i.e. we consider the determination of portfolio processes which yield the highest worst-case expected utility bound if the stock price may have uncertain (down) jumps. The optimal portfolios are derived as solutions of non-linear differential equations which itself are consequences of a Bellman principle for worst-case bounds. They are by construction non-constant ones and thus differ from the usual constant optimal portfolios in the classical examples of the Merton problem. A particular application of such strategies is to model crash possibilities where both the number and the height of the crash is uncertain but bounded. We further solve optimal investment problems in the presence of an additional risk process which is the typical situation of an insurer.

1 Introduction

Modelling stock prices at financial markets seems to be a classical field for the use of interacting particle systems. However, the most common stock price models do not contain explicit reference to the market participants, the traders. Even more, modern financial mathematics is based on the "small investor assumption" which requires that the action of the single trader has no impact to prices at all, an assumption which seems to contradict the idea of interaction at all.

The relation to interacting systems lies in a microeconomic modelling of financial markets. An excellent reference for this topic is [FS93]. Here,the authors show in particular how the usual assumption of stock prices following a geometric Brownian motion can be obtained via a limit argument out of a model where only a finite number of traders form the market and the stock prices are determined by supply and demand via the so-called market clearing condition.The geometric Brownian motion model is the limiting model that corresponds to the situation when only uninformed traders ("noise traders") are present.

Looking at the usual stock price models as limits that result from trading activities of many interacting traders, we are in a situation that is similar to limit considerations of particle models in statistical physics or biological applications. The main difference in financial mathematics is that the second step after the stock price modelling, the execution of tasks such as pricing of derivatives or of finding optimal investment straetgies is usually only done in the limit settings (such as the geometric Brownian motion model or other semi-martingale market models).

In this paper two of the main tasks of financial mathematics are touched. One is the modelling of stock prices and the other the determination of optimal investment strategies, the portfolio optimization problem. We will give a survey on the main results of the recently introduced approach of worst-case portfolio optimization (see [KW02] for its first introduction and [KM02], [KO03],[ME03] for generalizations). We specialize on portfolio optimization under the risk of market crashes but applications different from financial mathematics seem to be possible and should be considered in the future (examples could be the optimal control of a production line under the risk of a breakdown, optimal business strategies for food chains under the risk of sudden change of consumer behaviour (such as e.g.during the BSE crisis), evolution of populations/monocultures facing catastrophes). The basic model underlying our approach is worst-case modelling as introduced by Hua and Wilmott [HW97] where upper bounds on both the number of crashes until the time horizon and on the maximum height of a single crash are assumed to be known. Between the crashes the stock price is assumed to move according to a geometric Brownian motion. This makes the setting differ from classical approaches to explain large stock price moves such as e.g. described in [ME76], [EK95], [EKM97] where stock prices are given as Levy processes or other types of processes with heavy tailed distributions. As a second ingredient for our worst-case investment model we are more focused on avoiding large losses in bad situations via trying to put the worst-case bound for the expected utility of terminal wealth as high as possible.

In [KW02] this setup is introduced and the portfolio problem under the threat of a crash is solved in the case of a logarithmic utility function. Deriving systems of non-linear differential equations to characterize the optimal portfolio process for general utility functions and to allow the market parameters to change after each crash are the main achievements of [KM02]. Finally, in [KO03] the optimal investment problem of an insurer is considered who in addition faces a risk process which is non hedgeable in the financial market(a typical example is a life insurer that faces the biometric risk of the population getting older than estimated which seems to be uncorrelated (or at least not perfectly correlated) to the evolution of the financial markets).

The paper is organized as follows: Section 2 describes the set up of the model and contains the main theoretical results in the simple situation where at most one crash can occur. In Section 3 these results will be extended to the situation when the investor faces additional non-hedgeable risk. Finally, Section 4 contains various generalizations and states open problems.

2 The Simplest Set Up
of Worst-Case Scenario Portfolio Optimization

The most basic setup that we consider here consists of a riskless bond and a single risky security with prices during "normal times" given by

$$dP_0(t) = P_0(t)\,rdt, \ P_0(0) = 1 \tag{1}$$

$$dP_1(t) = P_1(t)\,(bdt + \sigma dW(t)), \ P_1(0) = p_1 \tag{2}$$

for constant market coefficients $b > r, \sigma \neq 0$ and a one-dimensional Brownian motion $W(t)$. At the "crash time" the stock price experiences a sudden relative fall which is assumed to be in the interval $[0, k^*]$ with $0 < k^* < 1$. Otherwise no further assumptions on both the crash size and time are made (we allow for changing market parameters and for multiple crashes in Sections 3 and 4).

We will assume that the investor is able to realize that a crash has happened and therefore introduce a process $N(t)$ counting the number of jumps (i.e. in our simple setting it is zero before the jump time and one from the jump time onwards). Let $\{f_t\}$ be the P-augmentation of the filtration generated by $W(t)$ and $N(t)$. We then define the set of admissible portfolio processes for our investor.

Definition 1. *Let $A(x)$ be the set of admissible portfolio processes $\pi(t)$ corresponding to an initial capital of $x > 0$, i.e. $\{f_t\}$ -progressively measurable processes such that*
a) the wealth equation in the usual crash-free setting

$$d\tilde{X}^\pi(t) = \tilde{X}^\pi(t)\left[(r + \pi(t)(b - r))\,dt + \pi(t)\,\sigma dW(t)\right], \tag{3}$$

$$\tilde{X}^\pi(0) = x \tag{4}$$

has a unique non-negative solution $\tilde{X}^\pi(t)$ and satisfies

$$\int_0^T \left(\pi(t)\,\tilde{X}(t)\right)^2 dt < \infty \ P - a.s. \tag{5}$$

i.e. $\tilde{X}^\pi(t)$ is the wealth process in the crash-free world.
b) the corresponding wealth process $X^\pi(t)$, defined as

$$X^\pi(t) = \begin{cases} \tilde{X}^\pi(t) \text{ for } t < \tau \\ (1 - \pi(\tau)\,k)\,\tilde{X}^\pi(t) \text{ for } t \geq \tau \end{cases}, \tag{6}$$

given the occurrence of a jump of height k at time τ, is strictly positive.
c) $\pi(t)$ has left-continuous paths with right limits.

This definition allows us to set up the worst-case portfolio problem we want to study:

Definition 2. *a) Let $U(x)$ be a utility function (i.e. a strictly concave, monotonously increasing and differentiable function). Then the problem to solve*

$$\sup_{\pi(.)\in A(x)} \inf_{0\leq\tau\leq T, 0\leq k\leq k^*} E\left(U\left(X^\pi\left(T\right)\right)\right) \tag{7}$$

(where the final wealth $X^\pi(T)$ in the case of a crash of size k at the (stopping) time τ is given by

$$X^\pi(T) = (1 - \pi(\tau)k)\tilde{X}^\pi(T) \tag{8}$$

with $\tilde{X}^\pi(\tau)$ as above) is called the worst-case scenario portfolio problem.
b) The value function to the above problem if one crash can still happen is defined as

$$v_1(t,x) = \sup_{\pi(.)\in A(t,x)} \inf_{t\leq\tau\leq T, 0\leq k\leq k^*} E\left(U\left(X^\pi\left(T\right)\right)\right). \tag{9}$$

c) Let $v_0(t,x)$ be the value function for the usual optimisation problem in the crash-free Black-Scholes setting, i.e

$$v_0(t,x) = \sup_{\pi(.)\in A(t,x)} E\left(U\left(\tilde{X}^\pi(T)\right)\right). \tag{10}$$

Under the assumption of $b > r$ a first fact which is very usefull and intuitively clear (note the requirement of left-continuity of the strategy !) is that it is optimal - with respect to the worst-case bound - to have all money invested in the bond at the final time (for a formal proof see [KW02]):

Proposition 1. *If $U(x)$ is strictly increasing then an optimal portfolio process $\pi(t)$ for the worst-case problem has to satisfy*

$$\pi(T) = 0. \tag{11}$$

We further require that the worst possible jump should not lead to a negative wealth process. Therefore, without loss of generality we can restrict to portfolio processes satisfying

$$1/k^* \geq \pi(t) \geq 0 \text{ for all } t \in [0,T] \text{ a.s.}. \tag{12}$$

which in particular implies that we only have to consider bounded portfolio processes. As after a crash it is optimal to follow the optimal portfolio of the crash-free setting, having a wealth of z just after the crash at time s leads to an optimal utility of $v_0(s,z)$. As $v_0(s,.)$ is strictly increasing in the second variable, a crash of maximum size k^* would be the worst thing to happen for an investor following a positive portfolio process at time s. As we only have to consider non-negative portfolio processes, and as by Proposition 1 we have

$$E\left(v_0\left(T, \tilde{X}^\pi(T)(1-\pi(T)k^*)\right)\right) = E\left(v_0\left(T, \tilde{X}^\pi(T)\right)\right) = E\left(U\left(\tilde{X}^\pi(T)\right)\right),$$

it is enough to consider only the effect of the worst possible jump. We have thus shown:

Theorem 1. *"Dynamic programming principle"*
If $U(x)$ and $v_0(t,x)$ are strictly increasing in x then we have

$$v_1(t,x) = \sup_{\pi(.)\in A(t,x)} \inf_{t\leq\tau\leq T} E\left(v_0\left(\tau, \tilde{X}^\pi(\tau)(1-\pi(\tau)k^*)\right)\right). \tag{13}$$

The dynamic programming principle will be used to derive a dynamic programming equation. A formal proof of the following result is again given in [KM02]. We will only sketch it.

Theorem 2. *"Dynamic programming equation"*
Let the assumptions of Theorem 1 be satisfied, let $v_0(t,x)$ be strictly concave in x, and let there exist a continuously differentiable (with respect to time) solution $\hat{\pi}(t)$ of

$$(v_0)_t(t,x) + (v_0)_x(t,x)(r + \hat{\pi}(t)(b-r))x + \tfrac{1}{2}(v_0)_{xx}(t,x)\sigma^2\hat{\pi}(t)^2 x^2$$
$$- (v_0)_x(t,x)x\frac{\hat{\pi}'(t)}{(1-\hat{\pi}(t)k^*)}k^* = 0 \text{ for } (t,x) \in [0,T[\times(0,\infty) \tag{14}$$

$$\hat{\pi}(T) = 0. \tag{15}$$

Assume further that we have:

(A) $f(x,y;t)$
$:= (v_0)_x(t,x)((y-\hat{\pi}(t))(b-r))x + \tfrac{1}{2}(v_0)_{xx}(t,x)\sigma^2\left(y^2-\hat{\pi}(t)^2\right)x^2$
is a concave fuction in (x,y) for all $t \in [0,T)$.
(B) $E^{0,x}\left(\hat{v}\left(t,\tilde{X}^\pi(t)\right)\right) \leq E^{0,x}\left(\hat{v}\left(t,\tilde{X}^{\hat{\pi}}(t)\right)\right)$ and $E^{0,x}(\pi(t)) \geq \hat{\pi}(t)$ for
some $t \in [0,T)$, $\pi \in A(x)$ imply

$$E^{0,x}\left(v_0\left(t,\tilde{X}^\pi(t)(1-\pi(t)k^*)\right)\right) \leq E^{0,x}\left(\hat{v}\left(t,\tilde{X}^{\hat{\pi}}(t)\right)\right).$$

Then, $\hat{\pi}(t)$ is indeed the optimal portfolio process before the crash in our portfolio problem with at most one crash. The optimal portfolio process after the crash has happened coincides with the optimal one in the crash free setting. The corresponding value function before the crash is given by :

$$v_1(t,x) = v_0(t,x(1-\hat{\pi}(t)k^*)) = E\left[v_0\left(s,\tilde{X}^{\hat{\pi}}(s)(1-\hat{\pi}(s)k^*)\right)\right]$$
$$\text{for } 0 \leq t \leq s \leq T. \tag{16}$$

Sketch of the proof:
Step 1: Derivation of (14)
The martingale optimality principle of stochastic control (see [KO03b] for a description of the martingale optimality principle) indicates that we obtain a martingale if we plug in the wealth process corresponding to the optimal

control into the value function. By using the Bellman principle (13), applying It's formula to the function inside the expectation of the right hand side and leaving aside the sup-opetator we obtain as a sufficient condition for the martingale property of the resulting process $v_0 \left(s, \tilde{X}^{\hat{\pi}} \left(s \right) \left(1 - \hat{\pi} \left(s \right) k^* \right) \right)$ that the portfolio process $\hat{\pi} \left(t \right)$ should satisfy the differential equation (14) with boundary condition

$$\hat{\pi} \left(T \right) = 0.$$

In particular, it should be differentiable.

Step 2: Optimality of $\hat{\pi} \left(t \right)$

The optimality proof for $\hat{\pi} \left(t \right)$ is motivated by the martingale optimality principle of stochastic control (see Korn (2003b)). We therefore introduce $\hat{v} \left(t, x \right) := E^{t,x} \left[U \left(\tilde{X}^{\hat{\pi}} \left(T \right) \right) \right]$. By considering $\hat{v} \left(t, \tilde{X}^{\pi} \left(t \right) \right)$ it will then be shown that under assumptions (A) and (B) all candidate processes $\pi \left(. \right)$ that could provide a higher worst case bound than $\hat{\pi} \left(t \right)$ do not deliver a higher one.

By verifying the requirements of Theorem 2 we obtain the central result of [KW02] as a special case:

Corollary 1. *There exists a strategy $\hat{\pi} \left(. \right)$ such that the corresponding expected log-utility after an immediate crash equals the expected log-utility given no crash occurs at all. It is given as the unique solution $\hat{\pi} \left(. \right) \in \left[0, \frac{1}{k^*} \right)$ of the differential equation*

$$\dot{\pi} \left(t \right) = \frac{1}{k^*} \left(1 - \pi \left(t \right) k^* \right) \left(\pi \left(t \right) \left(b - r \right) - \frac{1}{2} \pi \left(t \right)^2 \sigma^2 + \frac{1}{2} \left(\frac{b - r}{\sigma} \right)^2 \right) \quad (17)$$

with

$$\pi \left(T \right) = 0.$$

Further, this strategy yields the highest worst-case bound for problem (7). In particular, this bound is active at each future time point ("uniformly optimal balancing"). After the crash has happened the optimal strategy is given by

$$\pi \left(t \right) \equiv \pi^* := \frac{b - r}{\sigma^2}. \quad (18)$$

For numerical examples enlightening the performance of $\hat{\pi} \left(. \right)$ see [KW02] or [KM02].

Remark: a) The form of the differential equation for the optimal portfolio process in the above corollary in particular underlines that the differential equation in Theorem 2 is only an ordinary differential equation for $\hat{\pi} \left(. \right)$ and not for the value function $v_0 \left(t, x \right)$ of the crash-free setting. This value function is assumed to be known ! Further, the form of the differential equation (17) also implies that the fraction of wealth invested in the risky stock is continuously reduced over time if there is still the possibility of a crash to happen. This is in line with practitioners' behaviour.

b) In [ME03] the above situation is generalized to the case when the market coefficients after the crash depend on the crash size and crash time. This will introduce new cases that result in different optimal strategies. We will sketch one such situation in Section 4 below.

3 Optimal Worst-Case Investment with Non-hedgeable Risk

By introducing a non-hedgeable risk process into our scenario we arrive at a worts-case investment problem faced by an insurance company. This company invests at the stock market of the previous section (where for ease of notation we have set $r = 0$. The uncertainty of the insurance business is modelled via a risk process of diffusion type,

$$dR(t) = \alpha dt + \beta d\tilde{W}(t). \tag{19}$$

The additional one-dimensional Brownian motion $\tilde{W}(t)$ satisfies

$$\rho = Corr\left(W(t), \tilde{W}(t)\right). \tag{20}$$

The form of the above risk process is justified by a standard diffusion approximation argument (see [BR95]). The presence of this process however also introduces the possibility of bankruptcy. It is therefore convenient to consider the total amount of money $A(t)$ that the investor invests in the stock at time t instead of the portfolio process to describe the investor's activities. The corresponding wealth process $X^A(t)$ is then given by

$$dX^A(t) = A(t)(bdt + \sigma dW(t)) + \alpha dt + \beta d\tilde{W}(t) \tag{21}$$

in normal times. At the crash time it satisfies

$$X^A(\tau) = X^A(\tau-) - kA(\tau). \tag{22}$$

We now consider the worst-case problem of the form

$$\sup_{A(.)\in S(x)} \inf_{0\leq\tau\leq T, 0\leq k\leq k^*} E\left(-e^{-\lambda X^A(T)}\right) \tag{23}$$

where $S(x)$ consists of all deterministic strategies $A(t)$ which are left-continuous with right hand limits and almost surely square integrable with respect to time. The positive constant λ measures the investor's attitude towards risk. In the crash-free situation the optimal strategy is known from [BR95] as

Optimal Portfolios with a Crash

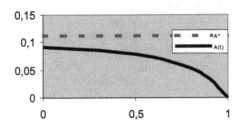

Fig. 1. Optimal investment for insurers with exponential utility and $b = 0.2, r = 0, \sigma = 0.4, k^* = 0.2, T = 1, \alpha = 0.3, \beta = 0.4, \lambda = 100, \rho = -0.1$

$$A(t) \equiv A^* = \frac{b}{\lambda\sigma^2} - \rho\frac{\beta}{\sigma}. \tag{24}$$

As in the setting of Section 2 this also forms the basis for the solution in the crash setting, a result proved in [KO03]:

Theorem 3. "*Optimal deterministic strategy with crash and risk process*"
If A^* is positive then the optimal deterministic amount of money invested in the stock before the crash is given by

$$A(t) = \frac{2k^*}{\lambda\sigma^2(t - T) - 2k^*/_{A^*}} + A^*. \tag{25}$$

The optimal amount of money invested into the stock after a crash equals A^*.

Remark: a) Theorem 3 differs from Corollary 1 by the fact that we now have an explicit expression for the optimal strategy. The reason for this is that the corresponding differential equation - obtained from the indifference argument mentioned in the sketch of the proof of Theorem 2 - can be solved explicitly. Indeed, this is the main difference in the proof of Theorem 3 which otherwise is very similar to the one of Theorem 2.
b) Note that one always invests less money in the stock than in the crash free model. The corresponding optimal wealth process is still a Brownian motion with drift (as in [BR95]) but now with a non-constant one. Figure 1 below shows the typical form of the optimal strategy before and after a crash. Note that the more negative the risk process is correlated with the stock price process the closer the optimal crash strategy approaches the one in the crash free setting.

4 Generalizations and Open Problems

a) *Changing market conditions after a crash*
Typically after a crash the market price of risk or some market coefficients

change as the expectations on the future perfomance of the stock price is then seen differently by the market participants. This feature is addressed in [KM02] in the stock market setting. It is extended to the insurer's case in [KO03]. The new aspect entering the scene is the fact that a crash need not necessarily be extremely disadvantageous if it happens, it can even be advantageous if it happens early when the market situation is better after the crash. To make things more precise, we assume that in normal times after the crash the stock price and the risk process follow

$$dP_1(t) = P_1(t)(b_1 dt + \sigma_1 dW(t)) \tag{26}$$

$$dR(t) = \alpha dt + \beta d\tilde{W}(t) \tag{27}$$

with $\rho_1 = Corr\left(W(t), \tilde{W}(t)\right)$. This leads to an optimal strategy after the crash of

$$A_1* = \frac{b_1}{\lambda \sigma_1^2} - \rho_1 \frac{\beta}{\sigma_1}. \tag{28}$$

This new aspect of the possibly advantageous crash leads to the following new optimality result given in [KO03]:

Theorem 4. *"Optimal deterministic strategy with crash, risk process, and changing market"*
Let A^* be positive.
a) If A_1^* is smaller than A^* then the results of Theorem 3 stay valid with A^* replaced by A_1^*.
b) If A_1^* is positive and bigger than A^* then the optimal strategy before the crash is given by

$$A(t) = \min\left(A^*, \frac{2k^*}{\lambda \sigma_1^2(t-T) - 2k^*/A_1*} + A_1*\right). \tag{29}$$

The optimal amount of money invested into the stock after a crash equals A_1^*.

An example illustrating Theorem 4 is given in Figure 2 where we have used the parameters $b = 0.2, r = 0, \sigma = 0.4, k^* = 0.2, T = 1, \alpha = 0.3, \beta = 0.4, \lambda = 100, \rho = -0.1, b_1 = 0.25, r_1 = 0, \sigma_1 = 0.3$. Note that due to the attractiveness of the crash we are allowed to follow the optimal strategy in the crash-free setting until $t = 0, 6$.

b)*n possible crashes*
Further aspects of the model such as the case of at most n possible crashes or more than one stock are considered in [KW02] and in [KM02]. As we now have to face n different crash scenarios we have to solve a system of n differential equations which however can be solved in an inductive fashion. Also it is shown in [ME03] that the above results are not changed if there is a probability distribution on the number of crashes that can still happen.

Optimal Portfolios with Crash and Change

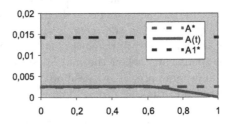

Fig. 2. Optimal investment with crash, changing coefficients, risk process

The worst case criterion is thus independent on the personal view of the probability for the worst case to appear as long as this probability is positive.

c) *Further aspects*
Interesting topics for future can be (among others):

– *including consumption to the portfolio problem*
– *use of options or option pricing under the threat of a crash*
– *of standard Hamilton-Jacobi-Bellman techniques that do not make use of the indifferernce argument but result in a Hamilton-Jacobi-Bellman equation (or more precisely into a variational inequality) for the value function before the crash.*

References

[BR95]   Browne, S. : Optimal investment policies for a firm with a random risk process: Exponential utility and minimizing the probability of ruin. Math. Op. Res., **20**(4), 937–957 (1995)

[EK95]   Eberlein, E., Keller, U. : Hyberbolic processes in finance . Bernoulli, **1**, 188–219 (1995)

[EKM97]   Embrechts, P., Klüppelberg, C., Mikosch, T.: Modelling Extremal Events. Springer, New York (1997)

[FS93]   Föllmer, H., Schweizer, M.: A microeconomic approach to diffusion models for stock prices. Mathematical Finance,**3**, 1–23 (1993)

[HW97]   Hua, P., Wilmott, P.: Crash course. Risk Magazine, **10**(6), 64–67 (1997)

[KM02]   Korn, R., Menkens, O.: Worst-case scenario portfolio optimization: A new stochastic control approach. Working paper, (2002)

[KO03]   Korn, R.: Optimal investment with crashes for insurers. Working paper, (2003)

[KO03b]   Korn, R.: The martingale optimality principle: The best you can is good enough. WILMOTT july issue, 61–67 (2003)

[KW02]   Korn, R., Wilmott, P.: Optimal portfolios under the threat of a crash. Int. J. Th. Appl. Fin., **10**(6), 171–187 (2002)

[ME03]   Menkens, O.: Crash hedging strategies and optimal portfolios. PhD Thesis, Technische Universität Kaiserslautern, Kaiserslautern (2004)

[ME76]   Merton, R.: Option pricing when underlying stock returns are discontinuous. J. Fin. Econ., **3**, 125–144 (1976)

Random Dynamical Systems Methods in Ship Stability: A Case Study

Ludwig Arnold[1], Igor Chueshov[2], and Gunter Ochs[1]

. Institut für Dynamische Systeme, Fachbereich 3, Universität
  Postfach 33 04 40, 28334 Bremen, Germany
  arnold@math.uni-bremen.de
. Department of Mechanics and Mathematics, Kharkov University
  4 Svobody Sq., 310077 Kharkov, Ukraine
  chueshov@univer.kharkov.ua

Summary. We first explain how to derive the archetypal equation describing the roll motion of a ship in random seaway from first principles. We then present an analytic and numerical case study of two simple nonlinear models of the roll motion using concepts of the theory of random dynamical systems. In contrast to the case of periodic excitation, the incorporation of noise leads to scenarios in which capsizing of the ship (i.e. the disappearance of the random attractor) is not preceded by a series of bifurcations, but happens without announcement "out of the blue sky".

Keywords and Phrases: Random seaway, random field, ship stability, ship capsizing, roll motion, random dynamical system, stochastic stability, stochastic bifurcation, random attractor, random invariant set, Conley index

MSC 2000: primary 34F05, 37H15, 37H20, 93E15; secondary 60H10, 70L05.

1 Introduction

The current regulations and criteria for assuring the stability of a ship and preventing it from capsizing (see the Codes of the International Maritime Organization (IMO) [7]) are empirical and are based on the properties of the righting lever of the ship, taking only hydrostatic forces into account. For details and criticism see Kreuzer and Wendt [9, p. 1836].

Those static criteria which neglect the motion of the ship as well as the impact of seaway and wind do obviously not guarantee total stability as they cannot prevent at least 30 ships yearly of tonnage greater than 500 GT from getting lost due to severe weather conditions. For recent figures of losses and their reasons cf. the publication [6] of The Institute of London Underwriters.

Hence researchers agree that those criteria have to be modified by using hydrodynamic models of the ship-sea system, by describing the sea as a random field and by analyzing the ship as a rigid body with 6 degrees of freedom using methods of nonlinear dynamics and the theory of random dynamical systems.

The present state of (deterministic as well as stochastic) ship dynamics research has been recently very well documented in the Theme Issue of the Phil. Trans. Royal Soc. London (series A) entitled "The nonlinear dynamics of ships", edited by Spyrou and Thompson [19]. The volume includes the extended overview by Spyrou and Thompson [18], presenting also the historical development of the field as well as future directions of research, the prime one being the "integration of the probabilistic character of the seaway into the study of nonlinear dynamics" (p. 1755) – which we have systematically addressed in the report [2] to which we refer the reader and of which this paper is a condensed version.

For further literature our default references are the volume [19] and the report [2]. We will quote additional papers only if needed.

The contents of this paper is as follows:

After this brief introduction (Sect. 1) we explain how the "archetypal" nonlinear random differential equation for the roll motion of a ship in random seaway can be obtained from "first principles" through a chain of controlled approximations and simplifications (Sect. 2).

We then give a brief introduction into the theory of random dynamical systems (Sect. 3). Our main contribution is an analytic and numerical case study of simple nonlinear models of the roll motion of a ship using concepts of the theory of random dynamical systems (Sect. 4).

Acknowledgement

This work grew out of the research project "Modellierung von Schiffsbewegungen durch zufällige dynamische System" (AR 137/15-1) in the framework of the DFG-Schwerpunktprogramm "Interagierende stochastische Systeme von hoher Komplexität" (SPP 1033). The support is greatfully acknowledged.

We also profitted from the collaboration with the project "Dynamik unendlichdimensionaler stochastischer Systeme" in the above DFG-Schwerpunktprogramm.

We finally would like to thank Professor Edwin Kreuzer (Technische Universität Hamburg-Harburg) and his collaborators for numerous valuable discussions.

2 The Motion of a Ship in Random Seaway

In this section we explain how the "archetypal" equation of the roll motion of a ship for which we perform the case study can in principle be derived from first principles by means of a chain of controlled approximations and simplifications.

2.1 The General Model

We will first briefly discuss in words the general equations of motion for a mechanical system consisting of a fluid (water) and of a partly immersed body B (ship). The fluid is assumed to be incompressible and to have an irrotational motion. The free surface of the fluid is assumed to extend to infinity in all horizontal directions. The body is assumed to be rigid and either describe a forced motion under the influence of external forces or be freely floating.

The "exact" equations of motion for a system of this type are well known (see e.g. John [8]).

The state of the fluid is completely described by a velocity potential function Φ satisfying Laplace's equation. The random seaway results in a component of Φ which is a Gaussian space-time stationary random field with a prescribed spectrum (see [2, Sect. 2]).

The boundary of the fluid consists of a fixed bottom surface, of the free surface S_A between the fluid and the atmosphere, and of the immersed surface S_B of the body B.

On each of these surfaces we have the condition that the normal velocity of the particles together with the pressure shall be continuous across the boundary surface (kinematic resp. dynamic boundary condition). On the free surface S_A it is assumed that the pressure is constant and equal to the atmospheric pressure. On the immersed surface S_B the *kinematic* condition that the normal velocity is continuous permits us to express the normal derivative of Φ in terms of the angular velocity of the rigid body B and of the velocity of its center of gravity. The *dynamic* condition along S_B, that the pressure is continuous, is taken into account by expressing that B moves under the influence of gravity, of the fluid pressure on S_B, and of other prescribed external forces. This yields the six differential equations for the motion of B (Newton's law and preservation of momenta). The six variables consist of the three translations x (surge), y (sway), z (heave) and the three angular coordinates ϕ (roll around the longitudinal x-axis, θ (pitch) and ψ (yaw).

For turning the above words into formulas see [2, Sect. 3].

However, in this generality of the problem little can be done in terms of rigorous mathematics either toward a discussion of the motion or toward an explicit solution of the equations. The difficulties arise from the fact that Φ is a solution of the potential equation determined by nonlinear boundary conditions on a variable boundary.

As a consequence of the analytical untractability of the general problem one has to resort to various approximation procedures, see [2]. One way of isolating three of the six variables of the full model is presented in the next subsection.

2.2 The Roll-Heave-Sway Interaction in Beam Sea

Aiming at the equations for roll ϕ, heave z and sway y only we adopt the following rather extreme version of strip theory: We assume that the ship is cylindrical and of infinite length. Hence the velocity potential Φ is a function of two spatial variables (width and height) only, and the free surface S_A is a function of one spatial variable (width).

We assume *beam sea*, i.e. the direction of waves is orthogonal to the ship's length axis.

In order to obtain the equations of motion of ϕ, y and z including hydrodynamic forces due to diffraction and radiation and the momentum of these forces we have to go through a long chain of simplifying hypotheses and approximations as done in [2, Sect. 4] (which we do not want to repeat here), and we finally obtain after lengthy calculations

$$m\ddot{y} = \xi(t)\eta_y'(y,t) + F_1^{\text{ext}}, \tag{1}$$

$$m\ddot{z} = \xi(t) - F_2^{\text{ext}} + F_1^{\text{ext}}\eta_y'(y,t) + 2\eta_{yt}''(y,t))\dot{y} + \eta_{tt}''(y,t), \tag{2}$$

$$I\ddot{\phi} = -\varepsilon\rho\xi(t)(\phi - \eta_y'(y,t)) + M^{\text{ext}}. \tag{3}$$

Here m is the mass and I is the second inertial moment of the ship, (Y_c, Z_c) are the coordinates of its center of gravity in a fixed frame, $y = Y_c$ denotes the sway, $z = \eta(Y_c, t) - Z_c$ is the *real heave*, ϕ is the roll angle, ε is a small parameter, $\xi(t)$ is a stationary stochastic process. and $\eta(y,t)$ is the equation for the free surface S_A of the sea. The latter can be assumed to be a stationary Gaussian random field as described in [2, Subsect. 2.3], with the fixed spatial travelling direction y, i.e. composed of infinitesimal beam waves of the form

$$\eta(y,t) = A\sin(\kappa y - \omega t + \alpha),$$

where the wave number κ is a function of the frequency ω through Airy's relation given by $\omega^2 = g\kappa\tanh(\kappa D)$, D being the depth of the sea.

2.3 The Equation of the Roll Motion

Rolling is probably the most obviously nonlinear and can be most realistically treated in isolation.

If we assume in the roll-heave-sway model that real heave and sway are small, we can put $y = 0$ in (3) and obtain (after rescaling) the equation of motion for the roll angle as

$$I\ddot{\phi} = -\varepsilon\xi(t)(\phi - \eta_y'(0,t)) + M^{\text{ext}},$$

where M^{ext} is to be calculated using deterministic theory.

For example, for a harmonic beam wave $\eta(y,t) = A\cos(\kappa y - \omega t)$ we obtain an equation of the form

$$I\ddot{\phi} + \beta_0\dot{\phi} + \beta_1|\dot{\phi}|\dot{\phi} - \alpha_1\phi^3 + (\alpha_2 + \varepsilon\xi(t))\phi = \varepsilon\kappa A\xi(t)\sin\omega t + F\sin\omega t. \quad (4)$$

Here we added

(1) a linear and nonlinear friction term $\beta_0\dot{\phi} + \beta_1|\dot{\phi}|\dot{\phi}$ (which accounts for the viscous roll damping not captured by potential theory),

(2) a restoring moment of the "softening spring" type $\alpha_2\phi - \alpha_1\phi^3$ which is zero at the critical roll angle $\phi^* := \sqrt{\alpha_2/\alpha_1}$ (locally approximating the righting lever curve),

(3) a further external harmonic forcing $F\sin\omega t$ (F small).

The additional friction and restoring terms of type (1) and (2) used in (4) are partly based on measurements and are widely used and accepted by the ship dynamics community, see e.g. Price and Bishop [15, p. 295], Faltinsen [5, p. 97–98] and Wendt [22, p. 44].

In deep water ($D = +\infty$) Airy's relation becomes the *dispersion relation* $\omega^2 = g\kappa$, and the assumption

$$\kappa A = \frac{\omega^2}{g}A \ll 1$$

is natural. Dropping the corresponding term in (4) we recover the model previously derived from "first principles" by Thompson, Rainey and Soliman [21].

We thus have (after some rescaling) arrived at what we call the *British model*

$$\ddot{\phi} + \beta_0\dot{\phi} + \beta_1|\dot{\phi}|\dot{\phi} - \phi^3 + (\alpha + \varepsilon\xi(t))\phi = F\sin\omega t, \quad (5)$$

where $\beta_0 > 0$, $\beta_1 \geq 0$, and $\alpha > 0$.

Investigating the roll motion of a ship with zero speed of advance and subject to harmonic longitudinal (following or head) waves leads to what we call the *Brazilian model* (see Neves, Pérez and Valerio [13])

$$\ddot{\phi} + \beta_0\dot{\phi} + \beta_1|\dot{\phi}|\dot{\phi} - \phi^3 + (\alpha + \varepsilon\xi(t) + A\sin\omega t)\phi = 0 \quad (6)$$

with parametric excitation.

We can hence say that beam sea leads to additive as well as parametric noise in the roll equation, while following and head sea causes parametric noise.

The task is now to perform a systematic study of the equations (5) and (6) in the framework of the theory of random dynamical systems which is partly done in Sect. 4.

Further studies of models of the roll motion including stochasticity were performed by Roberts and Vasta [17] and Moshchuk, Ibrahim and Khasminskii [11, 12].

3 Random Dynamical Systems Methods: A Brief Review

3.1 General Setup

We start with some basic definitions (for a systematic and detailed treatment of the general theory of random dynamical systems see [1]). Random dynamical systems are generated by differential equations with stationary noise input. A *random dynamical system* consists of two ingredients:

(i) The noise is modeled by a measure preserving flow $\theta = (\theta_t)_{t \in \mathbb{R}}$ on a probability space $(\Omega, \mathcal{C}, \mathbb{P})$.

(ii) The dynamics itself is given by a measurable mapping

$$f : \mathbb{R}^{(+)} \times \Omega \times \mathbb{R}^d \to \mathbb{R}^d, \quad (t, \omega, x) \mapsto f(t, \omega)x$$

such that
- $(t, x) \mapsto f(t, \omega)x$ is continuous for fixed ω,
- f satisfies the *cocycle property* $f(0, \omega) = \mathrm{id}_X$ and

$$f(t + s, \omega) = f(s, \theta_t \omega) \circ f(t, \omega)$$

for $t, s \in \mathbb{R}^{(+)}$ and $\omega \in \Omega$.

We say f is a random dynamical system on $\mathbb{R}^d$ over the *metric dynamical system* (Ω, θ).

The cocycle property implies that the *skew product*

$$\Theta_t(\omega, x) = (\theta_t \omega, f(t, \omega)x)$$

defines a measurable (semi–)flow on the product space $\Omega \times \mathbb{R}^d$.

There are two important ways of generating a random dynamical system: stochastic differential equations in the sense of Itô or Stratonovich, and so–called random differential equations.

A stochastic differential equation can be written in the form

$$dx = f(x)dt + \sum_{j=1}^{n} g_j(x)dW_j,$$

where $f, g_1, ..., g_n$ are Lipschitz continuous functions from $\mathbb{R}^d$ to $\mathbb{R}^d$ and $W_1, ..., W_n$ are independent d–dimensional Wiener processes. Then Ω is the path space of the underlying (two–sided) Wiener process, i. e. the space of continuous functions from $\mathbb{R}$ to $\mathbb{R}^{n \times d}$ which take value 0 at 0 equipped with the Wiener measure $\mathbb{P}$, which is invariant under the time shift θ_t defined by $(\theta_t \omega)(s) = \omega(t + s) - \omega(t)$.

The outcome $f(t, \omega)x$ of the cocycle mapping represents the solution of the equation with initial value x after time t under the particular realization

ω of the Wiener process. This "pathwise" existence of a solution is ensured by a perfection technique, see Arnold [1, Chap. 1.3].

A random differential equation is a family of non–autonomous ODE's, where the noise appears as a parameter on the right hand side, i. e. $\dot{x}(t) = F(x(t), \theta_t \omega)$ with $F : \mathbb{R}^d \times \Omega \to \mathbb{R}^d$. Here (Ω, θ) can be any measure preserving flow on a probability space. Again the cocycle mapping f is given by the solution operator for the corresponding integral equation

$$f(t, \omega)x = x + \int_0^t F(f(s, \omega)x, \theta_s \omega)ds.$$

Differential equations with periodic forcing fit also in this setup. Here θ is a rotation on the unit circle endowed with normalized Lebesgue measure.

3.2 Invariant Objects and Random Attractors

In general one can not expect that deterministic objects such as points or subsets of the state space $\mathbb{R}^d$ or measures on $\mathbb{R}^d$ are invariant under all the mappings $f(t, \omega)$. Therefore in the framework of random dynamical systems invariant objects are defined in a different manner.

A *random invariant set* is a family $(A(\omega))_{\omega \in \Omega}$ of subsets of $\mathbb{R}^d$ depending measurably on ω such that $f(t, \omega)A(\omega) = A(\theta_t \omega)$. Of particular interest are compact random invariant sets, i. e. random invariant sets where each $A(\omega)$ is a compact subset of $\mathbb{R}^d$. A random invariant set where each $A(\omega)$ consists of only one point is called a *random fixed point*.

A *random attractor* is a random invariant set which attracts trajectories from its neighborhood, more precisely: An compact random invariant set A is called a *local random attractor*, if there exists a measurable set $B \subset \Omega \times \mathbb{R}^d$, which is forward invariant under the skew product Θ_t and $B(\omega) := \{x \in \mathbb{R}^d : (\omega, x) \in B\}$ is an open neighborhood of $A(\omega)$, such that

$$\lim_{t \to \infty} \text{dist}(f(t, \omega)x(\omega), A(\theta_t \omega)) = 0$$

in probability for every random variable x with $x(\omega) \in B(\omega)$. The random set B is called the *basin of attraction* of A. If $B = \Omega \times \mathbb{R}^d$ is the whole space, then A is called a *global attractor*.

The existence of a random attractor is typically ensured by the construction of forward invariant absorbing random compact sets. That is, if $(C(\omega))_{\omega \in \Omega}$ is a measurable family of compact subsets of $\mathbb{R}^d$ such that $f(t, \omega)C(\omega)$ is contained in the interior of $C(\theta_t \omega)$ for $t > 0$, then

$$A(\omega) = \bigcap_{t \geq 0} f(t, \theta_{-t}\omega)C(\theta_{-t}\omega)$$

defines a random attractor with basin of attraction

$$B(\omega) = \{x : f(t,\omega)x \in C(\theta_t\omega) \text{ for some } t \geq 0\},$$

cf. Ochs [14, Theorem 17] and Ashwin and Ochs [3, Prop. 2.5].

The statistical behavior of random dynamical systems is described by *invariant measures*. An invariant measure is a probability measure μ on $\Omega \times \mathbb{R}^d$ with marginal $\mathbb{P}$ on Ω which is invariant under the skew product Θ_t. Via a disintegration $\mu(d\omega, dx) = \mu_\omega(dx)\mathbb{P}(d\omega)$ this corresponds to a family $(\mu_\omega)_{\omega \in \Omega}$ of probabilities on $\mathbb{R}^d$ with the invariance property $f(t,\omega)\mu_\omega = \mu_{\theta_t\omega}$.

Each invariant random compact set supports an invariant measure. If there exists a global random attractor, then it supports all invariant measures of the system.

3.3 Lyapunov Exponents

In the case of a differentiable random dynamical system (that is, all the mappings $f(t,\omega) : \mathbb{R}^d \to \mathbb{R}^d$ are differentiable) supporting an invariant measure, the asymptotic growth rate of the distance between neighboring trajectories is measured by Lyapunov exponents. Their existence is ensured by the

Multiplicative Ergodic Theorem (Oseledets, 1968). *Let f be a differentiable random dynamical system defined for two–sided time $\mathbb{R}$ with invariant measure μ. Denote the tangent mapping by $Df(t,\omega) : \mathbb{R}^d \to \mathbb{R}^d$ and assume $\int \log^+ \|Df\|^{\pm 1} d\mu < \infty$.*

Then there exist real numbers $\lambda_1 > \lambda_2 > ... > \lambda_p$ with $1 \leq p \leq d$ called Lyapunov exponents and subspaces $E_1(\omega,x), ..., E_p(\omega,x) \subset \mathbb{R}^d$ depending measurably on $(\omega,x) \in \Omega \times \mathbb{R}^d$ called Oseledets spaces, such that the following holds on a Θ_t–invariant subset of $\Omega \times \mathbb{R}^d$ with full μ–measure:

- $E_1(\omega,x) \oplus ... \oplus E_p(\omega,x) = \mathbb{R}^d$,
- $Df(t,\omega)E_k(\omega,x) = E_k(\theta_t\omega, f(t,\omega)x)$ *for $k = 1, ..., p$, and*
- $v \in E_k(\omega,x)$ *if and only if*

$$\lim_{t \to \pm\infty} \frac{1}{t} \log |Df(t,\omega)v| = \lambda_k.$$

Oseledets spaces can be used for the construction of invariant (stable, unstable etc.) manifolds, see [1, Chap. 7].

In the case $\lambda_1 < 0$, i. e. if all Lyapunov exponents are strictly negative, one can show, using an idea of Crauel [4, Proposition 2.1] (who considered the case of a system on a compact manifold), that the invariant measure μ is supported by an invariant random set A with $A(\omega)$ finite. Each element x of $A(\omega)$ has a local stable manifold which is an open neighborhood of x. This means, that $A(\omega)$ is a local attractor with basin of attraction the union of the stable manifolds of its elements.

4 The Roll Motion of a Ship: A Case Study

4.1 Numerical Studies

The Model

The basis of our study of the roll motion ϕ is the class of equations

$$\ddot{\phi} = -U'(\phi) - (\gamma + b|\dot{\phi}|)\dot{\phi} + (\delta_1 + \delta_2\phi)\sin \alpha t + (\varepsilon_1 + \varepsilon_2\phi)\xi_t, \qquad (7)$$

where $\gamma, b, \alpha, \delta_1, \delta_2, \varepsilon_1, \varepsilon_1$ are parameters and ξ_t is some stationary stochastic process, either white noise or sufficiently regular to define a random differential equation. In our numerical studies we use the arctan of a stationary Ornstein–Uhlenbeck process. The potential U is assumes to take the form

$$U(\phi) = \frac{1}{2}\phi^2 - \frac{1}{3}\phi^3 \text{ (one–sided case) resp. } U(\phi) = \frac{1}{2}\phi^2 - \frac{1}{4}\phi^4 \text{ (two–sided case).}$$

The two–sided model includes (5) and (6) and the one–sided model was proposed by Thompson as a generic model for studying one–sided capsize, see [20]. The interpretation of (7) is that 0 is the stable equilibrium of the ship in the absence of seaway, and capsizing occurs if the potential wall is crossed, i. e. if $\phi > 1$ in the one–sided and $|\phi| > 1$ in the two–sided case.

Written as a first order system

$$\dot{\phi} = y$$
$$\dot{y} = -U'(\phi) - (\gamma + b|y|)y + (\delta_1 + \delta_2\phi)\sin \alpha t + (\varepsilon_1 + \varepsilon_2\phi)\xi_t$$

(7) generates a random dynamical system on $\mathbb{R}^2$.

Attractors and Bifurcations

There are several studies of bifurcations for our model for the case of periodic forcing only, i. e. for $\varepsilon_1 = \varepsilon_2 = 0$, see Thompson and coworkers [16, 21, 20]. In the case of no forcing at all ($\delta_1 = \delta_2 = \varepsilon_1 = \varepsilon_2 = 0$) the origin is a local attractor. One can easily show that there exists an absorbing (deterministic) compact set C. This set C is still absorbing for small perturbations, i. e. in the case where $\delta_1, \delta_2, \varepsilon_1$ and ε_2 are sufficiently small and the noise process ξ is uniformly bounded.

In the case of purely periodic excitation $\varepsilon_1 = \varepsilon_2 = 0$ we can just calculate trajectories of the system and look at their asymptotic behavior in order to get information about the long term dynamics. For small δ_1, δ_2 one observes a stable periodic orbit with period equal to the period of the excitation. This corresponds to a "random" attractor with $A(\omega)$ a one–point set (a neighborhood of $A(\omega)$ in the (ϕ, y)–coordinates corresponds to a section in the space of (ϕ, y, t)–coordinates around a point on the limit cycle).

Increasing the magnitude of excitation the basin of attraction (considered as the "safe region") will shrink and develop a fractal boundary. Finally at a certain level of excitation the attractor disappears and all trajectories tend to ∞. For some parameter values one observes a series of bifurcations before the safe region vanishes. Examples are shown in Figures 1 and 2. The first bifurcations are period doublings, which should be also indicated by zeros of the top Floquet exponent on the periodic attractor. In the random case the attractor cannot be visualized as the limit set of a single trajectory, so we need a different method in order to "see" the random attractor. This method is based on heuristic considerations, namely, on the idea that for fixed sufficiently large time T and a fixed realization ω of the underlying stochastic process (which should be "typical" in the sense of the probability measure $\mathbb{P}$), the values of $f(T, \omega)x$ for different initial values x from the basin of attraction (the safe region) $B(\omega)$ should all lie almost on the attractor.

For this reason we do the following: We fix $T \gg 0$ and $h > 0$, which is smaller than one period of the deterministic excitation. Then we fix an $\omega_0 \in \Omega$ and caculate $f(t, \omega_0)x$ for $0 + h$ and $x \in X$, where X is a finite set of initial conditions distributed over the region where we expect the safe basin. Trajectories which escape to infinity are discarded, but for those trajectories which remain bounded, we draw the final part from time T to $T + h$, i. e. we obtain the set

$$L = \{f(t, \omega_0)x : t \in [T, T + h], x \in X^*\}$$

with X^* the set of those $x \in X$ which have bounded forward orbits.

If now the random attractor is a single point $a(\omega)$, we expect $f(t, \omega)x$ for t large to be close to $a(\theta_t\omega)$ for each $x \in X^*$, i. e. the set L will look like a piece $\{a(\theta_t\omega) : T \le t \le T + h\}$ of the orbit of $a(\omega)$.

In the case of a more complicated attractor we expect to see different orbit pieces in L depending on the different initial values $x \in X^*$.

Here we present the results of a case study with the system considered in Figures 1 with additional random forcing, see Figure 3. The calculations for $\varepsilon = 0$ are made in order to show that with our method it is indeed possible to detect bifurcations. For small additional noise $\varepsilon = 0.01$ we observe a similar bifurcation behavior as in the case with purely periodic forcing $\varepsilon = 0$. However, if the noise level increases to $\varepsilon = 0.05$, the random attractor seems to be a single point for all values of δ for which it exists, i. e. there are no bifurcations which "announce" capsizing, as commonly believed by engineers.

For the system $\ddot{\phi} = -\phi + \phi^2 - \dot{\phi}|\dot{\phi}| + \delta \sin 0.8t + \varepsilon \xi_t$ with asymmetric potential and additive forcing, see Figure 2 for the deterministic case, we did not observe any bifurcations for $\varepsilon > 0$, i. e. the attractor seems to be a single random point for all parameter values for which it exists.

Lyapunov Exponents

In deterministic systems bifurcations such as period doublings are typically indicated by a zero Lyapunov exponent (or eigenvalue resp. Floquet expo-

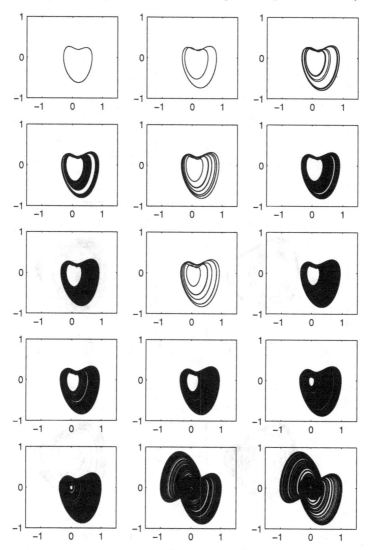

Fig. 1. The limit set of trajectories of $\ddot{\phi} = -\phi + \dot{\phi} - \dot{\phi}|\dot{\phi}| + \delta\phi\sin 0.8t$ with $\delta = 1.33$, 1.386, 1.421, 1.428, 1.43, 1.433, 1.438, 1.443, 1.448, 1.45, 1.451, 1.46, 1.465, 1.47, 1.473. Note that for $\delta < 1.47$ there are two local attractors symmetric to 0 which merge at $\delta = 1.47$. The pictures show only one of these local attractors

nent). In order to see what happens in the random case we have calculated the top Lyapunov exponent for our system (7) in different parameter regimes both with purely periodic and with random forcing. This was simply done by simulating a bounded trajectory and its tangent flow on a sufficiently long time interval and estimating the exponential growth rate of a tangent vector. The results for the system with symmetric potential and multiplicative pe-

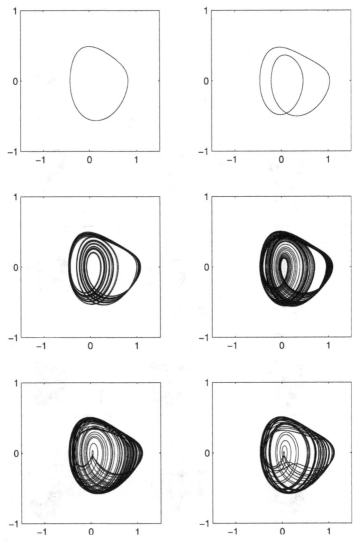

Fig. 2. The limit set of trajectories of $\ddot{\phi} = -\phi + \phi^{\bullet} - \dot{\phi}|\dot{\phi}| + \delta \sin 0.8t$ with $\delta = 0.24$, 0.25, 0.255, 0.2604, 0.2616, 0.2636

riodic forcing, see Figures 1 and 3 for the attractors, are shown in Figure 4. The curve for $\varepsilon = 0$ shows large regions with negative exponents separated by zeros at bifurcation values. If δ is close to the value ≈ 1.47 where capsizing occurs, the top Lyapunov exponent becomes positive, which indicates a chaotic attractor. This is consistent with the calculated attractors shown in Figure 1.

If noise is added the shape of the curve for the top Lyapunov exponent changes considerably. For $\varepsilon = 0.001$ we still observe a negative exponent in the

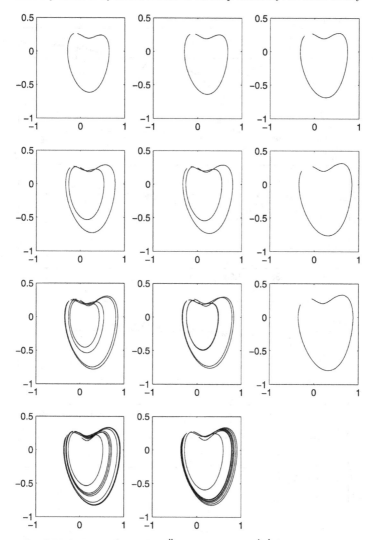

Fig. 3. The "random attractor" of $\ddot{\phi} = -\phi + \phi^{\cdot} - \dot{\phi}|\dot{\phi}| + \delta\phi\sin 0.8t + \varepsilon\xi_t$ with $\delta = 1.33$, 1.386, 1.421, and 1.448 (from top to bottom), and $\varepsilon = 0$, 0.01, and 0.05 (from left to right). For $\delta = 1.448$ and $\varepsilon = 0.05$ the attractor has disappeared, i. e. the ship has capsized

parameter regime between 1.37 and 1.42 after the first period doubling, where the attractor consists of two random points. The other regions with negative exponent corresponding to periodic attractors with higher period vanish. For large δ the exponent still becomes positive, so there seems to exist a chaotic random attractor before capsizing. For $\varepsilon = 0.05$ the top exponent increases almost monotonically and still seems to become positive before capsizing.

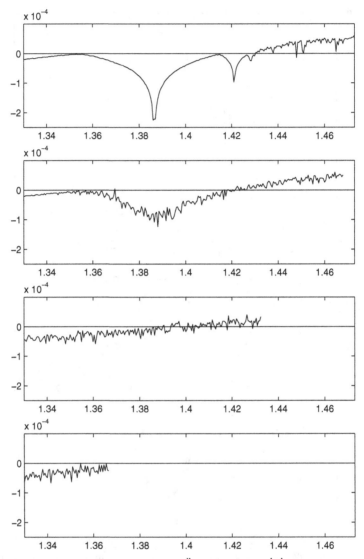

Fig. 4. The top Lyapunov exponent for $\ddot{\phi} = -\phi + \phi^{\cdot} - \dot{\phi}|\dot{\phi}| + \delta\phi\sin 0.8t + \varepsilon\xi_t$ as a function of δ with $\varepsilon = 0$, $\varepsilon = 0.01$, $\varepsilon = 0.05$, and $\varepsilon = 0.1$ (from top to bottom). The curves terminate at those δ values for which the attractor disappears, i. e. for which we have capsizing

Figure 5 shows the top Lyapunov exponent for the system with asymmetric potential and additive forcing. In the deterministic case it shows a similar behavior as in Figure 4. But here adding a small amount of noise makes the exponent remaining negative for all values of δ.

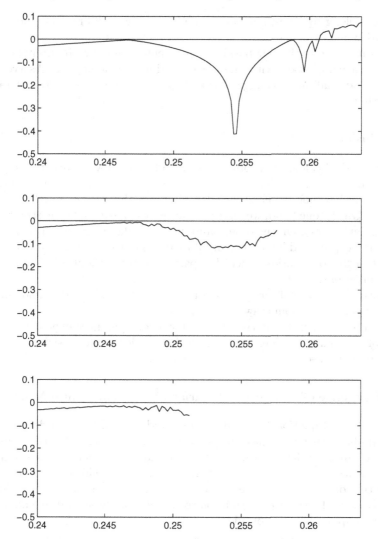

Fig. 5. The top Lyapunov exponent for $\ddot{\phi} = -\phi + \phi^{\cdot} - \dot{\phi}|\dot{\phi}| + \delta \sin 0.8t + \varepsilon \xi_t$ as a function of δ with $\varepsilon = 0$ (top), $\varepsilon = 0.01$ (middle), and $\varepsilon = 0.05$ (bottom). The curves terminate at those δ values for which the attractor disappears, i. e. for which we have capsizing

The Basin of Attraction

We also calculated the safe basin, which is a random set depending on the realization of the noise process, for certain parameter values. In order to do this we have to fix a "typical" realization ω of our stochastic process, which again we chose as the arctan of a stationary Ornstein–Uhlenbeck process, and calculate the forward trajectories for a grid of initial values. The initial values whose trajectories remain bounded for a certain (sufficiently large) amount of time belong by definition to the safe basin. Figure 6 presents an example for this type of calculations.

Figure 7 shows the safe basin in a parameter regime with purely periodic forcing in comparison with the corresponding attractor. We observe that there is no qualitative change of the shape of the basin when the attractor bifurcates.

Conclusions

The numerical calculations of attractors and Lyapunov exponents presented above show that the behavior of our toy model under the influence of (even small) random noise differs considerably from the case of purely periodic forcing. On the one hand, we observe that noise "smears out" the structure of random attractors. On the other hand, in cases where for periodic excitation capsizing is preceded by a series of bifurcations of the attractor, the addition of noise leads to scenarios for which capsizing occurs without being announced by bifurcations.

Future studies will have to take this "capsizing out of the blue sky" in the presence of random seaway into account.

Furthermore, it seems that the behavior of Lyapunov exponents in the case of random noise does not indicate bifurcations and capsizing as in the deterministic case.

4.2 Existence of a Compact Invariant Set in the White Noise Case

If $\xi = dW$ in Equation (7) is white noise and $\varepsilon_1 \neq 0$, then capsizing occurs with probability 1 regardless of how small the coefficients δ_1, δ_2, ε_1 and ε_2 are. We will show that, however, in the two–sided case there exists a compact invariant set, i. e. there is a set of initial conditions which may depend on the future of the noise, whose trajectories do not escape to ∞. For technical reasons we will restrict our attention to the case of purely additive noise and linear damping, i. e. we consider

$$\ddot{\phi} = -\phi + \phi^3 - \gamma\dot{\phi} + (\delta_1 + \delta_2\phi)\sin\alpha t + \varepsilon_1\dot{W} \qquad (8)$$

with $\gamma, \varepsilon_1 > 0$ and $\delta_1, \delta_2, \alpha \in \mathbb{R}$ arbitrary, where $\dot{W}$ denotes white noise.

Existence Proof via Conley Index Theory

The proof uses a topological argument and refers to Conley index theory.

In the first step we eliminate the white noise and transform (9) into a random differential equation. This is done by considering a stationary Ornstein–Uhlenbeck process ζ which solves

$$d\zeta = -\gamma\zeta dt + \varepsilon_1 dW$$

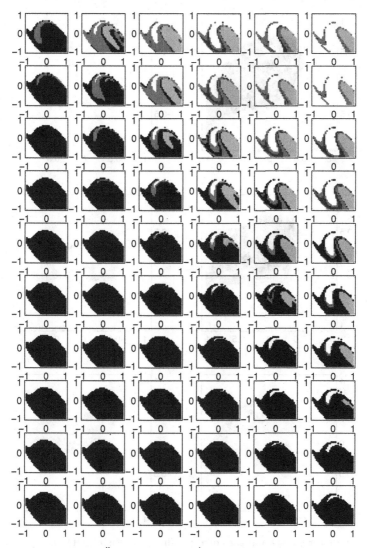

Fig. 6. The safe basin for $\ddot{\phi} = -\phi + \phi^{\cdot} - 0.5\dot{\phi} + \delta \sin 0.8t + \varepsilon \xi_t$ for $\delta = 0.25 + 0.005k$, $k = 1, ..., 10$ (from bottom to top) and $\varepsilon = 0.01n$, $n = 0, ..., 5$ (from left to right). The safe basin is the black region. The grey region consists of those initial values whose trajectories remain bounded a considerable amount of time but then tend to infinity. Note that the safe basin depends on the realization of ξ and moves in time

and setting

$$v = \dot{\phi} - \zeta \quad \Longrightarrow \quad \dot{v} = \ddot{\phi} + \gamma \zeta - \varepsilon_1 \dot{W}.$$

Using (8) we obtain

$$\dot{v} = -\phi + \phi^3 - \gamma v - \gamma \zeta + (\delta_1 + \phi \delta_2) \sin \alpha t + \varepsilon_1 \dot{W} + \gamma \zeta - \varepsilon_1 \dot{W},$$

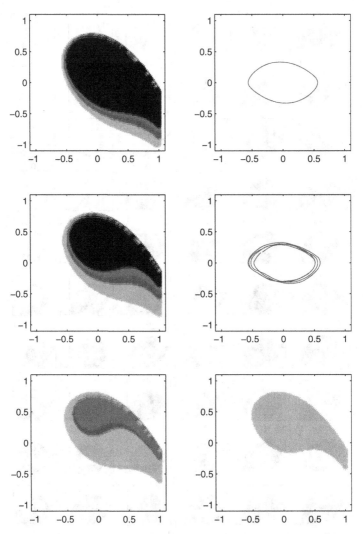

Fig. 7. Attractor (right) and the corresponding safe basin (left) for $\ddot{\phi} = -\phi + \phi^{\cdot} - 0.5\dot{\phi} + \delta \sin 0.8t$ for $\delta = 0.38$ (top) and $\delta = 0.39$ (center). For $\delta = 0.4$ and $\delta = 0.41$ (bottom, left and right) the attractor has disappeared, and the two pictures show the initial values with the longest life time

i. e.

$$\dot{\phi} = v + \zeta$$
$$\dot{v} = -\phi - \gamma v + \phi^3 + (\delta_1 + \phi \delta_2) \sin \alpha t \qquad (9)$$

In the next step we do some scaling. Therefore we choose a stationary stochastic process $\eta = e^{\beta}$ with differentiable trajectories and

$$\beta \le 0, \quad \eta \le \frac{1}{2\gamma|\zeta|}, \quad \text{and} \quad \left|\frac{\dot\eta}{\eta}\right| = |\dot\beta| \le \max\left\{\frac{1}{\delta_2}, \frac{1}{c}\right\}$$

with some sufficiently large constant $c > 0$ which will be specified later. Now introduce new coordinates

$$z = \eta\phi \quad \text{and} \quad w = \eta v.$$

Then we have

$$\dot z = \dot\eta\phi + \eta v + \eta\zeta$$
$$\dot w = \dot\eta v - \eta\phi - \gamma\eta v + \eta\phi^3 + \eta(\delta_1 + \phi\delta_2)\sin\alpha t$$

and hence

$$\dot z = w + \dot\beta z + \alpha\zeta$$
$$\dot w = -z - \gamma w + e^{-2\beta}z^3 + \dot\beta w + \eta(\delta_1 + e^{-\beta}z\delta_2)\sin\alpha t. \tag{10}$$

Now consider the compact rectangle N bounded by the lines

$$L_1 := \{(z, w) : z \ge 0, w \ge 0; \ 2\gamma z + w = c\},$$

$$L_2 := \{(z, w) : z \le 0, w \le 0; \ 2\gamma z + w = -c\},$$

$$M_1 := \{(z, w) : z \ge 0, w \le 0; \ 2\gamma z - w = c\} \text{ and}$$

$$M_2 := \{(z, w) : z \le 0, w \ge 0; \ 2\gamma z - w = -c\}.$$

We will show that on $L = L_1 \cup L_2$ the vector field (for all t and all realizations of the processes ζ and β) points outward of the set N, whereas on $M = M_1 \cup M_2$ it points inward. To show this consider

$$\frac{\partial}{\partial t}(2\gamma z + w)$$
$$= 2\gamma w + 2\gamma\dot\beta z + 2\gamma\eta\zeta - z + e^{-2\beta}z^3 - \gamma w + \dot\beta w + \eta(\delta_1 + e^{-\beta}z\delta_2)\sin\alpha t$$
$$= \gamma w + e^{-2\beta}z^3 - z + \dot\beta(2\gamma z + w) + 2\gamma\eta\zeta + \eta(\delta_1 + e^{-\beta}z\delta_2)\sin\alpha t$$

On L_1 we have

$$\gamma w + e^{-2\beta}z^3 - z = \gamma c + e^{-2\beta}z^3 - (1 + 2\gamma^2)z \ge \gamma c + z^3 - (1 + 2\gamma^2)z.$$

Since the function $z^3 - (1 + 2\gamma^2 + \delta_2)z$ is bounded from below on $\{z \ge 0\}$, we are free to choose $c > 0$ such that this term is greater than $4 + \delta_2 z$. Furthermore, by the choice of β, the term $|\dot\beta(2\gamma z + w)| = |\dot\beta c|$ can be estimated by 1, and $|2\gamma\eta\zeta|$ is also bounded by 1. Finally, we have

$$|\eta(\delta_1 + e^{-\beta}z\delta_2)\sin\alpha t| \le |\eta\delta_1| + |\delta_2 z| \le 1 + |\delta_2 z|.$$

Putting things together this yields

$$\frac{\partial}{\partial t}(2\gamma z + w) \geq 1$$

on L_1. By symmetry we have

$$\frac{\partial}{\partial t}(2\gamma z + w) \leq -1$$

on L_2.

On M_1 we consider analogously

$$\frac{\partial}{\partial t}(2\gamma z - w)$$
$$= 2\gamma w + 2\gamma\dot{\beta}z + 2\gamma\eta\zeta + z - e^{-2\beta}z^3 + \gamma w - \dot{\beta}w - \eta(\delta_1 + e^{-\beta}z\delta_2)\sin\alpha t$$
$$= 3\gamma w - e^{-2\beta}z^3 + z + \dot{\beta}(2\gamma z - w) + 2\gamma\eta\zeta - \eta(\delta_1 + e^{-\beta}z\delta_2)\sin\alpha t$$

Here we have

$$3\gamma w - e^{-2\beta}z^3 + z = -3\gamma c - e^{-2\beta}z^3 - (1 + 6\gamma^2)z \leq -3\gamma c - z^3 + (1 + 6\gamma^2)z,$$

which is smaller than $-4 - \delta_2 z$, if c is chosen large enough. Similar to the considerations on L_1 this leads to $\frac{\partial}{\partial t}(2\gamma z - w) \leq -1$ on M_1, such that the vector field points inward on M_1 and by symmetry the same holds on M_2.

This means that L is an exit set for N, and because the quotient set N/L has homotopy type of a circle, the random Conley index theory developed by Mischaikow and Ochs [10] yields the existence of a non–empty random compact set contained in C, which is invariant under the random dynamical system generated by the transformed equation in (z, w) coordinates. Transforming back into (ϕ, y)–coordinates, we obtain an invariant random compact set for our original equation. We have thus proved the following

Theorem 1. *The random dynamical system generated by the stochastic differential equation (8)*

$$d\phi = y\,dt$$
$$dy = (-\phi + \phi^3 - \gamma y + (\delta_1 + \delta_2\phi)\sin\alpha t)\,dt + \varepsilon_1 dW$$

with $\gamma, \varepsilon_1 > 0$ has a non–empty invariant random compact set. This set supports at least one invariant measure.

Visualization of the Invariant Set

We have no method to approximate the invariant set $A(\omega)$ directly. The main problem is that $A(\omega)$ depends on the past as well as on the future of the realization ω of the underlying white noise process dW.

For this reason we consider the *stable set*

$$S(\omega) = \{(\phi, y) \in \mathbb{R}^2 : f(t, \omega)(\phi, y) \nrightarrow \infty \text{ as } t \to \infty\}$$

(obviously depending on the future of the noise) and the *unstable set*

$$U(\omega) = \{(\phi, y) \in \mathbb{R}^2 : f(t, \omega)(\phi, y) \not\to \infty \text{ as } t \to -\infty\}$$

(depending on the past of the noise), where f denotes the random dynamical system generated by (8). Both $S(\omega)$ and $U(\omega)$ are closed and unbounded, and their intersection gives the random compact invariant set, i. e. $A(\omega) = S(\omega) \cap U(\omega)$.

The idea underlying our numerics is that $S(\omega)$ is some sort of random attractor (although not compact) in the sense that trajectories coming from the past will approach $S(\omega)$. More precisely, given a compact "initial set" $C \subset \mathbb{R}^2$ and a compact "window" $Q \subset \mathbb{R}^2$, we expect the semi–Hausdorff distance between $Q \cap f(t, \theta_{-t}\omega)C$ and $S(\omega)$ to be small for large t. Hence, if the initial set Q is chosen large enough, $Q \cap f(t, \theta_{-t}\omega)C$ will be an approximation for $Q \cap S(\omega)$.

In our numerical studies we chose $Q = [-2, 2]^2$, the region where the dynamics is most interesting for us and where we suppose the invariant set. The set C and its images under f are given by polygons. This proves to be an efficient method to describe the evolution of a subset of $\mathbb{R}^2$ under a dynamical system. Sets with non–empty interior can be represented by their boundary as a polygon.

A polygon P is represented by a sequence of knots (points in $\mathbb{R}^2$) and connecting lines between them. To calculate the image of P under the random dynamical system we have to fix a time interval Δt and realizations $\tilde{\omega}$ of the noise process. Then we calculate $x' = f(\Delta t, \tilde{\omega})x$ for all knots x with the same $\tilde{\omega}$ and define a new polygon P' with knots x'. Afterwards we delete some knots and create new ones adaptively in order to get an even distribution of knots on P'.

If we have $t = n \Delta t$ with $n \in \mathbb{N}$ and we repeat this procedure with $\tilde{\omega} = \theta_{-j\Delta t}\omega$ for $j = n, n-1, ..., 1$, we finally obtain an approximation for $f(t, \theta_{-t}\omega)P$, which in turn should approximate $S(\omega)$ for large enough t.

The unstable set $U(\omega)$ is approximated in a similar way with f replaced by its inverse, i. e. we approximate $f(-t, \theta_t\omega)P$ for some polygon P and t large enough.

An example of our calculations is shown in Figure 8. In Figure 9 we combine $S(\omega)$ and $U(\omega)$ in order to "see" the invariant set $A(\omega)$ as their intersection.

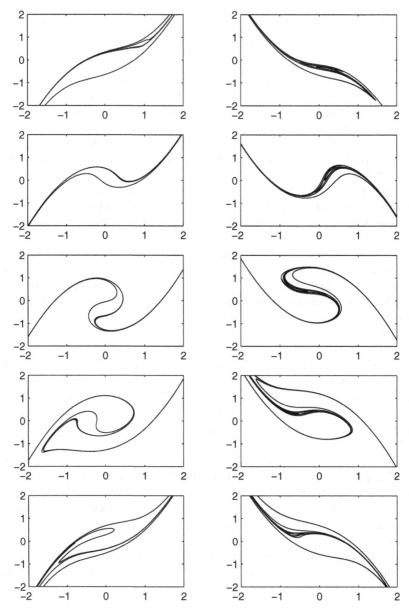

Fig. 8. The stable set $S(\omega)$ (left hand side) and the unstable set $U(\omega)$ (right hand side) for $\ddot{\phi} = -\phi + \dot{\phi} - 0.5\dot{\phi} + (0.3 + \phi)\sin(0.8t) + 0.2\dot{W}$ and its temporal development. The images from top to bottom show realizations for different ω of the form $\omega = \theta_{t_i}\omega$. with $0 = t. < t. < ... < t.$

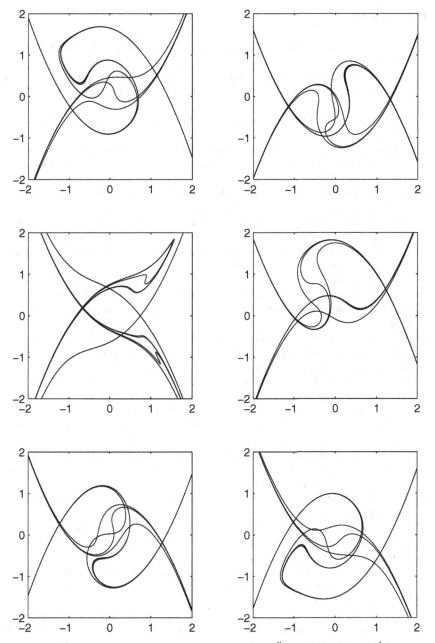

Fig. 9. Stable set $S(\omega)$ and unstable set $U(\omega)$ for $\ddot{\phi} = -\phi + \phi^{\cdot} - 0.5\dot{\phi} + (0.3 + \phi)\sin(0.8t) + 0.2\dot{W}$ and $\omega = \theta_{t_i}\omega$. with $0 = t. < ... < t..$. The intersection of both sets gives the invariant random compact set $A(\omega)$

References

1. L. Arnold. *Random dynamical systems.* Springer-Verlag, Berlin Heidelberg New York, 1998.
2. L. Arnold, I. Chueshov, and G. Ochs. Stability and capsizing of ships in random sea – a survey. Report No. 464, Institut für dynamische Systeme, Universität Bremen, June 2003.
3. P. Ashwin and G. Ochs. Convergence to local random attractors. *Dynamical Systems*, to appear, 2003.
4. H. Crauel. *Random dynamical systems*, PhD thesis, Bremen, 1987.
5. O. M. Faltinsen. *Sea loads on ships and offshore structures.* Cambridge University Press, 1998.
6. Institute of London Underwriters. *Hull casualty statistics. IUMI Conf. Paris.* The Institute of London Underwriters, 1997.
7. International Maritime Organization. *Code on intact stability for all types of ships covered by IMO instruments. Resolution A.749(18).* IMO, London, 1995.
8. F. John. On the motion of floating bodies. I. *Communications on Pure and Applied Mathematics*, 2:13–57, 1949.
9. E. Kreuzer and M. Wendt. Ship capsizing analysis using advanced hydrodynamic modelling. *Phil. Trans. R. Soc. Lond. A*, 358:1835–1851, 2000.
10. K. Mischaikow and G. Ochs. A Conley index for random homeomorphisms. Preprint, 2002.
11. N. K. Moshchuk, R. A. Ibrahim, R. Z. Khasminskii, and P. L. Chow. Asymptotic expansion of ship capsizing in random sea waves – I. First-order approximation. *Int. J. Non-Linear Mechanics*, 30:727–740, 1995.
12. N. K. Moshchuk, R. A. Ibrahim, R. Z. Khasminskii, and P. L. Chow. Asymptotic expansion of ship capsizing in random sea waves – II. Second-order approximation. *Int. J. Non-Linear Mechanics*, 30:741–757, 1995.
13. M. Neves, N. Pérez, and L. Valerio. Stability of small fishing vessels in longitudinal waves. *Ocean Engineering*, 26:1389–1419, 1999.
14. G. Ochs. Random attractors: Robustness, numerics and chaotic dynamics. In B. Fiedler (ed.): *Ergodic Theory, Analysis, and Efficient Simulation of Dynamical Systems*, pages 1–30, 2001
15. W. G. Price and R. E. Bishop. *Probabilistic theory of ship dynamics.* Chapman and Hall, London, 1974.
16. R. C. T. Rainey and J. M. T. Thompson. The transient capsize diagram – a new method of quantifying stability of waves. *Journal of Ship Research*, 35:58–62, 1991.
17. J. B. Roberts and M. Vasta. Markov modelling and stochastic identification for nonlinear ship rolling in random waves. *Phil. Trans. R. Soc. Lond. A*, 358:1917–1941, 2000.
18. K. J. Spyrou and J. M. T. Thompson. The nonlinear dynamics of ship motions: a field overview and some recent developments. *Phil. Trans. R. Soc. Lond. A*, 358:1735–1760, 2000.
19. K. J. Spyrou and J. M. T. Thompson, editors. *The nonlinear dynamics of ships*, volume 358 of *Phil. Trans. Royal Soc. London, Series A (Theme Issue)*. The Royal Society, 2000.
20. J. M. T. Thompson. Designing against capsize in beam seas: recent advances and new insights. *ASME Appl. Mech. Rev.*, 50:307–325, 1997.

21. J. M. T. Thompson, R. C. T. Rainey, and M. S. Soliman. Mechanics of ship capsize under direct and parametric wave excitation. *Phil. Trans. R. Soc. Lond. A*, 338:471–490, 1992.
22. M. Wendt. *Zur nichtlinearen Dynamik des Kenterns intakter Schiffe im Seegang.* VDI Verlag, Düsseldorf, 2000.

Analysis of Algorithms by the Contraction Method: Additive and Max-recursive Sequences

Ralph Neininger[*][1] and Ludger Rüschendorf[2]

- Department of Mathematics, J. W. Goethe University
  Robert-Mayer-Str. 10, 60325 Frankfurt a. M., Germany
  `neiningr@math.uni-frankfurt.de`
- Department of Mathematics, University of Freiburg
  Eckerstr. 1, 79104 Freiburg, Germany
  `ruschen@stochastik.uni-freiburg.de`

Summary. In the first part of this paper we give an introduction to the contraction method for the analysis of additive recursive sequences of divide and conquer type. Recently some general limit theorems have been obtained by this method based on a general transfer theorem. This allows to conclude from the recursive structure and the asymptotics of first moment(s) the limiting distribution. In the second part we extend the contraction method to max-recursive sequences. We obtain a general existence and uniqueness result for solutions of stochastic equations including maxima and sum terms. We finally derive a general limit theorem for max-recursive sequences of the divide and conquer type.

Keywords. Analysis of algorithms, parallel algorithms, limit laws, recurrence, probability metric, limit law for maxima.

1 Introduction to the Contraction Method

The analysis of algorithms is a rapidly expanding area of analysis. Since the introduction of the average case analysis in Knuth [4] there have been developed several approaches to limit laws for various parameters of recursive algorithms, random trees and combinatorial structures. The contraction method is a probabilistic technique of analysis with a broad range of applications which supplements the analytic techniques (generating functions) and the other probabilistic techniques like martingales or branching processes. The contraction technique was first introduced for the analysis of Quicksort in Rösler [15] and further on developed independently in Rösler [16] and Rachev and Rüschendorf [13], as well as in Neininger and Rüschendorf [9, 10], see also the survey article Rösler and Rüschendorf [18]. It has been successfully applied to a broad range of algorithms (see Neininger [6, 7], Rösler [17], Hwang and Neininger [2], and Neininger and Rüschendorf [9]).

---

[*]Research supported by an Emmy Noether Fellowship of the DFG.

The idea of the contraction method is to reduce the analysis of an algorithm to the study of contraction properties of transformations associated to the algorithm, and then to use some variant of the Banach fixed point theorem. We explain some general aspects of this method at the example of the Quicksort algorithm.

Let L_n denote the number of comparisons of the Quicksort algorithm to sort n randomly permuted real numbers, see, e.g. Mahmoud [5]. Then

$$\ell_n = EL_n = 2n \log n + (2\gamma - 4)n + O(\ln n), \tag{1}$$

$$\gamma \text{ the Euler constant}$$

$$\text{and} \quad \sigma_n^2 = \text{Var}(L_n) = \left(7 - \frac{2\pi^2}{3}\right) n^2 + O(n \ln n). \tag{2}$$

Regnier [14] established that $Z_n = \frac{L_n - \ell_n}{n+1}$ is a L_2-bounded martingale and, therefore, a.s. convergence to some rv's Z holds:

$$Z_n \to Z \text{ a.s.} \tag{3}$$

In order to determine the distribution of Z it is useful to consider the recursive structure of L_n. By an obvious simple argument we have

$$L_n \stackrel{d}{=} L_{I_n} + \bar{L}_{n-1-I_n} + n - 1, \tag{4}$$

where $\bar{L}_k$ are independent copies of L_k, I_n is uniformly distributed on $\{0, \ldots, n-1\}$ and independent of (L_k), (I_k). I_n is the size of the subgroup which is smaller than the first pivot element chosen by the Quicksort algorithm. After normalization, $Y_n = \frac{L_n - \ell_n}{n}$ satisfies the recursion

$$Y_n \stackrel{d}{=} \frac{I_n}{n} Y_{I_n} + \frac{n-1-I_n}{n} \bar{Y}_{n-1-I_n} + c_n(I_n), \tag{5}$$

$$\text{with } c_n(j) = \frac{n-1}{n} + \frac{1}{n}(\ell_j + \ell_{n-1-j} - \ell_n),$$

where $(\bar{Y}_n)$ is a distributional copy of (Y_n). With $c(x) := 2x \log x + 2(1-x) \log(1-x) + 1$ it is easy to see that: $\sup_{x \in (0,1]} |c_n([nx]) - c(x)| \leq \frac{4}{n} \log n + O\left(\frac{1}{n}\right)$. Choosing w.l.g. some version of I_n such that

$$\frac{I_n}{n} \to \tau \text{ a.s.}, \tag{6}$$

where τ is uniformly distributed on $[0, 1]$ we obtain from (3), (5) that the limit Y of Y_n exists a.s. and satisfies the *limit equation*:

$$Y \stackrel{d}{=} \tau Y + (1 - \tau)\bar{Y} + c(\tau). \tag{7}$$

There exists exactly one solution of the limiting equation (7) in the class $\mathcal{M}_2(0)$ of probability measures on $\mathbb{R}$ with mean zero and with finite variance. To that purpose we define the transformation $T : \mathcal{M}_2(0) \to \mathcal{M}_2(0)$ by

$$T(P) = \mathcal{L}(\tau Y + (1 - \tau)\bar{Y} + c(\tau)) \tag{8}$$

if $P = \mathcal{L}(Y)$. The operator T is closely related to the Quicksort algorithm. It is an asymptotic approximation of the recursion operator in (5). T is a contraction operator w.r.t. the minimal ℓ_2-metric defined for probability measures P, Q by

$$\ell_2(P,Q) = \inf \left\{ (E(X - Y)^2)^{1/2} ; X \stackrel{d}{=} P, Y \stackrel{d}{=} Q \right\} ; \tag{9}$$

$$\ell_2(TP,TQ) \le \sqrt{\tfrac{2}{3}} \ell_2(P,Q). \tag{10}$$

For the proof of (10) let $X_i \stackrel{d}{=} P$, $Y_i \stackrel{d}{=} Q$, $i = 1, 2$ be i.i.d. copies of P, Q such that $E(X_i - Y_i)^2 = \ell_2^2(P,Q)$. Then

$$\begin{aligned}
\ell_2^2(TP,TQ) &\le E(\tau X_1 + (1 - \tau)X_2 + c(\tau) - [\tau Y_1 + (1 - \tau)Y_2 + c(\tau)])^2 \\
&= E[\tau^2(X_1 - Y_1)^2 + (1 - \tau)^2(X_2 - Y_2)^2] \\
&= 2E\tau^2 \ell_2^2(P,Q) = \tfrac{2}{3} \ell_2^2(P,Q). \tag{11}
\end{aligned}$$

Thus by Banach's fixed point theorem the limiting equation (7) has a unique solution in $\mathcal{M}_2(0)$. The uniqueness of the solution of the limiting equation (7) implies that Y_n converges in distribution to Y

$$Y_n \xrightarrow{\mathcal{D}} Y, \tag{12}$$

where Y is the unique solution of the limiting fixed point equation (7) which is called the *Quicksort-distribution*.

The contraction method allows to extend this type of convergence argument to a general class of recursive algorithms. It simultaneously also allows to prove the essential convergence step (3) in the argument above without reference to a martingale argument as above. This is of considerable importance since a related martingale structure has been found only in few examples of recursive algorithms.

In section two of this paper we review some recent developments of the contraction method for additive recursive sequences of divide and conquer type. In the final part of the paper we develop some new tools which are basic for an extension of the contraction method to recursive sequences of divide and conquer type which are based on maxima (like parallel search algorithms). Similar as the additive recursive algorithms are 'relatives' of the classical central limit theorem for sums the max-based recursive algorithms can be considered as relatives of the classical central limit theorem for maxima.

2 Limit Theorem for Divide and Conquer Algorithms

In the recent paper Neininger and Rüschendorf [9] a general limit theorem has been derived for recursive algorithms and combinatorial structures by means of the contraction method. In comparison to the introductory example in section 1 the main progress in that paper is a general transfer theorem which allows to establish a limit law on the basis of the recursive structure and using the asymptotics of the first moment(s) of the sequence. Thus the strong information by the martingale structure can be replaced by the information on first moment(s). For a lot of examples of algorithms this information on moments is available by highly developed analytical methods.

A common type of univariate recursions (Y_n) of the divide and conquer type is of the following form:

$$Y_n \overset{d}{=} \sum_{r=1}^{K} Y^{(r)}_{I^{(n)}_r} + b_n, \quad n \geq n_0 \tag{13}$$

with $(Y_n^{(1)}), \ldots, (Y_n^{(K)})$, $(I^{(n)}, b_n)$ independent $Y_j^{(r)} \overset{d}{=} Y_j$, $P(I_r^{(n)} = n) \to 0$ and $\mathrm{Var}(Y_n) > 0$ for $n \geq n_1$. $I_r^{(n)}$ describe subgroup sizes of the divide and conquer algorithm and b_n is a toll function for the splitting into and merging of K smaller problems.

The analysis of the asymptotics of (Y_n) is based on the Zolotarev metric ζ_s on $\mathcal{M}$ the set of all probability measures on $\mathbb{R}^1$ defined by (see Zolotarev [19])

$$\zeta_s(P,Q) = \sup_{f \in \mathcal{F}_s} |Ef(X) - Ef(Y)|, \tag{14}$$

where $\mathcal{L}(X) = P$, $\mathcal{L}(Y) = Q$, and $\mathcal{F}_s = \{f \in C^{(m)}(\mathbb{R}); \|f^{(m)}(x) - f^{(m)}(y)\| \leq |x-y|^\alpha\}$ with $s = m + \alpha, 0 < \alpha \leq 1, m \in \mathbb{N}_0$. Finiteness of $\zeta_s(\mathcal{L}(X), \mathcal{L}(Y))$ is guaranteed if X, Y have identical moments of orders $1, \ldots, m$ and finite absolute moments of order s. Since ζ_s is of main interest for $s \leq 3$, we introduce the following subspaces of $\mathcal{M}_s$ – the set of measures with finite s-th moments – to obtain finiteness of ζ_s. Define $\mathcal{M}_s(\mu)(\mathcal{M}_s(\mu, \sigma^2))$ for $1 < s \leq 2$ $(2 < s \leq 3)$ to be the elements in $\mathcal{M}_s$ with fixed first moment μ (resp. also fixed variance σ^2) and define $\mathcal{M}_s^*$ to be identical to $\mathcal{M}_s$ for $0 < s \leq 1$, to $\mathcal{M}_s(\mu)$ for $1 < s \leq 2$ and to $\mathcal{M}_s(\mu, \sigma^2)$ for $2 < s \leq 3$, where μ, σ^2 are fixed in the context.

An important property of ζ_s for the contraction method is that

$$\zeta_s(X + Z, Y + Z) \leq \zeta_s(X, Y) \quad \text{and} \quad \zeta_s(cX, cY) = |c|^s \zeta_s(X, Y) \tag{15}$$

for all Z independent of X, Y and $c \in \mathbb{R} \setminus \{0\}$, whenever these distances are finite. ζ_s convergence implies weak convergence.

For the limiting analysis of Y_n we need a stabilization condition for the recursive structure and a contraction condition for the limiting fixed-point equation; in more detail: Assume for functions $f, g : \mathbb{N}_0 \to \mathbb{R}_0^+$ with $g(n) > 0$ for $n \geq n_1$ we have the following *stabilization condition in L_s*

$$\left(\frac{g(I_r^{(n)})}{g(n)}\right)^{1/2} \to A_r^*, \quad r = 1, \dots, K \quad \text{and}$$

$$\frac{1}{g^{1/2}(n)}\left(b_n - f(n) + \sum_{r=1}^K f(I_r^{(n)})\right) \to b^*, \tag{16}$$

as well as the *contraction condition*

$$E \sum_{r=1}^K |A_r^*|^s < 1. \tag{17}$$

Then the following limit theorem is obtained by the contraction method (see Neininger and Rüschendorf [9, Theorem 5.1]).

Theorem 1. *Let (Y_n) be s-integrable and satisfy the recursive equation (13) and let f, g satisfy the stabilization condition (16) and the contraction condition (17) for some $0 < s \leq 3$. Furthermore, in case $1 < s \leq 3$ assume the moment convergence condition*

$$EY_n = f(n) + o(g^{1/2}(n)) \quad \text{if } 1 < s \leq 2 \quad \text{and}$$

$$EY_n = f(n) + o(g^{1/2}(n)), \text{Var}(Y_n) = g(n) + o(g(n)) \quad \text{if } 2 < s \leq 3. \tag{18}$$

Then $\dfrac{Y_n - f(n)}{g^{1/2}(n)} \xrightarrow{\mathcal{D}} X$, *where X is the unique fixed-point of*

$$X \stackrel{d}{=} \sum_{r=1}^K A_r^* X^{(r)} + b^* \tag{19}$$

in $\mathcal{M}_s^*$ (with $\mu = 0, \sigma^2 = 1$), where $(A_1^*, \dots, A_K^*, b^*)$, $X^{(1)}, \dots, X^{(K)}$ are independent, $X^{(r)} \stackrel{d}{=} X$.

Remark 2. *a) For the proof of Theorem 1 one gets from the moment convergence condition (18) and the stabilization condition (16) the form of the limiting equation (19). The existence of a unique fixed point of (19) follows from the contraction condition (17) by Banach's fixed point theorem. From the regularity properties of ζ_s in (15) we can argue that the contraction property in the limiting equation can be carried over to the recursive sequence.*

b) Note that in the case that the conditions are satisfied for $0 < s \leq 1$ we do not need any information on the asymptotics of moments; for $1 < s \leq 2$ the asymptotics of the first moment is needed. The case $0 < s \leq 1$ arises for example for limit equations of the form

$$X \stackrel{d}{=} \tfrac{1}{\sqrt{2}} X + \tfrac{1}{\sqrt{2}} \mathcal{N}(0,1), \tag{20}$$

with the standard normal distribution as unique solution, or of the form

$$W \overset{d}{=} UW + U, \tag{21}$$

U uniformly distributed on $(0,1)$, with the Dickman distribution as unique solution. The Dickman distribution arises e.g. as a limit in the context of the Find algorithm. For normal limits, besides (20), the case $2 < s \leq 3$ is typical. Then typically the minimal ℓ_p-metrics (see (28)) cannot be used directly to derive normal limit laws directly.

c) *If the contraction method applies for $s = 1$ then Theorem 1 applied with ζ_1 yields the asymptotics of the first order moment. If it applies for $s = 2$, then one needs asymptotics of the first moment and obtains the asymptotics of the second moment.*

d) *A large class of examples for the application of Theorem 1 to the asymptotics of recursive algorithms has been established. Note that there are also several variants of this basic theorem (to the multivariate case, weighted recursions, random number of components, alternative contraction conditions, degenerative limits, ...). In particular one gets a classification of algorithms according to their contraction behavior. To get an impression of the range of application we give a list of established examples (without giving detailed references):*

1) *$\mathcal{M}_s, 0 < s \leq 1$. Examples contain: FIND comparisons, number of exchange steps, Dickman, Multiple Quickselect, Bucket selection, Quicksort with error (number of inversions), leader election (flips), skip lists (size), ideals in forest poset, distances in random binary search trees (rBST), minimum spanning trees in rBST, random split tree (large toll).*

2) *$\mathcal{M}_s(\mu), 1 < s \leq 2$. Quicksort (comparisons, exchanges), internal path length in quad trees, m-ary search trees, median search trees and recursive trees, Wiener index in rBST and recursive trees. Yaglom's exponential limit law, random split trees for moderate toll.*

3) *$\mathcal{M}_s(\mu, \sigma^2), 2 < s \leq 3$. Quicksort (rec. calls), patterns in trees, size in m-ary search trees, size and path length in tries, digital search trees, and Patricia trees, merge sort (comparisons top-down version), vertices with outdegrees in recursive trees, random split trees with small toll.*

For these and related examples see Neininger and Rüschendorf [9], Hwang and Neininger [2], and Neininger [6, 7].

e) *In Neininger and Rüschendorf [9] it has been shown that one can derive from the convergence results in the Zolotarev metric several local and global limit theorems. It is also possible to obtain rate of convergence results. In Neininger and Rüschendorf [8] it is shown that the convergence rate of the Quicksort algorithm is w.r.t. the Zolotarev metric ζ_3 of the exact order $\frac{\ln n}{n}$.*

3 Contraction and Fixed Point Properties with Maxima

In this section we extend the analysis of algorithms defined via sums as in (13) to recursive algorithms including maximum and sum terms. The analysis in section 2 based on the Zolotarev metric ζ_s would go through in this case if one could find a metric μ_s which is not only regular of order s for sums as in (15) but also simultaneously for maxima too, i.e.

$$\begin{aligned}
\mu_s(X \vee Z, Y \vee Z) &\leq \mu_s(X, Y) \quad \text{and} \\
\mu_s(cX, cY) &= |c|^s \mu_s(X, Y).
\end{aligned} \tag{22}$$

It was however shown in Rachev and Rüschendorf [12] that only trivial metrics may have this doubly ideal property.

For the central limit theorem for maxima the weighted Kolmogorov metric ϱ_s, defined by

$$\varrho_s(X, Y) = \sup_x |x|^s |F_X(x) - F_Y(x)| \tag{23}$$

is max-regular of order s for real rv's X, Y, i.e. it satisfies (22) and has been used for deriving limit theorems. But for recursions including also additive terms ϱ_s is not particular well suited (see Rachev and Rüschendorf [13] and Cramer [1]).

Limiting distributions of max-recursive sequences will typically be identified as unique solutions in some subclass of $\mathcal{M}$ of stochastic equations of the form:

$$X \overset{d}{=} \bigvee_{r=1}^{K} (A_r X_r + b_r), \tag{24}$$

where (X_r) are i.i.d. copies of X and $(A_r, b_r)_{1 \leq r \leq K}$ are random coefficients independent of (X_r). The right hand side of (24) induces an operator $T : \mathcal{M} \to \mathcal{M}$ defined for $Q \in \mathcal{M}$ and $X \overset{d}{=} Q$ by

$$TQ = TX \overset{d}{=} \mathcal{L} \left(\bigvee_{r=1}^{K} (A_r X_r + b_r) \right). \tag{25}$$

If A_r, b_r have absolute s-th moments and $\mathcal{L}(X) \in \mathcal{M}_s$ then also TX has absolute s-th moments. So T can be considered as operator $\mathcal{M}_s \to \mathcal{M}_s$ in this case.

We next establish that the minimal ℓ_s-metric is well suited for the analysis of equations as in (24) although it is not doubly ideal of order s. We need the following simple lemma.

Lemma 1. For all $a, b, c, d \in \mathbb{R}$ and $s > 0$ holds true:

$$|a \vee b - c \vee d|^s \leq |a - c|^s + |b - d|^s. \tag{26}$$

Proof. W.l.g. we assume that $b < a$, i.e., $a \vee b = a$ and $c < d$; otherwise we would have $c \vee d = c$ and so the left hand side of (26) is $|a - c|^s$ while the right hand side is $|a - c|^s + |b - d|^s$.

Furthermore, by symmetry we assume w.l.g. $a < d$. Then the left hand side of (26) is $|a - d|^s$. Noting that $|a - d|^s \le |b - d|^s$ since $b < a < d$ the result follows. $\qquad\square$

Define as usual the L_s-norm by

$$L_s(X,Y) = \begin{cases} (E|X - Y|^s)^{1/s}, & 1 \le s < \infty \\ E|X - Y|^s, & 0 < s < 1 \end{cases} \tag{27}$$

and the minimal L_s-metric ℓ_s by

$$\ell_s(P,Q) = \inf\{L_s(X,Y); X \overset{d}{=} P, Y \overset{d}{=} Q\}. \tag{28}$$

Then we obtain the following contraction property of T.

Proposition 3. *If (X_r) are i.i.d., $X_r \overset{d}{=} X_1$, (Y_r) are i.i.d., $Y_r \overset{d}{=} Y_1$ and A_r s-integrable, $1 \le r \le K$, then for $0 < s < \infty$*

$$a) \qquad L_s\left(\bigvee_{r=1}^{K}(A_r X_r + b_r), \bigvee_{r=1}^{K}(A_r Y_r + b_r)\right)$$

$$\le \left(E\sum_{r=1}^{K}|A_r|^s\right)^{1/s \wedge 1} L_s(X_1, Y_1). \tag{29}$$

b) For the operator T defined in (25) holds

$$\ell_s(TP, TQ) \le \left(E\sum_{r=1}^{K}|A_r|^s\right)^{1/s \wedge 1} \ell_s(P, Q). \tag{30}$$

Proof. a) Consider the case $1 \le s$. Then from induction we get by Lemma 1

$$L_s^s\left(\bigvee_{r=1}^{K}(A_r X_r + b_r), \bigvee_{r=1}^{K}(A_r Y_r + b_r)\right)$$

$$= E\left|\bigvee_{r=1}^{K}(A_r X_r + b_r) - \bigvee_{r=1}^{K}(A_r Y_r + b_r)\right|^s$$

$$\le \sum_{r=1}^{K}E|A_r(X_r - Y_r)|^s$$

$$= \sum_{r=1}^{K}E|A_r|^s L_s^s(X_1, Y_1).$$

The case $0 < s < 1$ is similar.

b) Choose (X_r) i.i.d., $X_r \overset{d}{=} P$ and (Y_r) i.i.d., $Y_r \overset{d}{=} Q$ such that $L_s(X_r, Y_r) = \ell_s(P, Q)$, $1 \leq r \leq K$. Then

$$\ell_s(TP, TQ) \leq L_s \left(\bigvee_{r=1}^{K} (A_r X_r + b_r), \bigvee_{r=1}^{K} (A_r X_r + b_r) \right)$$

$$\leq \left(\sum_{r=1}^{K} E|A_r|^s \right)^{1/s \wedge 1} L_s(X_1, Y_1)$$

$$= \left(\sum_{r=1}^{K} E|A_r|^s \right)^{1/s \wedge 1} \ell_s(P, Q). \qquad \square$$

Remark 4. *Note that inequality (29) holds more generally without any independence assumption on (X_r, Y_r) and thus may be used to analyze a more general class of stochastic equations. The b_r do not enter the contraction estimate in (29), (30).*

As a consequence we next obtain an existence and uniqueness result for the stochastic equation (24). For $\mu_0 \in \mathcal{M}$ define

$$\mathcal{M}_s(\mu_0) = \{\mu \in \mathcal{M}; \ell_s(\mu, \mu_0) < \infty\}, \qquad (31)$$

the equivalence class of μ_0 w.r.t. ℓ_s. If $\mu_0 \in \mathcal{M}_s$, then $\mathcal{M}_s(\mu_0) = \mathcal{M}_s$.

Theorem 5. *Let for some $s > 0$ the coefficients A_r, b_r be s-integrable and $\mu_0 \in \mathcal{M}$ such that $\zeta = E \sum_{r=1}^{K} |A_r|^s < 1$ and $\ell_s(\mu_0, T\mu_0) < \infty$. Then the stochastic equation $X \overset{d}{=} \bigvee_{r=1}^{K} (A_r X_r + b_r)$ has a unique solution in $\mathcal{M}_s(\mu_0)$.*

Proof. Define for $n \geq 1$, $\mu_n = T\mu_{n-1} = T^n \mu_0$. Note that, by induction, $\ell_s(\mu_0, T\mu_0) < \infty$ implies $\ell_s(\mu_n, T\mu_{n+p}) < \infty$ for all $n \geq 0$, $p \geq 1$. Then by Proposition 3 using the triangle inequality for ℓ_s we obtain

$$\ell_s(\mu_n, \mu_{n+p}) \leq \sum_{i=0}^{p-1} \ell_s(\mu_{n+i}, \mu_{n+i+1})$$

$$\leq \ell_s(\mu_0, \mu_1) \sum_{i=0}^{p-1} \zeta^{n+i} \leq \ell_s(\mu_0, \mu_1) \frac{\zeta^n}{1 - \zeta}$$

$$\to 0 \quad \text{as } \ell_s(\mu_0, \mu_1) < \infty.$$

Therefore, (μ_n) is a Cauchy-sequence in the complete metric space $(\mathcal{M}_s(\mu_0), \ell_s)$. Any limiting point is a fixed point of T by Banach's fixed point theorem. For the uniqueness let $\mu, \nu \in \mathcal{M}_s(\mu_0)$ be fixed points of T. Then $\ell_s(T\mu, T\nu) \leq \zeta^{1/s \wedge 1} \ell_s(\mu, \nu)$ and thus $\ell_s(\mu, \nu) = 0$ and $\mu = \nu$. $\qquad \square$

Remark 6. *a) Jagers and Röscler [3] recently obtained a general existence result for equations of the form $X \overset{d}{=} \vee_r A_r X_r$ by relating them to solutions of the additive form $W \overset{d}{=} \sum_r A_r^\alpha W_r$. This additive equation has been well studied.*

b) If $\mu_0 \in \mathcal{M}_s$ then the condition $\ell_s(\mu_0, T\mu_0) < \infty$ is fulfilled. So under the contraction condition $\zeta < 1$ there exists a unique fixed point of T in $\mathcal{M}_s$. But there may be further fixed points not in $\mathcal{M}_s$ but in some $\mathcal{M}_s(\mu_0)$ without finite absolute moments of order s. So, for example the stochastic equation

$$X \overset{d}{=} \tfrac{1}{2}X_1 \vee \tfrac{1}{2}X_2$$

has the (trivial) solution $X = 0$ which is in $\mathcal{M}_s$. The contraction factor is $\zeta = (\tfrac{1}{2})^{s-1}$ w.r.t. ℓ_s which is smaller than 1 for any $s > 1$. The extreme value distribution with distribution function $F(x) = e^{-x^{-1}}, x \geq 0$ is a further (nontrivial) fixed point of this equation without finite first moment. In fact a basic result of extreme value theory says that any nondegenerate max-stable distribution is one of the three classical types of extreme value distributions (Gumbel, Weibull, Fréchet). Recall that a distribution function G is called max-stable if for all $n \in \mathbb{N}$ there exist $a_n > 0$, $b_n \in \mathbb{R}$ such that $G^n(\frac{x}{a_n} + b_n) = G(x), x \in \mathbb{R}$ i.e., a random variable $X \overset{d}{=} G$ satisfies the stochastic equations of the form

$$X \overset{d}{=} \bigvee_{r=1}^{n}(a_n X_r - b_n), \quad n \in \mathbb{N}. \tag{32}$$

This characterization yields uniqueness without any moment considerations but uses a system of stochastic equations instead of only one equation as above.

c) Central limit theorem. *As consequence of Propositions 3 and Theorem 5 one gets an easy proof of the central limit theorem for maxima. (For a general discussion of this topic see Zolotarev [19] and Rachev [11]).*
Let $F(x) = F_{Y_1}(x) = e^{-x^{-\alpha}}, x \geq 0$, be an extreme value distribution of first type and let (X_r) be an i.i.d. sequence with tail condition $\ell_s(X_1, Y_1) < \infty$ for some $s > \alpha$. Then for the maxima sequence $M_n := \max\{X_1, \ldots, X_n\}$ holds:

$$\ell_s(n^{-1/\alpha}M_n, Y_1) \to 0. \tag{33}$$

For the proof note that Y_1 is a solution of the stochastic equation

$$Y_1 \overset{d}{=} n^{-1/\alpha} \bigvee_{r=1}^{n} Y_r. \tag{34}$$

This implies by Proposition 3

$$\ell_s(n^{-1/\alpha}M_n, Y_1) = \ell_s\left(n^{-1/\alpha}M_n, n^{-1/\alpha}\bigvee_{r=1}^{n} Y_r\right)$$

$$\leq (n \cdot n^{-s/\alpha})^{1/s \wedge 1}\ell_s(X_1, Y_1)$$

$$= (n^{1-s/\alpha})^{1/s \wedge 1}\ell_s(X_1, Y_1) \quad \to \quad 0 \ \text{as} \ s > \alpha.$$

For $s \to \infty$ the rate approaches the optimal rate $n^{-1/\alpha}$.

d) Transformation of the fixed point equation. *The fixed point equation*

$$X \overset{d}{=} \bigvee_{r=1}^{K}(A_r X_r + b_r) \tag{35}$$

can be transformed in various ways. Let, e.g., $Y = \exp(\lambda X)$, then (35) transforms to

$$Y = \bigvee_{r=1}^{K} e^{\lambda b_r}Y_r^{A_r}, \tag{36}$$

in particular, for $A_r = 1, \lambda = 1$,

$$Y = \bigvee_{r=1}^{K} e^{b_r}Y_r. \tag{37}$$

For $Z = Y^{(\alpha)} = |X|^\alpha \operatorname{sgn}(X)$ and $W = \frac{1}{X^{(\alpha)}}$ (35) transforms similarly to further equivalent forms, in particular in the case $b_r = b$.. In this way all possible extreme value distributions can be reduced to the case of extreme value distributions of type 1 considered in Remark 6a (see also Zolotarev [19]). Consider as example the stochastic equation:

$$X \overset{d}{=} \bigvee_{r=1}^{2}(X_r - \ln 2). \tag{38}$$

This equation cannot be directly handled w.r.t. the ℓ_s-metric. Using $Y = \exp(X)$ equation (38) transforms to

$$Y \overset{d}{=} \tfrac{1}{2}Y_1 \vee \tfrac{1}{2}Y_2. \tag{39}$$

A solution is the extreme value distribution $F(x) = e^{-x^{-1}}, x \geq 0$. The operator T corresponding to (39) has contraction factor $\zeta = (\frac{1}{2})^{s-1}$ with respect to ℓ_s. So for any $s > 1$ F is a unique fixed point in $\mathcal{M}_s(F)$ and the central limit theorem holds for (Z_r) with tail condition $\ell_s(Z_r, Y) < \infty$, i.e., $(1/n^{1/\alpha})\vee_{r=1}^{n} Z_r \overset{d}{\to} Y$, where $Y \overset{d}{=} F$ equivalently $\vee_{r=1}^{n}W_r - (1/\alpha)\ln n \overset{d}{\to} X$, where X is the corresponding solution of (38), $Y = \exp(X)$ and $Z_r = \exp(W_r)$.

4 Max-recursive Algorithms of Divide and Conquer Type

We consider a general class of parameters of max-recursive algorithms of divide and conquer type:

$$Y_n \stackrel{d}{=} \bigvee_{r=1}^{K} \left(A_r(n) Y_{I_r^{(n)}}^{(r)} + b_r(n) \right), \quad n \geq n_0 \tag{40}$$

where $I_r^{(n)}$ are subgroup sizes, $b_r(n)$ random toll terms, $A_r(n)$ random weighting terms and $(Y_n^{(r)})$ are independent copies of (Y_n) independent also from $(A_r(n), b_r(n), I^{(n)})$.

With normalizing constants ℓ_n, σ_n let X_n denote the normalized sequence $X_n = \frac{Y_n - \ell_n}{\sigma_n}$. Then

$$X_n = \bigvee_{r=1}^{K} \left(\frac{A_r(n) Y_{I_r^{(n)}}^{(r)}}{\sigma_n} + \frac{b_r(n)}{\sigma_n} \right) - \frac{\ell_n}{\sigma_n}$$

$$= \bigvee_{r=1}^{K} \left(\left(A_r(n) \frac{\sigma_{I_r^{(n)}}}{\sigma_n} \right) X_{I_r^{(n)}}^{(r)} + \frac{1}{\sigma_n} \left(A_r(n) \ell_{I_r^{(n)}} + b_r(n) - \frac{\ell_n}{\sigma_n} \right) \right)$$

$$= \bigvee_{r=1}^{K} \left(A_r^{(n)} X_{I_r^{(n)}}^{(r)} + b_r^{(n)} \right), \tag{41}$$

where $b_r^{(n)} = \frac{1}{\sigma_n} (b_r(n) - \ell_n + A_r(n) \ell_{I_r^{(n)}})$ and $A_r^{(n)} = A_r(n) \frac{\sigma_{I_r^{(n)}}}{\sigma_n}$. Thus we obtain again the form (40) with modified coefficients.

As in section 2 we need a stabilization condition in L_s:

$$\left(A_1^{(n)}, \dots, A_K^{(n)}, b_1^{(n)}, \dots, b_K^{(n)} \right) \rightarrow (A_1^*, \dots, A_K^*, b_1^*, \dots, b_K^*). \tag{42}$$

Thus we obtain as limiting equation a stochastic equation of the form considered in section 3:

$$X \stackrel{d}{=} \bigvee_{r=1}^{K} (A_r^* X_r + b_r^*). \tag{43}$$

For existence and uniqueness of solutions of (43) we need the contraction condition:

$$E \sum_{r=1}^{K} |A_r^*|^s < 1. \tag{44}$$

For the application of the contraction method let T be the limiting operator,

$$TX \stackrel{d}{=} \bigvee_{r=1}^{K} (A_r^* X_r + b_r^*). \tag{45}$$

Then $\ell_s(X, TX) < \infty$ if X, A_r^*, b_r^* have finite absolute s-th moments, X a starting vector. More generally finiteness also holds under some tail equivalence conditions for X and the corresponding TX. Finally, to deal with the initial conditions we need the nondegeneracy condition: For any $\ell \in \mathbb{N}$ and $r = 1, \ldots, K$ holds

$$E\left[1_{\{I_r^{(n)} \le \ell\} \cup \{I_r^{(n)} = n\}} |A_r^{(n)}|^s\right] \to 0. \tag{46}$$

Our main result gives a limit theorem for X_n.

Theorem 7 (Limit theorem for max-recursive sequences). *Let (X_n) be a max-recursive, s-integrable sequence as in (40) and assume the stabilization condition (43), the contraction condition (44), and the nondegeneracy condition (46) for some $s > 0$. Then (X_n) converges in distribution to a limit X^*, $\ell_s(X_n, X^*) \to 0$. X^* is the unique solution of the limiting equation*

$$X^* \overset{d}{=} \bigvee_{r=1}^{K} (A_r^* X_r^* + b_r^*) \quad \text{in } \mathcal{M}_s. \tag{47}$$

Proof. By our assumption we have $E|A_r^*|^s, E|b_r^*|^s < \infty$ and so for any s-integrable X_0 holds $\ell_s(X_0, TX_0) < \infty$. Define the accompanying sequence

$$W_n := \bigvee_{r=1}^{K} \left(A_r^{(n)} X_r^* + b_r^{(n)}\right), \tag{48}$$

where $X_1^*, \ldots, X_K^*$ are i.i.d. copies of the solution X^* of the limiting equation, which exists and is unique by the contraction condition and Theorem 5. Then

$$\ell_s(X_n, X^*) \le \ell_s(X_n, W_n) + \ell_s(W_n, X^*). \tag{49}$$

From the stabilization condition we first show that

$$\ell_s(W_n, X^*) \to 0. \tag{50}$$

Subsequently, we assume $s \ge 1$. For the proof of (50) we use the stabilization condition (42)

$$\ell_s(W_n, X^*) = \ell_s\left(\bigvee_{r=1}^{K} \left(A_r^{(n)} X_r^* + b_r^{(n)}\right), \bigvee_{r=1}^{K} (A_r^* X_r^* + b_r^*)\right) \tag{51}$$

$$\le \left(\sum_{r=1}^{K} L_s^s\left(A_r^{(n)} X_r^* + b_r^{(n)}, A_r^* X_r^* + b_r^*\right)\right)^{1/s}$$

$$\le \left(\sum_{r=1}^{K} \left[L_s\left(A_r^{(n)} X_r^*, A_r^* X_r^*\right) + L_s\left(b_r^{(n)}, b_r^*\right)\right]^s\right)^{1/s}$$

$$\le \left(\sum_{r=1}^{K} \left[L_s\left(A_r^{(n)}, A_r^*\right)(E|X^*|^s)^{1/s} + L_s(b_r^{(n)}, b_r^*)\right]^s\right)^{1/s}$$

$$\to 0.$$

Next let Υ_n denote the joint distribution of $(A_1^{(n)}, \ldots, A_K^{(n)}, I^{(n)}, b_1^{(n)}, \ldots, b_K^{(n)})$ and let $(\alpha, j, \beta) = (\alpha_1, \ldots, \alpha_K, j_1, \ldots, j_K, \beta_1, \ldots, \beta_K)$.

Then we obtain by a conditioning argument for $s \geq 1$

$$
\ell_s^s(X_n, W_n) = \ell_s^s\left(\bigvee_{r=1}^{K}\left(A_r^{(n)} X_{I_r^{(n)}}^{(r)} + b_r^{(n)}\right), \bigvee_{r=1}^{K}\left(A_r^{(n)} X_r^* + b_r^{(n)}\right)\right) \tag{52}
$$

$$
\leq \int L_s^s\left(\bigvee_{r=1}^{K}\left(\alpha_r X_{j_r}^{(r)} + \beta_r\right), \bigvee_{r=1}^{K}\left(\alpha_r X_r^* + \beta_r\right)\right) d\Upsilon_n(\alpha, j, \beta)
$$

$$
\leq \sum_{r=1}^{K} \int L_s^s\left(\alpha_r X_{j_r}^{(r)}, \alpha_r X_r^*\right) d\Upsilon_n(\alpha, j, \beta)
$$

$$
= \sum_{r=1}^{K} \int |\alpha_r|^s \ell_s^s(X_{j_r}, X^*) d\Upsilon_n(\alpha, j, \beta)
$$

$$
\leq p_n^s \ell_s^s(X_n, X^*) + \sum_{r=1}^{K} \int \mathbf{1}_{\{j_r < n\}} |\alpha_r|^s \ell_s^s(X_{j_r}, X^*) d\Upsilon_n(\alpha, j, \beta).
$$

where $p_n = \left(E \sum_{r=1}^{K} \mathbf{1}_{\{I_r^{(n)} = n\}} |A_r^{(n)}|^s\right)^{1/s}$. With the inequality $(a + b)^{1/s} \leq a^{1/s} + b^{1/s}$ for all $a, b > 0$ and $s \geq 1$ we obtain with (49), (51) and (52)

$$
\ell_s(X_n, X^*) \leq \frac{1}{1 - p_n}\left(\left(\sum_{r=1}^{K} E|A_r^{(n)}|^s\right)^{1/s} \max_{0 \leq j \leq n-1} \ell_s(X_j, X^*) + o(1)\right) \tag{53}
$$

Since, by (42), (44) and (46), we have $\left(\sum_{r=1}^{K} E|A_r^{(n)}|^s\right)^{1/s} \to \zeta < 1$ and $p_n \to 0$ as $n \to \infty$ it follows that the sequence $(\ell_s(X_n, X^*))_{n \geq 0}$ is bounded. Denote $\bar{\eta} := \sup_{n \geq 0} \ell_s(X_n, X^*)$ and $\eta := \limsup_{n \to \infty} \ell_s(X_n, X^*)$. Now we conclude that $\ell_s(\bar{X}_n, X^*) \to 0$ as $n \to \infty$ by a standard argument. For all $\varepsilon > 0$ there is an $\ell \in \mathbb{N}$ such that $\ell_s(X_n, X^*) \leq \eta + \varepsilon$ for all $n \geq \ell$. Then with (52), (49), and (51) we obtain

$$
\ell_s(X_n, X^*) \leq \frac{1}{1 - p_n}\left(\sum_{r=1}^{K} \int \mathbf{1}_{\{j_r \leq \ell\}} |\alpha_r|^s \ell_s^s(X_{j_r}, X^*) d\Upsilon_n(\alpha, j, \beta)\right.
$$

$$
\left. + \sum_{r=1}^{K} \int \mathbf{1}_{\{j_r > \ell\}} |\alpha_r|^s \ell_s^s(X_{j_r}, X^*) d\Upsilon_n(\alpha, j, \beta) + o(1)\right)^{1/s}
$$

$$
\leq \frac{1}{1 - p_n}\left((\bar{\eta})^s E \sum_{r=1}^{K}\left(\mathbf{1}_{\{I_r^{(n)} \leq \ell\}} |A_r^{(n)}|^s\right)\right.
$$

$$
\left. + (\eta + \varepsilon)^s E \sum_{r=1}^{K} |A^{(n)}|^s + o(1)\right)^{1/s}.
$$

With (46) and $n \to \infty$ we obtain

$$\eta \leq \zeta(\eta + \varepsilon) \qquad (54)$$

for all $\varepsilon > 0$. Since $\zeta < 1$ we obtain $\eta = 0$. The proof for $s < 1$ is similar. $\square$

Remark 8. *Theorem 7 is restricted to the case of solutions of the limit equation in $\mathcal{M}_s$. In the existence and uniqueness result in Theorem 5 also solutions have been characterized without finite s-th moments. For several applications it is of interest to extend Theorem 7 to this more general case. This is to be considered in a separate paper.*

References

1. Michael Cramer. Stochastic analysis of Merge-Sort algorithm. *Random Structures Algorithms*, 11:81–96, 1997.
2. Hsien-Kuei Hwang and Ralph Neininger. Phase change of limit laws in the quicksort recurrence under varying toll functions. *SIAM Journal on Computating*, 31:1687–1722, 2002.
3. P Jagers and Uwe Rösler. Fixed points of max-recursive sequences. Preprint, 2002.
4. Donald E. Knuth. *The Art of Computer Programming*, volume 3: Sorting and Searching. Addison-Wesley Publishing Co., Reading, 1973.
5. Hosam M Mahmoud. *Sorting*. Wiley-Interscience Series in Discrete Mathemstics and Optomization. Wiley-Interscience, New York, 2000.
6. Ralph Neininger. *Limit Laws for Random Recursive Structures and Algorithms*. Dissertation, University of Freiburg, 1999.
7. Ralph Neininger. On a multivariate contraction method for random recursive structures with applications to Quicksort. *Random Structures and Algorithms*, 19:498–524, 2001.
8. Ralph Neininger and Ludger Rüschendorf. Rates of convergence for Quicksort. *Journal of Algorithms*, 44:52–62, 2002.
9. Ralph Neininger and Ludger Rüschendorf. A general limit theorem for recursive algorithms and combinatorial structures. to appear in: Annals of Applied Probability, 2003.
10. Ralph Neininger and Ludger Rüschendorf. On the contraction method with degenerate limit equation. to appear, 2003.
11. Svetlozar T. Rachev. *Probability Metrics and the Stability of Stochastic Models*. Wiley, 1991.
12. Svetlozar T. Rachev and Ludger Rüschendorf. Rate of convergene for sums and maxima and doubly ideal metrics. *Theory Prob. Appl.*, 37:276–289, 1992.
13. Svetlozar T. Rachev and Ludger Rüschendorf. Probability metrics and recursive algorithms. *Advances Applied Probability*, 27:770–799, 1995.
14. Mireille Régnier. A limiting distribution for quicksort. *RAIRO, Informatique Théoriqué et Appl.*, 33:335–343, 1989.
15. Uwe Rösler. A limit theorem for Quicksort. *RAIRO, Informatique Théoriqué et Appl.*, 25:85–100, 1991.

16. Uwe Rösler. A fixed point theorem for distribution. *Stochastic Processes Applications*, 42:195–214, 1992.
17. Uwe Rösler. On the analysis of stochastic divide and conquer algorithms. *Algorithmica*, 29:238–261, 2001.
18. Uwe Rösler and Ludger Rüschendorf. The contraction method for recursive algorithms. *Algorithmica*, 29:3–33, 2001.
19. Vladimir M. Zolotarev. *Modern Theory of Summation of Random Variables*. VSP, Utrecht, 1997.